MODERN ASPECTS OF ELECTROCHEMISTRY

No. 30

LIST OF CONTRIBUTORS

T. N. ANDERSEN
Chemical Research and Development
Kerr-McGee Corporation
Oklahoma City, Oklahoma 73125

WAHEED A. BADAWY
Department of Chemistry
University of Kuwait
13060 Safat
Kuwait

M. I. ČEKEREVAC
Faculty of Technology and
Metallurgy
University of Belgrade
11000 Belgrade, Yugoslavia

M. ENYO
Hakodate National College of
Technology
Hakodate 042, Japan

W. JAEGERMANN
Hahn-Meitner-Institut
Bereich Physikalische Chemie
Abteilung Grenzflachen
14109 Berlin, Germany

N. V. KRSTAJIĆ
Faculty of Technology and
Metallurgy
University of Belgrade
11000 Belgrade, Yugoslavia

T. MIZUNO
Department of Nuclear Engineering
Hokkaido University
Sapporo 060, Japan

K. I. POPOV
Faculty of Technology and
Metallurgy
University of Belgrade
11000 Belgrade, Yugoslavia

A Continuation Order Plan is available for this series. A continuation order will bring delivery of each new volume immediately upon publication. Volumes are billed only upon actual shipment. For further information please contact the publisher.

MODERN ASPECTS OF ELECTROCHEMISTRY

No. 30

Edited by

RALPH E. WHITE
University of South Carolina
Columbia, South Carolina

B. E. CONWAY
University of Ottawa
Ottawa, Ontario, Canada

and

J. O'M. BOCKRIS
Texas A&M University
College Station, Texas

PLENUM PRESS • NEW YORK AND LONDON

The Library of Congress cataloged the first volume of this title as follows:

Modern aspects of electrochemistry. no. [1]
   Washington Butterworths, 1954–
   v. illus., 23 cm.
   No. 1–2 issued as Modern aspects series of chemistry.
   Editors: no 1– J. Bockris (with B. E. Conway, No. 3–)
   Imprint varies: no. 1, New York, Academic Press.—No. 2, London, Butterworths.
     1. Electrochemistry—Collected works. I. Bockris, John O'M.ed. II. Conway, B.E. ed.
(Series: Modern aspects series of chemistry)
QD552.M6   54-12732 rev

ISBN 0-306-45450-5

© 1996 Plenum Press, New York
A Division of Plenum Publishing Corporation
233 Spring Street, New York, N. Y. 10013

10 9 8 7 6 5 4 3 2 1

All rights reserved

No part of this book may be reproduced, stored in a retrieval system, or transmitted in
any form or by any means, electronic, mechanical, photocopying, microfilming,
recording, or otherwise, without written permission from the Publisher

Printed in the United States of America

Preface

This volume of *Modern Aspects of Electrochemistry* contains five chapters. In the first chapter, Jaegermann reviews the complexity of the structural and related electronic properties of semiconductor/electrolyte interfaces from a surface science point of view. A special emphasis on photoelectron spectroscopy is used to investigate the possible application of surface science techniques for investigating semiconductor/electrolyte interfaces. Model experiments are suggested as a very promising surface science approach to study semiconductor/electrolyte interfaces. These experiments allow one to investigate nearly all fundamental aspects of adsorbate/surface interaction. Case studies of contact formation with semiconductors are presented, with a special emphasis on comparing the surface science results to results obtained with electrolyte solutions.

The second chapter is by Badawy and includes a review of the basic principles of solar energy conversion by photovoltaic and photoelectrochemical systems and the improvement of the performance of these systems. The author investigates and discusses the photovoltaic and photoelectrochemical behavior of Schottky barrier photoanodes. He shows that the incorporation of foreign atoms, such as Sb, Zn, or Ru, in the oxide-film matrix increases the rate of charge-transfer reaction at the oxide/electrolyte interface and hence improves the electrochemical performance of the electrochemical system. The photoelectrochemical reactions in heterogeneous systems are compared with photoelectrochemical processes in homogeneous systems and natural processes. Conclusions are made concerning the use of such systems for practical applications.

Electrodeposition is the topic of Chapter 3. The authors, Popov, Krstajić, and Čekerevac, claim that morphology is probably the most important property of a electrodeposited metal. It depends mainly on the kinetic parameters of the deposition process and on the deposition overpotential or current density. The mechanisms of formation of different growth forms as functions of the above variables are elucidated. The authors claim that it is possible to correlate the morphology of a metal electrodeposit with deposition process conditions.

In Chapter 4 Andersen describes the MnO_2-electrode behavior in alkaline cells, the MnO_2 electrode in Leclanché and zinc chloride environments as modifications of alkaline discharge, and the MnO_2 structure and the effect of such structure on battery activity. The author also discusses the chemistry of recharging MnO_2 in alkaline media, which is currently a technological frontier of intense interest. Andersen ends the chapter with a focus on MnO_2 deposition, which, with its parametric variations, is responsible for the varied but controlled structural features on MnO_2.

The phenomena of adsorption and absorption of hydrogen are intimately related to each other in the cases of many hydrogen-absorbing metals. Mizuno and Enyo investigate this in the last chapter of this volume. A major portion of this chapter is devoted to describing means by which the concentration and the distribution of hydrogen are measured in hydrogen-absorbing metals, or possibly metal hydrides. Several views of the electrochemistry of hydrogen entry into electrode metal are first reviewed, and then the most plausible view is presented. Surface activity of hydrogen adatoms is discussed first on the basis of the mechanism of the electrochemical hydrogen evolution reaction, and the authors argue that the use of activity rather than surface coverage is preferable as it is a quantity that can be analyzed in terms of electrode kinetics. Mizuno and Enyo summarize that electrochemical techniques combined with other techniques provide a rather detailed picture of hydrogen adsorption and entry into hydrogen-absorbing metals.

                    Ralph E. White
                          University of South Carolina
                    John O'M. Bockris
                          Texas A & M University
                  Brian E. Conway
                          University of Ottawa

Contents

Chapter 1

THE SEMICONDUCTOR/ELECTROLYTE INTERFACE: A SURFACE SCIENCE APPROACH

W. Jaegermann

I. Introduction	1
II. The Electrochemical Potentials of Electrons	4
1. Metal Surfaces	4
2. Semiconductor Surfaces	9
3. Electrolyte Solutions	13
III. The Semiconductor Contact in the Schottky Limit	20
1. Metal/Semiconductor Contacts without Surface Dipoles	20
2. Surface Dipoles at Metal/Semiconductor Interfaces	25
3. The Semiconductor/Electrolyte Contact	28
IV. The Semiconductor Contact in the Bardeen Limit	35
1. Qualitative Classification of Surface States	36
2. Quantum Mechanics of Surface States	44
3. Fermi Level Pinning	53
V. Surface Science in Semiconductor Electrochemistry	61
1. Principal Aspects of Surface and Interface Analysis of Semiconductor/Electrolyte Junctions	61
2. Surface Science Techniques	67
3. Model Experiments in the Investigation of Semiconductor/Electrolyte Interfaces	80
VI. Case Studies on Clean Surfaces and Adsorbate and Electrochemical Interfaces	86
1. Silicon	87
2. Gallium Arsenide	113
3. Layered Chalcogenide Semiconductors	139
VII. Summary and Conclusion	166
References	170

Chapter 2

PHOTOVOLTAIC AND PHOTOELECTROCHEMICAL CELLS BASED ON SCHOTTKY BARRIER HETEROJUNCTIONS

Waheed A. Badawy

I. Introduction 187
II. Intrinsic and Extrinsic Semiconductors 189
III. The Photovoltaic Junction 191
IV. Photoelectrochemical Devices 193
V. Schottky Barrier Solar Cells (SBSC) 197
    1. Current Transport Mechanism in SBSC 199
    2. Role of Surface States 201
VI. Photoelectrochemical Processes in Solar Energy Conversion 203
    1. Regenerative Photoelectrochemical Cells 204
    2. Photoelectrolytic Cells 207
VII. Metal Oxide Films 211
    1. Preparation and Electrode Properties of SnO_2 211
    2. Homogeneity of the SnO_2 Films 214
    3. Conductivity of the SnO_2 Films 215
    4. Junction Characteristics of n-Si/SnO_2 216
    5. Improvement of the Characteristics of SnO_2 219
    6. The Electrochemical Behavior of the Oxide Film ... 229
    7. Titanium Oxide Films 239
VIII. n-Si/SnO_2 and n-Si/TiO_2 Photovoltaic and Photoelectrochemical Systems 240
    1. Effect of the Interfacial SiO_x Layer on the n-Si/Oxide Heterojunction 241
    2. Photovoltaic Characteristics 244
    3. Photoelectrochemical Characteristics 247
IX. Concluding Remarks 254
References 255

Chapter 3

THE MECHANISM OF FORMATION OF COARSE AND DISPERSE ELECTRODEPOSITS

K. I. Popov, N. V. Krstajić, and M. I. Čekerevac

I. Introduction... 261
II. Coarse Deposits 262
  1. Mathematical Model 262
  2. Physical Simulation............................... 266
  3. Real Systems 268
II. Spongy Deposits...................................... 283
  1. Mathematical Model 283
  2. Physical Model.................................... 285
  3. Real Systems 287
III. Dendritic Growth Initiation 294
  1. Mathematical Model 294
  2. Physical Simulation............................... 298
  3. Real Systems 300
References.. 311

Chapter 4

THE MANGANESE DIOXIDE ELECTRODE IN AQUEOUS SOLUTION

T. N. Andersen

I. Introduction.. 313
  1. Significance of Manganese Dioxide and Its Variations.... 313
  2. Brief Overview of Previous Literature 316
  3. Organization of Chapter............................ 316
II. Alkaline Discharge from Two-Step Viewpoint 317
  1. Discharge Curves and Steps for $\gamma/\varepsilon\text{-MnO}_2$ 317

	2. Step Mechanisms for γ/ε-MnO_2 318
	3. Experimental Support for Mechanism of γ/ε-MnO_2 Discharge 320
	4. Discharge of β-MnO_2 325
III.	Discharge in Leclanché and Zinc Chloride Electrolytes 326
	1. Overall Cell Chemistries.......................... 327
	2. Discharge Mechanism............................ 329
	3. Key Issues in Recent Studies 334
IV.	Structural Characterization of MnO_2.................... 339
	1. Crystal Structure 339
	2. Structural Water and Stoichiometry 347
	3. Surface Properties or Mesostructure.................. 352
V.	Effect of MnO_2 Structural Features on Battery Activity 355
	1. Battery Activity.................................. 355
	2. Comparison of Different Crystal and Generic Types 356
	3. Studies of Heat-Treated EMD (HEMD)................ 360
	4. Studies within CMDs and EMDs 361
VI.	Detailed Proton–Electron Insertion Mechanism 365
	1. Structural Evidence for Nonhomogeneous Reduction of γ-MnO_2.. 365
	2. Equilibrium Potentials of γ-MnO_2 368
	3. Modeling of γ-MnO_2 Discharge at Practical Rates 373
	4. Electrochemical Spectroscopy of γ-MnO_2 377
	5. Insertion in Modifications Other than γ/ε-MnO_2 380
VII.	Rechargeable Alkaline Electrode....................... 381
	1. Introduction: MnO_2 Rechargeability Problems 381
	2. Limitation of Depth of Discharge.................... 383
	3. Modified MnO_2 Structure with Two-Electron Capacity... 385
VIII.	MnO_2 Deposition................................... 391
	1. Technology 391
	2. Equilibrium Potentials 392
	3. Side Reactions in Deposition 394
	4. Deposition on Bare Substrate 395
	5. Deposition on Oxide-Covered Substrate 396
References 404

Chapter 5

SORPTION OF HYDROGEN ON AND IN HYDROGEN-ABSORBING METALS IN ELECTROCHEMICAL ENVIRONMENTS

T. Mizuno and M. Enyo

I. Introduction 415
II. Experimental Aspects of Tracing Hydrogen on and in Metals 419
   1. Hydrogen *Ad*sorption on Metal Electrodes 419
   2. Hydrogen *Ab*sorption into Metals 423
III. Absorption of Hydrogen in Metals 429
   1. Light-Hydrogen and Deuterium Absorption Isotherms for Pd 429
   2. Pressure versus Fugacity 432
   3. Temperature Dependence of Hydrogen Absorption in Pd .. 433
   4. Effective Charge of Hydrogen in Metals 434
   5. Mechanism of Hydrogen Embrittlement 434
   6. Confinement of Hydrogen in Metals and Possible Fracture 435
   7. Hydrogen Absorption in Fe and Ni 437
   8. Hydrogen Absorption in Ti and Zr 437
   9. Temperature Hysteresis in Hydrides of Ti and Zr 441
IV. Models of Hydrogen Entry into a Hydrogen-Evolving Electrode 441
   1. A Concentration Cell Model 441
   2. A Model of Hydrogen Entry Independent of the Hydrogen Evolution Reaction 442
   3. A Model of Hydrogen Entry via the Intermediate of the Hydrogen Evolution Reaction 443
   4. Effects of Catalytic Poisons 445
V. Activity of Hydrogen Adatoms as the Reaction Intermediate of the Hydrogen Evolution Reaction and the Equivalent Hydrogen Pressure 445
   1. Hydrogen Electrode Reaction and Hydrogen Permeation in Metals 445
   2. Hydrogen Electrode Reaction on Pd 447

      3. Activity of Hydrogen Adatoms and Hydrogen Pressure
         Equivalent to Hydrogen Overpotential................ 450
      4. The Cascade Model for Sequential Reactions 454
VI. Hydrogen Permeation in Metals 460
      1. Permeation of Hydrogen across the Metal Surface Layer
         and Effects of Poisons............................ 461
      2. Hydrogen Permeation into the Pd Electrode 463
      3. Hydrogen Permeation into Fe and Ni Electrodes 478
      4. Hydrogen Permeation into Ti and Zr Electrodes
         and Hydride Formation........................... 480
VII. Concluding Remarks 495
References 496

Cumulative Author Index 505

Cumulative Title Index 517

Subject Index 527

MODERN ASPECTS OF ELECTROCHEMISTRY

No. 30

1

The Semiconductor/Electrolyte Interface: A Surface Science Approach

W. Jaegermann

Hahn-Meitner-Institut, Physical Chemistry Department, Interfaces Section, 14109 Berlin, Germany

I. INTRODUCTION

The semiconductor/electrolyte interface is the essential part of semiconductor-based electrochemical devices, such as photoelectrochemical or photocatalytic cells for solar energy conversion[1-3] and sensors.[4-6] It is also of central importance in etching and processing steps in semiconductor microelectronics[7,8] in which a semiconductor comes into contact with a solution or a humid atmosphere. The basic principles that govern the contact formation between differently doped semiconductors immersed into electrolyte solutions containing redox couples of differing electrochemical potentials have been elaborated and are summarized in a number of review articles and books (e.g., Refs. 9–12). In order for electronic equilibrium to be achieved across the phase boundary, the difference in electrochemical potential of the electrons in the solid versus the electrolyte is compensated by charge (electron) transfer across the interface. The related changes in electrical interface potentials in combination with the formation of electrical double layers compensate for the original difference. However, a more detailed view of contact formation at semiconductor/electrolyte interfaces requires a detailed knowledge of the microscopic structure of the phase boundary—e.g., the structure of the semiconductor surface and the interfacial composition of the electrolyte—as well as of

Modern Aspects of Electrochemistry, Number 30, edited by Ralph E. White *et al.* Plenum Press, New York, 1996.

the spatial and energetic distribution of electron states from the interior of the semiconductor to the interior of the electrolyte. It has to be stated that for most systems this information is not available in adequate detail. However, these details, which are determined by the geometric structure and electronic structure of the semiconductor crystallographic planes, are indispensable for an understanding of the elementary steps of contact formation as well as of charge-transfer kinetics and electrocatalytic effects. This is even more true when strong chemical interactions of electrolyte components with the semiconductor surface or interfacial reactions (interphases) must be considered. Even very sophisticated electrochemical characterization techniques,[13,14] which are all based on the measurement of electric currents, potentials, and charges, do not provide this microscopic information. Thus, they must be complemented with interface- (surface-) sensitive diffraction, topographic, and spectroscopic techniques.

The solid/electrolyte interface is at present being intensively investigated with a variety of interface-sensitive spectroscopic techniques in order to complement standard electrochemical measurements. The aim of all these investigations is to elucidate the geometric and electronic structure of the interface on a molecular level and to reconsider the accepted electrochemical models of electric potential distributions and related charge-transfer processes. Many modern surface- and interface-sensitive techniques have been applied for the characterization of electrolyte interfaces.[15-18] They may be divided into *in situ* techniques, which may be applied within the electrolyte, and *ex situ* techniques, which can only be used in ultrahigh vacuum (UHV). The advantage of the *in situ* techniques (e.g., infrared spectroscopy, X-ray spectroscopy, diffraction experiments, or scanning tunneling microscopy) is their applicability under "real" conditions. The UHV surface science techniques offer the advantage that the interface may entirely be fully characterized as many complementary techniques are available.

Various UHV surface science techniques, especially low-energy electron diffraction (LEED) and electron spectroscopy (AES and XPS), have been applied for the *ex situ* analysis of electrochemical interfaces. The related problem of emersion and transfer into UHV has been discussed with respect to the potential and limitations of this approach.[19-22] Alternatively, electrolyte interfaces may be simulated in UHV to investigate fundamental properties of solid/electrolyte interfaces. These "model" experiments are based on the pioneering work of Sass and co-workers, who were the first to model electrochemical interfaces in UHV by adsorbing electrolyte components onto defined metal surfaces.[23,24] With this approach, all modern surface science techniques can be combined to investigate the chemical, structural, and electronic aspects of surface interactions. Sass and co-workers showed that a close correspondence could be obtained between the work-function change in such model experiments and the change in the point of zero charge (pzc) within the electrolyte.[25] Thus, these model experiments give significant results with regard to interfacial electrochemistry. Because of the success of this approach, other workers started to investigate the interaction of model electrolytes on inert metal substrates.[26–33] These experiments and the results obtained on metal substrates are summarized in Refs. 31–33.

Similarly, the basic interactions at semiconductor/electrolyte interfaces may be investigated in UHV using *ex situ* emersion and *in situ* model experiments. The model experiments, in which electrolyte components are adsorbed onto defined semiconductor surfaces to simulate the electrochemical interface, will be emphasized in this chapter. The preparation of semiconductor interfaces in UHV and their subsequent characterization by the application of surface science techniques is a standard approach for the investigation of solid-state semiconductor contacts (see, e.g., Refs. 34–38 and references therein). Such experiments have contributed considerably to a better understanding of the mechanisms of contact formation. Because semiconductor/metal (Schottky) junctions are, in many respects, equivalent to semiconductor/electrolyte contacts, they will be introduced as the basis for theoretical concepts in semiconductor electrochemistry. Wherever possible, the UHV surface science results for clean semiconductor surfaces and model interfaces will be critically compared to electrochemical investigations and to results obtained by *in situ* techniques. Only prototype semiconductors will be considered (i.e., Si as a representative elemental semiconductor, GaAs as representative of the important class of zinc blende type compound semiconductors, and layered metal chalcogenides, as e.g., WSe_2, because most of our investigations have been performed with these semiconductors).

The chapter is organized as follows. In the next section, the contribution of the dipolar surface potential to the electrochemical potential of electrons at metal, semiconductor, and electrolyte surfaces will be described based on a surface science viewpoint. The third section deals with the formation of semiconductor contacts to metals and electrolytes without the detrimental effect of surface states (Schottky limit of semiconductor junctions). In the fourth section, surface states (interface states) and their influence on contact formation are considered (Fermi level pinning or the Bardeen limit of semiconductor junctions). In the fifth section, the use of surface science techniques in interfacial electrochemistry will be introduced and the information obtained by photoelectron spectroscopy will be discussed. Case studies of contact formation with the above-mentioned semiconductors will be presented in the sixth section, with a special emphasis on comparison of the surface science results to results obtained in electrolyte solutions. As this chapter is directed toward a surface science approach to semiconductor/electrolyte interfaces, it will mostly emphasize the equilibrium situation; charge carrier transport and dynamics as well as charge-transfer processes will not be considered in depth. It should be noted, however, that the semiconductor/electrolyte interface in equilibrium is the starting point of any interface reactions, and its properties should be known in order to draw any conclusions on reaction mechanisms.

II. THE ELECTROCHEMICAL POTENTIALS OF ELECTRONS

When two phases are brought into contact and electrons are exchanged to achieve thermodynamic equilibrium (electronic equilibrium), as occurs, for example, at semiconductor/metal or semiconductor/electrolyte contacts, it is important to quantify the thermodynamic driving force for electron exchange. Solid-state physicists dealing with solid-state devices usually refer contact potentials to the position of the Fermi level, E_F, as an appropriate reference level, whereas electrochemists traditionally use electrode or redox potentials in describing the thermodynamics of electrochemical interfaces. This may lead to confusion, especially in semiconductor electrochemistry, where physical and electrochemical terminology are combined at the phase boundary. Therefore, it is very important to identify the appropriate reference level for the energy (potential) of electrons in the two phases and to reconsider its meaning in physical and electrochemical terms.

1. Metal Surfaces

(i) Bulk and Surface Potentials

The important quantities defining the potentials of electrons at metal surfaces are summarized in Fig. 1, which is based on Refs. 39–42. The

The Semiconductor/Electrolyte Interface

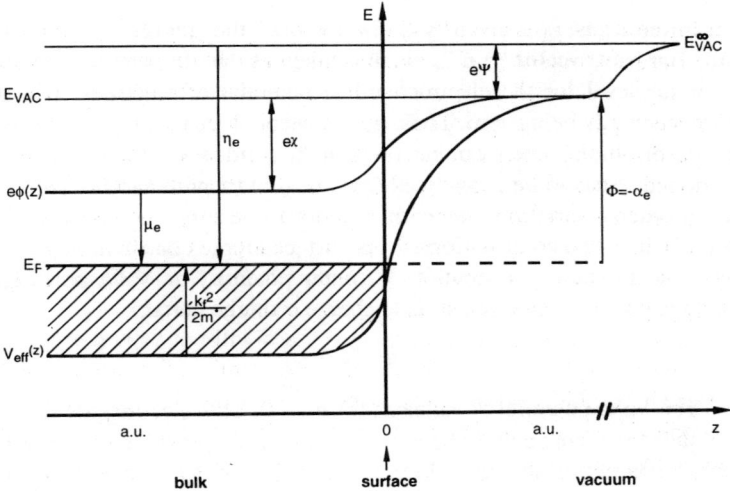

Figure 1. Schematic representation of the different surface potentials at metal/vacuum interfaces. (For details, refer to the text.)

effective one-electron potential $V_{\text{eff}}(z)$ is a function of the coordinate z normal to the surface. $V_{\text{eff}}(z)$ contains a bulk term, V_b, which describes the interaction of an electron with the nuclei and with all other electrons in the bulk, and a surface term, V_s $[V_{\text{eff}}(z) = V_b(z) + V_S(z)]$.[41,42] This figure is a conceptual diagram that separates in a gedankenexperiment the different electrostatic potentials that contribute to the electrostatic potential $\phi(z)$ of the electron inside the solid (inner potential or Galvani potential) and that define the surface (interface) potential $V_s(z)$. The electrochemical potential η_e of electrons inside the solid phase is defined according to the Gibbs (or Helmholtz) free energy G (or F) of the solid phase[43,44]:

$$\eta_e = \left(\frac{\partial G}{\partial n}\right)_{T,P} = \left(\frac{\partial F}{\partial n}\right)_{T,V} \quad (1)$$

with n being the number of electrons in the solid phase. The electron is considered to be transferred from a position at infinity without potential or kinetic energy (E_{vac}^{∞}) into the solid at electronic equilibrium. (For $T = 0$ K, the electron is added to the Fermi level E_F.[43]) E_F is the level of maximum kinetic energy of electrons inside the solid. (At $T = 0$ K for a free electron gas, E_F is given by $k_f^2/2m^*$, with m^* the reduced electron mass and k_f its wave vector.[43]) E_{vac}^∞ is thus taken as the ultimate (conceptual) reference level for the electrochemical potential of electrons with the electron energy being zero. Thus, η_e is a negative quantity that is equivalent to the binding energy of electrons in the solid at E_F. When a charging of the phase has to be considered (e.g., in contact with another phase or due to lateral variations in work function), the inner potential $\phi(z)$ is composed of two contributions,[45] the surface dipole contribution χ and a mean electrostatic contribution Ψ (outer potential or Volta potential), which is due to excess free surface charge σ on the metal:

$$\phi = \Psi + \chi \equiv V_s \qquad (2)$$

The Volta potential Ψ vanishes for $\sigma \to 0$. This electrostatic term Ψ will shift the whole energy band downward in the presence of a positive surface charge, as shown in Fig. 1. The value of $\Delta\Psi$ between the two phases depends on the charge and the distance to the contact phase. As the amount of charge σ is small compared to the metal density of states around E_F, it will hardly affect the position of E_F within the bulk, in contrast to the situation in semiconductors (see below). Thus, η_e, which defines the position of E_F, is composed of the chemical potential μ_e and electrostatic potential terms:

$$\eta_e = \mu_e - e\phi = \mu_e - e\chi - e\psi \equiv E_F \qquad (3)$$

The chemical potential of electrons, μ_e, contains no electrostatic surface term and only depends on bulk properties. It is not accessible by experiment and can only be calculated theoretically, based on simplified models,[39,46] by summing up the electron kinetic energy and the sum of core-level and electron exchange potentials.

For $\sigma \to 0$ and $\Psi = 0$, the electrochemical potential of a metal is given by

$$\eta_e^M = \mu_e^M - e\chi_M = \alpha_e^M = -\Phi_M \qquad (4)$$

Here, α_e is the so-called "real potential" of an electron inside the condensed phase, which is composed of the chemical potential of electrons, μ_e, and the contribution of the surface dipole, χ. It is equivalent to the work function Φ with a negative sign. Φ is defined as the work to be done to transfer an electron from the interior of the solid at E_F to outside the surface. The more precise definition is that the electron will be transferred to a position about 10^{-4} cm outside the solid ("near surface vacuum level"), which is outside the image charge of the surface (close to the break in distance units in Fig. 1). This is important as, in real cases, one has to account for extra excess charges σ even on surfaces in vacuum, which result from lateral charge transfer between different crystal planes. As the different planes of a crystal have different work functions, they will be charged by lateral charge transfer to compensate for this difference. The related outer potential Ψ will shift the low-work-function areas downward (as shown in Fig. 1), because they will be charged positively; the high-work-function areas will be charged negatively and thus will be shifted upward. As a consequence, there is no difference in the binding energy of electrons when they are transferred from E_F through any plane to E_{vac}^{∞}. For a more extended discussion of this point, see Ref. 40.

It should be noted at this point that the work function Φ, and thus the real potential α, is a measurable quantity. However, the relative contributions of μ and χ to α_e are not accessible by any experiment and can only be separated by theoretical calculations.[39,46]

(ii) Surface Dipoles

The surface dipole contribution to the work function, χ, warrants further analysis.[39,46,47] The surface dipole at the metal/vacuum interface contains, in general, several contributions, as may be deduced from a simple model (Fig. 2). First of all, there is the tail of conduction-band electrons spreading out from the bulk core potential into vacuum (also called spill-out). The distribution of atomic core potentials ρ^+ is represented by Wigner–Seitz spheres,[43,48] which also contain the electron density ρ^-. In the bulk the distribution of charge is symmetric, whereas at the surface the spreading of electron density leads to an asymmetric charge distribution, as schematically shown in Fig. 2. This effect was originally calculated for a jellium-like metal,[49] and the results are shown in the lower part of Fig. 2 (neglecting the oscillating behavior of the electron density in the surface near bulk, Friedel oscillations). The related surface potential drop due to the spread-out, χ_{spr}, is given by the Poisson equation across the vacuum interface and depends on the electron density and the bulk potential: $\chi = \phi(-\infty) - \phi(+\infty)$. In the expression

$$\frac{d^2}{dz^2}\phi = \frac{1}{\varepsilon \cdot \varepsilon_0} \cdot [\rho^+(z) - \rho^-(z)] \qquad (5)$$

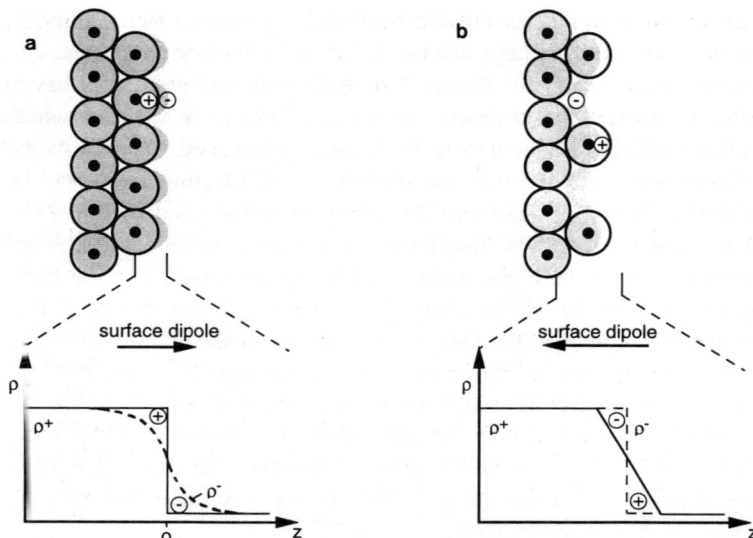

Figure 2. Schematic representation of different surface dipoles at metal/vacuum interfaces: (a) The spread-out of electrons leads to a positive dipole χ_{spr}. (b) The smooth-out of atomic surface roughness leads to a negative dipole χ_{sm}.

$\rho^+(z)$ is approximated by a step function, which is unity for $z < 0$ and 0 for $z > 0$, and $\rho^-(z)$ has to be calculated[39,46,47]; its schematic distribution is shown in Fig. 2.[49] The missing bulk electron density in the vicinity of the surface ($z < 0$) is separated from the extra electron density toward vacuum ($z > 0$) by atomic distances (typically 1 Å). The resulting positive dipole will increase the work function. This electronic effect is very much dependent on the atomic roughness of the surface, given by the actual positions of atomic core potentials (surface topography). For a more open surface, the electrons will smooth out the positive background potential, as also shown in Fig. 2.[50] This effect will lead to a positive (atom core related) charge toward vacuum, which will give rise to a negative surface dipole, χ_{sm}, reducing the dominant effect of spreading. The corresponding surface potential drop χ_{sm} may also be calculated according to Eq. (5) when the atom positions are known. The overall surface dipole then will be given by

$$\chi = \chi_{spr} + \chi_{sm} \qquad (6)$$

The formal separation of the two effects is helpful for understanding the variation in work function measured for different crystal planes. At a close-packed surface [e.g., (111) in face-centered cubic (fcc) crystal structures], the work function is high as χ_{spr} is large whereas χ_{sm} is small. In contrast, at the open (110) plane, χ_{sm} increases and Φ decreases. For transition metals, χ is usually larger than μ.[40] There are even cases, e.g., W(100), in which χ is calculated to be larger than ϕ,[51] which makes μ positive. For real metals the situation is considerably more complex due to the effects of surface relaxation on the observed atom positions of the topmost surface layers.[46,52]

As already mentioned above, the relative contributions of the chemical potential μ and the surface dipole χ to the work function Φ cannot be determined experimentally but can only be calculated theoretically. The relative contribution of χ to Φ may range from some tenths of an electron volt to a few electron volts; in some cases χ is even larger than μ (a listing of calculated values is given in, e.g., Refs. 40 and 41). As χ depends on the conditions of the phase boundary, it may be expected that on going from vacuum to a contact phase the related change of the surface dipole, $\Delta\chi$, will be considerable.

2. Semiconductor Surfaces

(i) Bulk and Surface Potentials

The important surface potentials of semiconductors are in close correspondence with those of metal surfaces (Fig. 1) and are schematically represented in Fig. 3. For simplicity, it is assumed that the semiconductor is under flatband conditions and that there is no excess charge σ on the surface: As a result, no outer potential contribution Ψ has to be considered. (The changes in surface potentials induced by charge transfer after contact formation will be discussed in detail in the next sections. The charging of surfaces due to different work functions of different surface planes is also not considered further but is the same as for metal surfaces.)

The most obvious difference with respect to metals is the lack of electron states around E_F, as E_F is situated in the band gap for moderate doping.[53,54] Therefore, the work function Φ is no longer equivalent to the electron affinity E_A, which is defined as the energy required for addition of an electron to the semiconductor conduction-band edge E_{CB} ($E_A = E_{vac} - E_{CB}$), and to the ionization potential I_p, which is defined as the energy required for removal of an electron from the semiconductor

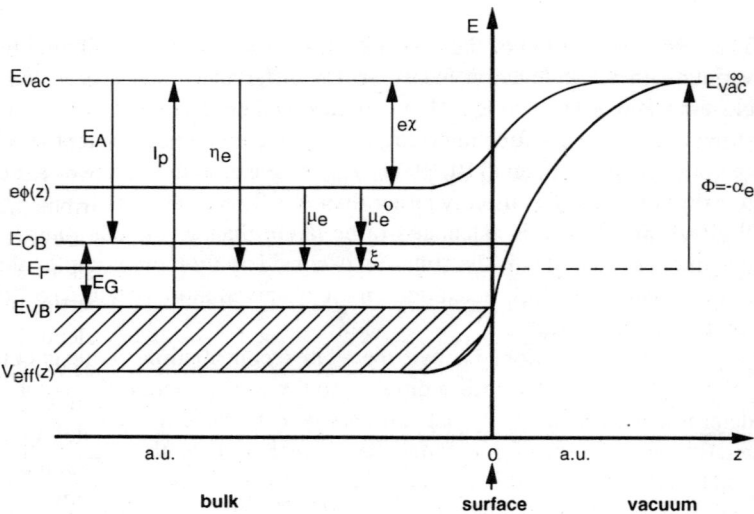

Figure 3. Schematic representation of the different surface potentials at semiconductor/vacuum interfaces. (For details, refer to the text.)

valence-band edge E_{VB} ($I_p = E_{vac} - E_{VB}$).[35] As the position of the Fermi level is dependent on the doping of the semiconductor, Φ changes with doping. It is larger (close to the value of I_p) for strong p-doping and smaller (close to the value of E_A) for strong n-doping. The occupation of conduction-band and valence-band states with electrons n and holes p, respectively, is defined by the Fermi distribution function $f(E)$, which for $E_i - E_F > 3kT$ may be approximated by the Boltzmann distribution[53,54]:

$$n = N_C \cdot f(E) \approx N_C \cdot \exp\left(-\frac{E_C - E_F}{kT}\right) \quad (7)$$

$$p = N_V \cdot [1 - f(E)] \approx N_V \cdot \exp\left(-\frac{E_F - E_V}{kT}\right) \quad (8)$$

$$f(E) = \frac{1}{1 + \exp\left(\frac{E_i - E_F}{kT}\right)} \quad (9)$$

N_C and N_V are the densities of conduction- and valence-band states at the band edges, approximated by $N_C(E) = 4\pi (2m_e^*/h^2)^{3/2} (E - E_C)^{1/2}$ and $N_V(E) = 4\pi(2 m_p^*/h^2)^{3/2} (E_V - E)^{1/2}$ for parabolic bands.

Nevertheless, the same definition for the electrochemical potential of electrons η_e (equivalent to E_F) is valid as for metals [see above, Eqs. (1)–(4)]: again a bulk term μ_e and a surface potential term χ contribute to η_e (and Φ). As for metals, the relative contributions of these terms are only accessible by theoretical calculation.[55-57] In addition, μ_e consists of an intrinsic part μ_e^0, which is only dependent on the chemical properties of the semiconductor, and a part ζ that depends on the additional electrons introduced or removed by bulk doping. The most reasonable choice would be to take the position of E_F for an intrinsic (undoped) semiconductor (intrinsic level E_i, close to midgap) as the reference level for defining μ_e^0. For convenience, we choose the conduction-band minimum E_{CB}; thus, the definition of the electrochemical potential of electrons in the semiconductor with respect to the vacuum reference level $E_{vac}^\infty = 0$ is

$$E_F^{SC} \equiv \eta_e^{SC} = \mu_e^0 + \zeta - e\chi_{SC} = \alpha_e^{SC} = -\Phi_{SC} \tag{10}$$

where ζ is given by

$$\zeta = E_{CB} - E_F = -kT\ln\frac{n}{N_C} \tag{11}$$

with n the concentration of electrons in the conduction band. It should be noted that some authors add μ_e^0 to the surface dipole contribution, which makes this value identical to the electron affinity, $E_A = \mu_e^0 - e\chi$. This has been justified by the large calculated surface dipoles, which even exceed the experimentally determined E_A values.[56] As this definition would neglect any bulk binding energy for electrons (the bulk system is thermodynamically unstable)[57] and the electrons are just confined to the solid by the surface dipole, we prefer our definition for the general case. Any change of the surface dipole $\Delta\chi$ (e.g., at different crystal planes or due to contact phases) will shift the energy bands and the related potentials parallel to the vacuum level, in close correspondence to the effect described for metals. These shifts do not alter the position of E_F relative to the band edges. Therefore, the work function Φ cannot be used for such a comparison, as at semiconductors the position of E_F within the band gap, and thus Φ, may also be changed independently, for example, due to band bending or doping. Only the determination of E_A or I_p will give reliable information on $\Delta\chi$.

(ii) Surface Dipoles

To our knowledge, there is only rather limited information in the literature on the relative magnitudes of surface dipoles at semiconductors. In principle, an electronic dipole contribution caused by the penetration of electronic charge density from the positive atom core bulk potential, as on the surface of metals, is also expected at semiconductor surfaces. However, the different character of bonding in semiconductors has to be considered, which could result in more directed dangling-bond "electron orbital" spatial distributions, which again should depend on the crystal planes. However, in addition, the structural surface dipole due to ionic charge distribution resulting from spatially different positions of differently charged atoms toward vacuum will dominate, especially for compound semiconductors. These effects are, of course, strongly influenced by the possible relaxations and reconstructions of surface atoms, which have been described for the different crystallographic planes of semiconductors (see, e.g., Refs. 58, 59, 60, 35, and 38). The different effects determine the value of the surface dipole changes $\Delta\chi$ and thus E_A and I_p. The work function Φ will also be affected by changed surface dipoles $\Delta\chi$, but Φ may also be independently modified by a shifted position of E_F in the band gap due to band bending induced by charge transfer from the space-charge layer to surface states. This complex interdependence may be the reason for the limited information available on the different dipole contributions at semiconductor surfaces. For example, for Si and Ge the photoemission threshold E_T, equivalent to I_p, has been determined to be rather independent of crystal orientation[61,62] (Fig. 4a). In contrast, the work function is considerably changed due to the influence of changed surface-state distributions inducing band bending (not shown). In Ref. 48 additional experimental and theoretical values of E_T are given; experimental values of E_T range from 4.75 eV for Si(100) to 5.15 eV for Si(111), and the theoretical values scatter between 4 and 5.9 eV. For GaAs, a prototype compound semiconductor, very different ionization potentials I_p have been determined, ranging from 5.1 eV (00$\bar{1}$) to 5.7 eV (110)[63] (Fig. 4b). Based on their experimental values, Ranke and co-workers[61–63] concluded that the ionic contribution to the surface dipole (structural dipole) is dominant and the electronic part is negligible. The contribution of surface dipoles and their changes upon contact formation with electro-

The Semiconductor/Electrolyte Interface

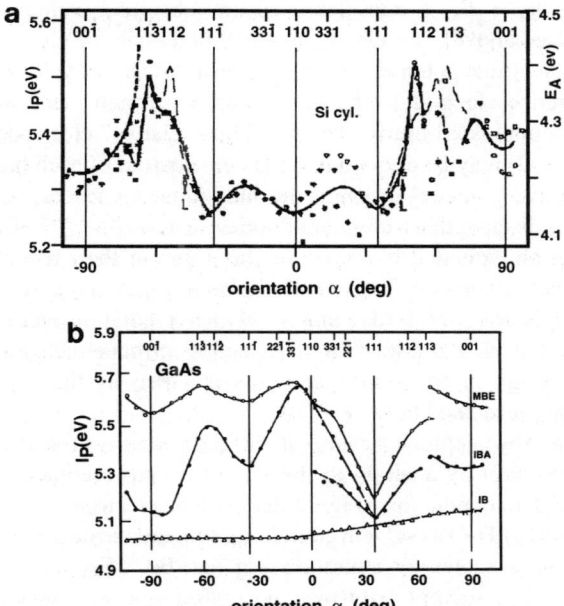

Figure 4. Variation of ionization potential I_p with crystal plane for Si (a) and GaAs surfaces (b), measured with a cylindrical single crystal. (The orientations are given in the figure; MBE, surface prepared by molecular beam epitaxy, IB, ion bombarded surfaces, IBA, ion bombarded and annealed surfaces. After Refs. 61–63.)

lytes and adsorbates, which will shift the relative positions of the energy bands in relation to the contact phase, will be further addressed later on in this chapter (Section III). It is also of general importance for solid-state semiconductor junctions,[35–38] e.g., for metal contacts (Section III.2).

3. Electrolyte Solutions

(i) Bulk and Surface Potentials and Electrolyte Density of States

The electrochemical potential of electrons in electrolyte solution, which may or may not contain a redox couple (reducible and oxidizable species in the potential range of interest), has been a matter of serious consideration and debate among electrochemists for a long time. Whereas for solids $\eta_e = E_F$ is considered as an "absolute" value referred to $E_{vac}^{\infty} = 0$ (see above), the electrochemical potentials of electrolyte solutions and electrode potentials are measured as "relative" values in relation to a reference electrode, e.g., the normal hydrogen electrode (NHE), whose potential is arbitrarily set as 0. This "relative" electrode potential (cell potential) may be divided into the contribution of each half-cell and the respective potential drops at the individual interfaces, as has been thoroughly discussed in a number of review articles.[41,44,64–67] Bockris and Khan give an extended overview of this topic in their recent book on surface electrochemistry,[68] to which the reader may refer. To our understanding, it is now well settled and widely agreed that the electrolyte may be considered as a phase with a defined electrochemical potential of electrons as given for solids, and this is taken as the basis for the description presented below.

In Fig. 5 we depict a scheme of surface potentials for an electrolyte solution ($\Psi = 0$) by analogy to the schematic representations used for metals and semiconductors (Figs. 1 and 3). The electrolyte is assumed to be composed of H_2O as solvent containing a reversible one-electron redox couple in about molar concentration (e.g., ox: Fe^{3+}, $Fe(CN)_6^{3-}$; red: Fe^{2+}, $Fe(CN)_6^{4-}$). The solvent H_2O may be considered as a wide-band gap amorphous semiconductor. The conduction-band and valence-band edges, E_{CB} and E_{VB}, with the density of state distributions $D_{unocc}^{H_2O}$ and $D_{occ}^{H_2O}$ result from the lowest unoccupied molecular orbital (LUMO) and highest occupied molecular orbital (HOMO) of the solvent.[69,70] For H_2O their positions are, respectively, about 1.5 eV and 9 eV below the vacuum level,[69,70,71,41] and they are derived from the empty $4a_1$ (LUMO) and occupied lone-pair $1b_1$ (HOMO) levels of molecular H_2O.[72] In the absence of additional redox-active species in solutions, free electrons or solvated electrons introduced into the solution may then be considered as dopants, which define the position of η_e close to E_{CB} (not shown here; see Ref. 41). The solvated electron experiences a different stabilization energy from that imported to the free electron in solution, which would lead to different electron work functions of the electrolyte, as discussed by Trasatti.[41] In general, semiconductor/electrolyte junctions are formed with added redox-active species in solution. Thus, additional electron states will contribute to the electrolyte density of states distribution D^{El} (shown as $(DOS)^{EL}$ in Fig. 5). These additional states may be understood as electronic doping of the wide-band-gap "semiconductor" formed by the solvent, the dopant concentration depending on the concentration of the

The Semiconductor/Electrolyte Interface

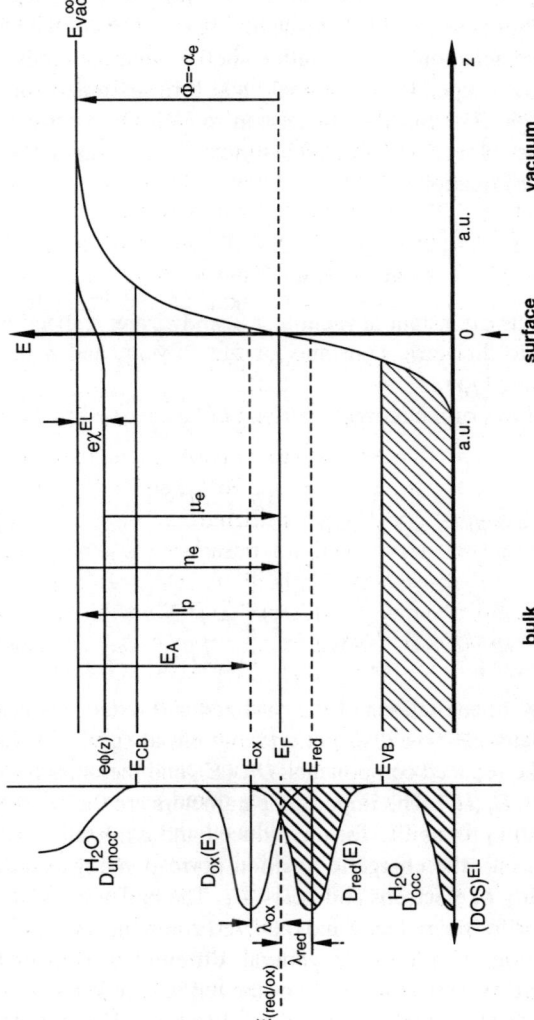

Figure 5. Schematic representation of the different surface potentials at electrolyte/vacuum interfaces. (For details, refer to the text.)

redox couple. We have chosen a representation of the D^{El} that has been derived from the solvent-fluctuation model.[9–12,68] It was originally derived from the Marcus theory of electron transfer[73,74] and was adapted to semiconductor electrochemistry by Gerischer.[75] It is based on the thermally activated reorganization of the solution shell around the redox-active species. Thus, D^{El} depends on the analytical form assumed for the reorganization energies. Within the fluctuation model, the electrostatic interaction due to the solvent is expressed in terms of a continuum theory given by a Born-type equation:

$$\lambda = \frac{e^2}{8\pi\varepsilon_0 \cdot a}\left(\frac{1}{\varepsilon_{opt}} - \frac{1}{\varepsilon_{st}}\right) \tag{12}$$

where ε_0 is the dielectric constant in vacuum, ε_{opt} and ε_{st} are, respectively, the optical and static dielectric constants of the solvent, and a is the diameter of the solvated ion.

The analytical form of the reorganizational D^{El} as derived by Morrison[10] is given by

$$D_{red}(E) = (4\pi kT \lambda_{red})^{-1/2} \exp\left[-\frac{(E_{red} - E)^2}{4\lambda_{red} kT}\right] \cdot c_{red} \tag{13}$$

$$D_{ox}(E) = (4\pi kT \lambda_{ox})^{-1/2} \exp\left[-\frac{(E_{ox} - E)^2}{4\lambda_{ox} kT}\right] \cdot c_{ox} \tag{14}$$

with c_{ox} and c_{red} the concentrations of the oxidized and reduced species, respectively. Two Gaussian-type distribution functions are given by these equations, one for the reduced components, $D_{red}(E)$, and the other for the oxidized component, $D_{ox}(E)$. The important parameters are the energetic positions of maximum probability for the reduced and oxidized components, E_{red} and E_{ox}, and the energetic position corresponding to equal occupation probability of electrons and holes, E_F. The half-widths of the distribution functions for the reduced and oxidized components are given by their reorganization energies λ. In general, different values for the reduced (λ_{red}) and the oxidized form (λ_{ox}) of the redox couple have to be taken into account [as shown in Fig. 5 and used in Eqs. (13) and (14)], owing to their different solution shells and the unequal reorganizational stabilization. However, very often λ_{red} and λ_{ox} are approximated by a single common value (as also assumed by Morrison[10]). Then the energetic distance between E_{red} and E_{ox} is 2λ, and that between E_{red} (E_{ox}) and E_F is λ.

The occupation of the electrolyte states by electrons n and holes h (equivalent to the occupation of the reduced and oxidized forms of a redox couple, respectively) may be expressed, by analogy to the equations for doped semiconductors (Eqs. 7–9), by

$$n(E) = D^{\text{El}}(E) \cdot f(E) \qquad (15)$$

$$p(E) = D^{\text{El}}(E) \cdot [1 - f(E)] \qquad (16)$$

$$f(E) = \frac{1}{1 + \exp\left(\dfrac{E - E_F}{kT}\right)} \qquad (17)$$

D^{El} is the sum of $D_{\text{red}}(E)$, $D_{\text{ox}}(E)$, and the solvent density of states $D^{\text{H}_2\text{O}}$, and $f(E)$ is the Fermi distribution function which governs the occupation of electron states of the phase under consideration, here the redox electrolyte.

It should be mentioned at this point that the general validity of the fluctuating-energy-level concept for a proper description of the electrolyte density of states distribution has been very controversial,[76–78] and other models have been suggested.[67,68,76,79,80] This search for alternative models was motivated by a detailed analysis of current–voltage curves of semiconductor electrodes and derived rate constants of charge transfer. In addition, it was stated that the thermal activation of the inner coordination shell around the redox species should lead to a Boltzmann-type activation of the vibrational modes and thus to a Boltzmann-type state distribution.[68,76] Unfortunately, this very important aspect of electrochemistry has not been settled so far and requires further experimental study. Moreover, it is evident that any simple model considering only bulk electrolyte species and their D^{El} is not appropriate when more complex redox reactions involving catalytic surface interactions are described. In these cases, electron states resulting from the interaction with the electrode surfaces will be involved and must be considered in the density of states distribution.[81,82,3]

However, independently of the exactly valid electrolyte D^{El}, η_e can be defined for the electrolyte as shown in Fig. 5. It is given by the energy level that governs the occupation of electron states by the appropriate thermodynamic distribution function [Fermi distribution function (Eq. 17); for the general case, E_F may be substituted by η_e]. As discussed by Reiss,[64,65] Fermi statistics are even valid when the occupation of states influences their density of states distribution, as found for redox couples in electrolytes.

(ii) Surface Dipoles and the Absolute Redox Potential

As the occupation statistics of D^{El} are given by the Fermi function for thermodynamic equilibrium, they correspond to an equilibrium redox reaction in the electrolyte:

$$\text{ox (solv)} + e^- \rightleftharpoons \text{red (solv)} \tag{18a}$$

such as

$$\text{Fe}^{3+}\text{ (aq)} + e^- \rightleftharpoons \text{Fe}^{2+}\text{ (aq)} \tag{18b}$$

If a Born–Haber cycle is applied to this redox reaction,[83] this translates to another expression for η_e:

$$\eta_e = \eta_{\text{red}} - \eta_{\text{ox}} = I_p + \lambda_{\text{ox}} - \lambda_{\text{red}} - e\phi^{El} \tag{19}$$

where η_{red} and η_{ox} are the electrochemical potentials of the reduced and oxidized forms of the redox couple, I_p is the ionization potential of the electrons, and ϕ^{El} is the inner (Galvani) potential of the electrolyte ($\phi = \Psi + \chi$). If the concentration dependence of η_{red} and η_{ox} is considered [64,65]:

$$\eta = \eta^0 + kT \ln c/c_0 \tag{20}$$

it follows that, with $\eta^0_{\text{red}} - \eta^0_{\text{ox}} = \eta^0_{\text{red/ox}}$,

$$\eta_e = \eta^0_{\text{red/ox}} + kT \ln \frac{c_{\text{ox}}}{c_{\text{red}}} \tag{21}$$

which is in analogy to the well-known Nernst equation for the concentration dependence of redox couples:

$$E(\text{red/ox}) = E^0 + \frac{kT}{e} \ln \frac{c_{\text{ox}}}{c_{\text{red}}} \tag{22}$$

Thus, it follows that the "relative" redox potential $E(\text{red/ox})$ of the electrolyte containing a redox couple can be made equivalent to the "absolute" electrochemical potential η_e of the electrons in the redox electrolyte by adding a constant reference value E^0_{abs}:

$$\eta_e = -eE(\text{red}/\text{ox}) - eE^0_{abs} \qquad (23)$$

The negative sign in Eq. (23) results from the different sign conventions of the "absolute" and "relative" electrochemical scales. A more strongly bound electron has a larger (negative) binding energy on the vacuum scale (but a larger positive work function), and this translates to a more positive redox potential.

In Fig. 5 a surface dipole for the interface between the electrolyte and vacuum has been included. Its value was estimated for aqueous electrolytes as 0.13 eV.[41] It is assumed that the oxygen of the H_2O dipole sticks out of the electrolyte, leading to the positive potential step shown in Fig. 5. Thus, the electrochemical potential of electrons η_e in electrolyte solutions (for $\psi = 0$) is composed of a bulk term μ describing the interaction of the redox species with the solvent and a surface dipole component χ:

$$\eta_e^{El} = \mu_e^{El} - e\chi_{El} = a_e^{El} = -\Phi_{El} \qquad (24)$$

Thus, the work function and real potential a_e of the electrolyte electron also contain a surface dipole χ, which cannot be determined experimentally. A precise number for the "absolute" value of the electrochemical potential of electrons in the electrolyte, equivalent to the "absolute" value or work function of the redox potential, cannot be given yet.[68] The electric surface potentials of solids as well as of semiconductor/adsorbate interfaces are referenced to the vacuum level (absolute scale vs. E^{∞}_{vac}). The electrochemical potentials are usually referenced to the redox potential of a reference electrode, for example, the saturated calomel electrode (SCE) or the normal hydrogen electrode (electrochemical scale vs. NHE). The two scales are related to each other by E^0_{abs} as discussed above [see Eq. (23)]. A scaling value of $E^0_{abs} = 4.5$ V is often used, but it is only an approximate number. Based on recent calculations,[84] a value of 4.44 V was suggested. However, experimentally determined values of 4.7 or 4.8 V have also been proposed in the literature.[85,86,87] Based on this scaling, it is, in principle, possible to relate energy values obtained from measurements in electrolytes to measurements in UHV. The different values suggested for E^0_{abs} result from the unknown contribution of χ and different estimations of the changes $\Delta\chi$ due to junction formation. Changes of solvent dipoles have been deduced from recent emersion experiments.[88]

Further experiments with different techniques, such as photoemission of solutions,[89–91] are urgently needed to improve the estimates of the surface dipole contribution for different electrolyte junctions and thus to achieve the precise determination of E_{abs}^0.

III. THE SEMICONDUCTOR CONTACT IN THE SCHOTTKY LIMIT

The contact formation at a semiconductor immersed into an electrolyte solution with a high concentration of the redox couple is often considered to be based on a similar mechanism as the formation of semiconductor/metal contacts (Schottky barriers). In addition, many similar concepts are used for the description of these two types of interfaces. Therefore, within this chapter, the theoretical framework and experimental procedures developed for metal interfaces will be transferred to electrolyte interfaces. However, in addition, it is important to consider the conceptual differences between semiconductor/electrolyte and semiconductor/metal contacts.

1. Metal/Semiconductor Contacts without Surface Dipoles

(i) Barrier Heights

The rectifying properties of metal/semiconductor contacts have been explained by Schottky[92] and Mott[93] in terms of the formation of a space-charge layer within the semiconductor due to different positions of the bulk Fermi levels E_F before electronic equilibrium is established. The schematic energy diagram of an idealized semiconductor/metal interface is shown in Fig. 6.[94–97] The semiconductor is considered to be free of surface states, which means that the bands are flat toward the surface (flatband conditions). Before the electrical contact between the two phases is established, the positions of their Fermi levels are given by the work functions (not considering different dipole contributions $\Delta\chi$ for different surfaces; see also the detailed discussion in Sections II and III.2). The corresponding work function difference $\Delta\Phi$ equivalent to the contact potential difference V_K leads to charge redistribution between the semiconductor and the metal and thus to band bending (formation of an extended space-charge layer) within the semiconductor, when electronic equilibrium is established. In the case of n-doped semiconductors in contact with high-work-function metals (as shown) in Fig. 6, a depletion

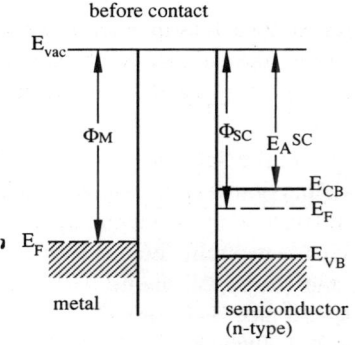 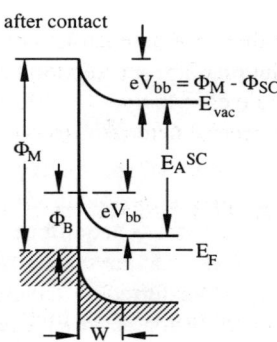

Figure 6. Formation of semiconductor/metal (Schottky) barriers: Ideal case without Fermi level pinning (Schottky limit). (Abbreviations are explained in the text.)

layer is formed. The band bending eV_{bb} or Schottky barrier height Φ_B is given by (see Fig. 6)

$$eV_K = eV_{bb} = \Phi_M - \Phi_{SC} = \Delta\Phi \quad \text{or} \quad \Phi_B = \Phi_M - E_A^{SC} \quad (25a)$$

For p-doped semiconductors, the following equations hold:

$$eV_K = eV_{bb} = \Phi_{SC} - \Phi_M = \Delta\Phi \quad \text{or} \quad \Phi_B = E_A^{SC} + E_G - \Phi_M \quad (25b)$$

Substituting the metal and thus changing the metal work function Φ_M also changes the band bending by the same amount. A plot of eV_{bb} versus Φ_M (or metal electronegativity X_M to avoid problems with surface dipoles) should give a straight line with a slope of unity, if ideal Schottky behavior is obtained. As the above derivation neglects the influence of surface dipole changes due to contact formation (see Section III.2 and Fig. 20), this slope S is usually taken as a criterion for classifying real (measured) semiconductor/metal contacts[98–100]:

$$\Phi_B = S(\Phi_M - E_A^{SC}) + \text{const.} \quad (26)$$

$S = d\Phi_B/d\Phi_M$ or $d\Phi_B/dX_M$ is a dimensionless factor ($0 < S \leq 1$–3).[98,100] It is considered to be unity[98] for "ideal" semiconductor/metal interfaces, which behave according to the Schottky limit. However, only rather ionic, large-band-gap semiconductors such as ZnS or TiO$_2$ have been found to follow the Schottky limit. For intimate metal contacts of most covalent semiconductors (e.g., Si, GaAs, ZnP), S is close to 0 (Bardeen limit[101]), which is due to the influence of a certain density of interface states (Fermi level pinning; see Section IV).

(ii) Space-Charge Layers

The actual space-charge layer formed in the Schottky limit depends on the relative positions of E_F or the work function Φ in the semiconductor and the metal. At equilibrium the Fermi levels in the semiconductor and the metal are at the same position, and charge neutrality demands that the charge transferred to the metal, Q_M, and the space charge within the semiconductor space-charge layer, Q_{SC}, balance each other[94–97,102,103]: $Q_M + Q_{SC} = 0$. The related potential drop (Galvani potential difference) $\Delta\phi$ compensates the contact potential difference $\Delta\Phi$. Due to the low concentration of mobile charge carriers within the semiconductor, the space-charge layer spreads into the bulk of the semiconductor. As a consequence, the related electrostatic potential φ within the semiconductor bulk varies continuously, leading to a bending of energy bands extending over a distance W of 100–10,000 Å. W is inversely dependent on the concentration of mobile charge carriers in the semiconductor (Fig. 7). As will be seen in Section III.2, $\Delta\varphi$ across the space-charge layer is equivalent to the Volta potential drop $\Delta\Psi$. Quantitative solutions for space-charge layers follow the one-dimensional Poisson equation [normal to the surface (z), neglecting lateral variations (x, y) of electric potentials] in conjunction with Fermi statistics for the occupation of electron states:

$$d^2\varphi(z)/d^2z = -\rho(z)/\varepsilon\varepsilon_0 \tag{27}$$

The density of charge $\rho(z)$ is given by the concentration of holes, p, electrons, n, charged donors, D^+, and acceptors, A^-, in the bulk of the semiconductor:

$$\rho(z) = e[D^+(z) - A^-(z) + p(z) - n(z)] \tag{28}$$

Three different kinds of space-charge layers may be classified, namely, accumulation layers, depletion layers, and inversion layers, as is schematically shown in Fig. 7 for an n-type semiconductor. Also included in this figure are schematic spatial distributions of charge $\rho(z)$ and electric potential $\varphi(z)$ and typical dimensions. The change of the electric potential (z) from the surface ($z = 0$) to the bulk ($z > W$) produces a potential difference eV_{bb} (equivalent to the contact potential difference eV_K, band bending) at the interface. As a consequence, the total energy of all

The Semiconductor/Electrolyte Interface

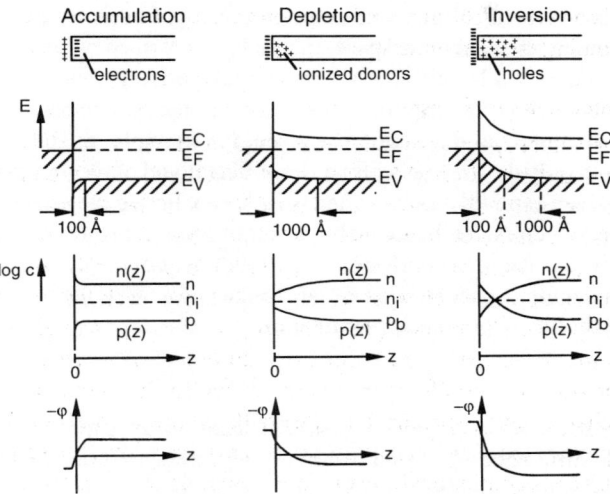

Figure 7. Different kinds of space-charge layers formed at n-type semiconductors. The changes in the energy scheme, the concentration of charge carriers, and the electric potential φ are shown as a function of the coordinate z normal to the surface (with reference to Ref. 104).

semiconductor electron states $E_i(z)$ (e.g., a core level or the conduction- and valence-band edges, E_{CB} and E_{VB}, respectively) becomes

$$E_i(z) = E_i(\text{fb}) + eV_{bb}(z) = E_i(\text{fb}) + e\varphi(z) \tag{29}$$

with $E_i(\text{fb})$ its energy for flatband conditions, which means that the electric potential has to be added to the energetic position of every electron state. (For positive potentials, this leads to an increase of the energetic position of the electron states.)

The type of space-charge layer that is formed depends on the relative positions of the bulk Fermi level and the surface Fermi level (given either by the contact phase or by surface/interface states). In the case of n-doped semiconductors (as shown in Fig. 7), the population of majority carriers is increased at the surface in accumulation layers. Depletion layers are characterized by reduced majority carrier concentrations, and for inversion layers the majority carrier type is reversed at the surface ($n \rightarrow p$ for the case shown). In the formation of accumulation and inversion layers, population changes of the band edges with a higher density of states (N_v or N_c) are involved, and the space-charge layers formed are less extended than the depletion layer (with N_D given by the doping concentration), as also shown in Fig. 7. A rigorous description of the procedure for calculation of the potential distribution is given, for example, in Refs. 102 and 103. Here, only the final result for a simplified model of the depletion layer will be given explicitly, as it is often used for analyzing the active junction of semiconductors. It is based on the assumption of only one type of charge carrier inside the semiconductor (e.g., oxidized donor levels for n-type semiconductors, concentration D^+) up to the width W of the space-charge layer (Schottky depletion approximation[105]):

$$\rho(z) = eD^+ \quad \text{for } 0 < z < W \quad (30)$$

Solving the Poisson equation (Eq. 27) leads to an expression for the potential difference V_{bb} (band bending), which corresponds to the difference between the electric potential at the surface, φ_s, and in the bulk, φ_b:

$$V_{bb} = \varphi_s - \varphi_b = (eD^+/2\varepsilon\varepsilon_0)W^2 \quad (31)$$

The spatial variation of the electric potential, $\varphi(z)$, is given by

$$\varphi(z) = -(eD^+/2\varepsilon\varepsilon_0)(W - z)^2 \quad (32)$$

With Eq. (31) a depletion layer width W of 1000 Å is calculated when a typical donor density of 10^{17} cm^{-3} and a barrier height of 1 eV are assumed. Space-charge layer widths in the inversion or accumulation region are considerably smaller due to higher concentrations of exchangeable charge carriers at the band edges ($N_c \approx N_v \approx 10^{19}$ cm^{-3} is typically assumed for the density of states at the band edge of semiconductors with parabolic energy bands [94]).

Within this approximation, the number of charge carriers Q_{SC} involved in the formation of space-charge layers and the differential capacity C_{SC} of the space-charge layer are given by:

$$Q_{SC} = (2\varepsilon\varepsilon_0 eD^+)^{1/2} V_{bb}^{1/2} \quad (33)$$

$$C_{SC} = [1/(2\varepsilon\varepsilon_0 eD^+)]^{1/2} V_{bb}^{-1/2} \quad (34)$$

Typically, 10^{11}–10^{13} charge carriers/cm^2 are involved in the formation of depletion layers, which corresponds to 0.1–10% of a surface monolayer.

It should be noted that in more elaborate treatments of space-charge layers,[102,103,94,95] the limiting case of a depletion layer leads to expressions similar to those presented in Eqs. (31)–(34), but V_{bb} has to be substituted by $(V_{bb} - kT/e)$.

2. Surface Dipoles at Metal/Semiconductor Interfaces

The contribution of surface and interface dipoles in contact formation has been neglected completely in the above derivation. However, the work function, which after Eqs. (25) defines the contact potential of two phases in intimate contact, contains a bulk term, the chemical potential μ, and a surface term χ, due to electrostatic dipole layers at the phase boundary to vacuum (see Section II). Generally, this surface dipole cannot be expected to remain unchanged when the phase boundary changes from vacuum to a contact phase. This is even more true when additional foreign atoms or molecules are adsorbed at the phase boundary. The conceptual differences in the representation of the interface potentials before and after contact formation are schematically summarized in Fig. 8 (cf. Fig. 6).

Figure 8a shows the energetic conditions before the contact is established: the electrochemical potentials of electrons in the two phases are $\eta_e^{SC} = E_F^{SC}$ and $\eta_e^M = E_F^M$ (by analogy to the definitions given in Section II; see Figs. 1, 2, and 4). For simplicity, it was assumed that no surface excess charge is found on both phases, i.e., $\sigma = 0$ (due to work-function differences of the different surface planes: see Section II and Ref. 40); thus, $\Psi = 0$ and η_e is equivalent to the inner potential α_e, with the contact potential difference V_K given by Eq. (25). However, when the two phases are brought into intimate contact (e.g., when semiconductor/metal junctions are formed), the surface dipole may change due to the changed boundary conditions:

$$\chi_M \rightarrow \chi'_M = \chi_M + \Delta\chi_M = g_M(\text{dip}) \tag{35a}$$

$$\chi_{SC} \rightarrow \chi'_{SC} = \chi_{SC} + \Delta\chi_{SC} = g_{SC}(\text{dip}) \tag{35b}$$

Here, $g(\text{dip})$ is the potential drop across the dipolar double layer of each phase, as in the electrochemical literature.[41,46,68] These dipole changes may result from changes of the electron polarization at the phase boundary of the solid (the free surface to vacuum leading to electron spill-out does not exist any more) or from changes of surface relaxation and reconstruction on the vacuum interface due to the influence of the contact phase (see

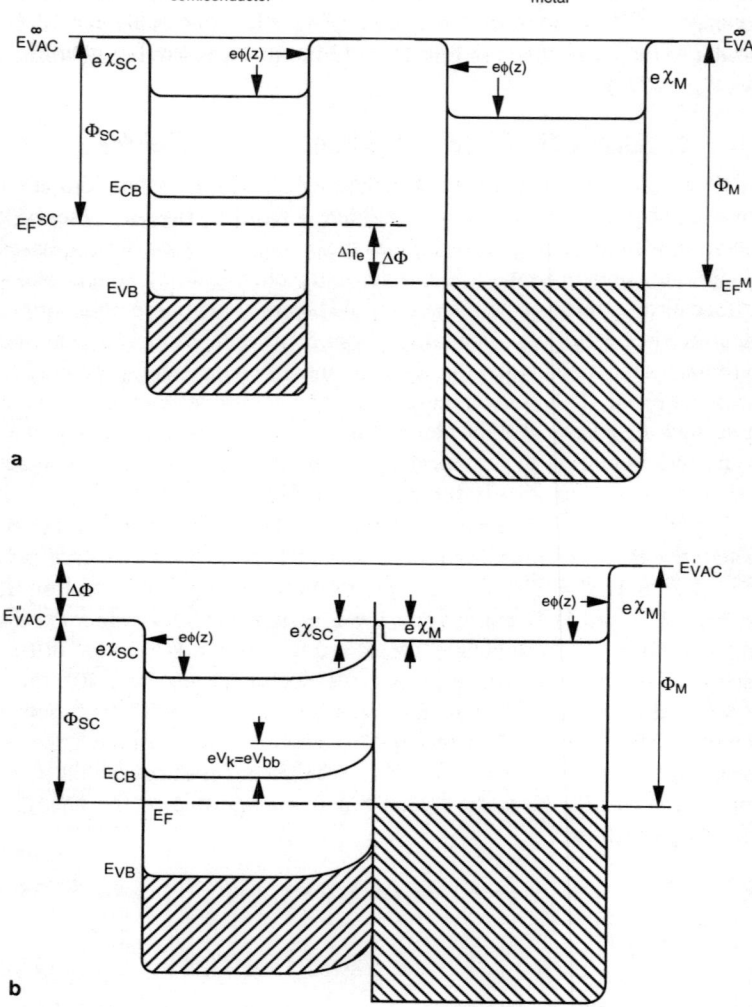

Figure 8. Schematic representation of the interface potentials at semiconductor/metal contacts: (a) Before contact formation; the difference in the positions of E_F or η_e is given by the difference in work function; (b) after contact formation; the changes in interface dipole potentials lead to modified values of eV_{bb}.

Section II). The two surface dipoles of the phases in contact may be combined into a new interface dipole (interface double layer) $^M\Delta^{SC}\chi$ representing the charge redistribution across the two phase boundaries on the atomic scale (about 2–3 Å):

$$^M\Delta^{SC}\chi = \chi'_M - \chi'_{SC} = (\chi_M + \Delta\chi_M) - (\chi_{SC} + \Delta\chi_{SC})$$

$$= g_M(\text{dip}) - g_{SC}(\text{dip}) = {^M\Delta^{SC}}g(\text{dip}) \qquad (36)$$

As a consequence, the contact potential at the phase boundary is composed of two parts. One part is due to charge transfer and thus changes the Volta (outer) potential Ψ of the phases. This part leads to an extended space-charge layer spreading into the bulk of the semiconductor ($\Psi \equiv \varphi$) (see Section III.1). The other contribution is due to the changed surface dipoles $\Delta\chi$. The "intimate" contact potential V^i_k may thus be written:

$$\Delta\Phi = \Phi_M - \Phi_{SC} = {^M\Delta^{SC}}\phi = {^M\Delta^{SC}}\Psi + {^M\Delta^{SC}}\chi \qquad (37)$$

$$eV^i_k = {^M\Delta^{SC}}\Psi = eV_{bb} = \Delta\Phi - {^M\Delta^{SC}}\chi = \Delta\Phi - (\Delta\chi_M + \Delta\chi_{SC}) \qquad (38)$$

It is assumed that the charge transfer σ, which changes the Volta potential Ψ of the phases in contact, can be treated independently from changes of dipolar surface potential $\Delta\chi$. In Fig. 8b the changed free surface dipoles of the semiconductor and the metal and the formation of the interface dipoles are shown at the phase boundary. In this example, they reduce the amount of band bending, as expected from Eq. (25). The hypothetical inner (Galvani) potential distribution $e\phi(z)$ across the interface is also illustrated in Fig. 8. The change in the Volta potential of the metal, $\Delta\Psi_M$, usually is negligible compared to the change in that of the semiconductor, $\Delta\Psi_{SC}$. The much lower charge density of the semiconductor implies a greatly enhanced Debye length ($d^{SC}_D/d^M_D = 10^4$ Å/0.1 Å), and as a consequence $^{SC}\Delta^M\Psi \to \Delta\Psi_{SC} = eV_{bb}$.[68,106] Thus, it follows that the original difference between the Fermi level positions (electrochemical potentials) of the semiconductor and the metal (given by, e.g., $\Delta\Phi$) leads to band bending within the semiconductor, eV_{bb} modified by the amount that is due to surface dipole changes ($\Delta\chi_{SC}, \Delta\chi_M$). The new positions of the near-surface vacuum levels E'_{vac} and E''_{vac} (Fig. 8b) are shifted from E^∞_{vac} by $e\Delta\phi_M$ and $e\Delta\phi_{SC}$, respectively; E'_{vac}, on the metal side, is shifted by $e\Delta\chi_M$ ($e\Delta\Psi_M \to 0$), and E''_{vac}, on the semiconductor side, by $e\Delta\phi_{SC} = e\Delta\Psi_{SC} + e\Delta\chi_{SC}$.

It should be noted at this point that the "pure" definition of the inner potential as a sum of outer potential and surface potential—in the Lange notation, $\phi = \Psi + \chi$ (presented in Section II)—is somewhat arbitrary, as the given distances at the solid/solid phase boundaries (a few angstroms) do not leave room for the outer potential to spread out (about 10^4 Å; see Fig. 1). For intimate metal/metal contacts, this partition is only of theoretical and conceptual value, as is also true for metal/electrolyte interfaces (see Ref. 31); the related double-layer dimensions due to charge transfer and surface dipole changes are comparable ($d = 1-3$ Å). However, for semiconductor/metal interfaces the changes in Volta potential due to charge transfer lead to the buildup of an extended space-charge layer (band bending) into the bulk of the semiconductor, whereas changes due to surface and interface dipoles lead to double-layer shifts on the atomic scale ($d = 1-3$ Å) at the phase boundary. The former effect shifts the surface Fermi level in the band gap (Fig. 8b), whereas the latter would lead to a relative shift of band-edge positions. In principle, these effects can be separated in model experiments of semiconductor junctions, e.g., using photoelectron spectroscopy (see Section VI).

The importance of such surface dipole changes (interface dipoles) in defining the barrier heights in Schottky barriers[95,97] and the band discontinuity in heterodevices[108,109] has been known for some time. For this reason, the Schottky limit is usually not defined as the absolute work-function difference [as given in Eq. (25)] but from the linear dependence of the barrier height on the metal work function or electronegativity [$S = 1$, Eq. (26)]. However, the relative contributions of interface dipole effects versus surface-state-induced Fermi level pinning effects have not been thoroughly differentiated in many cases. A controlled modification of band-edge positions (interface engineering) may be possible by controlling the interface structure and by selecting appropriate adsorbates.[37,107,111] The controlled modification of χ_{SC} and/or χ_M may induce a defined charge redistribution across the interface, which would lead to a defined dipolar double-layer potential drop $^{SC}\Delta^M\chi$ without introducing electronically active surface states into the band gap and thus inducing Fermi level pinning.

3. The Semiconductor/Electrolyte Contact

(i) Contact Potentials

The influence of surface dipole changes in contact formation is even of greater importance at semiconductor/electrolyte interfaces as an increased variety of contact species has to be taken into account. In addition

The Semiconductor/Electrolyte Interface

to the solvent molecules, the supporting electrolyte and the redox couple (and sometimes dissolved gases) may participate in specific interactions with the semiconductor surface, changing its surface dipole. In addition, the possibility of a reactive modification of the semiconductor surface must be considered. All these interactions may or may not be coupled to the formation of surface states in the semiconductor band gap.

The influence of surface dipoles leading to interface dipole potential drops must always be considered for semiconductor/electrolyte contacts, not only in the case of "Fermi level pinning" when surface (interface) states are involved. These interface dipoles may have drastic consequences for the electric potential distribution at electrolyte interfaces, which often leads to additional potential drops in electrochemical double layers (e.g., the Helmholtz layer across the interface).[9-12,67,68] The dipole effects and their spatial distribution will be described in this section; surface states and their consequences will be discussed in Section IV.

Before the contact between the semiconductor and the electrolyte is established, the electrochemical potential η_e^{El} or E_F^{El} of the electrons in the electrolyte solution and E_F^{SC} or η_e^{SC} of the electrons in the semiconductor (Fig. 9) are referred to the vacuum level and are given by the terms

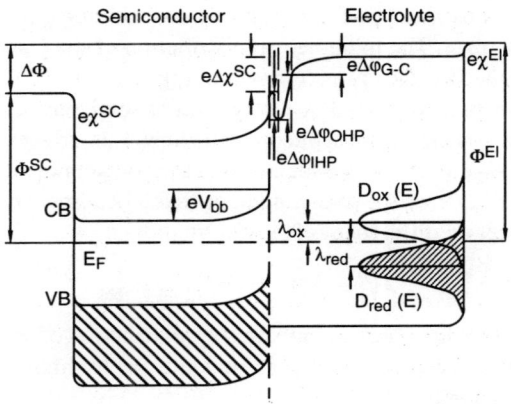

Figure 9. Schematic representation of the formation of Schottky barrier-like contacts at semiconductor/electrolyte interfaces. The outer boundaries refer to the vacuum interface. At the junction, the different potential contributions are schematically shown. In addition, the Gaussian distribution of occupied and unoccupied states of a redox couple is depicted.

introduced in Sections II.2 and II.3. In order to achieve thermodynamic (electronic) equilibrium between the phases, the original difference in electrochemical potentials, equal to the contact potential difference $\eta_e^{El} - \eta_e^{SC}$, is compensated by an electrostatic potential $^{El}\Delta^{SC}\Psi$ (Volta potential difference) due to exchanged excess charge on the semiconductor, Q_{SC} and the electrolyte, Q_{El}, and by changes of the dipole potentials $^{El}\Delta^{SC}\chi$ at the phase boundary, in close correspondence to what occurs at semiconductor/metal interfaces (see Section III.2); again the charge-neutrality condition holds:

$$Q_{SC} + Q_{El} = 0 \qquad (39)$$

$$\eta_e^{El} - \eta_e^{SC} = \Delta\Phi = eE(\text{red/ox}) - E_F^{SC} = {}^{El}\Delta^{SC}\Psi + {}^{El}\Delta^{SC}\chi \qquad (40)$$

The equivalence of $\eta_e^{SC} = E_F(\text{red/ox})$ to $E(\text{red/ox})$ has been shown in Section II.3. Even with the uncertainty in defining the absolute electrode potential for the electrolyte the original difference in electrochemical potentials, $\eta_e^{El} - \eta_e^{SC}$, can also be expressed by the difference in work functions, $\Phi^{SC} - \Phi^{El} = \Delta\Phi$.

The Volta potential difference $^{El}\Delta^{SC}\Psi$ can be measured by using an appropriate reference electrode. The Galvani potential ϕ of the different phases cannot be measured as the dipole contribution χ and its changes $\Delta\chi$ are not known. The hypothetical distribution of the Galvani potential $\phi(z)$ across the interface is also shown in Fig. 9.

In the originally defined Schottky limit of semiconductor/electrolyte junctions (by analogy to semiconductor/metal interfaces), the dipole potential change $e^{El}\Delta^{SC}\chi$ is neglected and the contact potential eV_K in the space-charge layer of the semiconductor (band bending eV_{bb}) is equal to the work function difference $\Delta\Phi$, given by

$$eV_K = \Delta\Phi = e^{SC}\Delta^{El}\psi = eV_{bb} \qquad (41)$$

As for metal/semiconductor contacts, the contribution of the Volta potential drop in the electrolyte $e\Delta\Psi_{El}$ can be neglected for high concentrations of the redox couple.

(ii) Band-Edge Shifts

In a more rigorous definition, the electrochemical double layers formed at the semiconductor/electrolyte interface will change the electron affinities of each phase by $\Delta\chi$ and will thus change the relative positions of the energy levels. As a consequence, the contact potential $^{SC}\Delta^{El}\psi$ inside the semiconductor space-charge layer (band bending eV_{bb}) is changed as shown in Fig. 9 [cf. Eq. (38)]

$$eV_{bb} = e^{SC}\Delta^{El}\psi = \Delta\Phi - (e\,\Delta\chi_{SC} - e\,\Delta\chi_{El}) \tag{42}$$

The changes in surface dipole contributions $\Delta\chi$ result from different dipole layers: (1) the change of the dipolar surface potential of the semiconductor from the vacuum interface to a contact phase, $\Delta\chi_{SC}$; and (2) the contributions of the inner Helmholtz (IHP), outer Helmholtz (OHP), and Gouy–Chapman (GC) $e\Delta\varphi_{IHP}$, $e\Delta\varphi_{OHP}$, and $e\Delta\varphi_{G-C}$, respectively, double layers, from adsorbed or electrostatically bound electrolyte components carrying ionic charge or dipoles. These double layers induce an additional potential drop of unknown magnitude across the semiconductor/electrolyte interface (Fig. 9). Little is known about the structure of the electrochemical double layer and the related potential distribution for semiconductor/electrolyte interfaces. Even for metal/electrolyte contacts, the microscopic structure of the double layer is still a subject of intensive research[41,68] involving electrochemical techniques,[13,14] various *in situ* spectroscopies,[15,16,18] and *ex situ*[16,18,19,21] and model experiments[20,23–33] with surface science techniques.

The contact potential difference is thus composed of several potential terms:

$$\eta_e^{El} - \eta_e^{SC} = \Delta\Phi = e^{SC}\Delta^{El}\psi + e^{El}\Delta^{SC}\chi$$

$$= eV_{bb} + e\Delta\chi_{SC} + e\Delta\varphi_{IHP} + e\Delta\varphi_{OHP} + e\Delta\varphi_{G-C} \tag{43}$$

Only the first two components are similar to solid-state semiconductor/metal Schottky barriers (Section III.2). The terms due to the inner and outer Helmholtz layer are equivalent to $\Delta\chi_M$ but are considered to be much more complex because of the more complex composition of the semiconductor/electrolyte phase boundary. Usually, $e\Delta\varphi_{G-C}$ may be neglected, when high ion concentrations are used in the experiments.

By analogy to metal/electrolyte interfaces, Eq. (43) may also be written as [see also Eqs. (35) and (36)]:

$$e^{SC}\Delta^{El}\phi = e^{SC}\Delta^{El}\Psi + e^{SC}\Delta^{El}\chi$$

$$= eV_{bb} + e^{SC}\Delta^{El}g(\text{ion}) + e^{SC}\Delta^{El}g(\text{dip}) \tag{44}$$

where $^{SC}\Delta^{El}g(\text{ion})$ accounts for the double-layer potential drop due to excess ionic charges of adsorbed ions across the phase boundary.[41,44]

The important difference with respect to metal/electrolyte interfaces is that the potential drop due to the electrochemical double layer including the ions in the electrolyte $^{SC}\Delta^{El}g(\text{ion})$ does not contain the complete Volta potential drop $e^{SC}\Delta^{El}\Psi$ (as for metals): band bending due to electron transfer, eV_{bb}, must also be considered. The positions of the band edges of semiconductors in contact with electrolyte solutions, which would give some information about $e^{SC}\Delta^{El}\chi$, when compared to work function measurements, e.g., in UHV, are usually obtained by capacity measurements.[10,11,112,113] Assuming the existence of a depletion layer, the following linear relationship between C_{SC}^{-2} and an externally applied potential E at the semiconductor is obtained [Mott–Schottky relation; see Eq. (34)]:

$$C_{SC}^{-2} = \frac{2}{\varepsilon \cdot \varepsilon_0 \cdot eN_d}\left(E - E_{fb} - \frac{kT}{e}\right) \quad (45)$$

where N_d is the bulk doping concentration. This simple relation is only valid when surface states and their influence can be neglected (see Refs. 10,11,112, and 113). E_{fb} is the electric potential of the semiconductor Fermi level, referred to the reference electrode, when no space-charge layer is formed (flatband potential). The difference between E_{fb} and the equilibrium redox potential E_0 in contact with the redox couple gives the value of band bending $eV_{bb} = E_{fb} - E_0$. When the energetic distance of E_F (the potential to which the externally applied potential is referred) to the conduction-band edge E_{CB} is known (ξ in Fig. 3 and Eq. (11)], the positions of the band edges in electrolyte solutions are determined. Due to the inherent problems with capacity measurements at semiconductor/electrolyte junctions, other techniques have also been applied to determine the flatband potential in electrolyte solution, such as electroreflectance measurements.[114–116] The flatband potential may also be inferred from transport experiments, but the results are less precise because carrier transport and charge-transfer processes may lead to nonequilibrium carrier-induced band-edge shifts. Examples of Mott–Schottky plots obtained for ZnO are shown in Fig. 10.[117,118] They indicate the well-known shifts of oxide semiconductor flatband potentials, which result from potential drops across the electrochemical double layer at the

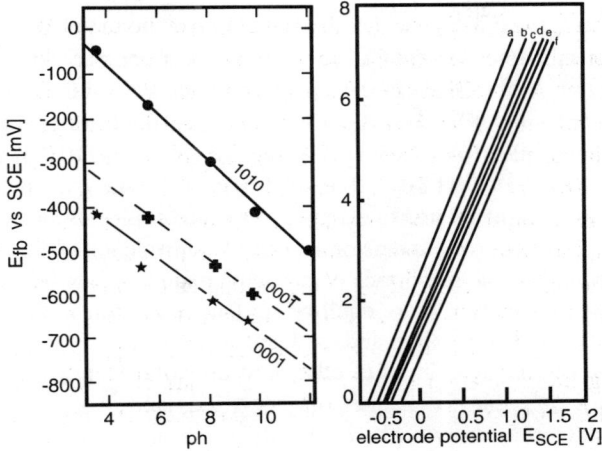

Figure 10. *Left*: Shift of E_{fb} as a function of pH for different ZnO surfaces. (After Ref. 118.) *Right*: Shifts of flatband potential E_{fb} (intersection with abscissa) of ZnO electrodes with decreasing pH (a–f).

solid/electrolyte phase boundary due to pH changes. The involved surface reactions are:

$$O_s + H^+ \rightleftharpoons O_sH^+ \tag{46a}$$

$$M_s + OH^- \rightleftharpoons M_sOH^- \tag{46b}$$

(O_s and M_s are surface sites for adsorption) and lead to a charging of the surface by adsorbed H^+ and OH^- ions. As a consequence, E_{fb} shifts by 59 mV/pH unit without introducing active surface states in the band gap. The charging of the double layer due to adsorbed ions changes $e^{SC}\Delta^{El}\chi$ by the contributions of $^{SC}\Delta^{El}g(\text{ion})$ (Eq. 44). There may still be an additional effect due to $^{SC}\Delta^{El}g(\text{dip})$. For no excess adsorbed ions, a point of zero zeta potential (pzzp) was defined[119]; this corresponds to the point of zero charge (pzc) of metal electrodes. For these conditions, no excess surface charge due to adsorbed ions is present at the semiconductor surface [$^{SC}\Delta^{El}g(\text{ion}) = 0$]. However, eV_{bb} is not necessarily zero as $^{SC}\Delta^{El}g(\text{ion})$ and eV_{bb} may approximately be considered to be independent of each other. The experimentally determined flatband potentials translate to an absolute work function with E^0_{abs}, which may be compared to the work functions measured in UHV. For ZnO as an example, this comparison gives the following[117,118,120]: (0001) Zn-terminated surface: $\Phi^{El} = 4.0$ eV (pH 10)–4.3 eV (pH 4), $\Phi^{vac} = 4.3$ eV; (000$\bar{1}$) O-terminated surface: $\Phi^{El} = 4.1$ eV (pH 10)–4.4 eV (pH 4), $\Phi^{vac} = 4.95$ eV; (10$\bar{1}$0) unpolar face: $\Phi^{El} = 4.3$ eV (pH 10)–4.6 eV (pH 4); $\Phi^{vac} = 4.6$ eV. The evident differences must be related to changes in surface composition and structure, which are not well understood so far. A dependence of E_{fb} on pH is only expected when the surface of the semiconductor may be treated in terms of Brønsted acid–base equilibrium (see, e.g., Ref. 121), which is usually only valid for oxides and oxidized surfaces.

However, the adsorption of other ions on surfaces may also lead to shifts of $^{SC}\Delta^{El}\chi$ due to changes of $^{SC}\Delta^{El}g(\text{ion})$ and $^{SC}\Delta^{El}g(\text{dip})$ at the semiconductor/electrolyte phase boundary. As an example of a pronounced shift of semiconductor band-edge positions, the energy diagram for CdSe in contact with the redox couples as Fe(CN)$_6^{3-/4-}$ and S^{2-}/S_n^{2-} is shown in Fig. 11.[66,122] When the positions of the band edges are compared to measurements in UHV (given by the threshold potential E_T), a strong interface dipole across the double layer must be assumed, which, in addition, is also very different for the two redox couples (see Fig. 11). Evidently, the contributions of interface double layers and/or dipoles to the band-edge positions may be rather large and are not well understood at the microscopic level. It should be noted at this point that in many investigations of semiconductor/electrolyte interfaces, well-defined single-crystalline surfaces are not used as in the case of, e.g., metals.[21] However, the use of such surfaces may be considered to be a precondition for a better understanding of interface effects (surface potential distributions) at semiconductor/electrolyte junctions.

As for metal/semiconductor contacts, a Schottky limit can be defined for the semiconductor/electrolyte interface by the slope $S = 1$ [$S = dV_{bb}/dE(\text{red/ox})$] for different redox potentials forming similar interface dipoles. In this case, the change in band bending eV_{bb} is given by

$$eV_{bb} = e\,[E_{fb} - E(\text{red/ox})] \cdot S + \text{const.} \tag{47}$$

with S approaching 1 and eV_0 approaching 0 (Schottky limit). In the case of Fermi level pinning, S approaches 0 and eV_0 gives the pinning position of surface states (see Section IV.3). However, there may be an experimental problem in determining eV_{bb} at electrolyte interfaces as E_{fb} is difficult

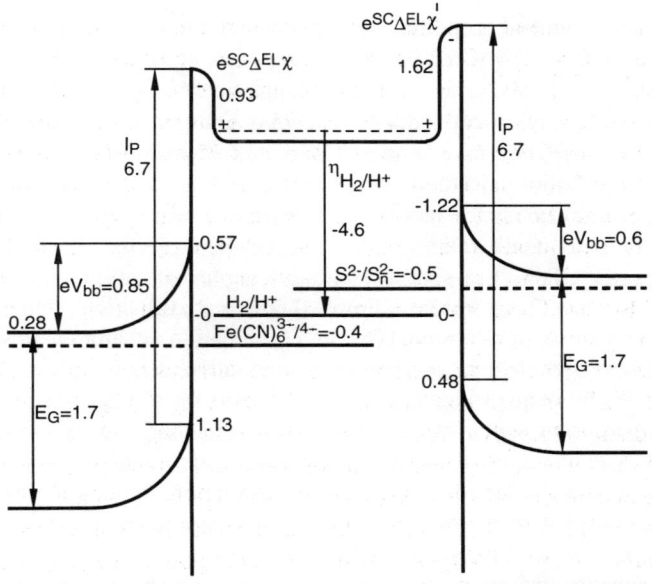

CdSe/Ferro-Ferricyanide   **CdSe/Sulfide - Polysulfide**

Figure 11. Energy diagram of CdSe electrodes in contact with two different redox couples, showing large (different) surface dipole potential drops at the semiconductor/electrolyte interface compared to that at the vacuum interface ($I_p = 6.7$ eV). (After Ref. 66; the diagram is not drawn to scale.)

to determine experimentally, especially for nonideal interfaces with surface states. In some cases, eV_{bb} may be derived from the saturation value of the photovoltage $U_{ph} \equiv -V_{bb}$.

IV. THE SEMICONDUCTOR CONTACT IN THE BARDEEN LIMIT

The Schottky limit of semiconductor junctions can usually only be achieved (see previous section) when active surface states, energetically situated in the band gap of the semiconductor, are not involved in junction formation. However, for many real semiconductor junctions this is not the case. The potential distribution across the interface and the charge-transfer processes are strongly modified by surface (interface) states and their physical and chemical properties. It is, in general, very easy to notice the electric influence and (opto)electronic response of surface (interface) states, and many deviations from ideal junction behavior have been assigned (especially in semiconductor electrochemistry) to often unspecified surface (interface) states. A more elaborate evaluation of their density of states distribution and their dynamic response is considerably more complex and is often not achieved with convincing accuracy; even more difficult is their identification on a microscopic, molecular level. This section is intended (1) to give a qualitative survey of different surface (interface) states, (2) to present a simple theoretical evaluation of surface states based on a tight-binding [linear combination of atomic orbital (LCAO)] approach, which may give some qualitative assignments but can also be made more quantitative, and, (3) to discuss the electronic equilibrium properties of semiconductor/metal and semiconductor/electrolyte junctions with a high concentration of interface states involved, which is classified as the Bardeen limit of junction formation. As this review is mostly restricted to semiconductor surfaces in electronic equilibrium, the influence of surface states on charge carrier dynamics and interfacial charge-exchange processes (nonequilibrium cases) will not be considered beyond some general remarks in Section IV.1.

1. Qualitative Classification of Surface States

In a very general viewpoint, all electronic eigenstates of the Schrödinger equation that have a maximum in their probability distribution function $|\psi_{SS}|^2$ at the semiconductor surface or interface may be classified as *surface states* (in contact with vacuum or with adsorbates) or *interface states* (in contact to a second bulk phase: solid state or electrolyte).[34-38,102-104] Usually only those states whose spatial distribution probability is completely restricted to the surface are called surface states (ψ_{SS} in Fig. 12a) whereas those with a considerable probability all over the semiconductor bulk are called *surface resonances* (ψ'_{SS} in Fig. 12b). The former are formed when the energy eigenvalues of the surface state E_{SS} have no corresponding solution of the bulk Bloch states: they are situated in a forbidden energy gap E_G of bulk states. The latter are formed when bulk Bloch states $\psi(k)$ hybridize with electron states of surface or adsorbate atoms (molecules) ψ_A. Their energy eigenstates E'_{SS} are in the energy range of bulk states: therefore, the occupation by electrons is identical to the bulk states. However, the spatial electron density distribution across the surface may deviate from the positive charge distribution of the ion

a) surface state

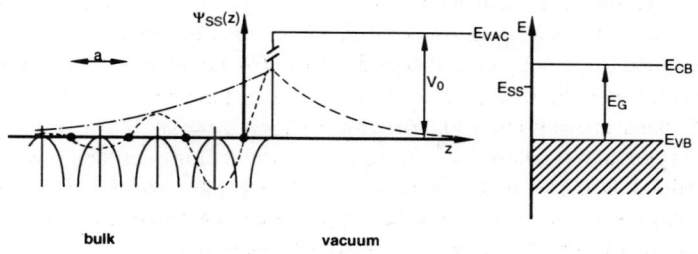

b) surface resonance state

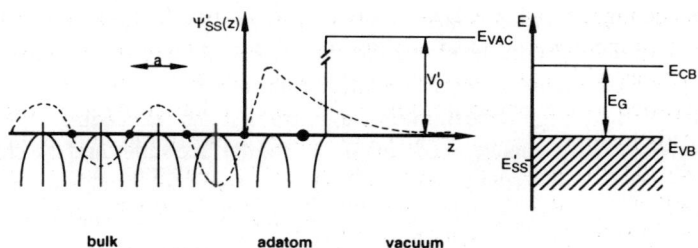

Figure 12. Schematic representation of the one-dimensional wave function ψ_{ss} of a surface state (a) and a surface resonance due to adatoms (b). Also shown are the energetic positions of these states in relation to the semiconductor band edges.

cores: as a consequence, a surface dipole layer is formed, which is responsible for the surface (interface) dipoles χ and their changes $\Delta\chi$ due to contact formation as discussed in the previous chapters.

(i) Intrinsic Surface States

Even for clean semiconductor surfaces in contact with vacuum, electronic surface states may exist within the band gap as a result of the broken periodic crystal symmetry at the surface. In a gedankenexperiment, one may separate the surface as a phase with its own Fermi level (charge neutrality level) for the surface states, E_F^{SS}, dependent on their density of states distribution D_{SS} and their intrinsic occupation, given by the number of electrons originally occupying the valence states of the surface atoms. In a first step no electronic equilibrium with the bulk of the solid is assumed. Then the occupation with electrons n and holes p is given by [by analogy to Eqs. (7)–(9)]:

$$n(E) = D_{SS}(E) \cdot f(E) \qquad (48a)$$

$$p(E) = D_{SS}(E) \cdot [1 - f(E)] \qquad (48b)$$

with $f(E)$ the Fermi distribution function. The position of E_F^{SS} is defined from the integral over the surface states with no excess positive or negative electron charge: the surface is electrically neutral.

If the surface Fermi level E_F^{SS} deviates from the bulk position E_F, electrons will be exchanged between the bulk and the surface, and the occupation of surface states will change (Fig. 13). In electronic equilibrium, the bulk Fermi level and the surface Fermi level are the same, and D_{SS} is changed in accordance with the number of exchanged electrons. However, for high concentrations of surface states (exceeding 10% of a monolayer), the change in occupation of the surface states and the shift of E_F^{SS} remain small (the number of charge carriers for establishing band bending is only about 10^{12} cm^{-2} compared to the surface state density of a monolayer of about 10^{15} cm^{-2}; see Fig. 13). As a consequence, a bending

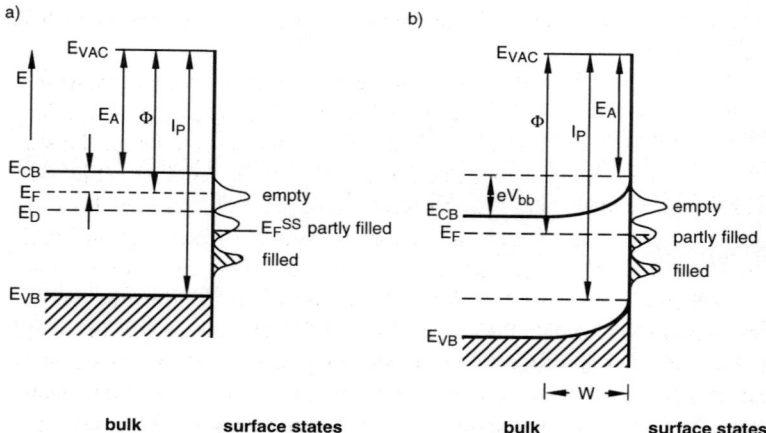

Figure 13. Formation of band bending eV_{bb} at n-type semiconductors due to the existence of surface states. The changed work function Φ and the unchanged electron affinity E_A and ionization potential I_p are indicated. E_D denotes donor energy levels.

of energy bands and an electric potential distribution will be induced at the surface, in a similar manner as with metal contacts (cf. Fig. 8). The band bending leads to a change of the work function Φ by eV_{bb}, whereas the ionization potential I_p and the electron affinity E_A remain unaffected. In general, compound semiconductors of the type AB may be classified according to their bonding character. A sharp transition occurs in their junction formation properties at an electronegativity difference between elements A and B of $\Delta X = 0.8$.[98-100] The more ionic compounds show more ideal behavior, e.g., in Schottky barrier formation (S approaches 1) whereas covalent semiconductors tend to show Fermi level pinning (S approaches 0). Accordingly, two types of intrinsic surface states are identified based on the type of core potential termination at the surface (Fig. 14): ionic surface states (Tamm states)[123] and covalent surface states (Shockley states).[124] The Tamm states result from an asymmetric Madelung potential at the surface. In chemical terms, this corresponds to a heteronuclear splitting of bonding interactions. These states have energetic positions close to the semiconductor band edges and are of cationic origin for states close to the conduction band and of anionic character for states close to the valence band (Fig. 14). Originally, these states are either completely empty or filled with two electrons. Thus, they will usually not induce band bending, and flatband conditions are expected, at least for the vacuum interface. Typical examples of semiconductors with Tamm states are oxides, e.g., ZnO and TiO_2, and wide-band-gap chalcogenides such as CdS and ZnS.

Shockley states result from a symmetric potential termination, which, in chemical terms, corresponds to a homonuclear splitting of covalent bonds, leading to unsaturated radical states (dangling bonds) at the surface. They are usually situated near the center of the band gap (Fig. 14). These surface states are very reactive and, without external bonding partners, may bond to each other, forming states of bonding and antibonding character (with respect to intersurface bonding interactions). Typical semiconductors forming such surface states are elemental semiconductors such as Si and Ge. The 3-5 semiconductors, e.g., GaAs and InP, are on the borderline between Shockley states and Tamm states as their A-B bonds are heteronuclear covalent bonds with some ionic character. Therefore, the distinction between Tamm and Shockley states may be overestimated in some cases and should be considered with care. Several review articles have been published on experimental[125-127] and theoretical[128-134] aspects of intrinsic surface states of "classical" semicon-

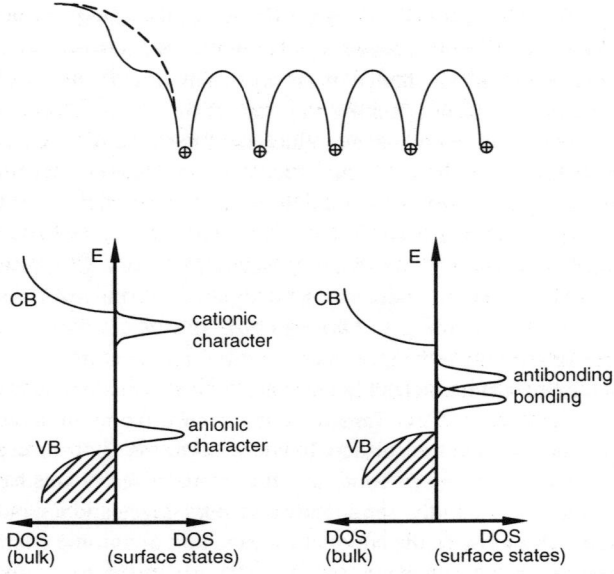

Figure 14. Schematic density of states distribution of Tamm (*left*) and Shockley (*right*) surface states.

ductors such as group 4 and 3–5 compounds and oxides (see also Refs. 34, 35, 38, and 46).

The electronic surface states of real semiconductor surfaces deviate in most cases from the theoretical models outlined above due to the possible relaxation and reconstruction of the surface.[34,35,60,134] The termination of the surface, which keeps the positions of atoms in the bulk unchanged, leads to a strong increase in Gibbs free energy (surface tension). As a consequence, the surface atoms tend to rearrange to find a relative minimum of the surface Gibbs free energy. The process that most easily occurs is a change of bonding distances and/or bonding angles without migration of surface atoms, a process that is already effective at room temperature (relaxation). The thermodynamically most stable surface reconstruction may be kinetically hindered, and heating of the sample will be necessary to allow reasonable mobility of surface atoms. Typical activation energies are in the range of 1–2 eV, which requires temperatures well above 500 K.[104]

Semiconducting transition-metal chalcogenides have surface states of considerably different character. They may best be described in terms of surface coordination chemistry by analogy to the chemistry of transition-metal coordination complexes.[82] The existence of bonding via directed metal d-states in the compounds may result in dangling bonds of metal d-character at the surface for certain crystal plane terminations. Unfortunately, no theoretical calculations of the electronic structure of transition-metal chalcogenide surfaces have been performed to our knowledge. In addition, only very few experimental results are available for semiconductor surfaces with nonsaturated metal d-orbital coordination.

(ii) Extrinsic Surface States

The distinction between intrinsic and extrinsic surface states at semiconductor surfaces is based on the source of the new electronic states. Extrinsic surface states result from interactions with extrinsic sources, such as strongly interacting adsorbates but also bulk phases, such as metals or electrolytes, in contact with the semiconductor, which lead to the formation of new electronic states.

These extrinsic surface states may be generated intentionally by adsorption of a particular species in order to study the properties of interfaces or to modify the junction properties in a controlled way (interface engineering).[36-38,111] In many cases, the formation of extrinsic surface states may be an unwelcome, unavoidable result of surface preparation and contact formation which is strongly dependent on compound properties. The action of extrinsic surface states strongly impairs the junction character of the idealized interfaces. Many of the fundamental investigations on semiconductor junction properties deal with the underlying processes, which determine interface formation in real semiconductor devices. Also of interest are the mechanism of barrier formation and its dependence on the physical and chemical properties of the contact phases. For the classical semiconductors and their solid-state junctions (semiconductor/metal and semiconductor heterojunctions), a large number of experimental investigations have been reported, and there is a reasonable understanding of the mechanisms involved[34-38,94-97] (even if this has been questioned in some recent investigations[135,136]). However, for novel semiconductors, such as the transition-metal chalcogenides, or for electrolyte interfaces, only a few experimental results on the molecular arrangement of surfaces and interfaces are available so far, and only tentative models may be proposed.

It should be emphasized that intrinsic and extrinsic surface states mutually exclude each other. Intrinsic surface states, and especially Shockley states, are generally related to surface sites of increased reactivity. Therefore, adsorbates preferentially interact with these states, and, as a consequence, new (extrinsic) electronic states are formed. The energetic positions of the extrinsic surface states are given by a bonding–antibonding coupling of intrinsic surface states and adsorbate electron states (see Section IV.2). This is also true when interfaces to contact phases are established. The resulting interface states and their density of states distribution D_{SS} are strongly influenced by bonding–antibonding interactions with the interface atoms of the contact phase. Their physical and chemical character is thus different from that of intrinsic surface states at vacuum interfaces.

Finally, some remarks should also be made regarding surface states in nonequilibrium cases. Generally, at solid-state junctions, e.g., to wide-band-gap semiconductors or oxides, with localized electron states, the occupation of surface states by electrons may also influence their energetic position. Only in cases in which the surface states have extended (itinerant) wave functions along the surface, the electron–electron correlation energy U (Hubbard correlation energy) may be neglected, and the position of the state is unaffected by its occupation. The electronic properties of solids can only be treated by normal occupation statistics, as given by Eqs. (7)–(9), when the width of the bands exceeds the Hubbard correlation energy.[134,46] The contribution of the electron coupling term to the change in the position of surface states with changes in occupation number has been elaborated for surfaces and bulk states of transition-metal solids, e.g., oxides[137–139] and defect states in amorphous Si,[140] but it must also be considered for localized states at semiconductor interfaces.[141–143] In these cases it is still possible to define a surface Fermi level for the occupation statistics of the surface states [as for electron states in electrolyte solutions; cf. Section II.3, Eqs. (15)–(17)], but the Fermi level may not necessarily be situated within D_{SS}, as shown in Fig. 13. It is expected that such localized states may easily be formed at semiconductor/electrolyte interfaces owing to strong bonding of surface atoms to adsorbates, which reduces the interaction between neighboring sites on the surface.

In addition, it should also be emphasized that for the description of nonequilibrium cases, e.g., reactions at semiconductor/electrolyte interfaces, a special kind of surface state must be considered which (probably) has no analog in solid-state junctions. Especially in more complex multistep redox reactions, for which a strong interaction of chemical intermediates with the semiconductor surface has to be taken into account, e.g., in photoelectrocatalysis or surface etching, *transient surface states* may be formed. Transient surface states are defined as short-lived transition states in multistep redox reactions. Their influence in photoelectrochemical reactions may be "bad," when they function as recombination centers, or "good," when they function as "catalysts" for charge transfer.[68,144,145] Chemical reaction steps may be involved in the formation and forward reactions of these transient surface states, and they usually do not exist under equilibrium conditions. As a special illustration, the hypothetical reaction sequence of the $FeS_2(100)$ surface with H_2O is shown in Fig. 15. It shows the assumed changes of extrinsic surface states due to adsorbed H_2O and $OH\cdot$ formation based on a ligand-field treatment of the coordination interaction.[3] It is extremely difficult to obtain information on such transient surface states as the response of the majority carriers in capacity measurements may be different from the response of the minority carriers in photocurrent or photocurrent modulation experiments. Some aspects of

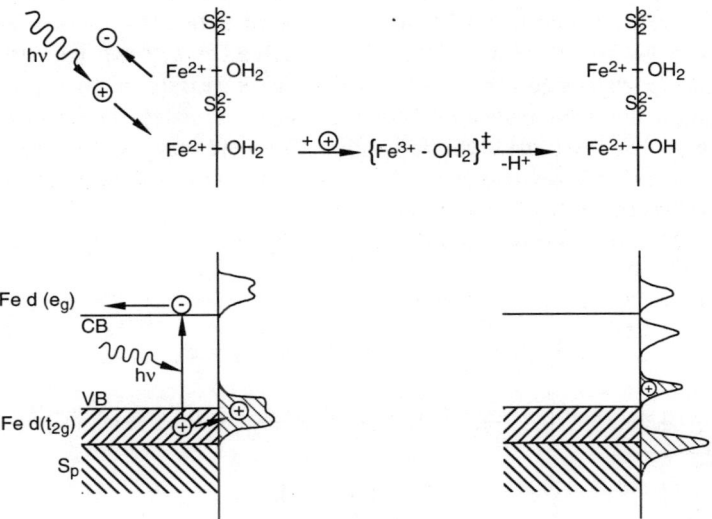

Figure 15. Transient surface states formed by trapping of light-induced holes in extrinsic surface states and hypothetical reaction sequence on $FeS_2(100)$ surfaces interacting with H_2O.

the problems involved are discussed in several recent reviews.[68,144,145] However, many of the assignments of electronically detected states to specific surface moieties on a microscopic level are very speculative as it is considerably more difficult to get spectroscopic information on these transient surface states compared to the static ones dealt with in this chapter. However, the identification of the molecular structure of the static surface states is a precondition for "serious speculation" on transient states and their role in multistep charge-transfer reactions, e.g., in photoelectrocatalysis. The quantum-mechanical treatment of electrochemical processes has been discussed in, e.g., Refs. 68 and 146.

2. Quantum Mechanics of Surface States

The quantitative quantum-mechanical treatment of semiconductor surface states is described in detail in a number of review articles and monographs.[128–134] However, these calculations are mostly restricted to clean surfaces in contact with vacuum and to solid contact phases such as metals and semiconductors. To our knowledge, no quantitative theoretical calculations have been performed on possible surface states that may be formed at semiconductor/electrolyte interfaces. For this reason, an introduction to a tight-binding (LCAO) treatment of surface states will be given here. This treatment may be used in a qualitative way but may also be extended to semiquantitative calculations. With the tight-binding approximation, the semiconductor bulk as well as surface states can be treated in a way that most chemists are familiar with, and many of the principal properties of these states can be deduced on the basis of simple symmetry and two-center bonding considerations.

The one-electron Schrödinger equation of the solid may be written as[43,150,151]

$$H\psi_{nk}(\mathbf{r}) = \left(-\frac{\hbar^2}{2m}\nabla^2 + U(\mathbf{r})\right)\psi_{nk}(\mathbf{r}) = E_{nk}\psi_{nk}(\mathbf{r}) \tag{49}$$

with a periodic potential $U(\mathbf{r})$ reflecting the periodicity of the Bravais lattice,

$$U(\mathbf{r}) = U(\mathbf{r} + \mathbf{R}) \tag{50}$$

which accounts for the translational symmetry of the three-dimensional solid along the fundamental directions x, y, and z by the Bravais lattice vectors $\mathbf{R} = n\mathbf{a} + m\mathbf{b} + n\mathbf{c}$ with $\mathbf{a}$, $\mathbf{b}$, and $\mathbf{c}$ as basis vectors.

Within the solid, the atom core potentials reflect the translational symmetry of the solid, and thus the solutions of the Schrödinger equation for the electrons also show this translational symmetry:

$$\psi_{n\mathbf{k}}(\mathbf{r}) = \psi_{n\mathbf{k}}(\mathbf{r}+\mathbf{R}) \tag{51}$$

Solutions of the electron eigenstates $\psi_{n\mathbf{k}}(\mathbf{r})$ (Bloch functions) have the form of planar waves times a function reflecting the periodicity of the crystal (Bravais lattice):

$$\psi_{n\mathbf{k}}(\mathbf{r}) = u_{n\mathbf{k}}(\mathbf{r})\exp(i\mathbf{kr}) = u_{n\mathbf{k}}(\mathbf{r}+\mathbf{R})\exp(i\mathbf{kr}) \tag{52}$$

On the surface the translational symmetry of the potential $U(\mathbf{r})$ is broken along **c** (normal to the surface) but retained along **a** and **b** (parallel to the surface):

$$u(\mathbf{r}_p + \mathbf{R}_p) = U(\mathbf{r}_p); \quad U(\mathbf{r}_n + \mathbf{R}_n) = U(\mathbf{r}_n) \quad \text{for } z < 0$$

$$u(\mathbf{r}_n + \mathbf{R}_n) \neq U(\mathbf{r}_n) \quad \text{for } z > 0 \tag{53}$$

The phase boundary is set at $z = 0$.

As a consequence, the wave functions of a surface state may be split into one component with a wave vector normal to the surface, $\mathbf{k}_n$, and a second component with wave vectors parallel to the surface, $\mathbf{k}_p$, retaining the symmetry of the bulk. The solution of the Schrödinger equation normal to the surface leads either to wave functions with exponentially decaying parts into the solid (with complex wave vectors **k**) and into vacuum (surface states) or wave function, which exhibit a maximum at the surface (surface resonances), as shown in Fig. 12. The solution of the Schrödinger equation parallel to the surface leads to (two-dimensional) Bloch waves with energy eigenvalues that are a function of the wave vector $\mathbf{k}_p$ parallel to the surface (with band dispersions along $\mathbf{k}_p$).

$$\psi_{SS}(\mathbf{r}_p, z) = u_{\mathbf{k}_p} \exp(i\mathbf{k}_p \mathbf{r}_p)\, \psi_{SS}(z) \tag{54}$$

The calculation of $\psi_{SS}(z)$ for a one-dimensional lattice leads to solutions of the form[152]

$$\psi_{SS}(z) = e^{\kappa z} \cos\left(\frac{1}{2}\mathbf{g}z + \delta\right) \quad z < 0 \tag{55}$$

$$\psi_{SS}(z) = e^{-qz} \qquad z > 0 \qquad (56)$$

where **g** is the reciprocal lattice vector, κ is a complex wave vector, and $q = (V_0 - E)^{1/2}$ with V_0 the surface potential step to vacuum. Equation (55) gives exponentially decaying and modulated wave functions into the solid, and Eq. (56) gives exponentially decaying wave functions into vacuum. Real surface states of energy E_{SS} exist when both parts of the surface-state wave function can be matched at $z = 0$.[128–134]

Surface states for three-dimensional solids are shown schematically in Fig. 16.[34] A surface state is described by its energy level E_{SS} and its wave vector $\mathbf{k}_p$ parallel to the surface. For bulk states, the wave vector $\mathbf{k}_n$

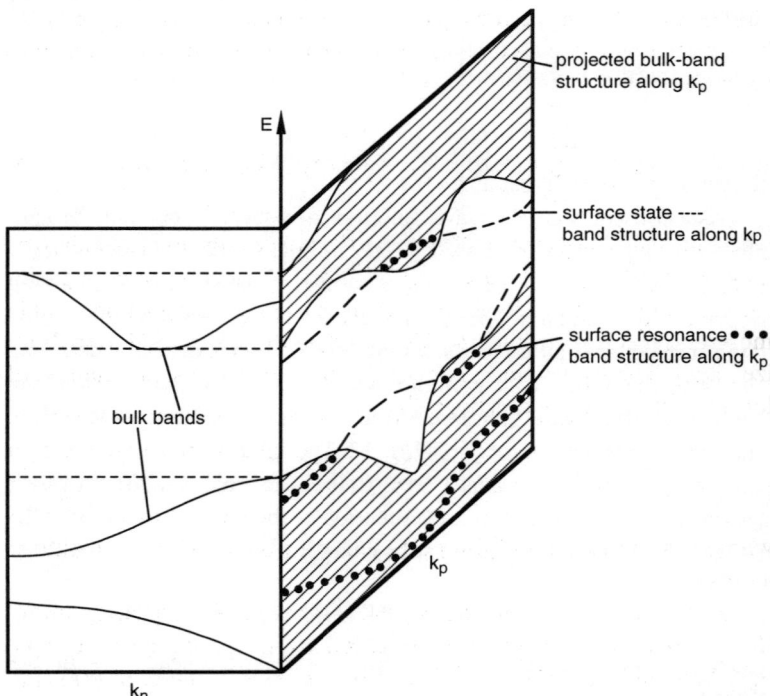

Figure 16. Hypothetical electronic band structure of a three-dimensional semiconductor. The bulk bands and the projected bulk-band structure are shown in relation to surface-state and surface-resonance bands.

normal to the surface is also defined. In Fig. 16 the bulk bands are shown together with their projections onto a special surface plane (hatched area). True surface states ψ_{SS} have an energy level E_{SS} in a fundamental gap of the projected bulk band structure. They give rise to wave functions normal to the surface as shown in Fig. 12a. Surface resonance states are formed for surface states that for certain $\mathbf{k}_p$ values are degenerate with bulk Bloch states and will hybridize with these, forming wave functions as shown in Fig. 12b.

(i) LCAO Approach for Intrinsic Surface States

In the tight-binding approach, which may easily be used for qualitative estimation but also for quantitative calculations of surface states and the band structure of complex solids (see, e.g., some recent reviews by Hoffmann[148,149]), the periodic part $u_\mathbf{k}$ in the Bloch functions (Eq. 52) can be replaced by a linear combination of localized atomic wave functions (orbitals):

$$\Psi_k(\mathbf{r}) = \sum_{n=1}^{N} c_n \cdot \phi(\mathbf{r} - \mathbf{R}_n) = \sum_{n=1}^{N} \exp(i\mathbf{k}\mathbf{R}_n) \cdot \phi(\mathbf{r} - \mathbf{R}_n) \qquad (57)$$

where the index n runs over all atoms in the lattice. For simple solids, $\phi(\mathbf{r})$ may be an atomic wave function.[148,149] In more general cases, e.g., for several atoms in the unit cell, for stronger chemical interactions within molecular species or within the first coordination sphere, and for hybridized atomic wave functions, one may expand $\phi(\mathbf{r})$ in a small number of localized atomic wave functions[151]:

$$\phi(\mathbf{r}) = \sum_{\mu,i} b_i^\mu \chi_i^\mu (\mathbf{r}) \qquad (58)$$

with $\chi_i^\mu(\mathbf{r})$ the μth atomic wave function (orbital) of the constituent atoms i.

For the calculation of surface states, the exponent $\exp(i\mathbf{k}\mathbf{R}_n)$ in Eq. (57) is replaced by the product $\exp(i\mathbf{k}_p \cdot \mathbf{R}_n) \cdot \exp(i\mathbf{k}_n z)$. With the use of the tight-binding or LCAO approach and by a proper selection of $\phi(\mathbf{r})$ (Eq. 58), the electronic interaction of electron eigenstates of surface atoms with the atoms of the first coordination sphere, with the other atoms of the solid, and with adsorbates can be described in a straightforward manner.

For general cases, the energy eigenvalues ε of the one-electron Schrödinger equation are given by solution of the secular determinant

$$\det [H_{ij}^{\mu\nu} - \varepsilon s_{ij}^{\mu\nu}] = 0 \tag{59}$$

where the overlap matrix $s_{ij}^{\mu\nu} = \langle \chi_i^\mu | \chi_j^\nu \rangle$ contains the overlap of the atomic orbitals χ_i^μ, χ_j^ν, and the energy matrix contains the Coulomb integrals $H_{ii}^{\mu\mu} = \langle \chi_i^\mu | H | \mu \rangle$ and the exchange integrals $H_{ij}^{\mu\nu} = \langle \chi_i^\mu | H | \chi_j^\nu \rangle$ of the atomic orbitals. The value of electron coupling is mainly determined by:

1. the relative positions (Coulomb potentials or atomic valence shell ionization potentials) of the atomic valence orbitals, $H_{ii}^{\mu\mu}$, and the changes in the relative positions induced by the net charge on the atoms due to bonding interactions

2. the magnitude of intra- and interatomic electron–electron interactions (electron resonance integrals), $H_{ij}^{\mu\nu}$

Quantum-mechanical calculations of surface-state energies and their corresponding wave functions depend on the quality of the approximations used for solving the secular determinant (Eq. 59). In many cases, reasonable results can be obtained by evaluating two-center resonance integrals of the atomic bonds in the solid and at the surface as introduced by Harrison.[147] As an example, the bulk and surface properties of zinc blende-type semiconductors such as Si and GaAs will be deduced (Fig. 17). In the bulk, the atoms are coordinated tetrahedrally: thus, the valence band is composed of four sp^3-hybrid orbitals $|h^i\rangle$ of the type

$$|h^i\rangle = \frac{1}{2} [\langle s^i| + |p_x^i\rangle + |p_y^i\rangle + |p_z^i\rangle] \tag{60}$$

with the sign of the p atomic orbitals being either positive or negative for the different tetrahedral directions (for the element i). The energy eigenvalues of these hybrids are given by

$$\varepsilon_h^i = (\varepsilon_s^i + 3\varepsilon_p^i)/4 \tag{61}$$

where $\varepsilon_{s,p}^i$ are the atomic energy eigenvalues. With these hybrid orbitals, bonding and antibonding combinations are formed in the same way as for simple diatomic molecules:

$$\varepsilon_h^{b,a} = [(\varepsilon_h^i + \varepsilon_h^j)/2] \pm \{ [(\varepsilon_h^i - \varepsilon_h^j)^2/4] + H_{ij}^2\}^{1/2} \tag{62}$$

The positive sign is for the antibonding state (a), and the negative sign is for the bonding state (b); $H_{ij} = -V_2$ is the so-called hybrid covalent

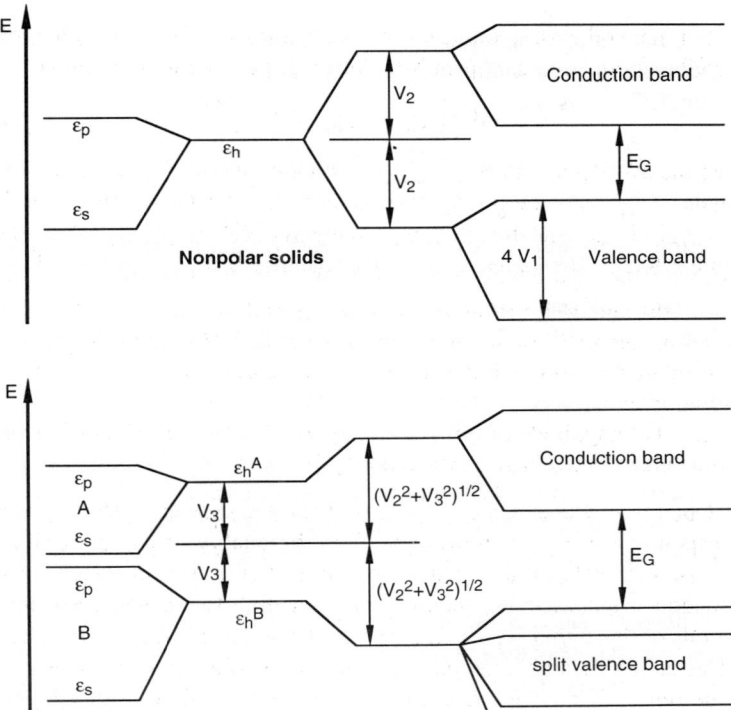

Figure 17. Successive band formation in the LCAO scheme for a tetrahedral semiconductor such as Si (*top*) and GaAs (*bottom*). Hybrid orbitals are formed, which are combined to bonding and antibonding states; these are taken as basis of the Bloch sums and form band states.

energy,[147] and the energy difference $(\varepsilon_h^i - \varepsilon_h^j)/2 = V_3$ is the hybrid polar energy (it is larger for more ionic semiconductors). The resonance integrals H_{ij} for orbitals of neighboring atoms pointing toward each other are evaluated from matrix elements of the atomic orbitals involved[147,35,46]:

$$-H_{ij} = V_2 = \langle h_i | H | h_j \rangle = (-V_{ss\sigma} + 2\sqrt{3}\,V_{sp\sigma} + 3V_{pp\pi})/4 \quad (63)$$

The interatomic matrix elements $V_{ll',\lambda}$ for the different orbitals $l\,(= s, p, d)$ and bonding types/interactions $\lambda\,(= \sigma, \pi, \delta)$ are evaluated from atomic or corrected atomic values (for details, refer to Refs. 147, 35, and 46).

The bulk Bloch states are formed by combining the hybrid states of the antibonding and bonding combinations as Bloch sums in accordance with Eq. (57):

$$\psi_k(\mathbf{r}) = \sum_i \sum_n e^{i\mathbf{k}\mathbf{R}_n} \cdot |h_i(\mathbf{r} - \mathbf{R}_n)\rangle \qquad (64)$$

where the outer sum is over the four bonding and antibonding hybrid orbitals of the zinc blende-type semiconductor. As a consequence, the bond orbitals and energy levels split into bands (Fig. 17). The width of the bands is given by $4V_1 = 1/(\varepsilon_p^i - \varepsilon_s^i)$ and depends upon the *sp* splitting and the interatomic interaction.

The above scheme of bulk electron state construction is represented in Fig. 17 for a nonpolar semiconductor like Si and a polar semiconductor like GaAs. It should be noted at this point that the absolute energies obtained by this procedure deviate from experimental values by about 3.8 eV.[147] This offset may be eliminated by subsequent calibration procedures or by the use of more complicated procedures for solution of the secular determinant.[128–134,147]

Within the surface, the formation of intrinsic surface states corresponds to missing bonding partners in the standard bulk coordination and bonding geometry. The positions of the hybrid orbitals before forming bonding and antibonding combinations may be taken as the approximate position of dangling-bond surface states (compare Fig. 17 to Fig. 14). Relaxation of the surface atom positions changes the bonding distances and thus the distance-dependent resonance integrals: as a consequence, the surface-state energies shift with respect to the band edges. Reconstruction of the surface is often due to extra intersurface bonding interactions that have different geometries and usually smaller resonance integrals than the bulk bonding. The intrinsic surface-state energies of many semiconductors surfaces may be found in e.g., Refs. 34, 128–134, 125, and 126.

(ii) LCAO Approach for Extrinsic Surface States

The bonding interaction with adsorbates (contact phases) is also of special interest for semiconductor/electrolyte interfaces. It is described as the formation of a surface molecule based on a two-center bonding interaction of the adsorbate electron states $|A\rangle$ with the hybrid orbitals $|h^i\rangle$ introduced in the previous section or with a small number of LCAO–Bloch states. The contribution of all other atoms representing the solid can

The Semiconductor/Electrolyte Interface

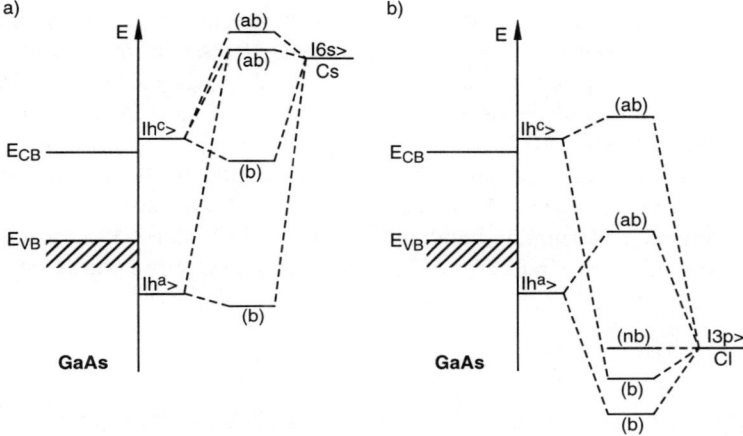

Figure 18. LCAO energy level diagram for adatom bonding of Cs and Cl on GaAs surfaces. The energetic positions of the surface states and surface resonances and their origin are schematically shown.

be considered as a perturbation of the resulting molecular eigenstates. This approach is exemplified by the LCAO energy level diagram shown in Fig. 18 for the formation of extrinsic surface states (surface resonances) by adatom bonding on polar semiconductors such as GaAs.[35] In this case the bonding and antibonding energy levels of the surface molecule are given by [by analogy to Eq. (62)]:

$$\varepsilon_{b,a} = [(\varepsilon_h^i + \varepsilon_A)/2] \pm \{[(\varepsilon_h^i - \varepsilon_A)^2/4] + H_{iA}^2\}^{1/2} \qquad (65)$$

The positive and negative signs correspond to the antibonding and the bonding state, respectively. As the adatom (adsorbate) energy level ε_A will be changed from the atomic energy value by some charge transfer from the solid, i.e., $\varepsilon_A \to \varepsilon_A'$, one needs to perform a tight-binding Hartree–Fock approximation for the Coulomb integrals $\langle A|H|A\rangle = \varepsilon_A$ and $\langle h^i|H|h^i\rangle = \varepsilon_h^i$. The resonance integral $\langle A|H|h^i\rangle = H_{iA}$ may again be evaluated in the simplest case based on simple two-center atomic matrix elements [as given in Eq. (63)]. The example given in Fig. 18 shows the interaction of Cs and Cl with the cation $|h^c\rangle$ and anion $|h^a\rangle$ dangling-bond states of GaAs(110). Assuming the hybrid orbitals $|h^c\rangle$ to be empty and $|h^a\rangle$ to be doubly occupied, the bonding combinations ($|h^c\rangle - |6s\rangle$) of the Ga–Cs bond are donor states, whereas the antibonding combinations ($|h^a\rangle + |3p\rangle$) of the As–Cl bond are acceptor states. (The bonding states are occupied by two electrons and the antibonding states are empty when bonding of two adsorbate atoms to both cation and anion hybrid states is considered.)

In more elaborate approaches to the calculation of adsorbate interactions with semiconductor surfaces, $|h_i\rangle$ has to be replaced by a linear combination of dangling-bond states and bulk Bloch states $|\mathbf{k}_j\rangle$ to account for the influence of bulk atoms. The semiconductor electron states are then of the type

$$\psi_{sc}^{ij} = a_i |h_i\rangle + \sum_j b_j |\mathbf{k}_j\rangle \tag{66}$$

The coefficient a_i may also be zero when no hybrid dangling bonds are involved [$\psi_{SC} \to \psi_k(\mathbf{r})$]. The formation of such surface resonance states is well known for adsorbates on metal surfaces.[153–155,149,146,134] As a consequence of interaction, the adsorbate states broaden and shift in energy.

For surface sites on transition-metal compounds, the adsorbates can be considered as extra ligands of differing ligand field strength. The electronic character and energetic position of a surface coordination site of reduced coordination symmetry (intrinsic surface state) will be changed by the adsorption of an extra ligand X' (MX$_5$X'). The strength of the bonding interaction $H_{Ld} = \langle d|H|A \rangle$ determines the energetic position of the resulting extrinsic surface states. Strongly interacting ligands may even push the surfaces states beyond the band edges, thus passivating them. In a first approximation, the bonding character can qualitatively be estimated from coordination compounds of comparable local symmetry. For an octahedrally coordinated compound, e.g., FeS$_2$, the intrinsic surface states result from the reduction in symmetry on going from the idealized bulk octahedral coordination of the metal to a square pyramid (idealized C_{4v} symmetry) at the surface, which results from the absence of a ligand at one chalcogenide coordination site. As a consequence the degeneracy of the t_{2g} levels and the e_g levels is removed.[3] The extrinsic surface states result from the adsorption of additional ligands, e.g., H$_2$O, at these coordinatively unsaturated sites. The resulting change in metal d-electron state splitting (energy positions of extrinsic states) may be estimated from the ligand-field strength or from semiquantitative extended Hückel calculations of the exchange integrals H_{ij} of neighboring sites[148,149] (cf. Fig. 15).

3. Fermi Level Pinning

For many semiconductor/metal and semiconductor/electrolyte interfaces, the ideal junction behavior defined by the Schottky limit (Section III) is not observed experimentally (see, e.g., Refs. 94–96, 34–36, 115, 145, 156, and 157). The reason is the influence of a high concentration of interface (surface) states in the semiconductor band gap on contact formation. In many cases the term surface state is even used when the interaction with the contact phase has been established and interface states have been formed. In the following, we will use the term "interface states" for solid-state contacts but will use the somewhat incorrect term "surface states" for electrolyte contacts, as this term is well established in literature.

(i) Semiconductor/Metal Contacts

In the nonideal case of Fermi level pinning, the charge redistribution between the semiconductor and the metal, accounting for the difference in electrochemical potentials $\Delta\Phi = \Delta\eta = eV_k$ (contact potential difference), also involves the charging Q_{IS} of the interface states: $Q_{SC} + Q_{IS} + Q_M = 0$. As a consequence, an additional double-layer potential drop $\Delta\Psi_{dl}$ may exist between the semiconductor surface and the metal contact. Its distance d_{dl} is on the order of atomic dimensions, i.e., some angstroms. For very high concentrations of interface states, the Fermi level in the semiconductor bulk remains pinned by these states, and the contact potential leads only to a double-layer potential drop $\Delta\Psi_{dl}$ between the metal surface and the semiconductor surface (Fig. 19). In these cases the band bending is constant for different metals with changed work functions, and the slope factor S approaches zero [see Eq. (26) and Fig. 20]. The interface (surface) state distribution is characterized by a neutrality level Φ_0, which is defined by charge neutrality of the interface states, being occupied by holes and electrons up to E_F^{SS}.[94,95,35] The position is usually measured from the top of the valence band. When the concentration of interface states is assumed to be considerably larger than the number of charge carriers exchanged for the double-layer potential drop $\Delta\Psi_{dl}$, the position of E_F^{SS} remains nearly constant ($E_F^{SS} = \Phi_0$) even after contact formation. Thus, the barrier height is given by:

$$\Phi_B \approx E_G - \Phi_0 \quad (n\text{-doped}) \quad (67a)$$

$$\Phi_B \approx \Phi_0 \quad (p\text{-doped}) \quad (67b)$$

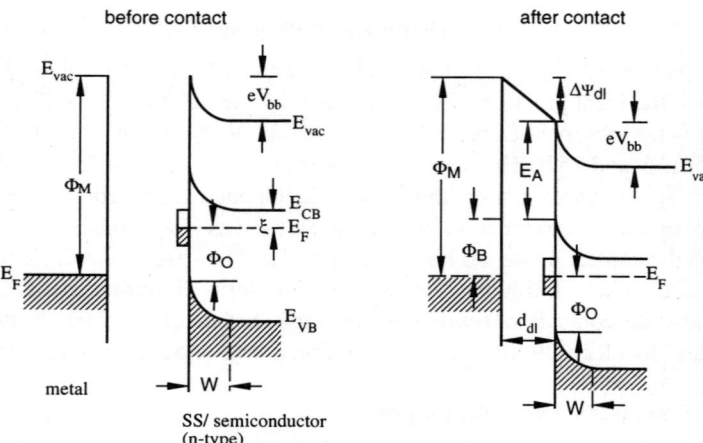

Figure 19. Formation of semiconductor/metal (Schottky) barriers with Fermi level pinning (Bardeen limit). (The abbreviations are explained in the text. The double-layer distance d_{dl} in atomic distances is not drawn to scale.)

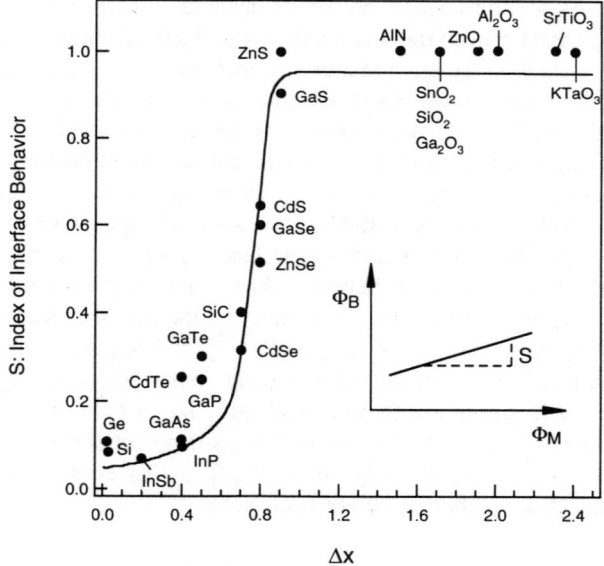

Figure 20. Index of interface behavior S ($S = 0$: Bardeen limit; $S = 1$: Schottky limit) for various semiconductors as a function of their electronegativity difference ΔX. (After Ref. 94.)

The Semiconductor/Electrolyte Interface

This behavior is defined as complete Fermi level pinning or the Bardeen limit of semiconductor/metal junctions (see also Section III). The degree of Fermi level pinning in real devices may vary (S is between 0 and about 1) and is very much dependent on the concentration and energy (density of states) distribution of interface states, D_{is}. The following mathematical expression for pinned metal/semiconductor contacts was derived by Cowley and Sze[98] (neglecting the originally considered image-charge lowering of the barrier height). Assuming a continuous and nearly constant distribution of interface state D_{is}, its charging is given by

$$Q_{is} = -e \cdot D_{is} (E_G - \Phi_0 - \Phi_B) \tag{68}$$

The term in parentheses is a measure of the shift in Fermi level position from Φ_0 to a new value as determined by Φ_B due to the metal contact. Its product with D_{is} gives the number of additionally occupied states due to contact formation.

The charge in the space-charge layer is given by Eq. (33). For thin interfacial layers (<10 Å) with no space-charge effects, the charge on the metal, Q_M, is opposite to Q_{SC} and Q_{is}:

$$Q_M = -(Q_{is} + Q_{SC})$$
$$= eD_{is} \cdot (E_G - \Phi_0 - \Phi_B) - \left[2e \cdot \varepsilon\varepsilon_0 \cdot D^+\left(\frac{\Phi_B}{e} - \frac{\xi}{e} - \frac{kT}{e}\right)\right]^{1/2} \tag{69}$$

As a consequence, the potential drop $\Delta\Psi_{dl}$ across the interfacial layer is given by

$$e\Delta\Psi_{dl} = -\left(\frac{e}{\varepsilon_i\varepsilon_0}\right) \cdot Q_M \cdot d_{dl} \tag{70}$$

where ε_i is the permittivity of the interfacial layer.

If the surface dipole contributions χ_M and χ_{SC} are assumed to be unchanged by the interfacial dipole layer, it follows from the energy-band diagram of Fig. 19 that

$$e\Delta\Psi_{dl} = \Phi_M - E_A^{SC} - \Phi_B \tag{71}$$

Combining Eqs. (70), (71), and (69) leads to

$$(\Phi_M - E_A^{SC}) - \Phi_B =$$
$$-\frac{e^2 \cdot D_{is}}{\varepsilon_i\varepsilon_0} \cdot d_{dl}(E_G - \Phi_0 - \Phi_B) + \left[\frac{2e^3 \cdot \varepsilon}{\varepsilon_i^2 \cdot \varepsilon_0} \cdot D^+ \cdot d_{dl}^2 \left(\frac{\Phi_B}{e} - \frac{\xi}{e} - \frac{kT}{e}\right)\right]^{1/2} \tag{72}$$

The second term on the right-hand side of Eq. (72) may be neglected for reasonable values of ε (≈ 10 for the semiconductor bulk), ε_i (1–10), and D^+ ($\approx 10^{17}$ cm^{-3}) and $d_{dl} < 5$ Å; then we obtain for the barrier height:

$$\Phi_B = S \cdot (\Phi_M - E_A^{SC}) + (1 - S)(E_G - \Phi_0) \tag{73}$$

with $S = [1 + (e^2/\varepsilon_i \varepsilon_0) D_{is} \cdot d_{dl}]^{-1}$. Two limiting cases can be considered:

1. $D_{is} \to 0$; for this case, S goes to 1 and the barrier height Φ_B is given by the Schottky limit (Eq. 25).
2. $D_{is} \to \infty$; for this case, S goes to 0 and Φ_B is given by $E_G - \Phi_0$ [see also Eq. (67)].

For an intermediate density of interface states, the potential drop is distributed between the space-charge layer, V_{bb}, and the surface double layer, $\Delta \Psi_{dl}$. The density of interface states D_{is} for real semiconductor/metal contacts can be determined from the slope S of a plot of Φ_B versus Φ_M in accordance with Eq. (73). The reduced value of band bending eV_{bb} or barrier height Φ_B is dependent on the total charge $Q'_{is} = Q_{is} + Q_{SC}$ (due to the formation of the space-charge layer and the metal contact) residing on the interface states[95]:

$$eV_{bb} = \Phi_M - \Phi_{SC} - \left(\frac{e}{\varepsilon_0 \varepsilon_i}\right) d_{dl} \cdot Q'_{is} \tag{74a}$$

$$\Phi_B = \Phi_M - E_A^{SC} - \left(\frac{e}{\varepsilon_0 \varepsilon_i}\right) d_{dl} \cdot Q'_{is} \tag{74b}$$

The last term accounts for the potential drop $\Delta \Psi_{dl}$ across the surface double layer of width d_{dl} and permittivity ε_i. The interface charge Q'_{is} is given by the integral over $e \cdot D_{is}(E)$, as given by Eq. (48a or 48b), from Φ_0 up to the equilibrium position of $E_F \equiv \Phi_B$, approximately determined by Φ_M:

$$Q_{is} = e \cdot n_{is} = e \cdot \int_{\Phi_0}^{\Phi_M} D_{is}(E) \cdot dE \tag{75}$$

The degree of Fermi level pinning is thus dependent on the density and energy distribution of the interface states. A deviation from the Schottky limit (S considerably smaller than 1) can only be expected when the surface concentration of interface states is larger than 10^{13} eV^{-1} cm^{-2}.

It should be mentioned that a similar expression can be derived when donor and acceptor surface state distributions D_{is}^D and D_{is}^A are assumed, which may be separated by a gap as in the case of polar semiconductors[95] (compare Figs. 17 and 14). In this case the plot of barrier height Φ_B versus metal work function Φ_M would lead to areas with S approaching 0, when the Fermi level intersects with D_{is}^A or D_{is}^D. In other cases Φ_B follows the Schottky limit ($S \to 1$).

As already mentioned above, the magnitude of the electronic barrier Φ_B formed at the interface varies between the extremes of the Schottky limit ($S = 1$) and the Bardeen limit ($S = 0$) (Fig. 20). This dependence on the chemical nature of the semiconductor clearly indicates that the type of interface interaction and the related nature of interface states play an important role. A more ideal behavior ($S = 1$) is observed for ionic semiconductors, such as ZnSe and ZnO (closer relation to the Schottky limit), whereas a higher tendency for Fermi level pinning ($S = 0$) is observed for covalent semiconductors, such as Si, GaAs, and InP (closer relation to the Bardeen limit). A sharp transition between the two extreme cases occurs at the point at which the electronegativity difference between the constituent elements, ΔX, is 0.7–0.8.[99,100] For example, S is about 0.6 for CdS, 0.3 for CdSe, and 0.25 for CdTe.

Surface science techniques have been used to elucidate the microscopic details of semiconductor/metal interface formation and to determine the decisive factors governing the barrier height. The results indicate that the "interphase" may be of considerable complexity in both structure and composition and is strongly influenced by the interface chemistry of the semiconductor/metal combination.[94–96,34–36] For this reason, a number of different theoretical models are still being discussed in the literature to account for the experimental facts regarding nonideal Schottky barrier heights. The original model of Bardeen attributes the pinning effect to intrinsic surface states of clean semiconductor surfaces.[55] In more recent models, extrinsic surface states resulting from the interaction with the contact metal are favored.

In the defect model the interfacial pinning states are due to near-surface lattice defects induced by the deposited contact metal, which only reflect the intrinsic bulk properties of the semiconductor.[158,159] In the metal-induced gap states (MIGS) model, the surface states result from the penetration of metal electron states into the semiconductor band gap.[160–162]

An interfacial reaction layer creating interface states responsible for Fermi level pinning has also been discussed as a model for strongly reacting semiconductor/metal combinations.[163] A general unified model combining the various theoretical approaches has been suggested for group 4 and 3–5 semiconductors[159,163,164] but may not be appropriate for semiconductor compounds of completely different chemical character. In the low-coverage regime of metal deposition, the interface states are closer in their properties to localized extrinsic surface states,[165,166,141,148] and this situation resembles closely that found for semiconductor/electrolyte interfaces (see below). One of the major problems in understanding barrier formation is the inhomogeneity due to the different local barriers that may be found at the interface.[135,167] These local barriers are very important for transport experiments[167] but may also effect the results obtained from surface science spectroscopies.[168–170]

(ii) *Semiconductor/Electrolyte Contacts*

As for metal contacts, many semiconductor/electrolyte contacts show deviations in their experimentally observed behavior from the idealized picture of contact formation discussed above (Section III.3). In both cases the deviation is related to the electronic influence of extrinsic surface states. The term Fermi level pinning, introduced into the field by Bard, Wrighton, and co-workers,[171–173] derives from the nonlinear response of photovoltage to changes in redox potentials. As the conclusions drawn from this work are based on nonequilibrium properties involving charge-transfer rates, they may be misleading.[174] A large number of subsequent investigations (see, e.g., Refs. 3, 9–12, 115, 145, 156, and 157) have shown that, for most semiconductor/electrolyte interfaces, surface states and their response must be considered and may lead to deviations from the idealized Schottky limit. As discussed above, surface states are expected to have an influence only in the energy range in which E_F intersects the energetic position of the surface states E_{SS} due to contact formation with the electrolyte (Fig. 21). In this case the charge-neutrality condition for electrochemical contacts is modified to take into account the charge stored in the surface states: $Q_{SC} + Q_{SS} + Q_{El} = 0$.

A schematic energy diagram for semiconductor/electrolyte junctions with two different electrochemical potentials for the case of Fermi level pinning is shown in Fig. 21. For simplicity, it was assumed that no surface dipolar effects, as discussed in Section III.3, must be considered. In principle, the charging of surface states and the related double-layer

The Semiconductor/Electrolyte Interface

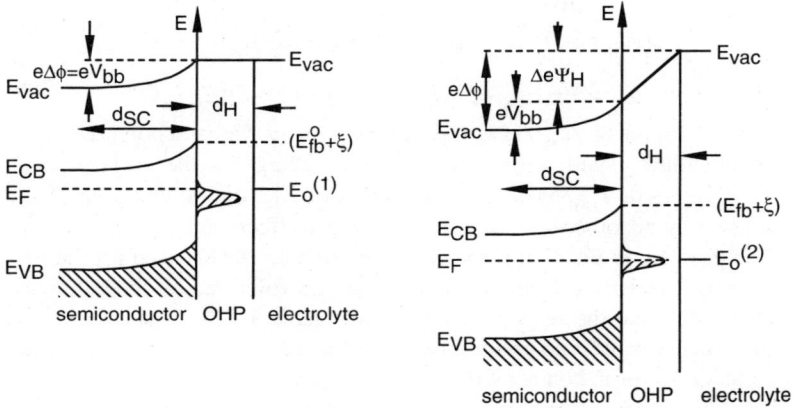

Figure 21. Influence of surface states on electric potential drop at semiconductor/electrolyte interfaces (Fermi level pinning). For nonactive surface states (*left*) $\Delta\Phi$ just leads to band bending. For active surface states (*right*), an additional double-layer potential drop $\Delta\Psi_H$ across the Helmholtz layer leads to a shift of band edges. (The influence of other double-layer contributions is assumed to be zero in this case.)

potential drops must be added to the changes in surface dipoles. It cannot be expected that these changes can be treated independently of each other, which makes the interface charge and electric potential distribution extremely complex. When the available surface state density far exceeds the typical charge carrier numbers of 10^{11}–10^{13} cm^{-2} exchanged for the formation of space-charge layers, the band bending remains fixed (the semiconductor Fermi level pinned versus the surface) when electronic equilibrium is established. The contact potential drops across the electrostatic double layer (Helmholtz layer) between the surface and the electrolyte ($\approx \Delta\Psi_H$) by changing the occupancy of surface states. In reality, the situation is more complex as different types and lower concentrations of surface states may be involved and the contact potential difference is distributed inside the semiconductor and across the double layer.

Semiconductor/electrolyte interfaces may be treated by the Cowley–Sze formalism,[98] derived above for semiconductor/metal contacts. It is assumed that no kinetic overpotentials develop at the phase boundary to the electrolyte in the case of a fast redox couple. Then Eq. (73) may be rewritten also in accordance with Eq. (47):

$$eV_{bb} = S \cdot e[E_{fb}^0 - E(\text{red/ox})] + (1 - S)(E_G + \xi - \Phi_0) \quad (76a)$$

$$\Phi_B = S \cdot e[E_{fb}^0 - \xi - E(\text{red/ox})] + (1 - S)(E_G - \Phi_0) \quad (76b)$$

where $S = [1 + (e^2/\varepsilon_H \varepsilon_0) \cdot D_{ss} \cdot d_H]^{-1}$ and E_{fb}^0 is the flatband potential of the semiconductor with no excess electronic charge in the surface states ($\sigma = Q_{SS} \to 0$, $\Psi_{dl} \to 0$). However, shifts of E_{fb}^0 due to charging of the surface by adsorbed ions and due to dipolar effects must be considered. The problem in Eq. (76) is the different reference levels used for the first term in brackets (electrochemical scale, in volts versus the reference electrode), and the second term (absolute scale, in electron volts versus the valence-band maximum, considered as an intrinsic property of the semiconductor in contact with the electrolyte.

The shift of the band edges due to surface-state charging, $E_{fb}^0 - E_{fb} = \Delta \Psi_H$, is given by [see also Eqs. (74) and (75)]:

$$\Delta \Psi_H = \left(\frac{1}{\varepsilon_0 \varepsilon_H}\right) \cdot Q'_{SS} \cdot d_H \quad (77)$$

$$Q'_{SS} = \int_{\Phi_0}^{\Phi_B} D_{SS}(E) \, dE \quad (78)$$

Expressing Φ_B and Φ_0 in electrochemical terms and assuming a uniform distribution of D_{SS} leads to

$$Q'_{SS} = eD_{SS} \cdot [E_{SS} - E(\text{red/ox})] \quad (79)$$

with $E_{SS} = E_{fb} - \xi - e^{-1} \cdot E_G + e^{-1} \Phi_0$ and $e^{-1} \Phi_B \approx E(\text{red/ox})$. Similar expressions are given by van Meirhaeghe et al.[175] The term Fermi level pinning in the original sense is related to an equilibrium situation. Therefore, its definition in terms of deviations of the linearity of photovoltage versus redox potential is, at least, misleading and may also lead to incorrect results, as the influence of charge-transfer rates in nonequilibrium may lead to band-edge shifts due to kinetic charging of the semiconductor surface. As for semiconductor/metal contacts, the barrier properties should also be investigated with n- and p-doped samples of the same material, as the effect of Fermi level pinning is considered to be independent of doping. The barrier heights for n- (Φ_B^n) and p-doped samples (Φ_B^p) should add up to the band gap E_G: $\Phi_B^n + \Phi_B^p = E_G$. A deviation may be related to different values of the surface dipoles $^{SC}\Delta^{El}\chi$.

There are only a limited number of detailed spectroscopic investigations on the physical and chemical nature of surface states at semiconductor/electrolyte contacts. As in the investigation of semiconductor/metal interfaces, surface science techniques may provide valuable additional information on the structural and electronic properties of these extrinsic surfaces states, as will be discussed in detail in the next section. It can be expected that, as in the case of metal contacts, the more ionic semiconductors are less affected by surface states in the center of the band gap, and therefore a tendency to exhibit Fermi level pinning is less probable for these semiconductors (cf. Fig. 20).

However, one important difference with respect to semiconductor/metal contacts should be emphasized again. When the semiconductor band bending of a metal/semiconductor Schottky barrier is changed, the occupancy of surface states is affected in accordance with Fermi statistics. For electrolyte contacts, however, also the type and concentration of surface states at the surface may change due to electrochemical reactions occurring at certain threshold electrochemical potentials. This problem is even more severe when currents flow due to illumination. Photocorrosive decomposition reactions of semiconductors are in most cases multistep electron-transfer processes. The intermediates may form different distributions of transient surface states (see also Section IV.1) with electron occupations not in equilibrium with bulk states, which in turn will change the electric potential distribution at the interface. This eventual instability of surface conditions always has to be kept in mind when the properties of semiconductor/electrolyte contacts are analyzed according to concepts of solid-state junctions (Schottky barriers). However, this problem of dynamic Fermi level pinning involving transport of charges and reactants from and to the semiconductor surfaces cannot easily be approached by UHV investigations.

V. SURFACE SCIENCE IN SEMICONDUCTOR ELECTROCHEMISTRY

1. Principal Aspects of Surface and Interface Analysis of Semiconductor/Electrolyte Junctions

The semiconductor surface (interface) and especially the electronic band-gap states formed at the phase boundary strongly influence the optoelectronic properties of the prepared junction. The response of band-gap states is easily observed in electric and optoelectronic measurements, but it is often very difficult to identify the microscopic, molecular origin of these states. A deeper scientific understanding, which also enhances the possibilities for junction optimization, requires spectroscopic identification of the surface or interface states.

Many optical and X-ray spectroscopies can only be made interface-sensitive by special experimental procedures (e.g., modulation techniques). They offer the advantage that the junctions can principally be analyzed in their "real" conditions: they allow an *in situ* characterization of electrolyte interfaces. However, their applicability is often limited to special cases (see Refs. 15–18, 115, and 116). As these techniques are outside the scope of this chapter, they will not be considered any further. Electron spectroscopies used in UHV are specifically surface-sensitive and, in principle, allow the structural, chemical, and electronic properties of surfaces to be analyzed. However, they do not allow the investigation

Table 1
Application of Ultrahigh-Vacuum (UHV) Techniques in Investigations of Energy-Converting Interfaces[a]

Type of investigation	Type of information	Problem(s)
Analysis of "real" interfaces	Chemical composition (XPS, AES, ISS, SIMS) Topological structure (LEED, ISS, SIMS)	*Ex situ* experiment → emersion of sample → change of surface
Electrochemistry with UHV-prepared surfaces (well-defined surface structure and electronic structure)	Electronic structure [(S)XPS, UPS, ELS] Topological structure (LEED, ISS) Chemical analysis (XPS, AES)	Preparation of clean, defined surfaces; transfer to electrochemical cell
Modeling of the active interface (adsorption of contact-phase adsorbates onto defined semiconductor surfaces)	Complete characterization of interface possible: Electronic structure [(S)XPS, UPS, ELS] Topological structure (LEED, ISS) Chemical analysis (XPS, AES) Surface potentials (UPS, SPV)	Model experiment → relation to real interface not clear

[a]For definitions of abbreviations, see Table 2.

of the interfaces of electrochemical junctions under "real" conditions in contact with the electrolyte. For this reason, special experimental procedures have to be developed in order to obtain relevant information. Table 1 summarizes the possible applicability of UHV techniques for electrochemical interface problems. The types of information provided by these techniques are indicated in Table 2.

The analysis of the junction properties in *ex situ* experiments requires that the thickness of the contact phase (electrolyte) be less than the information depth (see Table 2) of the applied technique. Therefore, a transfer from the electrolyte into UHV is necessary. However, the emersion procedure and the subsequent transfer process may considerably change the conditions of the interface. The *ex situ* use of surface science techniques has been mostly restricted to metal electrodes and is summarized in a number of review articles (see, e.g., Refs. 17–22 and 177–179). Generally, it is assumed that strong chemical interactions at the interface are not changed substantially during transfer, so that the surface species formed in the electrochemical environment can still be analyzed. The electrochemical double layer may also be preserved during the emersion experiment as in most cases the shear plane is assumed to occur outside the Helmholtz layer. However, the possibility that surface transformations may occur due to dissolution or evaporation of interface species must always be considered. Therefore, the analytical results obtained must be interpreted with care. Usually, a certain amount of impurities (adsorbed CH compounds and water) cover the surface, restricting the applicable surface techniques (mostly to LEED, ESCA, and AES). Reviews on experimental setups for the coupling of electrochemical cells to UHV chambers are given in Refs. 22 and 177. As is evident from Table 2, ESCA and AES are less surface-sensitive than some of the others (UPS, SXPS, ISS, HREELS). Therefore, low concentrations of contaminants do not prevent the analysis of surface composition and structure. However, a more detailed analysis of the potential distribution at the interface (e.g., the determination of work functions and their changes due to contact with the electrolyte) is not possible as undetectable contaminations in fractions of a monolayer may completely falsify the results.

There is a distinctly smaller number of papers on *ex situ* surface science analysis of semiconductor surfaces published in the literature (see Section VI), and nearly no systematic studies besides our experiments on transition-metal chalcogenides.[3,180] In a number of recent investigations, Lewerenz and co-workers[181–183] have shown for different Si planes that

Table 2
Ultrahigh-Vacuum (UHV) Surface Science Techniques[a]

Technique	In/out	Information depth (Å)	Type of information	Key reference(s)
UV photoelectron spectroscopy (UPS)	UV/electrons	5–20	Electronic structure of valence-band region, interfacial energetics	193–197
X-ray photoelectron spectroscopy (XPS, ESCA)	X rays/electrons	10–30	Elemental composition, valence states of elements	198–201
Inverse photoemission (IPE, BIS)	Electrons/photons	5–30	Electronic structure of conduction-band region	202, 203
Electron loss spectroscopy (ELS, HREELS)	Electrons/same electrons	5–20	Electron excitation, vibronic excitation	204–206
Auger electron spectroscopy (AES, SAM)	Electrons/Auger electrons	10–20	Elemental composition, valence state of elements	207–209
Low-energy electron diffraction (LEED)	Electrons/same electrons	5–10	Geometric surface structure	210, 211
Secondary-ion mass spectroscopy (SIMS)	Ions/ions	50	Elemental composition	212
Ion scattering spectroscopy (ISS, LEISS)	Ions/same ions	3	Elemental composition, structural information	213–215
Stimulated desorption (ESD, ESDIAD)	UV, electrons/ions	3	Composition, bond direction	216, 217
Extended X-ray absorption fine structure (EXAFS)			Structural information	218, 219
Tunneling electron microscopy			Surface structure, electronic structure	220, 221
Electron microscopy with microprobe (SEM)	Electrons/electrons, X rays	10^4	Surface topography, composition	222

[a]Refs. 192 and 193.

very clean surfaces can also be prepared in electrochemical cells and by transfer into UHV, which allows the application of UPS and HREELS as extremely surface-sensitive techniques. The cleanliness obtained by their procedure is among the best that may be found in the literature: there is practically no contamination of carbon present on their surfaces.

Clean and/or defined surfaces (preferentially of single crystals) are a precondition for the characterization of semiconductor bulk and surface properties and for the reproducible and interpretable performance of electrochemical experiments. For many semiconductor materials, this may cause serious problems as even the surface condition after electrode preparation is often not known. Therefore, it seems to be a good starting point to perform electrochemical experiments with clean surfaces for a microscopic understanding of semiconductor surface properties, even if they have limited relevance for many applications of semiconductor electrochemistry. Clean surfaces can only be obtained and conserved under UHV conditions with residual gas pressures in the range of 10^{-10} mbar. It is immediately evident that it is generally not an easy task to obtain clean surfaces that subsequently can be studied experimentally. There are three types of contaminations that must be removed, namely, surface reaction layers, adsorbates, and crystal damage. The type and degree of contamination is dependent on the quality of pretreatment and the chemical nature of the semiconductor of interest. Also, the suitability of various methods of surface preparation depends on the chemical properties of the semiconductor. Thus, the experience obtained with a certain material cannot automatically be transferred to other semiconductors. The methods used, for example, for the preparation of oxides,[184–186] 3–5 compounds and elemental semiconductors,[34,35,38,187,188] or transition-metal compounds[3,180] must always be checked with respect to the structural, chemical, and electronic properties of the surfaces obtained. In many cases, the preparation of a certain surface with the desired properties is a scientific problem in its own right.

The most versatile technique, *a priori* leading to surfaces bearing the closest relationship to the bulk, is cleaving, crushing, or scratching of the sample. The advantage of this technique is its applicability to most semiconductors. However, defined, ordered surfaces can only be prepared for cleavage planes, and some crystal planes of interest are not accessible at all. If no specific cleavage planes exist, only atomically rough surfaces which contain a mixture of most low-index planes are obtained. The cleavage plane of 3–5 and 2–6 semiconductors with the zinc blende structure is the (110) plane, and most surface science studies have been performed with this plane. On the other hand, most electrochemical experiments have been performed with the (100) and (111) planes. For the transition-metal chalcogenides, cleaving is by far the most promising technique of surface preparation. Especially for two- or one-dimensional compounds, the van der Waals faces can easily be prepared by cleaving in vacuum.

The simple procedures of heating the samples in vacuum or ion bombardment with subsequent annealing (IBA), which often provide clean surfaces for metals and elemental semiconductors such as, e.g., Si and Ge, cannot be applied easily for 3–5 compounds, oxides, and transition-metal chalcogenides. These compounds decompose at higher temperatures and form new phases of different stoichiometry and very different electronic and structural properties. The same effect occurs during ion bombardment, which usually leads to a preferential sputtering of one component.

A promising method is the *in situ* preparation of the semiconductor of interest within the UHV system by chemical vapor deposition, preferentially by molecular beam epitaxy (MBE) or alternative atomic layer-by-layer epitaxy (ALE). Also, annealing of heated or sputtered surfaces in combination with evaporation of chemical reactants is a possible means of obtaining clean surfaces of compound semiconductors.

Chemical methods of surface preparation can be applied in some cases. Especially for the preparation of semiconductor electrodes, empirically developed etching techniques are used which provide surfaces of better electronic properties.[7,8] In many cases, however, the surface chemical composition has not been analyzed after chemical treatment, and new interfacial phases without electronically active surface states may equally well be responsible for the observed improvements. Differences in surface composition are suggested by the high concentrations of surface states obtained for many semiconductor surfaces prepared in UHV. The surface concentration is drastically reduced after (electro)chemical treatment. Evidently, adsorbates still reside on the surfaces and have beneficial effects due to the formation of extrinsic surface states (or, more correctly, surface resonances), which are situated outside the semiconductor band gap.

It seems therefore very interesting to study systematically the influence of different methods of surface preparation, either in UHV or by (electro)chemical means, on the electronic response of semiconductor/electrolyte interfaces. For such experiments, the surface chemical composition after the chemical treatment must be carefully analyzed by surface science techniques following the above-described procedure. In parallel, such surfaces, prepared and characterized in UHV, must be analyzed by electrochemical means. To our knowledge, there are only very few electrochemical studies of semiconductors using defined surfaces such as cleavage planes or UHV-prepared surfaces,[189–191] besides the extensive literature on layered chalcogenides.[2]

The most complete set of results may be obtained by model experiments to simulate the active junction. Such experiments are performed in UHV by adsorbing contact-phase constituents onto defined semiconductor surfaces. In this way, any contact phase that can be evaporated by any means can be deposited *in situ* and analyzed. Basically all surface science techniques may thus be applicable, and a rather complete characterization of the interface is possible. It may, however, be a problem to relate these model junctions to the real interface prepared and analyzed under completely different pressure and temperature conditions. Especially all surface transformations that involve transport of reactants are hardly accessible by such model experiments. It seems reasonable to assume that model experiments allow the investigation of the fundamental equilibrium properties of surface interaction and the initial steps of surface reactions. A more detailed description of this experimental approach will be given in Section V.3.

2. Surface Science Techniques

UHV surface science techniques[192,193] use probes that strongly interact with matter—i.e., electrons or ions—to achieve high surface sensitivity. Usually, a combination of techniques is applied to characterize the surface as completely as possible. The most important surface science techniques were presented above in Table 2, which includes the abbreviation, the type of information available and the typical surface sensitivity for each technique. The UHV techniques that are most often used to characterize the electrochemical interfaces of semiconductors, as discussed later on, will be described here in terms of their basic physical principles.

(i) Photoelectron Spectroscopy (Photoemission, XPS, UPS, SXPS)

In photoelectron spectroscopy (UPS, XPS) high-energy photons are used to eject electrons from the valence-band states or core levels of the sample (Fig. 22). Usually, a division is made between XPS or ESCA

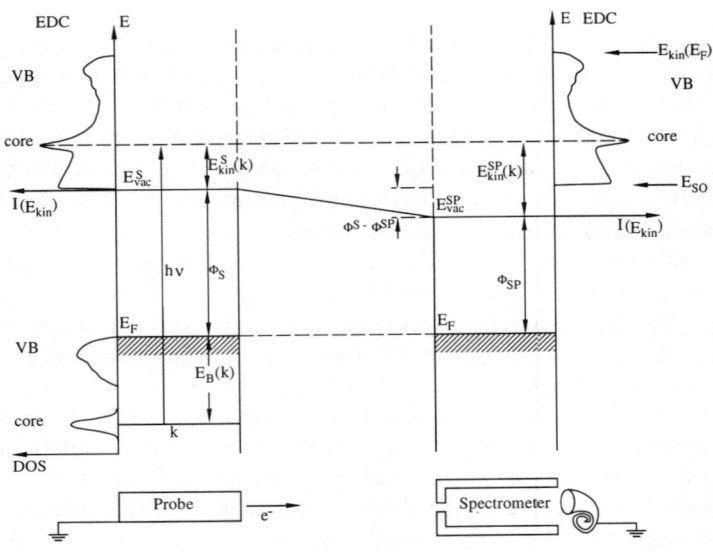

Figure 22. Schematic representation of the experimental and the energetic conditions of photoelectron spectroscopy. The initial excitation of the occupied electron states (DOS) in the solid leads to an electron energy distribution curve (EDC) in the sample, which is changed in energy in the spectrometer by the sample–spectrometer contact potential. The reference level is the identical Fermi level in equilibrium.

(excitation by X rays; most often used sources: Mg $K\alpha$ = 1253.6 eV and Al $K\alpha$ = 1486.6 eV), probing inner orbitals and the valence-band region, and UPS (excitation by UV light; most often used sources: He I = 21.22 eV and He II = 40.82 eV), probing mostly the valence-band region. Such a division has become outdated with the development of continuous light sources from synchrotrons which provide photons over a wide energy range typically 10–1000 eV (SXPS).[197,223] The current density due to the emitted photoelectrons is analyzed as a function of their kinetic energy E_{kin} and emission direction $I(E_{kin,\Theta,\varphi})$. Such a spectrum is called an electron energy distribution curve (EDC). The EDC is a superposition of directly excited primary photoelectrons $I_p = I_p[E_{kin}(i)]$ from the energy states of the probe and a background of secondary electrons I_b, which result from inelastic scattering processes (Fig. 22): $I = I_p + I_b$. For solids, $E_{kin}(i)$ is given by the equation

$$E_{\text{kin}}(i) = h\nu - I_p(i) = h\nu - E_B(i) - \Phi \tag{80}$$

$h\nu$ is the photon energy of the excitation source, $I_p(i)$ denotes the ionization potential and $E_B(i)$ the binding energy of the electronic state (i) versus E_F, and Φ is the work function of the sample.[192,194,196] As $E_B(i)$ of the core levels is sensitive to the chemical environment of the specific atom (chemical shift), XPS provides a powerful analytical technique [hence the acronym ESCA (electron spectroscopy for chemical analysis)].

The electronic structure of the valence-band region is probed by UPS; thus, angle-resolved measurements (ARUPS) give information on the band structure $E(\mathbf{k})$ of the solid.[194,195] During the photoemission process, only the electron momentum $\mathbf{k}_p$ parallel to the surface is conserved:

$$\mathbf{k}_p = (2\pi/h)(\sin \Theta)[2mE_{\text{kin}}(\Theta,\varphi)]^{1/2} \tag{81}$$

$E_{\text{kin}}(\Theta,\varphi)$ gives the kinetic energy of photoelectrons along a certain direction from the surface determined by the polar angle Θ to the surface normal and the azimuthal angle φ, which represents a line in the surface Brillouin zone. For surface states the spatial extension is restricted to the surface region (a few atomic layers). A dispersion of photoemission peaks is thus only expected along $\mathbf{k}_p$ and not along $\mathbf{k}_n$ (momentum normal to the surface). Therefore, the lack of dispersion obtained in normal emission with changes in excitation energy is an important criterion for the identification of surface states.[194,195]

For samples in electronic equilibrium with the spectrometer (Fig. 22), the photoelectron energy distribution curve (EDC) is referred to the Fermi level E_F, and the binding energy E_B^F is given by the relation

$$E_B^F = h\nu - E_{\text{kin}} - \Phi_{\text{sp}} \tag{82}$$

where $h\nu$ is the energy of the excitation source, E_{kin} is the measured kinetic energy of the photoelectrons, and Φ_{sp} is the spectrometer work function.[194,198] The spectrometer work function Φ_{sp} can be calibrated via the Fermi edge of clean metal surfaces and with the known value of Φ_M. Referencing the binding energy to E_F eliminates the contribution of different sample work functions Φ_s to E_B. The sample work function Φ_M of metallic samples can be determined from the width of the EDC, $\Delta E = E_{\text{so}} - E_F$ (where E_{so} is the secondary electron onset, binding energy scale):

$$\Phi_M = h\nu - \Delta E \tag{83}$$

For semiconductor samples, E_F is situated in the forbidden energy gap and therefore is not defined in the spectrum via the onset of photoemission as in the case of metallic samples. The first expected contribution to the emitted photoelectrons usually originates from the valence-band maximum E_{VB}. For this reason, in many cases E_{VB} has been chosen as a reference level for semiconductor studies, as spectral shifts due to changes of E_F (e.g., because of different doping levels or band bending) are avoided. However, when the energetic condition of the interface (especially band bending) is of interest, E_F is an adequate reference level. The position of E_F can be defined via the Fermi edge of metallic samples if electrical contact between the semiconductor and the reference is established. In this way the energetic distance of the valence-band maximum to E_F can also be determined, which for flatband conditions allows the doping of the semiconductor to be determined.

The work function of the semiconductor Φ_{SC} is defined by the following equation, corresponding to Eq. (83), with $\Delta E = E_{SO} - E_{VB}$:

$$\Phi_{SC} = h\nu - (E_{SO} - E_F) = h\nu - \Delta E - (E_{VB} - E_F) \tag{84}$$

where the term $(E_{VB} - E_F)$ gives the distance of E_F from the edge of the valence band, E_{VB}. As the position of E_F at the semiconductor surface is dependent on the energetic condition at the interface, changes of Φ_{SC} and, according to Eq. (85) of E_B^F are observed in the photoelectron spectrum (see Fig. 23):

$$\Delta\Phi = \Delta\chi - eV_{bb} - (E_{VB} - E_F) \tag{85}$$

$\Delta\chi$ represents changes due to surface dipoles altering the electron affinity or ionization potential I_p and thus only the secondary electron onset (Fig. 23b). The term eV_{bb} represents the contribution of band bending, which shifts the whole EDC and thus changes the binding energy E_B^F and the secondary electron onset E_{SO} (corresponding to the work function) by the same amount (Fig. 23c). These values can be measured due to the high surface sensitivity of the photoemission techniques (10 Å vs. typically 1000 Å–10,000 Å for the width of the space-charge layers). The last term $(E_{VB} - E_F)$ represents the contribution due to redoping of the semiconductor bulk (for example, due to intercalated atoms) which shifts the Fermi

The Semiconductor/Electrolyte Interface

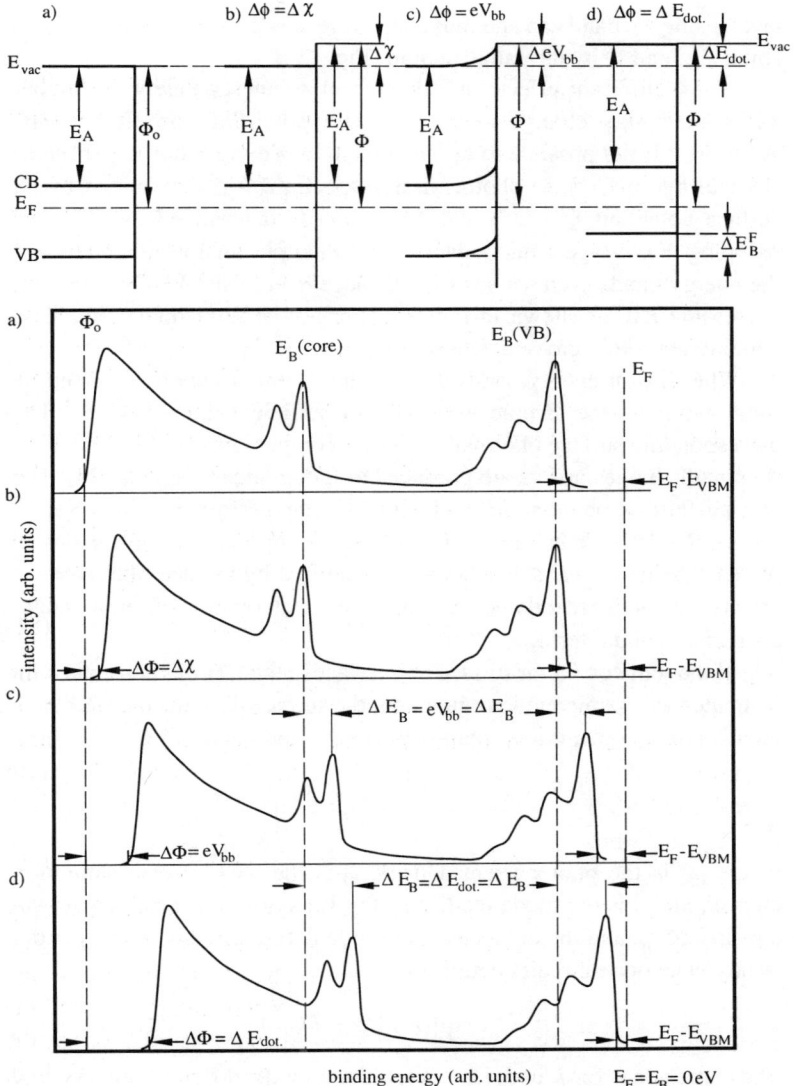

Figure 23. Schematic changes of energetic conditions at semiconductor interfaces and their consequences for the valence-band spectra. The change of the surface dipole only shifts E_A and therefore Φ. Band bending, eV_{bb}, and changes of doping, ΔE_{dot}, shift both E_B^F and Φ.

level inside the band gap and thus also influences the whole EDC and thus shifts E_B^F and Φ by the same amount (Fig. 23d).

These different effects usually do not occur separately, but rather adsorbates may change several surface potentials simultaneously. Whereas it is not possible to separate them by Kelvin probe experiments (measuring only Φ), in photoemission spectra $\Delta\Phi$ is determined by all surface potentials (Eq. 85) and ΔE_B^F is only determined by eV_{bb} and redoping. For n-type semiconductors, for example, an upward bending of the energy bands corresponds to a displacement of the EDC toward E_F, decreasing E_B^F as shown in Fig. 23. (For p-type semiconductors, band bending leads to increased values of E_B^F.)

The shift of energy bands due to band bending may be reversed by illumination of the sample with light of sufficient intensity.[224,225] The corresponding surface photovoltage (U_{ph}) compensates the band bending if enough charge carriers are produced to flatten the bands (Fig. 24). The original surface photovoltage experiments were performed with a Kelvin probe. Recently, it has been shown that the binding energy values in photoemission experiments can also be shifted by surface photovoltage effects, which can readily be used for the characterization of semiconductor surfaces and interfaces.[226-230]

If the semiconductor surface or interface in the UHV experiments can be treated as a semiconductor junction, the surface photovoltage (SPV) is identical to an open-circuit photovoltage U_{ph} and is given by[231,94]

$$U_{ph} = \frac{n \cdot kT}{e} \ln\left(\frac{j_{ph}}{j_0} + 1\right) \tag{86}$$

where j_{ph} is the photocurrent density, j_0 is the dark reverse saturation current, and n is the diode quality factor. Equation (86) results from the equality of j_{ph} and the compensating diode dark current density j_d under steady-state open-circuit conditions:

$$j_d = j_0\{\exp[(eU_{ph})/nkT] - 1\} \tag{87}$$

The expressions for j_0 and n are determined by the different processes of transport and charge-carrier injection into the contact phase and/or surface (interface) states. When the dark current is dominated by the thermionic emission of the majority carriers over the barrier Φ_B[94-96] (Schottky diode), n is equal to 1 and j_0 is given by

The Semiconductor/Electrolyte Interface

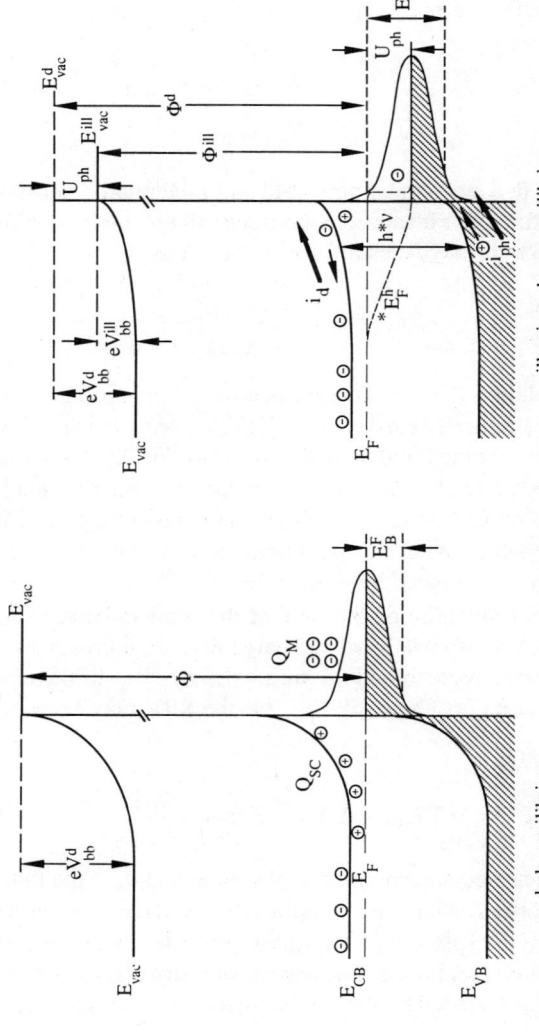

Figure 24. Surface-photovoltage-induced rebending of energy bands. The surface photovoltage U_{ph} and the effect of U_{ph} on the binding energy E_B^F are indicated. Steady-state conditions are obtained for $i_{ph} = i_d$.

$$j_0 = A\left(\frac{m^x}{m}\right)T^2 \exp(-\Phi_B/kT) \tag{88}$$

where m^x is the effective mass and $A = 4\pi mkT/h^3 = 120$ A/cm^2 is the Richardson constant.

The photocurrent density is given by

$$j_{ph} = e \int \eta(h\nu)\cdot I(h\nu)\, d\nu \tag{89}$$

with η the energy-dependent quantum yield and I the photon density per unit area per unit time. For idealized conditions (no recombination in the space-charge layer), j_{ph} may be estimated by the Gärtner model[233]:

$$j_{ph} = e\cdot I_0 \left(1 - \frac{\exp(-\alpha W)}{1 + \alpha L}\right) \tag{90}$$

I_0 is the integral photon flux, α is the mean absorption constant, L is the diffusion length of minority carriers [$L = (D\cdot\tau)^{1/2}$, with D the diffusion constant and τ the lifetime], and W is the extension of the space-charge layer. The expected photocurrents for semiconductors with different band gaps are given in Ref. 232. In cases in which the photovoltage is induced by high-energy photons from the photoemission source, the effective quantum yield can be estimated by $\eta = h\nu/3E_G$.[234,235]

A quantitative evaluation of the SPV at the semiconductor surface may be very complex, especially when charge exchange processes with surface states (surface recombination) are involved.[236,237] In the case of n-type semiconductors (as shown in Fig. 24) the SPV may be approximated by[238–240]

$$\text{SPV} = -kT \ln(1 + \Delta_p) = -kT \ln\left(1 + \frac{\delta_p}{p_b}\right) \tag{91}$$

p_b is the minority carrier concentration in the bulk, and δ_p is the light-induced excess minority carrier concentration at the surface. Equation (91) describes basically the splitting of the quasi-Fermi levels at the surface under nonequilibrium conditions. The excess minority carrier concentration Δ_p is strongly influenced by the surface recombination velocity S[241]:

$$\Delta_p = \frac{\eta(h\nu)\cdot I(h\nu)\cdot[(1 - R(h\nu)]}{(D/L) + S}\frac{\alpha(h\nu)L}{1 + \alpha(h\nu)\cdot L} \tag{92}$$

For large values of S, Δ_p as well as U_{ph} goes to 0. If S is small, Δ_p and U_{ph} may change considerably for small light intensities. S is strongly dependent on the concentration and properties of surface (interface) states. Thus, U_{ph} may show large differences in UHV experiments. It is, in general, not easy to discriminate between charge transfer to the adsorbate phase and recombination in adsorbate-induced surface states. However, the measured value of SPV in UHV experiments is in any case equivalent to the photovoltage U_{ph} of semiconductor junctions[240] when equivalent interfaces have been prepared.

This simplified view of bias-light-induced SPV is only valid when the thermodynamic equilibrium is not noticeably disturbed by the light source in the photoemission experiment; this is usually the case when room-temperature experiments are performed. However, recent results have shown that the photoemission light source, particularly with synchrotron radiation and at low temperatures, may introduce enough electron–hole pairs to disturb the electronic equilibrium and thus induce a large surface photovoltage.[242–247] For most of the excited electrons, which are not emitted as photoelectrons, the photoemission experiment is an open-circuit experiment with a low-intensity but high-brilliance and high-energy light source. The magnitude of the surface photovoltage effect is, as discussed above, dependent on the intensity of the light sources used (including additional bias light), which determines the excess carrier flow, and the surface and interface properties, which determine the dark current flow.

The magnitude of the SPV induced by the photoemission light source can be checked by measurements with a Kelvin probe and may reach considerable values for good junctions even at room temperature.[247,248] As an example, the SPV produced with an α-Si:H p–i–n solar cell at room temperature with bias light as compared to monochromatized synchrotron light may be mentioned. The SPV induced by the soft X-ray photons (0.6 V) is close to the white-light SPV of 0.8 V, which is identical to U_{ph} measured with electrical contacts. The SPV values measured for adsorbate interfaces (e.g., p-GaSe/H_2O) may be considerably smaller (see Fig. 25). Nonetheless, SPV-induced binding energy shifts as a function of, e.g., sample temperature may give valuable information on the device properties of semiconductor interfaces prepared and characterized in UHV.[168,169,249–251]

Bias-light-induced nonequilibrium conditions may also provide an appropriate means of inducing charge-transfer processes at semiconduc-

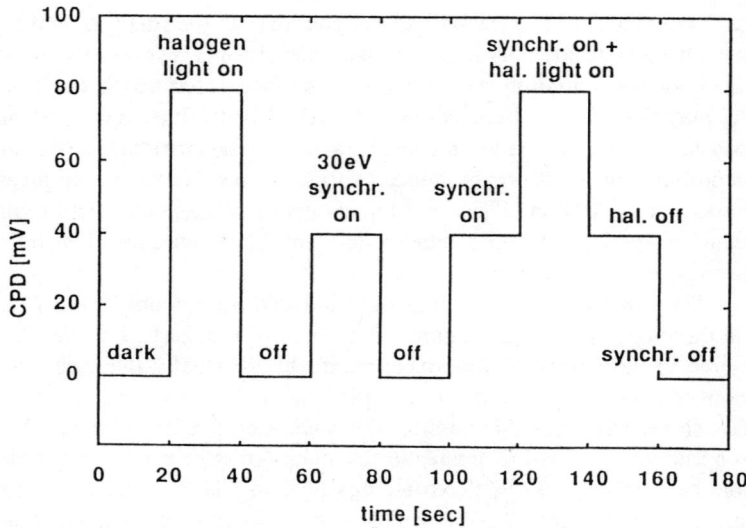

Figure 25. Surface photovoltage (SPV) experiments performed with the p-GaSe/H_2O interface at low temperature. The changes in the Kelvin contact potential difference (CPD) induced by 30-eV synchrotron light and bias light (W/halogen; 100 mW/cm^2) are shown for comparison.

tor/adsorbate interfaces in UHV, which would allow elementary reaction steps to be investigated in UHV experiments.[252]

(ii) Low-Energy Electron Diffraction (LEED)

Low-energy electron diffraction (LEED) is the standard technique for analyzing the geometric structure of surfaces.[210,211] A low-energy electron beam (20–500 eV) is backscattered from a crystalline surface and analyzed with respect to the symmetry of the diffracted beams (Fig. 26). At normal incidence, constructive interference of scattered electrons is described by two Laue equations:[104]

$$h_1 \cdot \lambda = a \cdot \sin \alpha \qquad (93a)$$

$$h_2 \cdot \lambda = b \cdot \sin \beta \qquad (93b)$$

where h_1 and h_2 are the diffraction orders (integers), λ is the de Broglie wavelength of the X rays, and a and b are the lattice parameters of the

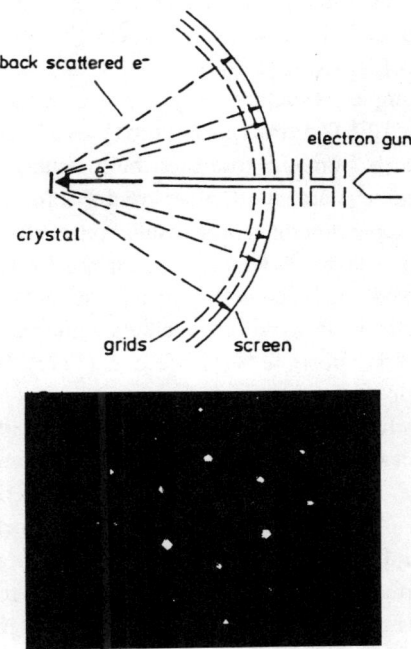

Figure 26. Experimental arrangement for low-energy electron diffraction (LEED) experiments (*top*) and LEED pattern of a cleaved WSe$_2$(0001) surface (*bottom*).

surface. The angles α and β give the directions at which the diffraction maxima will be obtained. For a complete surface structure analysis, the intensity variations with beam energy are recorded and compared to theoretical calculations (i–E curves).[210,211] The intensities of the diffraction spots are proportional to the square of the structure factor $F(h_1,h_2)$, given by (kinematic approximation)

$$F(h_1,h_2) = \sum_j f_j t_j \exp\{2\pi i[h_1 x_j + h_2 y_j + (1 + \cos \Theta)z_j]/\lambda\} \quad (94)$$

with f_j and t_j the atomic scattering factor and transmission factor, respectively. The positions of the atoms j are specified by the coordinates x_j, y_j, and z_j, with z_j normal to the surface plane defined by the vectors **a** and **b**. The appearance of defects on surfaces is accessible from the analysis of the LEED spot profiles (SPA-LEED).[253]

Whereas a complete structure analysis by a quantitative evaluation of the intensity of the LEED diffraction pattern takes a larger effort, changes of the surface mesh from the truncated bulk structure can readily be discerned from the appearance of superstructures in the diffraction pattern.[210,211] These superstructures may result from surface reconstructions or ordered adsorbate layers A which form surface Bravais lattices with a defined orientation to the basis mesh (given by the bulk truncated lattice). Thus, the superstructure designation is related to the basis mesh found for a certain surface of the substrate M (hkl) [e.g., Si(111)]. The superstructure is given as the ratio of the overlayer mesh unit vectors **a***, **b*** to those of the substrate, **a**, **b**. In addition, the angle γ between the substrate mesh and the overlayer mesh and the type of lattice mesh (primitive: α = p; centered: α = c) may be specified. Thus, the designation is M(hkl) – $\alpha(a^*/a \times b^*/b)\gamma$ – A. The occurrence of certain superstructures gives directly valuable information on the structure of the surface. For example, on the Si(100) surface a pairing of the surface atoms to reduce the number of dangling bonds gives an Si(100)(2 × 1) LEED pattern. For the H-terminated surface, it may transform to an Si(100)(1 × 1)-H surface, which readily indicates the loss of pairing (for details, refer to Section VI). Thus, simple inspection of the LEED pattern and measurements of the geometrical spot positions give valuable information on the structural conditions of semiconductor surfaces and interfaces.

(iii) Low-Energy Ion Scattering Spectroscopy (LEISS)

Low-energy ion scattering spectroscopy (LEISS) is a very interesting technique for analyzing site-specific interactions on semiconductor surfaces.[213–215] An ion beam (usually He^+ or Ne^+ or alkali ions, E_0 = 0.5–3 keV) is analyzed with respect to its kinetic energy E_1 after interacting with the surface of interest (Fig. 27). The technique is extremely surface-sensitive for noble-gas ions due to their high neutralization probability (monolayer range). The kinetic energy ratio of the scattered to the initial ion beam is given in a simple binary collision model by:

$$E_1/E_0 = [\cos \delta + (M_2/M_1 - \sin^2 \delta)^{1/2}]^2 (1 + M_2/M_1)^{-2} \quad (95)$$

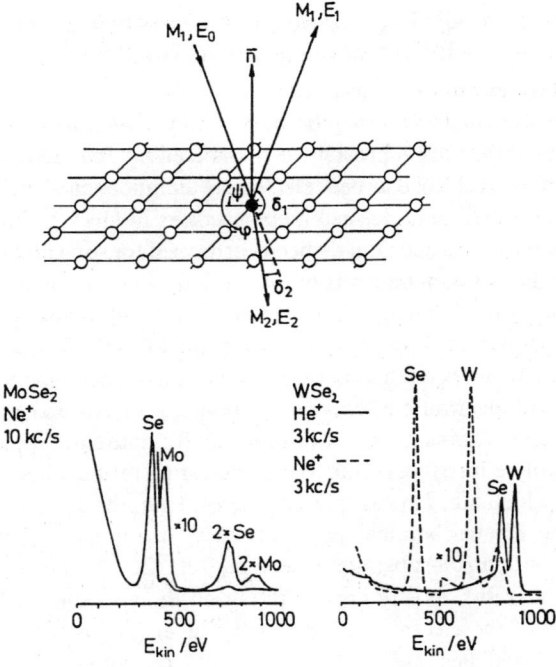

Figure 27. Geometric arrangement for low-energy ion scattering spectroscopy (LEISS) experiments (*top*) and ion scattering spectra for UHV-cleaved layered semiconductors with different scattering ions ($\delta = 142°$, $\psi = 72°$).

As δ, E_0, and the mass of the scattering ion, M_1, are defined by the experimental conditions and the kinetic energy of the backscattered beam E_1 is measured, the mass M_2 of the target atom on the surface can be calculated. The scattering cross section is given by a nucleus–nucleus coulomb repulsion potential which is modified by the screening electron cloud (Thomas–Fermi–Moliere potential[213–215]). Using this potential, a shadowing cone is defined. As a consequence, adsorbates specifically interacting with specific surface atoms reduce their respective backscattering intensities. A detailed analysis of the scattering intensity as a function of scattering geometry also gives information on structural parameters. A disadvantage of the technique is the lack of information on the chemical state of the surface atoms.

3. Model Experiments in the Investigation of Semiconductor/Electrolyte Interfaces

The most extensive set of information about elementary processes at solid/electrolyte interfaces may be obtained by modeling the electrolyte interface in UHV. This approach was pioneered by Sass and co-workers, who adsorbed and coadsorbed electrolyte components, such as H_2O, halogens, or alkali, onto defined metal surfaces in UHV.[23-25,254-256] Adsorption experiments had already been performed for some time in surface science studies of metal/adsorbate interactions, e.g., in investigations of H_2O adsorption,[257] halogen adsorption,[258,259] and alkali (co)adsorption.[260,261] However, Sass was the first scientist to relate these adsorbate studies, which had been directed toward a fundamental understanding of structural and electronic effects on the bonding of adsorbates and their consequences for surface reactions, to fundamental problems of solid/electrolyte interfaces. Due to the interesting results obtained by this approach, which are, in principle, of interest from the surface science as well as the electrochemical point of view, many more studies were performed with metal substrates (see, e.g., Refs. 26–32 and 260–267 and references therein), mostly concentrating on the structure and potential distribution of the electrochemical double layer. The approach has also been questioned because the experimental conditions are completely different from those of standard solution electrochemistry (for a discussion of concerns about this approach, see, e.g., Ref. 31). However, a similar discussion on the relevance of surface science studies in relation to problems of catalysis due to the differences in temperature and pressure has been settled: the fundamental interactions in model experiments are comparable to those under real-world conditions, but the more complex composition of the systems and the difference in state conditions must be considered when the results are transferred from one environment to the other.

There is a considerably smaller number of model adsorption experiments directed toward the investigation of semiconductor/electrolyte interfaces (see Section VI). However, many of the fundamental surface science investigations performed with semiconductor surfaces and different adsorbates[34,35,38,268] may be of importance for an understanding of semiconductor/electrolyte interaction, even when the authors did not have this interface in mind. The investigation of semiconductor/model electrolyte interfaces may offer specific advantages compared to the study of metal substrates. As discussed extensively in Sections III and IV, the formation of the solid/electrolyte contact leads to an equilibration of the electrochemical potentials of the two phases forming the contact, which changes the Volta potential $\Delta\Psi$ due to charge transfer and changes of the surface dipole $\Delta\chi$ due to contact formation. At metal surfaces, these two effects generally cannot be separated as they occur over comparable atomic distances. (There may be a chance of separating them by following binding energy shifts of adsorbate levels compared to the overall work function change; see Ref. 40 for an extended discussion of this point.) At semiconductors, charge transfer and the related Volta potential drop $\Delta\Psi$ lead to an extended space-charge layer in the semiconductor. The surface dipole changes $\Delta\chi$ again only occur over atomic distances, shifting the surface position of the energy bands (change of electron affinity or ionization potential). As discussed in Section V.2(i), these two contributions can be separated in a straightforward manner in photoelectron spectra. In addition, intrinsic and extrinsic surface states induced by the adsorbate phase can be directly studied from the extra emission intensity in the valence-band region and the shifts of electrostatic potentials. However, there is also a serious drawback, as the sensitivity for direct spectroscopic examination of surface species with most surface science techniques is restricted to about 1% of a monolayer, depending on the conditions and systems used for the investigation. However, measurable changes in surface potentials, especially the formation of band bending, may already be induced by less than 0.1% of a monolayer.

In principle, experiments with model semiconductor/electrolyte interfaces allow simulation of the fundamental interactions of electrolyte species with defined semiconductor surfaces. They will, in our opinion, contribute a lot to a better understanding of electrochemical elementary processes at semiconductor surfaces. They are of similar relevance as the surface science investigations of semiconductor/solid interfaces prepared in UHV, which model buried interfaces of semiconductor solid-state junctions such as Schottky barriers and semiconductor heterojunctions.[34-38,95,96] The model electrolyte interface may be built up with rather high complexity, approaching that of real electrolyte contacts, by coadsorption of several electrolyte species. Nevertheless, they should be complemented with *ex situ* experiments and electrochemical studies of UHV defined surfaces as suggested in Table 1. We are convinced that this combination of different surface science approaches is a possible and necessary route to an understanding of semiconductor electrochemistry on the molecular level.

Typically, the model experiments are performed in a UHV system such as that schematically shown in Fig. 28. It contains two preparation chambers: one is used for the preparation of clean semiconductor surfaces, preferentially by cleavage; the second is used for the adsorption of the electrolyte species (in the example shown, Br_2 and H_2O are adsorbed here, and Na may be codeposited in the analyzer chamber). The sample manipulator allows the transfer of the substrate from the preparation chamber to the analyzer chamber under controlled temperature conditions. The analyzer chamber is equipped with an angle-resolving hemispherical analyzer, LEED, and a Kelvin probe; as excitation sources, synchrotron light

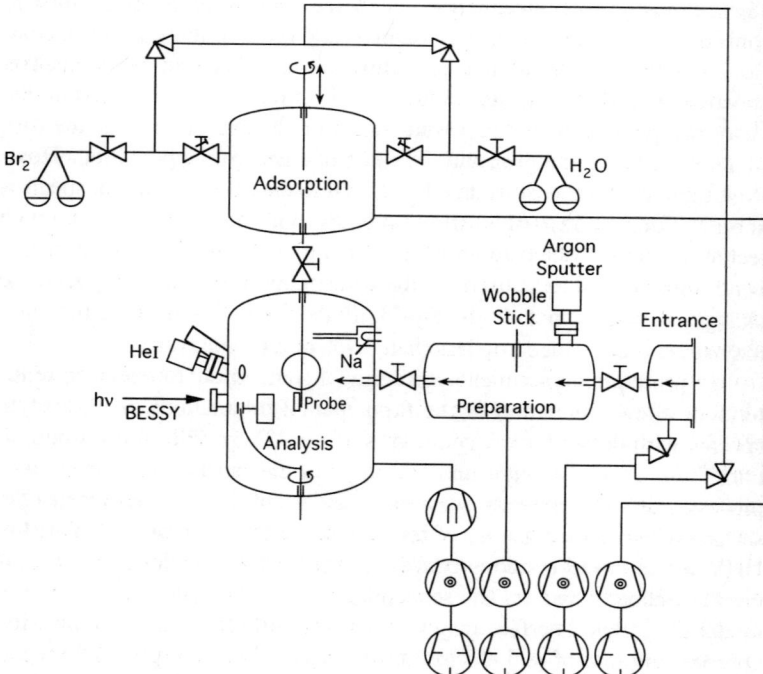

Figure 28. Experimental setup for performing model experiments with semiconductor/electrolyte interfaces. The main analyzer chamber (analysis) contains the spectrometer and ports for different excitation sources. The preparation chamber is used for sample cleavage. The electrolyte components are adsorbed in the adsorption chamber, where the temperature of the probe can be controlled.

or laboratory sources such as X-ray guns or discharge lamps may be used. A large variety of experiments are possible with this equipment. The adsorbate model electrolytes can thus be prepared in a way that allows high-surface-sensitivity and high-resolution photoemission experiments using synchrotron light. A typical model experiment involves the following steps: (1) Preparation of clean surfaces; (2) structural and electronic characterization of the surface; (3) stepwise adsorption or coadsorption of the electrolyte species from very low coverages to several monolayers with systematically varied experimental parameters; (4) after each step characterization of the interface with the available surface science techniques; and (5) as most of the adsorption experiments require lower temperature, heating of the samples to room temperature and above. The experimental conditions are, of course, adapted to the specific problem under investigation.

The type of information that can be expected from model electrolyte adsorbate interfaces is schematically shown in Fig. 29. It shows a hypothetical case of an n-type semiconductor in contact with a complex adsorbate phase, which contains a strongly chemisorbed adsorbate layer directly in contact with the semiconductor surface and another adsorption layer on top of this. In addition, coadsorbed solvent molecules like H_2O interact with the adsorbates and the surface. The first reactive adsorbate layer produces surface states and surface resonances; their respective density of states distributions, $(DOS)^{SS}$ and $(DOS)^{SR}$, are schematically shown in Fig. 29. Their energy position and chemical state can be inferred from the binding energy positions E_B^F of the adsorbate-induced extra photoemission intensity [neglecting a strong contribution of final-state effects (relaxation energy) due to changing photo hole screening].[40,192,194–195] Also, the second adsorbate layer, which is assumed to interact with the solvent molecules, will give rise to extra emission intensity that can be directly measured. In the figure the $(DOS)^{ads}$ of the HOMO state has been drawn in accordance with the Gaussian type of distribution expected for the electron-exchanging electron state of a redox couple with the expected reorganization energy of the solvent [see Eqs. (13) and (14) and Fig. 5]. Whether this distribution will be valid for the adsorbate phase may also be directly addressed with photoemission experiments.

Finally, the changes in electric potential distribution, which give rise to the energy diagram, as shown in Fig. 29, can also be deduced from the photoemission experiment. On the left-hand side of the diagram, the

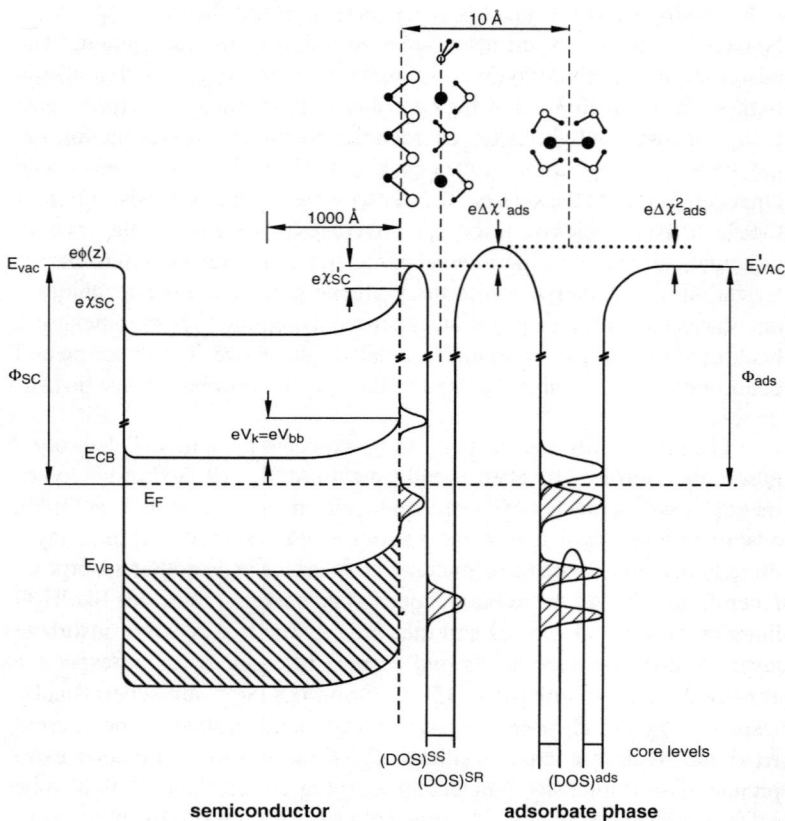

Figure 29. Schematic representation of a semiconductor/model electrolyte interface. The adsorbate is composed of a reactively chemisorbed layer and a second layer interacting with coadsorbed H_2O. The extra adsorbate-induced density of states (DOS) is shown as well as changes in surface potentials of the different double layers (for details refer to the text).

energy potentials of the semiconductor, assuming no Fermi level pinning, are shown (the situation before starting the adsorption experiment). The position of the Fermi level versus the valence-band edge and the work function Φ_{SC} are directly given by the experiment [cf. Section V.2(i)]. Due to the adsorbate phase, a band bending is induced and also the work function is changed to a final value given by Φ_{ads}. In the case shown in

Fig. 29, the adsorbate phase is composed of two layers, which will be formed step by step in the adsorption experiment. Therefore, it may be possible to deduce band bending, electron affinity changes at the semiconductor surface, due to the reactively chemisorbed layer, $\chi_{SC} \rightarrow \chi'_{SC}$ and $\Delta\chi^1_{ads}$, and changes induced later on by the second adsorbate phase, $\Delta\chi^2_{ads}$ (the work function sums up $eV_{bb} + e\Sigma\Delta\chi_i$). With the given estimated value of the absolute redox potential (see Section II.3), this model electrolyte interface can then be compared to real semiconductor/electrolyte interfaces.

Usually, the change of the electron affinity due to adsorbed or adsorbate-induced dipole layers is given by the Helmholtz formula[270,271]:

$$e\Delta\chi = \frac{4\pi e}{\varepsilon_i \cdot \varepsilon_0} \cdot N_{ad} \cdot \mu_n(\theta) \qquad (96)$$

where N_{ad} is the surface concentration of adsorbed species ($N_{ad} = N_{max} \cdot \theta$, with θ the fractional surface coverage) and $\mu_n(\theta)$ is the dipole contribution normal to the surface.

The adsorbate-induced surface dipole $\mu_n(\theta)$ normal to the surface depends on the nature of the surface dipole. It may be formed by a charge-transfer to the substrate corresponding to a polarization of chemisorption bonds by adsorbed atoms (typical cases are adsorbed alkali or halogen atoms) or by adsorbed bipolar molecules (typical cases are adsorbed H_2O or CO molecules). The surface dipole normal to the surface, $\mu_n(\theta)$, is given by the orientation of the dipoles and is influenced by the depolarization of the dipoles for higher surface coverages θ, induced by the dipole-induced electric field on itself. Thus, $\mu_n(\theta)$ may be written as

$$\mu_n(\theta) = \mu_n^0 - \mu_d(\theta) \qquad (97)$$

where $\mu_d(\theta)$ depends on the arrangement of dipoles and the lateral interdipole interaction. In the simplest case, following a treatment by Topping[272] for an array of parallel point dipoles, it follows that

$$\mu_n(\theta) = \frac{\mu_n^0}{(1 + 9\alpha_{ad} \cdot N_{ad}^{3/2})} \qquad (98)$$

where α_{ad} is the polarizability of the surface dipoles. In the special case discussed above (Fig. 29), there is the contribution of different dipole layers:

$$e^{SC}\Delta^{Ads}\chi = e(\Delta\chi_{SC} + \Delta\chi^1_{ads} + \Delta\chi^2_{ads}) \qquad (99)$$

with $\Delta\chi_{SC}$ being due to changes of the semiconductor surface dipole ($\chi_{SC} \rightarrow \chi'_{SC}$), $\Delta\chi^1_{ads}$ being due to the strongly chemisorbed first adsorption layer, and $\Delta\chi^2_{ads}$ being due to the ionosorbed second adsorption layer. The terms $\Delta\chi^1_{ads}$ and $\Delta\chi^2_{ads}$ may be identified with the inner Helmholtz the outer Helmholtz layers, respectively, in electrolyte contacts [cf. Fig. 9 and Eq. (43)]. The sum of all contributions will give rise to the overall change in the electron affinity contribution to the new value of the work function measured for the adsorbate-covered surface. As mentioned above, it may be possible to deduce the contribution of the different dipole layers by the step-by-step formation of the adsorbate phase. In addition, the adsorbate emission lines may shift differently for different spatial positions within the overall potential distribution.[40] As the binding energy values are referred to the position of the Fermi level, double-layer potential drops that completely occur in a plane closer to the solid also will affect the adsorbate binding energy values E^F_B (ads). In other words, if the adsorbate is outside a double-layer potential drop (closer to vacuum), its binding energy position E^F_B will see the contribution of inner surface dipoles $\Delta\chi$; as a consequence, it will shift nearly in parallel with the work function: It is coupled to the vacuum level. If, however, the adsorbate is part of the dipole, the changes in binding energy position E^F_B (ads) will occur nearly in parallel with changes of substrate levels (governed by changes in band bending eV_{bb}): It is coupled to the Fermi level.

VI. CASE STUDIES ON CLEAN SURFACES AND ADSORBATE AND ELECTROCHEMICAL INTERFACES

In the following sections of this chapter, typical surface science results on surface and interface properties of prototype semiconductors will be presented. We will only present typical case studies, which serve as examples in support of the theoretical framework presented above on the importance and influence of surface and interface effects on semiconductor/electrolyte contacts. The nature and distribution of surface states of clean and adsorbate-covered semiconductor surfaces will be related to changes of surface potentials such as band bending and shifts of band-edge positions and will be compared to electrolyte contacts. We will take Si and GaAs as examples of established semiconductors, and layered chalcogenides will serve as examples of more "exotic" semiconductors, as these

The Semiconductor/Electrolyte Interface

were the subject of our investigations. We will not attempt to give a complete survey of semiconductor electrochemistry as a number of review articles are available.[2,3,115,145,273-277] We also do not want to give a complete survey of UHV studies on semiconductor surfaces and interfaces; the reader may refer to Refs. 34, 35, 38, 46, 60, 126, and 278-280. It should be noted in this context that valuable information on the structure of semiconductor surfaces and interfaces has been obtained using scanning tunneling microscopy (STM) and vibrational spectroscopic techniques such as infrared reflection–absorption spectroscopy (IRAS) or HREELS. These techniques can, in principle, be combined for the complementary analysis of adsorbate interfaces in UHV and electrolyte interfaces within the electrolyte. However, there is still a lack of systematic investigations on semiconductor/electrolyte interfaces; for this reason and because of necessary restrictions on the size of this chapter, we refer to the general references cited above and will not specifically address this topic.

The electronic properties of semiconductor surfaces are always extremely sensitive to the quality of surface preparation, which determines the concentration of defects and extrinsic surface states. In the following sections we will only describe the intrinsic properties of idealized terminated surfaces, unless stated otherwise.

1. Silicon

(i) Clean Surfaces

Si crystallizes in a cubic diamond structure. The most important low-index planes in technological applications are the (100) and the (111) planes, which therefore are also the most intensively investigated surfaces in surface and interface studies. The schematic survey of surface properties presented below relies on extensive theoretical work on surface electronic band structure and on numerous experimental surface science results. The numbers given are mean values as reported in Refs. 35 and 279.

(a) The (100) plane

The (100) plane is usually prepared by ion bombardment and annealing at elevated temperatures (IBA) of oriented single-crystalline wafers that have been polished and treated with chemical etchants. After cleaning of the chemically treated surfaces [their surface composition is discussed in more detail in Section VI.1(iii)] with several IBA steps, a contamination-free Si(100) surface is obtained. As a bulk-like termination would result in two dangling bonds on every surface atom, neighboring Si atoms form dimers to reduce the surface free energy (see Fig. 30). As a consequence, clean Si (100) planes show a 2×1 superstructure in the LEED pattern. Other superstructures have also been reported in the literature; these are related to a superposition of 2×1 domains and single-atom defects at the domain boundaries.[35,60,38] The actually measured LEED pattern depends on the details of the preparation and the cleaning procedure.

On the 2×1 plane every outer-plane Si surface atom forms two bonds to the next inner plane and one σ-type bond to its surface neighbor, leaving one dangling bond toward vacuum. As these dangling bonds are parallel to each other, they may be stabilized by π bonding.[280,281] In Fig. 30 a symmetric dimer and an asymmetric dimer are shown, as both structures have been reported in the literature. The asymmetric dimer would be stabilized by a Jahn–Teller distortion of the degenerate dangling-bond state.[35] This distortion leads to rehybridization of the surface bonds. The inward-displaced Si atom is closer to an sp^2 hybridization, and its dangling bond has an increased p character. The dangling bond of the outward-displaced atom will have a larger s character. As a consequence, electron

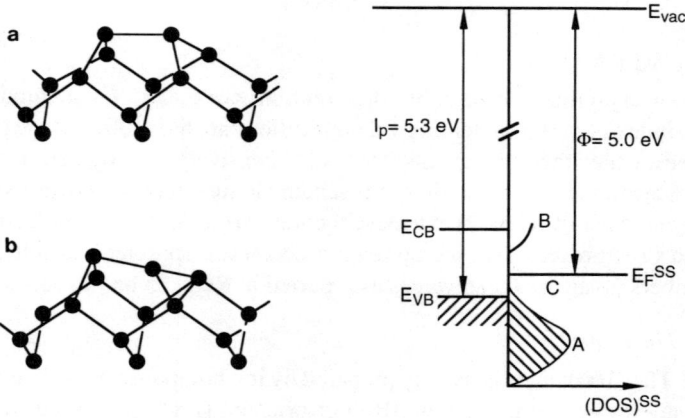

Figure 30. Schematic representation of the Si(100)(2 × 1) reconstructed surface with symmetric (a) and asymmetric dimers (b). Also shown are the surface potentials and a schematic representation of the density of surface states.

density will be transferred to the outer Si atom, which leads to a completely filled s-like dangling bond on this atom and a completely empty p-like dangling bond on the inner Si atom. The differently charged Si surface atoms can be directly identified by high-resolution core-level photoelectron spectroscopy,[282] as they will exhibit different binding energy values. The schematic density of surface states (DOS)^{SS} is also included in Fig. 30 in relation to the bulk band edges.[279,35] The electronic surface structure is of semiconducting nature for both dimer models with a gap in the range of 0.6 eV.[283] The occupied dangling bond gives rise to a strongly dispersing band from the valence-band maximum down to about 1.5 eV (feature A in Fig. 30). The unoccupied dangling bonds are represented by feature B. There are additional states tailing into the bulk band gap which may be related to defect states (feature C). Also included are the experimental values for the threshold energy (ionization potential), $I_p = 5.3$ eV,[279,35] and for the position of the surface Fermi level (given by the work function), $E_F^{SS} = \Phi = 5.0$ eV. E_F^{SS} is shifted from this mean value by about ±0.15 eV for strongly p- and n-doped samples.

(b) The (111) plane

The (111) plane of Si may be prepared by cleavage in UHV, but it can also be prepared by IBA of polished and etched samples. After cleavage in UHV, a metastable (2 × 1) reconstructed surface is formed. There have been several models proposed for this structure; an in-depth discussion of these models by Mönch[35] includes the theoretical and experimental facts in support of each of them. The chain model suggested by Pandey[285] and presented in Fig. 31 seems to be well established now. The structure is determined by π-bonded zigzag chains of Si atoms along the (110) direction forming alternating five- and seven-membered rings. These rings result from a bond rearrangement of the originally present six-membered rings of the outermost Si double layer of the (111) surface. As there are three equivalent directions on the (111) plane and usually a nonuniform density of steps, several domains and a number of defects are formed on the (2 × 1) reconstructed surface, which leads to surface defect atoms in the range of some percent.[279]

The schematic density of states in relation to the band gap and the surface potentials of the Si(111)(2 × 1) reconstructed surface is also shown in Fig. 31. The π bonding leads to an occupied bonding π-band (A) and an unoccupied antibonding π^*-band (B), both about 1 eV wide, which are separated by a gap of about 0.4 eV. The ionization potential of the (2 × 1)

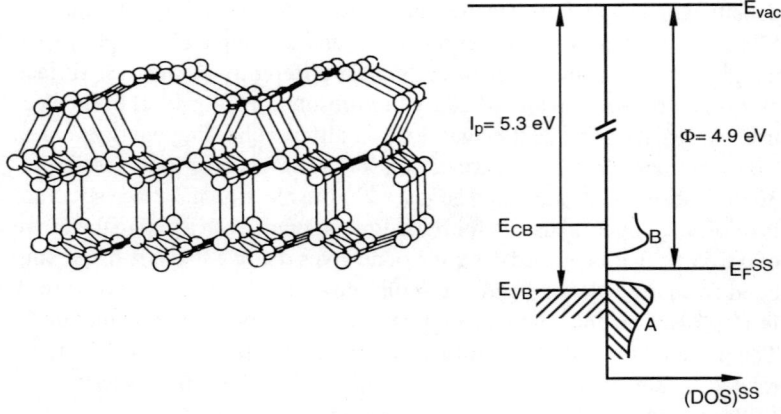

Figure 31. Schematic representation of the Si(111)(2 × 1) reconstructed surface after the π-bonded chain model of Pandey.[285] Also shown are the surface potentials and a schematic representation of the density of surface states.

reconstructed (111) surface is again about 5.3 eV, and the surface Fermi level position E_F^{SS} (given by the measured work function Φ) is about 4.9 eV. E_F^{SS} may shift as a function of doping by ±0.1 eV.

When a cleaved (2 × 1) reconstructed surface or a polished and etched sample is annealed at elevated temperatures (ranging from 500 to 1200 K, depending on the pretreatment), the famous (7 × 7) reconstructed surface forms. There have been a number of different structure models proposed since its discovery almost 40 years ago.[286] The dimer–adatom–stacking fault (DAS) model[287] is well accepted based on a number of refined structure determinations.[35,38,60] This model of the structure is schematically shown in Fig. 32. The driving force for this rather complex reconstruction, which involves the movement of a large number of surface atoms, is the minimization of the number of surface dangling bonds by the formation of different Si adatoms, dimers, and stacking faults. There are the topmost surface adatoms A with three back bonds and one remaining dangling bond. In addition, there are the so-called rest atoms R in the layer below, also with one dangling bond. There is another dangling bond of the Si atom at the bottom of the corner hole. The dimers D are formed along the boundary of the triangular subunit.

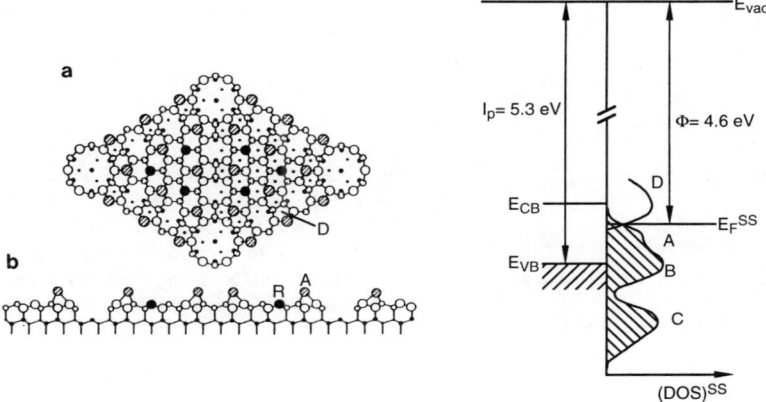

Figure 32. Schematic representation of the Si(111)(7 × 7) reconstructed surface after the dimer–adatom–stacking fault model a, top view; b, side view.[287] Also shown are the surface potentials and a schematic representation of the density of surface states.

As a consequence of this complicated structure, many different Si surface atoms with different charge compared to that of the bulk atoms are expected, and their presence can be inferred from, e.g., high-resolution core-level photoelectron spectra.[288,289] Three surface core-level shifted Si lines S_1–S_3 can be clearly identified (see Fig. 33), and another line S_4 is located close to the bulk line S_0. S_2 is assigned to the rest atoms, and S_1 to those Si atoms bound to the rest atoms and the adatoms, respectively. The adatoms are identified with S_3. As a consequence, the surface band structure is rather complex. A schematic DOS is also shown in Fig. 32. The A and C surface states are assigned to the adatom dangling bond and adatom back bonds, respectively, B belongs to the rest atom dangling bonds, and D corresponds to the empty adatom states.[35,280] The (111) surface with the (7 × 7) reconstruction turns out to be metallic due to remaining dangling bonds, which are not saturated by adatom bonding or dimer formation. This metallicity can be directly measured by photoemission intensity at E_F or by metallic-like STM current–voltage curves. Therefore, strong Fermi level pinning occurs with a position of the surface Fermi level E_F^{SS} (given by its work function Φ) of 4.6 eV, which is nearly independent of bulk doping. The mean value for the ionization potential (see also Fig. 4) is given as 5.3 eV.

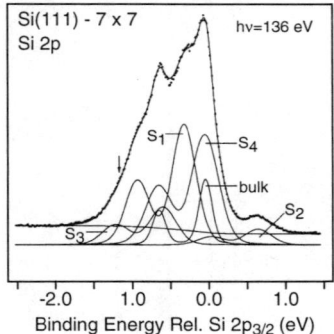

Figure 33. High-resolution SXPS core-level spectra of the Si $2p$ line of the Si(111)(7×7) reconstructed surface, (excitation energy: 136 eV). In addition to the bulk line, four surface core-level peaks (S_1–S_4) are resolved. The assignments of the peaks are given in the text. (After Ref. 289.)

The low-index planes (110) and (113) of Si are of less importance and therefore have been less well studied.[279] Both surfaces exhibit a number of different surface reconstructions. The surface Fermi level position (work function) is 4.7 and 4.8 eV, respectively, and the ionization potential is in both cases also close to 5.3 eV.

In summary, the clean Si surfaces of the different orientations show complex surface reconstructions. These are due to surface bond or adatom formation, which reduces the surface free energy by minimizing the number of dangling bonds. The resulting electronic band structure of the surface may be semiconducting or metallic. As a consequence, the position of E_F^{SS}, as given by Φ, and the degree of Fermi level pinning turn out to be different for the different surfaces. However, all surfaces show nearly the same value of the ionization potential (about 5.3 eV; see also Fig. 4), which indicates a nearly constant surface dipole χ_{SC}.

The measured difference between Φ and I_p (or between E_A and the known band gap of Si) directly gives the barrier height Φ_B^n or Φ_B^p for n- or p-doped samples. The values of the band bending eV_{bb} (and thus Φ_B) can also be directly measured by the saturation value of source-induced SPV shifts in photoelectron spectra at low temperatures[290] (see Section V.2, Fig. 24); e.g., the shifts measured for Si(111)7×7 are -0.65 and $+0.46$ eV for p- and n-doped samples, respectively, which is equivalent to the expected band bending.

(ii) Adsorbate Interfaces

We will restrict our discussion of adsorbate interfaces to a few examples involving oxygen, hydrogen, water, and halogens, which are of importance in Si photoelectrochemistry and chemical surface modification and passivation.

(a) Oxygen adsorption

The adsorption of oxygen has been investigated in order to gain information on the initial steps of silicon oxide formation and its electronic surface properties. Silicon oxide is a well-known passivation layer used in Si-based electronic devices (see, e.g., Ref. 291). The interaction of oxygen with a clean Si surface depends on the temperature, oxygen partial pressure, orientation and pretreatment of the Si surface, and the excitation state of the oxygen.[279,35] However, independent of the conditions and the surface orientation, a disordered oxidized layer is always formed, which leads to a rapid loss of the LEED pattern on all surfaces. Except in a low-temperature precursor state, the oxygen always ends up dissociated and bonded between two Si atoms. The precursor state has been identified for the (111)[292,293] and (100)[294] oriented Si surfaces. It is related to an initial rapid adsorption with a high sticking probability, which is followed by a saturation regime. The nature of the precursor state has been a subject of controversy,[35,292,293] but it can now be identified as a chemisorbed molecular dioxygen species.[293–296] Different proposed bonding sites on Si surfaces are shown in Fig. 34. At saturation all Si surface atoms are bound to oxygen; already at an early stage the O atoms penetrate the lattice into

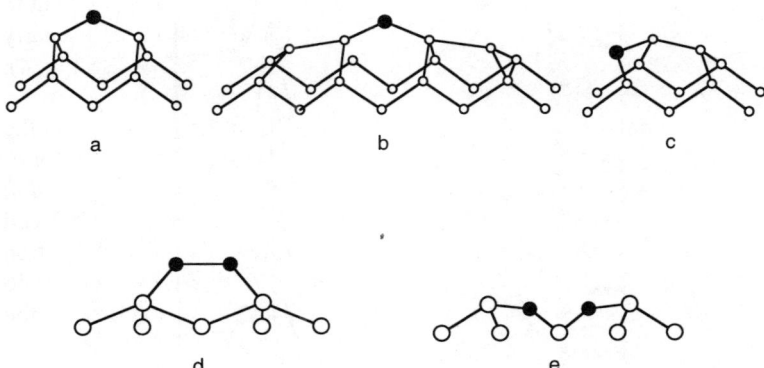

Figure 34. Schematic representation of O coordination sites on Si surfaces. O bonding to Si(100): (a) broken dimer, (b) bridged dimers, (c) broken back bond (from Ref. 279); O bonding to Si(111): (d) molecular precursor, (e) broken back bond (from Refs. 292 and 293).

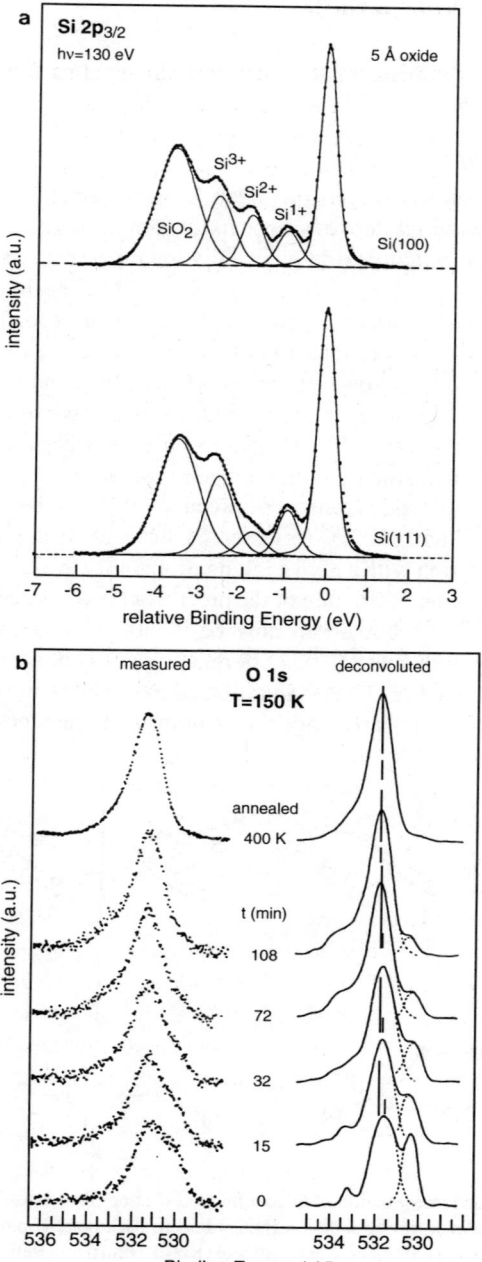

the bulk, forming an SiO_{2-x} layer, with x mostly depending on temperature and oxygen pressure. In the Si/SiO_2 interface layer, nearly every O atom is coordinated to two Si atoms, and, therefore, all Si dangling bonds and π bonds formed by reconstruction should be saturated. Thus, for an ideal interface, no interface states should remain in the band gap. But owing to the disorder in the oxide layer, different defect states are formed with different bonding geometries and numbers of Si and oxide bonds,[297] which leads to a U-shaped distribution of interface states. Its density depends very much on the preparation method used but is considerably lower than the intrinsic density of surface states [see Section VI.1(i)].

The SXPS of Si during oxygen exposure and stepwise oxidation have been intensively investigated by Hollinger, Himpsel, and co-workers.[298–300] Typical core-level spectra of an oxygen-exposed Si surface are shown in Fig. 35a. The silicon $2p$ line (the Si $2p$ multiplex splitting has been removed by a curve-fitting routine) shows several shifted lines depending on the amount of oxidation,[300] which originally was attributed to the chemical shifts induced by different oxidation states of Si. More recent studies have questioned the original assignment[292,293,301] and relate the different emission lines to specific chemical moieties at the oxidized Si interface. The typical binding energy of the Si bulk line is $E_B(2p_{3/2}) = 99.3$ eV; the different Si^{n+} lines ($n = 1-4$) are shifted by about 0.9 eV per unit change in oxidation state.[279] The binding energy value of thick, bulk-like SiO_2 is 104 eV. Also shown (Fig. 35b) are typical X-ray photoelectron spectra obtained for the O $1s$ core line after exposure to 2.5 L of O_2 at 150 K.[293] The line at 530.5 eV has been assigned to the precursor state of adsorbed oxygen. The main peak at 531.5 eV is also present in the case of room-temperature adsorption ($E_B = 532$ eV) and is due to the stable broken back bond.[292] The O $1s$ core line of thick, bulk-like SiO_2 has a binding energy of 534 eV. As the O $1s$ line is outside the energy range of many synchrotron monochromators, the O $2s$ line may be used instead as a core-level line for chemical identification of surface reactions (see the collection of $2s$ line chemical shifts compiled by Ranke[302]).

Figure 35. Si $2p$ (a) and O $1s$ (b) core-level spectra of Si surfaces exposed to oxygen. (a) Si $2p_{3/2}$ line after deconvolution referred to the bulk line Si° (SXPS, $hv = 130$ eV; after Ref. 300). (b) O $1s$ line after exposure to 25 l of O_2 at 150 K for different times (XPS, raw data and deconvoluted spectra; after Ref. 293).

Valence-band spectra of oxygen-exposed Si surfaces have not been investigated by angular-resolved photoelectron spectroscopy due to the disorder of the oxygen-covered surface. The spectral series, which can be measured, is strongly dependent on external experimental parameters but rather similar for the different Si surfaces.[292–294,298] Angle-integrating valence-band spectra measured for Si(111)(7 × 7) surfaces,[292] e.g., those presented in Fig. 36, show the loss of surface-state-related emission intensity at E_F and the growth of extra lines at around $E_B = 7$ eV and 11 eV, which are related to the oxygen 2p-based σ-bonding (p_z) and π-bonding (p_x, p_y) states. There are also some other features measured with different intensities in other experiments, which are due to the contribution of varying amounts of precursors and defect states. As already discussed above, the σ- and π-bonding interaction of O with Si atoms removes the intrinsic surface states found on the clean surfaces from the fundamental

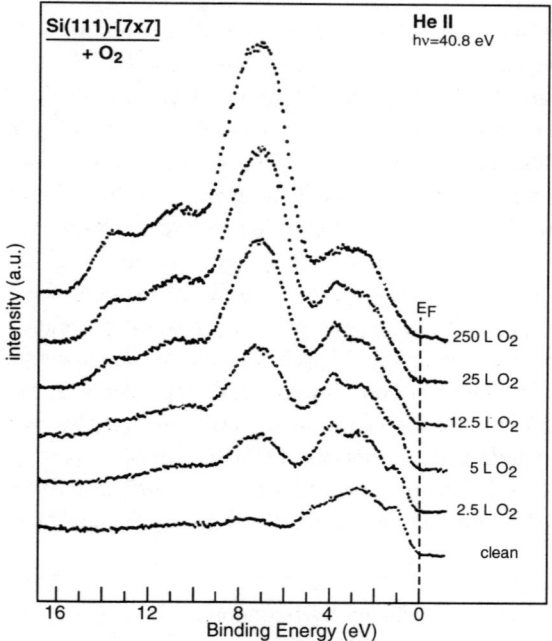

Figure 36. He II valence-band spectra of Si(111)(7 × 7) surfaces exposed to oxygen as indicated on the figure.[292]

energy gap. This is expected based on a simple LCAO estimation of the energetic position of bonding–antibonding states formed between Si sp^3 hybrids and O $2s$ and $2p_{x,y,z}$ atomic states[303] but is also suggested by more elaborate electronic structure calculations.[304,305]

As many defects are formed due to the disorder of the oxygen adsorbate interface and of the subsequently formed oxide layer, the interface states are not entirely removed, as already discussed above. As a consequence, the oxygen-treated samples show Fermi level pinning even when the interface density of states in the range of the band gap is drastically reduced. High-quality Si/SiO_2 interfaces with extremely low interface density of states are only expected for nearly perfect oxide interfaces grown, for example, at elevated temperatures.[291]

The surface potentials of oxygen-covered Si surfaces are given as[61]:

Si(100): $\Phi \approx 4.7$ eV, $I_p \approx 5.3$ eV

Si(111): $\Phi \approx 4.8$ eV, $I_p \approx 5.5$ eV

These values lead to a nearly constant position of the surface Fermi level close to midgap, which is the center of the U-shaped interface density deduced for Si/SiO_2 interfaces. The values given above are supported by other studies which indicate that the pinning position 0.6 eV above the valence-band maximum can be removed by subsequent H_2 treatment, which leads to flatband conditions for n-doped samples ($E_F - E_{VBM} = 0.9$ eV).[300]

Somewhat surprising is the fact that the ionization potential is hardly affected by the adsorption of such an electronegative species as O_2. However, the values for I_p given above are values obtained after a complete rearrangement of the surface structure, which involves the penetration of O atoms into the Si back bonds. When the changes in surface potentials are measured for the precursor state, which corresponds to an adsorbed oxygen species on top, a drastic increase of the work function is reported (in these references ΔI_p and $\Delta \Phi$ are not discriminated)[296,293,306]: $\Delta \Phi$ is reported to range between 0.65 for $\Theta \approx 0.1$ of a monolayer[293] to about 1.6 eV for saturation coverage. During annealing, $\Delta \Phi$ is reduced again to the values given above as the oxygen is now incorporated into the lattice. An extensive review of oxygen adsorption experiments on Si surfaces is given by Engel[307] with special emphasis on the primary steps of oxide formation.

(b) Monovalent adsorbates (halogens, hydrogen)

The adsorption of monovalent adsorbates such as halogens and hydrogen on Si surfaces has been intensively investigated because these adsorbates are expected to chemically saturate and thereby passivate the intrinsic surface states. They form chemical single bonds to the Si dangling bonds or their weaker intersurface bonds formed during reconstruction. Detailed reviews on these adsorbates are found in Refs. 35, 187, 278, 279, and 308. The expected structures of such monovalent adsorbates X bound to the Si (100) and (111) surfaces are schematically shown in Fig. 37.[279] On the (100) plane an adsorbate site bridging two Si surface atoms has been assumed as an intermediate species (Fig. 37a). The other structures may be stabilized under certain experimental conditions, which depend on the adsorbate under study. There is a monoadatom species, which leaves the Si dimers intact. The LEED pattern still shows the (2 × 1) superstructure (cf. Fig. 37b). For larger adsorbed halogens such as Br and I, also a bridging configuration between two dimer blocks has been assumed. In addition, at higher coverages diadatom species (Fig. 37c) are

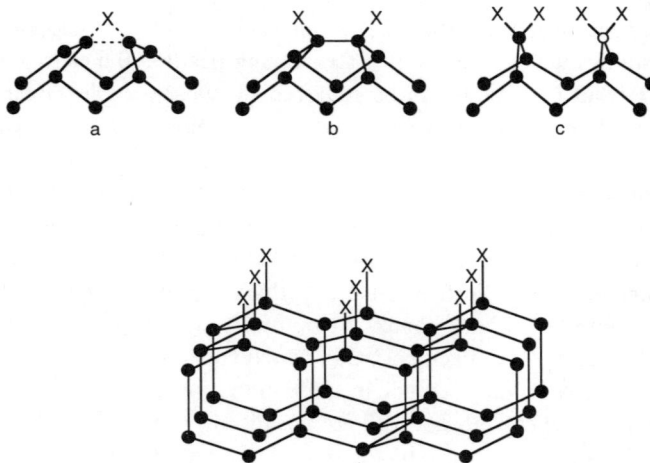

Figure 37. Schematic representation of the monovalent adsorbate bonding geometry on Si surfaces: (a) intermediate bridging site; (b) monoadatom bonding (2 × 1); (c) diadatom bonding (1 × 1) on Si(100); (d) monoadatom bonding on Si(111)(1 × 1).

formed. Here also the Si dimer bonds are broken, and the two possibly formed dangling bonds per Si atom are saturated by strong covalent bonds to the adatoms. As a consequence, a (1 × 1) LEED pattern is observed, indicating the loss of surface reconstruction. The situation for ideally terminated (111) surfaces is depicted in Fig. 37d. Here all dangling bonds pointing to vacuum are saturated with one adatom. As a consequence, the different complex reconstructions [see Section VI.1(i)] should be lost, and a (1 × 1) LEED pattern is expected. However, it turns out that for most monovalent adsorbates (see below and Refs. 35, 187, 278, 279, and 308) a number of different complex adsorption stages exist, which also lead to complex distributions of adsorbate bonding sites and consequently to different LEED patterns. Only under special conditions is the (1 × 1) pattern of the (111) surface obtained. The reason is evidently the complex reconstruction of the surface after sample preparation. Therefore, the adsorbates face many different adsorption sites with different bonding properties. It has been noticed that the idealized surface structures given above are indeed formed under appropriate experimental conditions. However, in general, the adsorption usually does not stop after monolayer coverage but may proceed by forming bonds to Si back bonds, depending on experimental parameters and the properties of the adsorbates.

The electronic properties of Si surfaces covered with monovalent adatoms should also be very simple for the idealized structural conditions shown in Fig. 37. The interaction of dangling bonds with the adsorbates to form Si–X surface molecules can be described in terms of a simple LCAO diagram as shown in Fig. 38 (see also Section IV.2). The position of the ionization potential of the adsorbate varies systematically with the adatom electronegativity and is given by Hartree–Fock atomic terms [H($1s$): 13.6 eV, F($2p$): 19.9 eV, Cl($3p$): 13.8 eV, Br($4p$): 12.4 eV, and I($5p$): 11.0 eV].[35] These values must be compared to the dangling-bond energy of the Si sp^3 hybrid orbitals ($|h\rangle$: 9.4 eV).[35] In Fig. 38, only the simple case with one dangling bond is considered. For the reconstructed surfaces, the intersurface Si bonding may be considered, which would lead to an energetic splitting of the surface states.

In any case, a strong covalent single bond (σ-type) is formed between the Si sp^3-hybrid orbital and the adatom s (H) or p_z (X = halogen) atomic orbital. In addition, for the halogens, nonbonding (or weakly π-bonding) states ($p_{x,y}$) are formed close to the initial ionization potentials. The antibonding states of Si character may be driven out of the semiconductor band gap. The precise energetic positions have been calculated or meas-

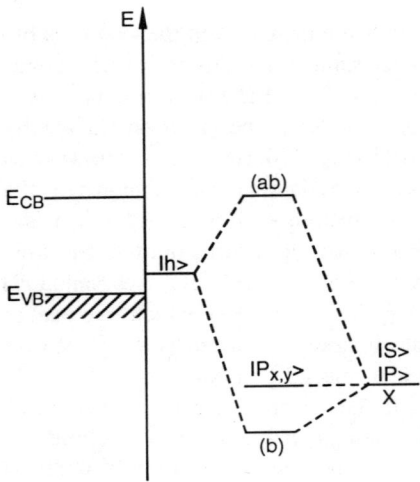

Figure 38. LCAO diagram of Si–X bonding interaction. The position of the adatom states (H: $|s\rangle$; halogens: $|p\rangle$) vary for the different adsorbates, and therefore so do the energetic positions of the bonding (b), antibonding (ab), and nonbonding states $|p_{x,y}\rangle$ (the latter not existing for H).

ured by photoelectron spectroscopy and will be given below for typical monovalent adsorbates. As the position of the adatom-induced states is outside the semiconductor fundamental gap, they have to be classified as surface resonances (see Section IV.1). As a consequence, no surface states exist in the band gap for perfectly terminated Si surfaces, and no surface-state-induced band bending or Fermi level pinning is expected. However, due to charge transfer from the Si surface atoms to the more electronegative adsorbates, the ionization potential I_p, and thus the position of the energy band, should be changed.

Hydrogen adsorption. Molecular hydrogen does not adsorb at room temperature on Si surfaces because of its high activation barrier for bond breaking. The dissociation energy of H_2 is given as 4.52 eV, and the Si–H bond energy is 3.9 eV.[35] Therefore, the adsorption is thermodynamically favored but evidently kinetically hindered. Assuming reversibility in the elementary steps involved, this would indicate that the formation of H_2 molecules is kinetically hindered as well, which is also documented by the high overpotentials of H_2 evolution on bare Si surfaces in electrochemical cells.[309] Therefore, in UHV adsorption experiments atomic hydrogen is produced by precracking the H_2 molecule (e.g., by a hot tungsten filament), and H atoms are easily adsorbed on the different Si surfaces. On the other hand, atomic H may also be inserted into back bonds. The formed SiH$_x$ species, with x approaching 4, will be desorbed from the surface, which leads to dry chemical etching of Si.[310,311] Atomic H may also easily diffuse into the bulk and interact with dopants.[312] As a consequence, the type of doping and the dopant concentration may change, which may also induce changes in the surface potentials. H-covered Si surfaces with defined LEED patterns have been prepared by careful control of the experimental parameters, e.g., Si(100):H (2 × 1) and Si(100):H (1 × 1)[313–315] (with surface monohydride and dihydride structures as shown in Fig. 37b, c) as well as Si(111):H δ(7 × 7) or Si(111):H (7 × 1).[310,311,316–318] A very perfect Si(111):H (1 × 1) surface can be prepared by chemical etching[288] and also, as has recently been shown, by H adsorption at elevated temperatures.[319]

The electronic properties of such defined H-covered surfaces are well studied. The H-induced surface resonances [bonding state (b); cf. Fig. 38] are found in a binding-energy range of about 4.5–6.5 eV for the different surfaces.[288,314,316–318] The number of H-induced bands depends on the number of H adatom sites. These states are found to be nearly nondispersive (the bands are flat as a function of wave vector $\mathbf{k}_p$), which indicates weak lateral interaction.

The related changes in surface potentials are close to the expectations for the idealized case as derived from Fig. 38. Without surface states in the band gap, the work function Φ now depends on the doping of the semiconductor.[312,321,327] As for perfect terminated Si:H surfaces, flatband conditions are expected; the position of E_F in the semiconductor band gap is close to the value given by the bulk doping. Φ is therefore also given by the value of $E_F - E_{VB}$ and by the absolute position of the valence-band edge versus the vacuum level, given by I_p. The ionization potential I_p seems to be not very much affected by adsorbed H, which would indicate only a small negative charge on the H atom. The reported values for Si(100):H are $I_p = 5.3$ eV[312] and 4.9 eV,[322] and for Si(111):H $I_p = 5.2$ eV to 5.0 eV.[323,321] For electrochemically treated (111) surfaces, the values are similar, but also a surprisingly large change ($I_p = 6.6$ eV) has been reported[35,324] (see also Section 1.3).

Halogen adsorption. The adsorption of halogens may also be rather complex under many experimental conditions. Especially F_2 strongly reacts with Si due to the large Si–F bond energy (the bond strength for Si–F bonds is 5.9 eV) and the low dissociation energy of F_2 (1.65 eV). F is readily adsorbed when XeF_2 is used as a source. For molecular F_2 as the source, the adsorption of F depends very much on the experimental conditions.[325] As F is also the most electronegative element, it strongly polarizes the Si–F bond as well as the Si back bonds, which makes a further attack kinetically favorable. Therefore, after adsorption various SiF_x species are formed until volatile SiF_4 is formed and Si is etched. The SiF_x species have been identified by core-level photoelectron spectroscopy[326,327] (Fig. 39). Their relative concentrations depend on the surface

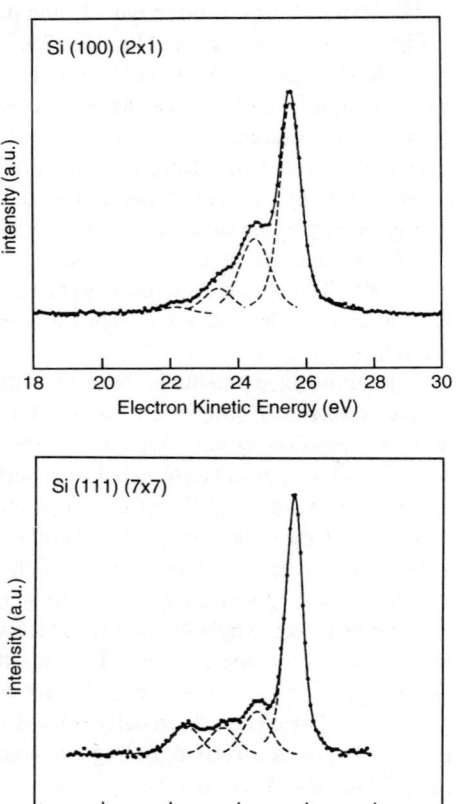

Figure 39. SXPS $2p$ core-level photoemission spectra of different Si surfaces after exposure to 50 L of XeF_2 (hv = 130 eV, deconvoluted spectra with the $2p_{3/2}$ component only; after Refs. 326 and 327).

orientation and the experimental conditions of the surface pretreatment. Owing to the strong Si–F bond, the surface states of the clean surface are expected to be passivated (as shown in Fig. 38), and the strong dipolar polarization of the bond is expected to produce a large dipolar shift of the ionization potential. However, as the adsorption can hardly be restricted to a monolayer, the idealized surface terminations shown in Fig. 37 are not achieved in most cases. However, after F_2 adsorption the surface coverage saturates at 1.5 ML (ML = monolayer) coverage and a defined Si(100):F (2 × 1) surface has been identified by LEED.

The other halogens (Cl_2, Br_2, and I_2) behave much more conveniently due to their weaker Si–X bond strengths. They saturate the dangling bonds on the different Si surfaces, and ideal surface termination can be obtained as indicated in Fig. 37. Thus, their saturation coverage is close to a monolayer. A number of detailed investigations on structural and electronic properties have been performed with stable adsorbate-covered Si surfaces.[35,187,278,279,329] The best defined and most thoroughly investigated surfaces are Si(111):Cl (1 × 1)[330] prepared from Si(111)(2 × 1), Si(111):Cl δ(7 × 7)[316] prepared from Si(111)(7 × 7), and Si(100):Cl (2 × 1)[331] prepared from Si(100)(2 × 1) by Cl_2 adsorption. The reported superlattices in the LEED patterns indicate that the idealized surface structures shown in Fig. 37 represent the surface bonding geometry in a correct way. The bond lengths and bonding geometry have been deduced from a detailed analysis of surface-extended X-ray adsorption fine structure (SEXAFS) data.[332,333] The bonding distances are close to the values expected for a polarized σ-type single bond.

The electronic properties of the halogen-covered surfaces can be discussed based on the simple LCAO diagram shown in Fig. 38. For Cl-covered surfaces, the bonding state (b), which results from the overlap of the Si sp^3 hybrid $|h\rangle$ with the p_z level of the Cl adatoms, is found at a binding energy of about 8 eV for all surfaces.[334–337] This state shows no dispersion. In addition, the quasi-nonbonding chlorine states $|p_{x,y}\rangle$, often also referred to as π states, are situated at a binding energy between 4 and 7 eV. These states show a very weak dispersion. In contrast to the simplified picture of Fig. 38, the degeneracy of these states is not observed in more elaborate theoretical calculations and angle-resolved photoelectron spectra (for a detailed discussion of the electronic structure of Si adsorbate surfaces, refer to Ref. 278). The antibonding states (ab) are considered to be situated above the semiconductor conduction-band edge. For Br- and I-covered Si surfaces, all these states are shifted upward (to lower binding energies) as the ionization potential of the halogen p states is decreased. Owing to the large electronegativity of the halogens, the surface resonance states formed should lead to a strong dipole-induced increase of the ionization potential. Indeed, the Si(100):Cl-(2 × 1) surface shows an increase of I_p to 6.1 eV,[334] which is 0.8 eV above the clean-surface value. Also, the surface Fermi level of n^+-doped Si shifts from the pinning position at midgap ($E_F - E_{VB} = 0.6$ eV, corresponding to $\Phi = 4.7$ eV) for the clean surface to a position close to the conduction-band edge for the Cl-covered surface ($E_F - E_{VB} = 1.1$ eV, with an experimentally determined value $\Phi = 5.1$ eV), which would indicate a complete passivation of surface states. For the Si(111):Cl-(7 × 7) surface,[336] with nearly intrinsic Si, the position of the Fermi level moves from $E_F - E_{VB} = 0.63$ to 0.74 eV. Here the work function was not given.

Evidently, the experimentally determined surface energy diagrams of monovalent adsorbates are well described by the theoretical expectations. For perfect surface terminations without any Si back bond attack, the dangling-bond states (surface states) of the clean surfaces are removed from the band gap (Fig. 38) and flatband conditions may be obtained. However, the ionization potential I_p, and thus the work function Φ, is changed by the dipole layer $\Delta\chi$ induced by the surface resonances formed. The increase in I_p is proportional to the electronegativity of the adsorbed adatom.

(c) H_2O adsorption

The adsorption of H_2O on Si surfaces is of special interest from the viewpoint of semiconductor electrochemistry, as H_2O is the most familiar solvent used for electrolyte solutions. H_2O may also be adsorbed from wet air or as a contaminant in organic electrolytes. In addition, there is also a technological interest in wet oxidation of Si because of its use in the Si electronics industry for producing good SiO_2 passivation layers.[338,339] Despite this importance, the number of fundamental studies of Si/H_2O adsorbate interfaces in UHV is considerably smaller than the number of studies for the adsorbates discussed above; good reviews on the results obtained before 1987 are given in Refs. 72 and 308. Most of these studies were related to the question of whether H_2O is molecularly or dissociatively adsorbed. This problem seems to be settled now, and the consensus is that the type of adsorption depends mostly on the adsorption temperature and the surface orientation. On Si(100)(2 × 1) the initial sticking coefficient at room temperature is very large (close to 1), and the adsorption saturates at about 0.5 ML.[340,341,344] The 2 × 1 reconstruction is not changed.[341,342] The H_2O is dissociatively adsorbed at low and room temperature (RT), forming $OH_{ad} + H_{ad}$ surface species, as shown in Fig. 40. This was proven by vibrational spectroscopy (see Refs. 72 and 308) and was corroborated in a recent study of valence-band photoemission spectra.[343] On Si(111)(2 × 1) the sticking coefficient is smaller by a factor of 10, and the (2 × 1) surface reconstruction transforms to a (1 × 1).[341,342,344] There is agreement on the fact that H_2O is dissociatively adsorbed for all adsorption temperatures investigated.[72,308] Evidently, similar species to those depicted in Fig. 40 are formed but with a statistical distribution of adsorption sites. In contrast, Si(111)(7 × 7) shows no dissociative adsorption at low sample temperatures, but dissociated H_2O is found at elevated temperatures.[343,72,308] The sticking coefficient is extremely low at RT (in the range of 0.01), and the surface retains the (7 × 7) surface periodicity on an increased diffuse background.[344,345] These discrepant findings regarding the adsorption behavior have been based on the use of different techniques, such as UPS and vibrational spectroscopies (IRAS, HREELS). It has been suggested that the energy required to surmount the activation barrier for H_2O dissociation at low sample temperatures may be provided by the electron beam of the HREELS technique.[346] The type of adsorption found may also be influenced by the quality of surface preparation.

The expected surface molecular structures suggested for dissociative adsorption are shown in Fig. 40. The adsorbed H and OH species on the surfaces are, in principle, both able to saturate the Si dangling bonds, and good electronic surface properties are expected without high concentrations of surface states. The electronic properties of the H-terminated

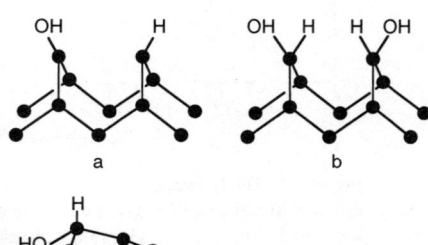

Figure 40. Schematic representation of H_2O adsorbate bonding geometry on Si(100)(2 × 1): (a) Dissociative monoadatom type bonding; (b) dissociative diadatom type bonding; (c) dissociative back-bond bonding.

surfaces have been discussed above, and the OH moieties should be qualitatively comparable in their chemical properties to monovalent halogens. Therefore, one would expect filled surface states due to H with a binding energy around 6 eV and a bonding σ and nearly nonbonding π surface resonances due to chemisorbed OH. Indeed, the valence-band spectra show adsorbate-induced emission intensity at E_B = 12 eV due to Si–O bonds and at E_B values of 7.5 and 6.5 eV, which have been lately assigned to OH σ bonds and OH π bonds, respectively.[343] For cases in which molecularly adsorbed H_2O is assumed, three emission features are observed at E_B = 6.2, 7.2, and 13.5 eV; these are assigned to the $1b_1$, $3a_1$, and $1b_2$ molecular orbitals of H_2O.[346,347] Originally, all three emission lines were attributed to the molecular orbitals of H_2O, shifted from their gas-phase energy differences due to the interaction with the surface.[346,347] Typical He I valence-band spectra obtained for the different adsorbate stages of H_2O are shown in Fig. 41.

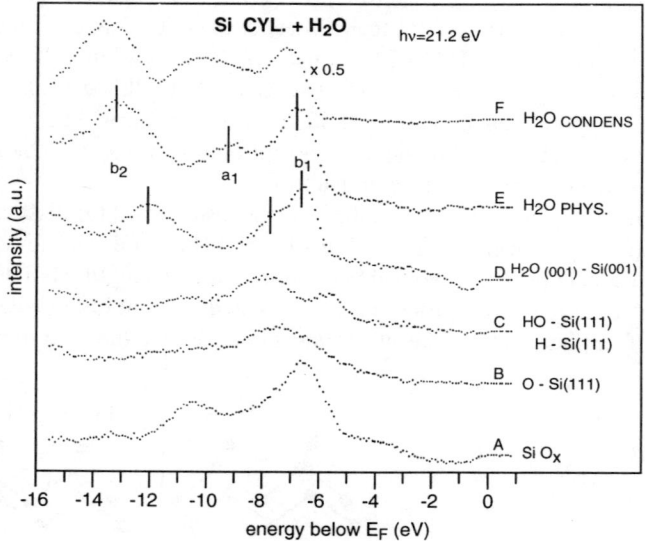

Figure 41. He I photoelectron difference spectra of H_2O adsorption on different Si surfaces. (The assignments are given on the figure; after Ref. 347, Reprinted from *Surf. Sci.* **152/153**, W. Ranke, D. Schmeisser, and Y.R. Ying, Orientation dependence of H_2O adsorption on a cylindrical Si single crystal, p. 1103, 1985, with kind permission of Elsevier Science-NL, Sara Burgerhartstraat 25, 1055 KV Amsterdam, The Netherlands.)

The ionization potential should also be affected by the adsorbate phase. It may be expected that the ionization potential would be strongly increased by adsorbed OH (adsorbed H moieties have only a small effect). However, the reported values for the different surface orientations show a nearly constant decrease in ionization potential of 0.2 eV.[61] For the molecularly adsorbed H_2O, the work function is decreased by up to 1 eV.[347] This value is in good agreement with that for metal substrates ($\Delta\Phi = 0.4-1.2$ eV).[72] There are no consistent data on the changes in band bending except for those published in Ref. 61; however, in the latter study nearly intrinsic Si was used, and thus a final conclusion about the H_2O-induced modification of surface states cannot be drawn.

When the OH- and H-covered surfaces are heated to elevated temperatures (500 K), dissociation of adsorbed OH_{ad} to O_{ad} and H_{ad} occurs.[345,348-350] At even higher temperatures (770 K), only adsorbed O remains.

(iii) Electrolyte Interfaces

Si is the best known semiconductor material to date, which has led to a broad and detailed knowledge of its bulk and interface properties. Therefore, its electrochemical properties can be analyzed without uncertainties regarding its bulk properties, which is an important precondition for fundamental studies of (photo)electrochemical charge-transfer processes at semiconductors.[351] The electrochemical properties of Si electrodes have been studied extensively and are summarized by Lewis and Bocarsly.[352] The specific aspect of oxide growth and etching is discussed in Refs. 353 and 354. However, there is a considerable scatter in the evident electrochemical response reported by different authors, which is mostly due to differences in the surface pretreatment before the start of the electrochemical experiments. This may be expected based on the survey given above on Si surface and adsorbate interface properties [Sections VI.1(i) and VI.1(ii)], as there are many different adsorbate sites (on top and in back bonds), and these may be differently occupied after subtle changes in pretreatment conditions. As a consequence, the surface state distribution and the interface dipoles will be changed.

One of the most important steps in Si-electrode conditioning is the cleaning and etching of its surfaces. Various procedures have been presented in the literature that are used to remove surface impurities and, subsequently, the native oxide layer.[352,355,356] The wafer cleaning typically involves oxidizing solutions—e.g., in the "RCA standard clean,"

$H_2O/H_2O_2/NH_4OH$ or $H_2O/H_2O_2/HCl$.[355] The resulting Si-oxide layer is removed by forming soluble oxides. The most often used treatment involves HF and leads mostly to H-passivated surfaces.[356] With proper experimental conditions, such surfaces are nearly free of surface states and are optimized with respect to their electronic properties.[357-360] The concentrations of surface states can be as low as 1×10^{10} eV^{-1} cm^{-2}, as measured for the Si(111):H (1 × 1) surface.[361]

This etching process and the surface conditions obtained are an intensively investigated field of Si electrochemistry. Many investigations deal with the electroless "chemical etch," a process in which no current is drawn through an external circuit.[288,355,357-359,362-365] For a more detailed analysis of the etching mechanism, a number of experiments have also been performed in standard electrochemical setups, which allow control of the potentials or the currents (potentiostatic or galvanostatic control) and also investigation of photoexcitation processes.[181-183,366,367] Both etching processes lead to H-terminated surfaces, which are comparable in their structural and electronic properties to the H-adsorbate interfaces discussed above [Section VI.1(ii)]. However, as chemical and electrochemical treatments easily allow the transport of reactants and products, the surface properties may even be better than those of the UHV-prepared surfaces. The dangling bonds of the Si(111) surface are terminated by H bonds (cf. Fig. 37). Owing to the perfect H termination, the LEED pattern shows a clear (1 × 1) structure.[181,353,359,368] The existence of terminal Si–H bonds has been identified by vibrational spectroscopy (IRAS, HREELS)[182,358,363-365,369,370] and also by photoelectron spectroscopy (SXPS, UPS).[181,288,359,362-365] For the Si(100) surface, dihydride surface molecules, as shown in Fig. 37, are formed. Also, a clear square LEED diffraction pattern, as expected for the (1 × 1) unreconstructed bulk termination, is obtained.[359] The terminal Si–H$_2$ moieties have again been identified by their vibrational spectra.[363,364]

Extensive investigations of Si electrodes after electrochemical treatment have been performed by Lewerenz and co-workers using an electrochemical setup directly coupled to a UHV system (see Fig. 42A).[181-183] Typical photoemission results obtained after electrochemical etching in a NH$_4$F solution before and after an oxidation transient current is passed are presented in Fig. 42B. In more recent investigations, these authors have also shown by STM studies that the smoothness as well as the electronic properties can be considerably improved when current oscillations occurring during the oxidation etch process[371] are utilized in the Si surface

The Semiconductor/Electrolyte Interface

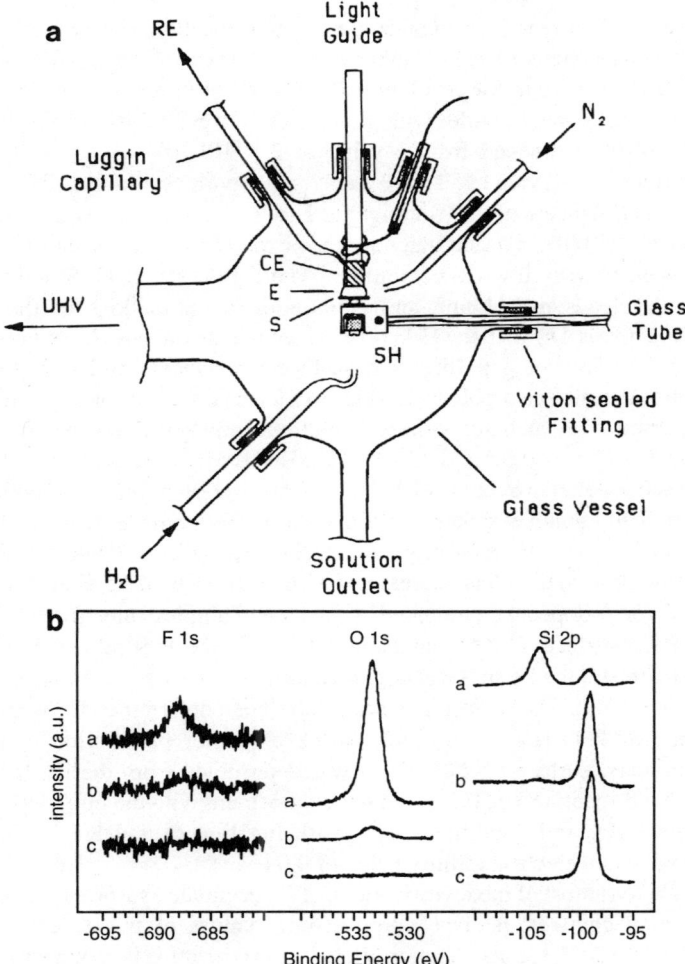

Figure 42. (a) Experimental setup for performing (photo)electrochemical experiments directly coupled to a UHV analysis system. (After Refs. 181 and 182.) (b) XPS core-line spectra of Si(111) after etching in 0.2M NH$_4$F solution obtained before (a), during (b), and after (c) anodic current transient.[181] (Reprinted from *Surf. Sci.* **269/270**, T. Bitzer and H.J. Lewerenz, Preparation of hydrogen-terminated Si single-crystal surfaces, p. 886, 1991, with kind permission of Elsevier Science-NL, Sara Burgerhartstraat 25, 1055 KV Amsterdam, The Netherlands.)

passivation. Whereas for normal etching conditions the surface roughness is rather large (steps 15 Å in height on 200-Å terraces), it is considerably reduced after the current oscillations (atomically flat terraces of several hundred angstroms in width with single atomic steps). Also, the surface state density is reduced from a value of 2×10^{11} eV^{-1} cm^{-2} without oscillations[360] to a value of 1×10^{10} eV^{-1} cm^{-2} with oscillations.[361]

In accordance with the findings for H-adsorbate interfaces [reported in Section VI.1(ii)], surface states are only expected at defects and should be present in very low concentrations. The formation of H–Si surface molecules leads to bonding and antibonding combinations of the Si dangling bonds $|h\rangle$ with the H $1s$ orbitals, which are outside the semiconductor fundamental gap (they are surface resonances) and may only influence the surface dipole layer (Fig. 38). This surface dipole layer may be expected to be small due to the low electronegativity difference between Si and H. The electronic states of H adsorbates on Si surfaces were discussed in detail in Section VI.1(ii). Unfortunately, the surface potentials of chemically and eletrochemically treated surfaces have only been analyzed in a very few cases in *ex situ* UHV experiments. Thornton and Williams observed a clear increase of Si core-level binding energies by 0.8 eV for p-doped as compared to n-doped samples only for the H-terminated surface.[359] After annealing at 520°C, the H is removed from the surface and the Fermi level is pinned again 0.3 eV above the valence-band maximum. The surface potentials have been determined for weakly n-doped Si(111) [$E_F - E_{VBM}$(bulk) = 0.8 eV] after preparation of the H-terminated surface.[372] $E_F - E_{VBM}$ was determined from the He II valence-band spectra to be 0.8 eV, in good agreement with the bulk doping, indicating flatband conditions. The work function Φ is 4.3 eV, which translates to an electron affinity value of 4.0 eV.

Whereas most of the investigations of H-terminated surfaces indicate only small changes in the electron affinity values of the different Si surfaces [≤0.2 eV; see also Section VI.1(ii) (the ionization potentials given there differ from the electron affinity by the Si band gap of 1.1 eV)], a rather large value of $\Delta I_p = +1.3$ eV has also been reported in the literature[35,324] and explained by semiquantitative calculation of the expected surface dipole [using Eq. (96)]. We are not able to resolve this controversy regarding the values reported for the surface dipole changes due to H-termination. However, it has been shown that for slightly different etching conditions there is also a small fraction of F-terminated Si bonds, up to several tenths of a percent of a monolayer.[364,365,181] As the Si–F moiety would lead to a strong surface dipole, it may be expected that a small extent of surface contamination with Si–F may explain the large deviation reported above.

When electrochemical investigations are performed with Si surfaces, it must also be considered that the Si–H terminated surface will be oxidized again. It has been shown that this oxidation process is very slow in contact with air.[359,373] However, in contact with H_2O[374,375] and with aqueous electrolytes and especially for positive potentials drawing anodic currents, the surface oxidation is faster and depends on the composition of the electrolyte.[376,354] As a consequence, Si-oxide/hydroxide mixtures of various compositions are formed. The initial steps and the surface and near-surface moieties formed are equivalent to those formed in the adsorption experiments with O_2 and H_2O, as presented in Section VI.1(ii). As already discussed there, the experimental conditions during the adsorption experiments determine the amount of surface and back bond adsorbate bonding. Therefore, also different distributions of surface states and values of the interface dipole are expected. Surface states should not exist in the band gap for a perfect termination of all Si dangling bonds by H-, O-, or OH-adsorbate termination. However, the surface resonances formed lead to interface dipoles, changing the electron affinity. Surface states, which are often found for imperfect surface oxidations in such low-temperature processes, have to be assigned to interface defects with incomplete coordination numbers and to polarized bonds due to electronegative ligands and/or stressed bond angles. The amount of surface oxide/hydroxide after surface treatment may be deduced from *ex situ* XPS analysis. For example, it has been shown that only minor traces (≤0.4% of a monolayer) of $Si-O_x$ and $Si-F_x$ species are found when organic electrolytes and efficient charge transfer to the redox couple is utilized.[376] On the other hand, anodic currents in aqueous electrolytes may lead to porous Si layers.[354]

Unfortunately, in many cases the conditions of the surface are not properly analyzed by surface analytical means. Therefore, the flatband potentials that are reported for Si in different electrolytes and after different surface pretreatments may deviate from each other considerably.[352] These deviations may be attributed to different amounts and different adsorption sites of adspecies X, with on-top bonding and Si back bond bonding (cf. Figs. 37 and 40). The dipole potential drop $^{SC}\Delta^{El}\chi$ at the phase boundary depends also on the nature of X and may be qualitatively deduced from the adsorbate interfaces described in Section VI.1(ii).

The flatband potentials in nonaqueous electrolytes may be considered to be less affected, even though there is considerable scatter in the reported data.[352] $E_{fb} = -0.85$ V vs. SCE is given as a reasonable value for n-doped samples in CH_3CN. For p-doped samples, E_{fb} is placed at +0.05 V vs. SCE. These values may be translated to positions of the valence-band and conduction-band edges at +0.25 V and −0.85 V vs. SCE, respectively. In ethanol the conduction-band edge is placed at −0.4 V vs. SCE,[352] and in methanol at about −0.55 V vs. SCE.[377] In aqueous electrolytes the flatband potentials shift with pH as expected for oxide-type surfaces.[379] For comparison of band positions to the results obtained from UHV experiments, the point of zero zeta potential (pzzp) of the adsorbed ions must be taken into account (cf. Section III.3). Madou et al.[378] reported flatband potentials for n- and p-doped Si with a very thin but defined oxide layer as a function of pH value. The position of the conduction band was given as −0.47 V vs. SCE. Other reported values of E_{fb} in aqueous electrolytes are −1.68 V (for highly p-doped) and −1.43 V vs. SCE (for weakly p-doped) Si(100) after HF etch in $2M$ KOH solution[380] and −0.74 V vs. NHE for Si(111) in HF (1%) solutions.[381] The position of E_{CB} was determined to be at −1.2 V vs. SCE[382] after HF etch and measurement in NaOH solutions and also at −1.2 V vs. SCE in HF solutions.[383] Also, a pH-dependent change of flatband potential has been reported for F^--containing solutions with a Nernstian dependence of 59 mV/pH unit.[384] Here E_{CB} ranges from 0.0 V vs. SCE (pH = 2) to −0.6 V vs. SCE (pH = 12), with $E_{CB} - E_F = 0.3$ eV, which is considerably more positive than the values given above. This shift of E_{fb} with pH is unexpected as properly treated Si surfaces should be H-terminated, and thus E_{fb} should not show a pH shift. These values of the flatband potential may be compared to the ionization potentials (or work functions) of UHV-treated Si surfaces and their changes due to the presence of adsorbates (cf. preceding sections). With a scaling term of $eE^0_{abs} = 4.5$–4.7 eV, a flatband potential of −0.5 V vs. SCE translates to an absolute value of the electron affinity of about 4.2–4.0 eV. These values may be compared to the electron affinities E_A (or ionization potentials $I_p = E_A + E_G$) of clean Si surfaces [$E_A \approx 4.2$ eV; Section VI.1(i)] and their changes due to the different adsorbates [Section VI.1(ii)]. Evidently, the band-edge positions of thin oxide-covered surfaces in electrolytes are in good agreement with the values obtained for surfaces prepared and measured in UHV.

If the flatband potential shows the Nernstian shift of 59 mV per decade of pH change, an oxide/hydroxide layer on the surface is very

The Semiconductor/Electrolyte Interface

probable. Unexpectedly, a pH-dependent shift of the flatband potential has also been reported for an H-terminated surface.[384] Positive shifts of the flatband potential (increased ionization potentials or electron affinities) can easily be explained in terms of adsorbed electronegative species X and a double-layer drop across the inner Helmholtz plane. A negative shift cannot be as readily understood. Whether the solvent of the electrolyte leads to an additional dipole or whether adsorbed negative ions of the supporting electrolyte lead to a double-layer potential drop across the outer Helmholtz plane is just a matter of speculation and has to be further examined.

Generally, it seems that defined and constant electrochemical properties of Si surfaces may only be expected when a rather perfect thin oxide layer is formed on the surface, such that the surface may still be efficiently tunneled. In this case the semiconductor/electrolyte interface may be better described as a semiconductor/oxide/electrolyte junction. It is also known that metal/oxide/semiconductor (MOS) devices give more reliable results in solid-state contacts than simple semiconductor/metal junctions.[94] Otherwise, the preparation and conservation of a perfect H- (or halogen-) terminated surface is necessary. The Si–X adatom surface molecules remove the dangling bonds (surface states) from the fundamental energy gap. Thus, only dipolar changes of the band-edge position are expected (changes of ionization potentials). However, as the Si–X bonds facilely undergo further reactions, transformation of the Si surface to a less defined condition easily occurs, which will lead to deviations in measurable equilibrium and nonequilibrium properties. Existing surface states can be assigned to defects due to structural disorder at the interface (broken Si back bonds).

2. Gallium Arsenide

(i) Clean Surfaces

Like most 3–5 semiconductors, GaAs crystallizes in a cubic zinc blende structure, which is similar to the structure of Si (see above).[94] Both structures can be considered as two interpenetrating fcc lattices. Whereas in Si all atoms are identical, in GaAs the existence of cation and anions in the lattice leads to sublattices (alternating layers) of Ga and As atoms. The cleavage plane of zinc blende semiconductors is the nonpolar (110) plane, and most of the UHV investigations of surface and interface properties have been performed with cleaved surfaces. The more important planes in technological applications and also in electrochemical studies are the (100) and (111) planes; however, the preparation of well-defined surfaces is more complex in their case. The structural and electronic properties of the different GaAs surfaces have been reviewed in a number of articles, with special emphasis on their dependence on surface treatment.[35,38,60,278–280]

After cleavage in UHV, a clean and contamination-free GaAs(110) surface is obtained. It shows a (1 × 1) structure in LEED and is free of surface states in the band gap for perfect cleaves.[385,386] The cleavage should be performed along the (100) direction.[34] The bulk truncated (110) plane without surface reconstruction is schematically shown in Fig. 43a. As already shown in Fig. 14, two sets of surface states may be expected for this structure; these are due to the dangling bonds situated at the Ga and As surface atoms. As the LEED measurements do not indicate any superstructure lattice spots, a reconstruction of the surface can be excluded, but the surface atoms relax to new positions, as also indicated in Fig. 43a.[35,385,386] This relaxation involves a tilt of the topmost surface layer, which leaves the Ga–As bond distances mostly unaffected. The As atoms are pushed outward and the Ga atoms inward by a rotation angle of about 30°. This rotation leads to a rehybridization of the surface atoms from the original sp^3 hybridization to approximately sp^2 hybridization for Ga, as expected from its nearly planar geometry. The surface As atoms form s-like nonbonding and p-like bonding orbitals, with nearly pyramidal coordination of the back bonds, approaching 90° angles.

As a consequence of this relaxation, the dangling-bond states are pushed out of the fundamental gap as shown schematically in Fig. 43a. The changes in the electronic structure of the GaAs(110) plane with changed surface geometry have been analyzed by a number of theoretical calculations.[35,126,387,388] In addition, the electronic structure of the perfect cleavage plane has been thoroughly characterized by photoelectron and inverse photoelectron spectroscopy.[35,126,280] The data show conclusively that the surface states of the relaxed As and Ga atoms are situated below the valence-band edge and above the conduction-band edge, respectively (see Fig. 43a). There are additional surface resonances identified in the spectra at lower and higher energy values (further away from the band edges), which are due to disturbed back-bonding states of the second and third layer.

As the fundamental band gap of perfectly cleaved GaAs(110) planes is free of surface states, the position of the Fermi level (flatband potential), and thus also the work function Φ, is dependent on the bulk doping. The

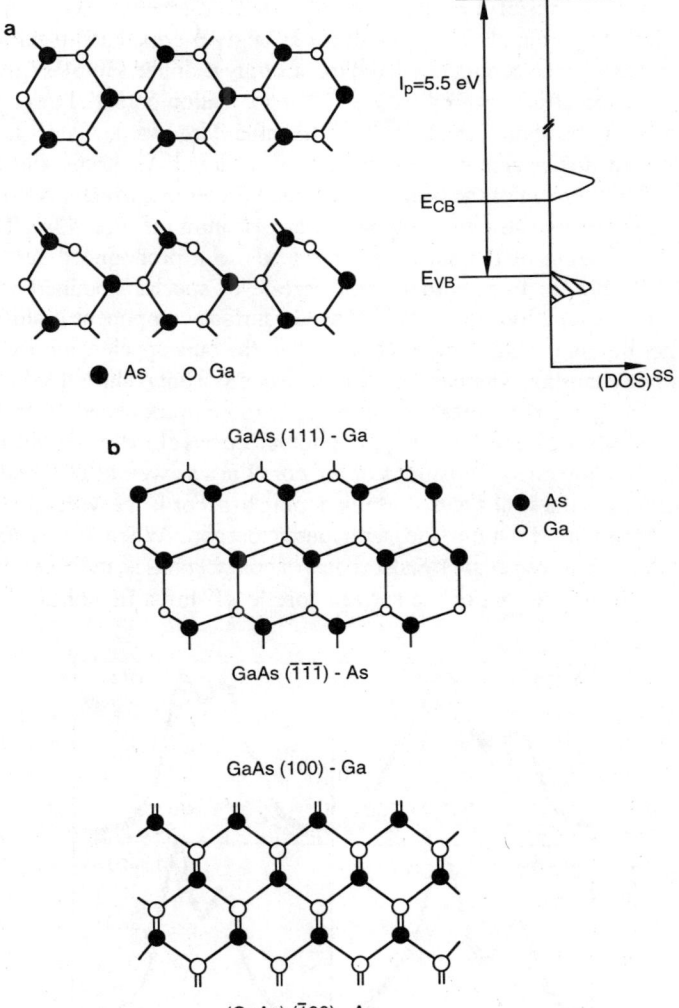

Figure 43. (a) Schematic representation of the GaAs(110) surface in its bulk truncation and after relaxation (tilt of Ga–As surface pairs). Also shown are the surface potentials and the density of surface states. (b) Schematic side view of the GaAs(111) and GaAs(100) bulk truncated surface structures (to the top the Ga-terminated and to the bottom the As-terminated plane).

Fermi level may shift from a position close to the valence band (given by the ionization potential, $\Phi = 5.5$ eV) for strongly p-doped samples, to a position close to the conduction band for strongly n-doped samples (given by the electron affinity, $\Phi = 4.0$ eV).[279] The ionization potential has been determined to be about 5.5 eV.[63,389-391] Its rather large value results from the outward displacement of the negatively charged As atoms and the inward displacement of the positively charged Ga atoms, which gives rise to a strong positive structural dipole toward vacuum (cf. Fig. 43a). This different charging of the surface atoms leads to a pronounced surface core-level shift in the photoelectron core-level spectra obtained with synchrotron excitation (Fig. 44).[392] The Ga surface component is shifted to higher binding energy values compared to the bulk species, indicating its positive charging, whereas the As surface component is shifted to lower binding energies. This surface component is more pronounced when the surface sensitivity is increased (E_{kin} of photoelectrons close to 50 eV). The absolute binding energy positions of the core lines as well as of all other photoemission lines shift with the bulk doping in accordance with the shift of the Fermi level within the semiconductor gap. When the surface relaxation is removed again by adsorbates or contact phases, the core-level spectra indicate the loss of the surface core-level shifts. In addition, the

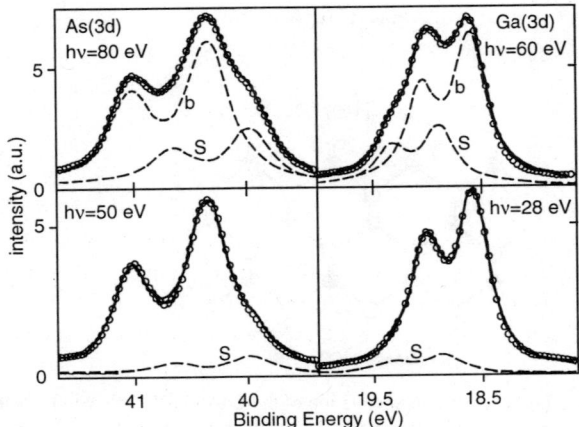

Figure 44. Surface core-level shifts of GaAs(110) cleavage planes. The Ga(3d) and As(3d) core levels were measured with different excitation energies in order to change the surface sensitivity. (Binding energies are referred to E_{VB}; after Ref. 392.)

surface resonances of the Ga and As dangling bonds may again move into the band gap and form real surface states. This effect must always be considered in addition to the possible modification of the surface hybrid states by the interaction with the adatom (adlayer) electronic states. The bulk truncated, unreconstructed (100) and (111) surfaces of GaAs formally contain a full layer of either Ga or As atoms, respectively, and this should lead to very polar surfaces. Their structure is shown schematically in Fig. 43b. In practice, these surfaces do not exist, and even the (1×1) LEED pattern is usually not found. Instead, many complex ordered structures are observed, which depend on the conditions of surface preparation and vary in their structure and stoichiometry. The surface preparation involves ion bombardment (sputter) and annealing cycles of pretreated surfaces (IBA) and molecular-beam epitaxy (MBE). Especially the sample temperature in the annealing step as well as in the epitaxy experiments plays a crucial role, because it determines the surface stoichiometry. As the number of known structures and their investigations exceed the scope of this chapter by far, the interested reader may refer to review articles covering the subject; general overviews on GaAs surface structures are given, e.g., by Mönch[35] and Duke.[385] More detailed overviews on the (100) plane are given in Refs. 393–396 and on the (111) plane in Refs. 397 and 398. The stable reconstructions of the (100) plane can be understood based on simple energy arguments and by counting valence electrons. The electronic energy will be lowest when the surface atoms in the top layer form dimers [similar to the case of Si(100) surfaces (Fig. 30)] and when the dangling bonds are filled on surface anions and are empty on surface cations. The different superstructures thus contain different local reconstructions which are often stabilized by As adatoms.[393-396] The main superstructures and the related surface stoichiometries are summarized in Table 3. Much more information on the details of the atomic arrangements has been obtained in recent investigations using STM.[396] The different superstructures also lead to different surface-state (surface resonance) density distributions. These have been studied by photoelectron spectroscopy and by electronic structure calculations and are summarized in Refs. 35 and 126. However, the understanding of the electronic properties of the different superstructures seems to be still not very well settled. Therefore, no additional details are given here. There are also different surface core-level shifts reported for the different superstructures, which may be taken as calibration of the surface composition. There is a clear indication of Fermi level pinning close to the midgap position

Table 3
Surface Reconstructions on GaAs(100) Surfaces and the Related Surface Potentials

Reconstruction	Ga(3d)/As(3d) intensity ratio[a]	$E_F - E_{VBM}$ (eV)[b]	I_p (eV)[c]
$c(4 \times 4)$	0.65	0.7	5.3–5.65
$c(2 \times 8)$	0.72	0.7	5.2–5.6
4×6	1.17	0.6	5.05–5.44
$c(8 \times 2)$	0.98	0.7	5.2–5.3

[a]Ref. 394.
[b]Mean value, Refs. 399–402.
[c]Refs. 399–402.

($E_F - E_{VB}$ = 0.7 eV) for all reconstructions. The ionization potential is found to vary among the different superstructures.[399–402] The published data on ionization potentials are also included in Table 3.

The UHV-prepared (111) plane is also characterized by a number of reconstructions. As for the (100) surfaces, the structure and composition depend on the orientation and preparation, which again may be by IBA or MBE.[35,385,397,398] The bulk truncated (111) plane is either As-terminated or Ga-terminated (Fig. 43b). The corresponding surfaces are denoted as ($\bar{1}\,\bar{1}\,\bar{1}$) and (111), respectively. The Ga-terminated (111) (2 × 2) surface reconstruction has been well characterized. It is formed by an ordered array of missing Ga atoms (one out of four) in the topmost layer.[403–405] The remaining topmost Ga atoms with threefold coordination move inward, and their hybridization changes toward sp^2 hybridization. On the other hand, the As atoms of the layer below move upward and outward, and their bonds change to s- and p-type bonds. Thus, the bonding geometry is similar to the tilted chains on the (110) plane. As a consequence, the surface is considered to be semiconducting.

The As-terminated ($\bar{1}\,\bar{1}\,\bar{1}$) surface also forms a (2 × 2) reconstruction, which is assigned to an ordered (2 × 2) array of As trimers as adatom clusters.[404,406] As this structure contains completely occupied As dangling bonds, the surface should be semiconducting, in agreement with the photoelectron spectra. For the (111) surface, additional reconstructions may be formed, depending on the sample temperature[397]:

$$(2 \times 2) \rightarrow (\sqrt{3} \times \sqrt{3} - R30°) \rightarrow (\sqrt{19} \times \sqrt{19} - 23.4°)$$

and also a (3 × 3) and a (1 × 1) reconstruction.

The electronic properties of these surfaces are not well studied. The ionization potential for the Ga-terminated (111) surface is given as 5.1 eV, and for the As-terminated ($\bar{1}\bar{1}\bar{1}$) surface as 5.35–5.6 eV.[63] The position of the Fermi level is situated in the middle of the band gap.

In summary, the polar (100) and (111) surfaces of GaAs (prototype of a 3–5 semiconductor) are rather complex and depend on the preparation conditions. Only the nonpolar (110) plane is well defined. It is expected that for semiconductor/electrolyte interfaces the surface structures, and thus the electronic properties, may be altered considerably from those reported above for GaAs surfaces prepared and characterized in UHV. These changes due to the contact with ambient air or the electrolyte may also be rather complex and are expected to depend on the surface pretreatment. Therefore, the electrochemical interfaces may vary considerably in their interface properties. Unfortunately, only a few systematic electrochemical studies have been performed with (110) cleavage planes; additional studies of this nature would allow a better comparison with the UHV results.

(ii) Adsorbate Interfaces

Whereas for the different Si surfaces many adsorbates tend to passivate the dangling-bond-induced surface states (see Section VI.1) and remove Fermi level pinning, for the GaAs(110) cleavage plane the adsorbates have an opposite effect. The dangling-bond states, which are removed from the semiconductor band gap on perfect (110) surfaces, are shifted again into the band gap by the adsorbates and lead to Fermi level pinning. For the (100) and (111) GaAs surfaces, which are already pinned due to their complex reconstructions with different surface stoichiometries, the adsorbates do not change the pinning effect: There is no efficient surface passivation process identified so far as for Si with adsorbate or oxide layers.

The reason for this completely different behavior is related to the fact that different surface atoms with cationic (Ga) and anionic (As) character are involved in the surface interaction. This may be exemplified by a comparison of the LCAO energy-level diagrams of the surface molecules formed by monovalent, more electronegative adsorbates, X (such as H, halogens, and O) with Si (Fig. 38) and GaAs (Fig. 18).[35] On Si the dangling-bond hybrid states $|h\rangle$ are removed from the semiconductor band gap by forming bonding (b) and antibonding (ab) states due to Si–X bond formation. On GaAs the interaction with the Ga hybrid states $|h^c\rangle$ leads to a similar effect: the bonding (b) and antibonding (ab) states are outside the band gap. The interaction with the As hybrid states $|h^a\rangle$ leads to bonding states (b) (and nonbonding states $|p\rangle$ in the case of halogens and O; these states are missing for H) which are below E_{VB} and therefore also leads to surface resonances. However, the antibonding states (ab) are situated in the band gap. These states have acceptor character and lead to Fermi level pinning. This general argument is basically valid for all 3–5 semiconductors interacting with electronegative adsorbates, even though the details may be modified for a specific surface/adsorbate interface (see below). This can be verified by noting the linearity of a plot of acceptor pinning position ($E_F^{SS} - E_{VB}$) for GaAs(110) surfaces with different adsorbates X (Fig. 45) as a function of adatom atomic electron affinity (as a measure of the electron energy level of the adsorbate). The position of the antibonding state (ab) of the As hybrid $|h^a\rangle$ with $|p\rangle$ is closer to the valence-band edge in the band gap for more deeply bound adatom states.[35,407] (having larger electron affinities; cf. Fig. 45).

The exposure of GaAs surfaces with many adsorbates having different bonding properties shows pronounced differences in terms of adsorbate interactions, compared to elemental semiconductors, e.g., Si. In

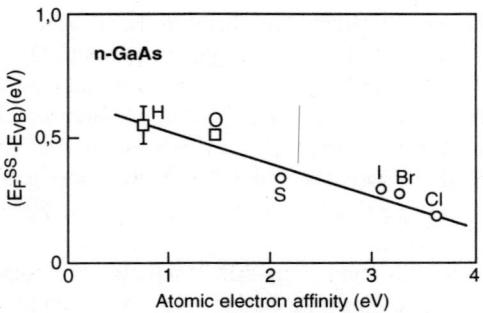

Figure 45. Surface Fermi level position ($E_F^{SS} - E_{VB}$) as determined by the position of surface states of acceptor character induced by different adsorbates of increasing atomic electron affinity. ○, RT, room temperature (~300 K); □, LT, low temperature (~140 K). (After Ref. 35, *Semiconductor Surfaces*, W. Monch, Fig. 14.37, 1993, © Springer-Verlag.)

general, the sticking coefficient (reactivity) to adsorbates is very small, especially for the (110) planes. Therefore, defects and different reconstructions found on the noncleavage planes [GaAs (100) and (111)] may drastically enhance the adsorbate interaction. Thus, the preparation and pretreatment of the GaAs surface drastically influences the results obtained. There is a strong tendency to form disordered surfaces as, for many adsorbates, the coordination numbers and bonding geometries are different for Ga and As, as may be expected from their general chemical bonding properties (see, e.g., Ref. 408). The specific surface interactions of GaAs with adsorbates such as O, H, halogens, and H_2O will be further considered below as these are of interest for electrochemical interfaces.

The numerous investigations on O_2 interaction with GaAs surfaces are summarized in e.g., Refs. 35, 279, and 409–413. We will therefore not concentrate on specific effects in relation to UHV adsorption experiments, e.g., the adsorption mechanisms as a function of surface pretreatment, but will specifically address the type of surface oxide formed and stabilized at room temperature and its consequence on interface potentials. The room-temperature adsorption of molecular O_2 on high-quality GaAs(110) cleavage planes has an extremely low sticking coefficient: The oxygen uptake slows down at a coverage of about one to two monolayers and an exposure of 10^{10} L.[411,412,414–417] The adsorption can be considerably enhanced by surface defects or by using external excitation mechanisms such as photoexcitation with light of energy greater than the band gap,[411,413,414,418] predissociation or excitation of the O_2 molecule,[417] or coadsorption of metals, e.g., alkalis.[419] The excitation process has been related to the enhanced formation of adsorbed O_2^- molecules by band gap excitation[414] or also to a direct (by light)[418] or indirect (by secondary electrons)[413] dissociation. In any case, the surface species finally formed are of a similar type and are characterized by chemically shifted core-level emission lines,[35,411,412] as shown in Fig. 46. The As $3d$ line shows more pronounced shifts, with four high-binding-energy components [$\Delta E_B = 0.8$ (As^1), 2.3 (As^2), 3.1 (As^3), and 4.2 (As^4) eV, relative to the bulk line]. The Ga $3d$ line shows two high-binding-energy components ($\Delta E_B = 0.45$ (Ga^1) and 1.0 (Ga^2) eV]. Their relative intensities depend on the degree of surface oxidation. The main chemically shifted lines, As^3 and Ga^2, are assigned to the formation of a disordered surface-oxide layer containing Ga_2O_3 and As_2O_3. Additional evidence for this assignment comes from Ga M1M45V Auger lines of the Ga-site density of states, which also indicates the formation of Ga_2O_3.[417] The As^2 and Ga^1 components may be

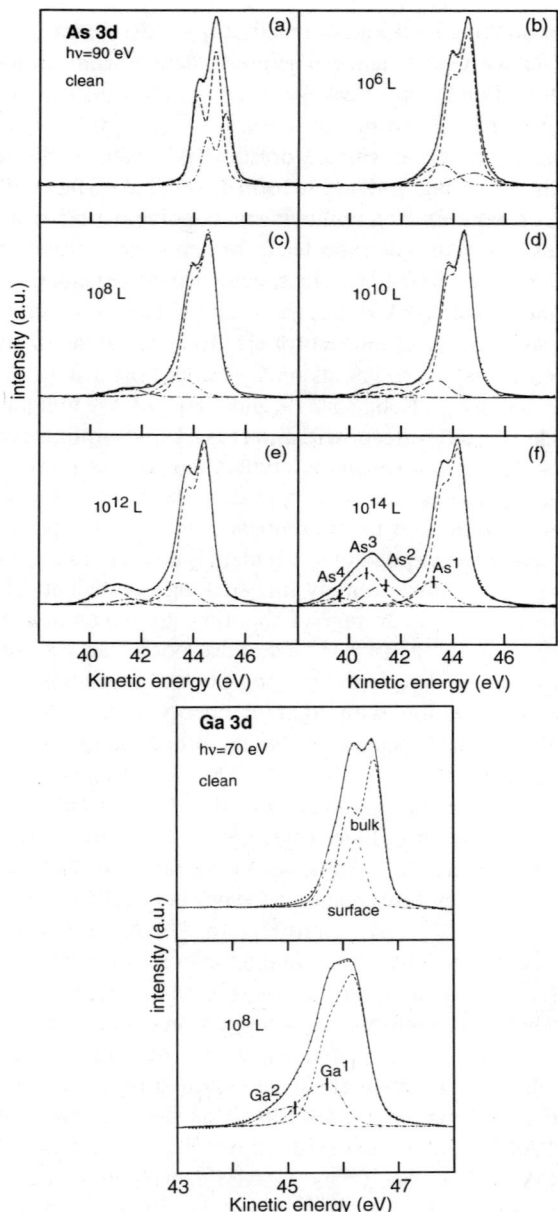

Figure 46. High-resolution photoemission core-level spectra (Ga $3d$ and As $3d$ lines) of GaAs(110) for different O_2 exposures. The assignment of the peaks are given in the text. (After Refs. 411 and 412.)

assigned to a partly oxidized GaAs layer with O introduced into the Ga–As bonds (see Figs. 46 and 47), which is related to the initial chemisorption stage. As[4] is evidently due to higher As oxidation states (As_2O_5 or $GaAsO_4$), and As[1] may be assigned to elemental As. The presence of elemental As is expected for very thick oxide films and long-term oxidation, as the reaction of GaAs with As_2O_3 to form Ga_2O_3 is thermodynamically favored[419] (see also Section VI.3). For a coverage of up to one monolayer, the adsorbed O is proposed to initially occupy twofold coordination sites between As surface atoms and—though less probable—between Ga and As surface atoms (Fig. 47).[420] Subsequently, a completely oxidized $GaAsO_4$ layer may be formed. Similar bridging sites have also been proposed based on valence-band[416] and high-resolution core-level

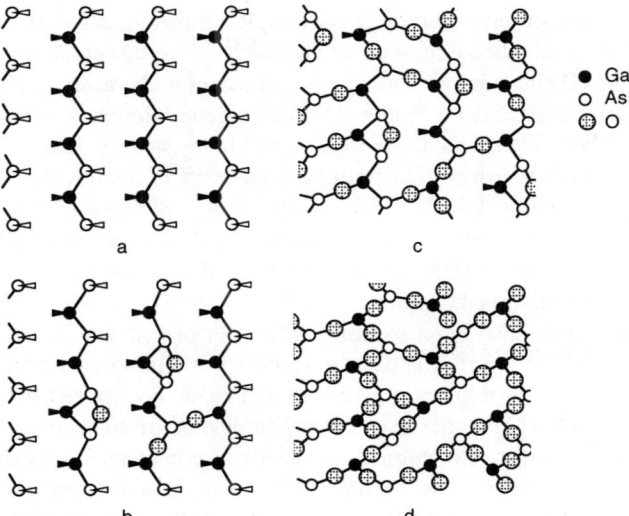

Figure 47. Oxygen adsorption model for GaAs(110) as proposed by Ludeke[420]: (a) clean surface; (b) initial O adsorption; (c) about half a monolayer coverage; (d) fully oxidized surface layer. (After Ref. 420, Reprinted from *Solid State Commun.* **21**, R. Ludeke, The oxidation of the GaAs (110) surface, p. 815, © 1977, with kind permission from Elsevier Science Ltd., The Boulevard, Langford Lane, Kidlington OX5 1GB, UK.)

spectra.[411] In addition, it was assumed that adsorbed O reacts to form a terminal double-bonded As=O unit. The O-induced valence-band features in the photoelectron spectra are a prominent peak at 4.6 eV and a shoulder at 6.9 eV superimposed on a broad background.[416,421] They are assigned to weakly bonding O π-bands [(nb) in Fig. 18)] and strongly bonding σ-bands [(b) in Fig. 18].

The thicker oxides are considered to be formed from this precursor GaAs-oxide layer by a field-assisted transport of O^{2-} species through the oxide film.[422] The LEED diffraction pattern is readily lost after oxygen adsorption[423,424] due to the disordered oxygen adsorbate bonding sites in the submonolayer regime and due to the disordered structure of the monolayer oxide and thicker oxide mixtures.

As a consequence of the complex surface structure, a large number of defects are expected, which also have considerable influence on the surface potentials. It is generally agreed that at room temperature the Fermi level is pinned at $E_F - E_{VB} = 0.6 \pm 0.1$ eV for n-type doping and at $E_F - E_{VB} = 0.9 \pm 0.1$ eV for p-type doping.[412,413,425–428] The deviations of the pinning positions observed at low sample temperatures must be related to SPV-induced rebending of energy bands,[413,426] as discussed in Section V. The As–O single bonds should possess acceptor character, as has been discussed above based on simple LCAO arguments. However, donor states have also been identified; these are assigned to the surface defects formed at the disordered surface. The ionization potential of the oxygen-covered surface is given as 5.4 eV.[429] This surprisingly small value can be explained by the insertion of the electronegative O into the GaAs bonds, as this does not produce a large dipole layer normal to the surface as expected for an on-top adsorbed oxygen.

There have also been several studies performed with other GaAs surfaces.[393,397,409,410,429] The basic oxidation properties are comparable to those of the cleavage plane, and disordered oxide layers are formed. As the (100) and (111) planes exhibit very complex surface reorganizations and different surface stoichiometries, the oxidation mechanisms are even more complex than those for the (110) plane. There is experimental evidence that the "Ga-rich" surfaces exhibit a faster oxygen uptake than the "As-rich" surfaces.[431] The Fermi level is reported to be pinned close to the midgap position.[279]

A very interesting result has been obtained by UV photoelectron spectroscopy with pulsed laser irradiation.[433] In this experiment a nonequilibrium occupation of conduction band and surface states was produced by a short "pump" laser pulse. Subsequently, the energetic distribution of the excited electrons, and thus also the density of states distribution of the surface states, was measured as a photoelectron spectrum by a second "probe" laser pulse with a defined time delay (the results presented were obtained with overlapping time pulses). In Fig. 48 the photoelectron spectra of clean and O-covered GaAs(110) surfaces are shown. As expected for the clean GaAs(110) surface (Fig. 48a), only conduction-band photoelectrons are obtained in addition to the valence-band electrons after a nonequilibrium situation has been produced. The O-exposed surface (Fig. 48b) directly

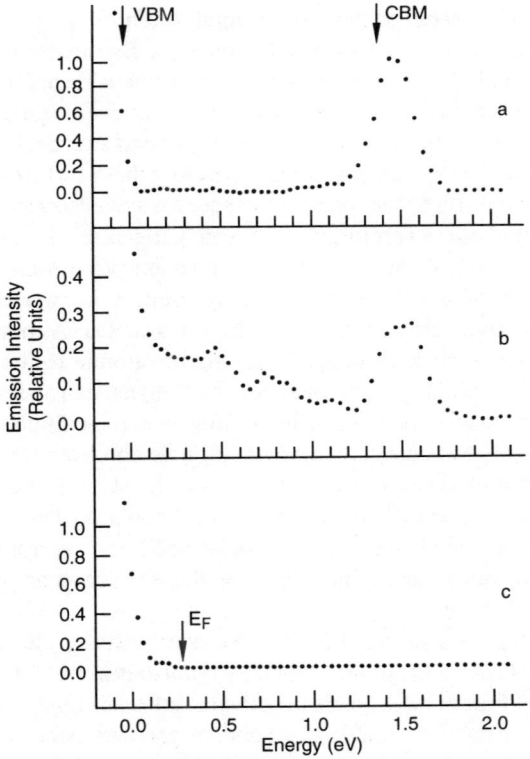

Figure 48. Gap region in the photoemission spectra obtained by laser pulse "pump-probe" experiments for GaAs(110) surfaces: (a) clean surfaces; (b) oxygen-covered surfaces; (c) oxygen-covered surfaces without "pump" excitation. (After Ref. 433.)

shows a broad distribution of surface states, which is formed by the surface reaction of GaAs(110) with adsorbed excited oxygen. Without "pump" excitation, no surface states are observed above the Fermi level E_F (Fig. 48c). Such experiments are of great value for the direct identification of the type and distribution of surface and interface states, but, unfortunately, they are very sophisticated.

In contrast to oxygen, with its tendency for divalent bonding, the interaction of monovalent adsorbates such as H and halogens on the GaAs cleavage plane (Fig. 43a) may theoretically be considered as an ideal procedure for transforming the remaining dangling bonds into chemically saturated surface molecules. However, as in the case of Si, the adsorption is very much dependent on the experimental conditions (e.g., pretreatment of the sample) as well as adsorbate excitation. Especially for the polar GaAs (100) and (111) surfaces, for which different reconstructions and surface stoichiometries have been reported (see above), a general adsorption scheme cannot be given, beyond some general similarities in surface reactivity to the (110) cleavage plane. Therefore these surfaces will not be considered any further; the interested reader may refer to some published adsorption studies concerning H[434–437] and halogens.[438–442]

As in the case of Si, molecular H_2 does not adsorb on GaAs(110) surfaces.[443,279] When dissociated H, e.g., from a plasma is used, the increased reactivity leads to adsorption as well as to Ga–As bond breaking and subsequent surface etching. Therefore, the atomic H concentrations (produced preferentially by thermal dissociation) have to be kept very low to get optimized adsorption conditions. However, in addition to adsorbed H, there will always be a certain non-negligible concentration of surface defects formed by H-induced back-bond cleavage. Unfortunately, there is no chemical or electrochemical procedure known so far that leads to perfect H-terminated surfaces as in the case of Si.[279] Therefore, there still remain some open questions regarding the surface properties of the GaAs(110):H surface.

At low H coverages the LEED (1×1) pattern remains intact, indicating that the basic GaAs(110) surface geometry is not destroyed.[444] It is generally agreed that H forms single bonds to both Ga and As dangling bonds, based on HREELS data[445–447] and high-resolution photoemission results.[448–451] Thus, the LCAO scheme presented in Fig. 18 is a good description of the surface electronic structure. Detailed theoretical electronic structure calculations[452,453] as well as the experimental results mentioned above indicate that the tilting of the topmost GaAs surface pairs found for the clean surface is removed. The Ga and As surface atoms move back to their bulk truncated atom positions (Fig. 43a). The driving force is the increased bonding strength of both As- and Ga-based dangling bonds $|h\rangle$ to the adsorbed H. With longer exposures to H plasmas, a preferential bonding to As sites is observed, and subsequently As is removed from the surface.[449,444,454] For even more extended exposures, etching of GaAs is obtained.

The valid surface potentials for the GaAs(110):H surface are still unclear. The Fermi level is pinned close to the midgap position $E_F^{SS} - E_{VB} = 0.6 \pm 0.1$ eV for n- and p-doped samples, as reported in Refs. 444 and 450. This pinning position was confirmed for n-doped samples by Mönch and co-workers,[455,456] but they found no pinning for p-doped semiconductors at exposures exceeding 10^2 L of H_2. The acceptor pinning position may be due to the surface acceptor state formed by the As–H antibonding level (see the LCAO scheme, Fig. 18), as suggested by Mönch and co-workers.[455,456] On the other hand, the back-bond surface defects formed, as described above, may induce less well defined surface states in the band gap which also exhibit donor character. There is also a striking discrepancy in the values reported for the ionization potential by Mönch and co-workers.[455,456] For both n- and p-doped samples, the ionization potential measured after a surface exposure to 10^5 L of H_2 at a sample temperature of about 140 K was about 1.2 eV lower than the value for the clean surface ($I_p = 5.5$ eV). This large difference cannot be explained in terms of the surface dipole estimated for the Ga–H and As–H[35] bonds. The electronegativity difference is small, with H being more electronegative, which would lead to a positive shift of I_p based on an estimation of the surface dipole (Eq. 96). The evident discrepancy was attributed to the removal of the GaAs surface relaxation. As already discussed above [Section VI.2(i)], the outward and inward displacements of As and Ga surface atoms of the tilted Ga–As surface pairs lead to an increase in I_p for the clean surface, which is again removed for the H-covered surface. The effect is considered to be even overcompensated by a Ga outward displacement on the GaAs(110):H plane. In contrast to these results, M'Hamedi et al.[444] reported an unchanged ionization potential for room-temperature conditions ($I_p = 5.4$ eV), and Santoni et al.[450] have found a small decrease ($\Delta I_p = -0.2$ eV). These large discrepancies in the reported values of the surface potentials indicate the difficulties in preparing a well-defined and reproducible GaAs(110):H surface. The polar GaAs surfaces is even expected to be less defined after exposure to activated H.

The interaction of halogens with GaAs surfaces has been intensively investigated in recent years because of the potential applicability of halogens as dry etching reagents. Most of the work has been directed toward Cl_2 adsorption,[457–463] but F_2,[464] Br_2,[460,465,407] and I_2[466] have also been investigated. For all halogens investigated, the adsorption has been found to be dissociative at room temperature. For low-temperature adsorption, a molecular precursor adsorption stage has been identified. It has been shown that the adsorption of Cl_2 saturates very fast at room temperature (about monolayer coverage),[463] and the (1 × 1) surface reconstructions remain unaltered.[458] Based on angle-integrated as well as angle-resolved UPS results in comparison with theoretical calculations of the possible surface molecule electronic structure, it was originally concluded that the Cl forms bonds only to the As dangling bonds.[457] Five nearly nondispersive to weakly dispersive Cl-induced bands are found at around 2.0, 4.5, 5, 6–7, and 8 eV and were attributed to a monohalogenide surface molecule only bound to As. The As–Cl σ-bonding states were identified with the band at 8 eV, and the nonbonding (weakly π-bonding) Cl $p_{x,y}$ states with the band at 5 eV. All other features were assigned to disturbed back bonds. Also, high-resolution core-level spectra[458] of the Ga $3d$ and As $3d$ lines gave support to this monochloride adsorption site model, as only one chemically shifted As line (ΔE_B = 0.82 eV) was measured, and nearly no change was observed in the Ga $3d$ line. Later, this assignment was questioned as SXPS[459] core-level spectra as well as data from HREELS[462] and STM[459,460] indicated that a dichloro adsorption geometry (as for H) is formed (see above). The Cl-induced valence band features (mentioned above) were then assigned in a different way,[463] with the features at 5 and 8 eV belonging to the Ga–X and As–X σ-bonding states (Fig. 18) and the nonbonding $|p_{x,y}\rangle$ states being found at about 4 eV. Similar adsorption geometries have been reported for Br_2 and I_2.[407,460,465,466] Only F_2 seems to be extremely reactive, leading to immediate etching of the GaAs surfaces without a defined adsorption phase.[464]

The surface potentials of halogen-covered GaAs surfaces follow the trend expected from the decrease in electronegativity ongoing from Cl to I. As suggested by Fig. 18, the electron acceptor position should be shifted closer to the valence-band maximum for Cl compared to I, as has also been verified in a number of experimental studies (cf. Fig. 45). The following values have been reported for the position of the surface Fermi level relative to the valence-band maximum ($E_F^{SS} - E_{VB}$) for n-doped samples: GaAs(110):Cl[461]: 0.2 eV (for low coverage, <1 ML) and 0.55 eV (for higher coverages, >1 ML); GaAs(110):Br[407]: 0.3 eV (<1 ML) and 0.55 eV (>1 ML); GaAs(110):I[466]: 0.3 eV (1 ML) and 0.5 eV (>1 ML). The values obtained at coverages below 1 ML are taken as the basis for the LCAO calculation of the pinning position (Figs. 18 and 45). The second value, which is independent of the halogen adsorbed, is evidently due to surface defects formed by Ga–As bond breaking when surface etching begins. The large electronegativity of the halogens should also lead to an increased value of the ionization potential, with an increasing trend from I to Cl. Indeed, the reported values qualitatively follow these expectations (ΔI_p = 0.8 eV for I_2[466]; ΔI_p = 0.8 eV for Br_2[407]; ΔI_p = 1.4 eV for Cl_2[461]). However, as discussed in detail for the H-terminated surfaces, the possible relaxation of surface geometry (removal of the Ga–As tilt angle, Fig. 43) may reduce the absolute value of the surface dipole, which is calculated for the Ga–X and As–X bonds based on the electronegativity difference.

There is only very limited information on the interaction of H_2O with GaAs surfaces. The adsorption of H_2O on (110) cleavage planes has been investigated in a set of experiments using UPS, SPV spectroscopy, and LEED as well as thermal desorption spectroscopy (TDS).[467–469] It has been concluded that H_2O is physisorbed at low sample temperature (below 180 K) and undissociatively chemisorbed at 300 K. This conclusion is based on the adsorbate-induced additional emission features in the UPS data, which show the typical threefold emission pattern of physisorbed H_2O (cf. Fig. 41) and shifted emission lines at 300 K.[467] TDS shows a constant-desorption-temperature peak at 350 K with an asymmetric shoulder extending up to 650 K.[469] In our opinion, and as also mentioned in the references cited above, the data may also suggest a certain fraction of dissociatively adsorbed H_2O. The UPS and XPS results of Webb and Lichtensteiner[470] clearly indicate dissociative adsorption and the formation of OH adsorbates, which preferentially interact with Ga sites and not with As sites, as is deduced from the XPS core-level shifts. Contrary to these findings, a strong interaction of H_2O with Ga sites has been suggested based on Auger line shape measurements, and the spectra were found to resemble those obtained with O_2 adsorption.[471] The LEED (1 × 1) pattern of GaAs is conserved after H_2O adsorption. The analysis of the intensity variation of the peaks indicates that the tilt of the Ga–As surface pairs is removed by the adsorbate.[469] H_2O adsorption and desorption experiments using photoreflectance give evidence for a physisorbed and chemisorbed state on GaAs(100).[472]

The reported data on the surface potentials do not seem to be consistent. Surprisingly, an upward band bending of 0.7 eV has been reported for n-doped samples,[467] which would make molecularly adsorbed H_2O on GaAs(110) an electron acceptor. On the other hand, the work function is reduced by a maximum value of 1 eV, which would translate to a 1.7-eV reduction in ionization potential. There is a definite need for more investigations on the interaction of H_2O with GaAs surfaces.

In summary, the results obtained with the different adsorbate interfaces indicate that the GaAs surfaces are much more complex in their structural and electronic properties than, for example, the Si surfaces. This is true for the GaAs(110) cleavage plane, which loses many of its preferential electronic properties after interaction with adsorbates. This is even more true for the polar (100) and (111) surfaces, whose complexity is further increased in contact with adsorbates. As a consequence, the surface potentials (distribution of surface states and surface resonances, surface Fermi level position, and ionization potentials) may differ considerably for the different surfaces and depend very much on the details of surface preparation and treatment.

(iii) Chemically Treated Surfaces and Electrolyte Interfaces

Owing to the importance of GaAs in optoelectronic solid-state devices, there has been a considerable amount of work on surface passivation, e.g., by forming a surface oxide layer as for Si, and on surface modification for its use as a substrate in the MBE type of growth experiments. Unfortunately, these attempts have been not very successful so far, as the density of surface and interface states remains high. The composition and thickness of the interface layer depends very much on the procedure used: oxidation by different oxidants and with different experimental parameters or by electrochemical means (anodic oxidation in aqueous electrolyte) leads to different compositions of the oxide layer.[473-476] As the surface oxide layer is an amorphous phase, nearly no structural information can be gathered except for EXAFS data on the local environment.[477] Therefore, most of the information on the morphology and stoichiometry of the oxide layer has been obtained by XPS and AES. The surface composition have been analyzed after different pretreatments, and the main species detected are summarized in Refs. 473 and 474, covering the work before 1984.

The oxide composition has to be separated into that of the "bulk" oxide and that of the interface layer. Wilmsen[473] reported that long-term oxidation with O_2, air, H_2O, and As_2O_3 produced a Ga_2O_3-rich "bulk" oxide while the interface contained Ga_2O_3 and As. This is in reasonable agreement with the results of the oxygen adsorption experiments (see above), in which the formation of Ga_2O_3 and As was also detected. The anodic oxides contain a mixture of Ga_2O_3 and As_2O_3 in the bulk and at the interface, when they are analyzed after formation. In more recent studies with XPS,[475] the GaAs oxide layer has been analyzed as "native oxides," "non-oxidized," "chemical oxide," and "thermal oxide." In all cases, Ga_2O_3 and As_2O_3 has been detected as the main species together with varying degrees of Ga and As suboxides. Even the "non-oxidized" samples (after HCl rinse) have oxide layers of more than 7-Å thickness. The assignment of the XPS core-level spectra to defined bulk-like phases such as Ga_2O_3 and As_2O_3 based only on the measured binding energy values has been questioned by Hollinger et al.[476] They concluded, based on studies of thermal oxides, that a single phase of a nonstoichiometric compound is formed, which contains only Ga–O and As–O subunits of the respective oxidation states but not the binary oxides. The most efficient way of achieving GaAs surface modification is the treatment of GaAs with S_x and Se_x [or S- (Se-) containing precursors].[478,479]

The surface composition that may be obtained after the different pretreatments (surface etching) is of special interest for the questions raised in this chapter as the surface conditions of GaAs electrodes are thereby defined for electrochemical investigation. Unfortunately, most of the electrochemical studies have been performed with the polar (100) and (111) surfaces of GaAs, which, based on the UHV studies, are considerably more complex than the (110) cleavage plane. On the other hand, most of the better understood UHV adsorption studies have been performed with the (110) plane. Therefore, the comparison of adsorbate interfaces in UHV with electrochemically characterized GaAs interfaces must be done with great care. A summary of GaAs electrochemistry and photoelectrochemistry is given in Ref. 480 (see also Refs. 7, 8, 115, and 145). We will not cover the changes in interface properties that are obtained by metallization of GaAs electrodes. There has been a considerable amount of work in the latter area due to the beneficial effect of metals on the catalytic properties of semiconductor electrodes. A summary of the electronic and electrochemical properties has recently been given by Allongue,[309] but many of the obvious beneficial effects seem to be still unclear because of the lack of consistent surface analytical investigations. On the other hand, the GaAs/metal interface is also known to be extremely complex from

UHV investigations of Schottky barrier properties.[35–38,95–97] A comprehensive comparison of UHV and electrochemically prepared metal/interfaces from a surface science perspective is not within the scope of this chapter.

The pretreatment of GaAs electrodes usually involves a combination of electropolishing and chemical etching.[480,481] The chemical etches typically contain an oxidant in acidic or basic solutions, which leads to dissolution of the surface reaction products. Typical etch solutions are 1% Br_2 in CH_3OH for electropolishing and $H_2SO_4/H_2O_2/H_2O$ (acid) or NaOH (or NH_4OH)/H_2O_2/H_2O (basic) solutions. These etchants are used to produce mirror-like planes of the different GaAs surfaces with minimum oxide coverage. There are a number of surface science analytical studies published on the surface composition as analyzed by XPS after application of the surface etches.[482–488] All these investigations propose that nearly stoichiometric GaAs surfaces, which are mostly oxide-free and contain only minor amounts of elemental As, can be obtained. Lu et al.[482] used aqueous acid mixtures containing either H_3PO_4 or H_2SO_4 in addition to H_2O_2 to produce nice surface morphologies of (100) surfaces. The remaining surface oxides—Ga_2O_3 and As_2O_3 in different ratios—were removed by subsequent rinsing in hot HCl, and As was removed by rinsing in NH_4OH solutions. For their best etch (1. $H_3PO_4/H_2O_2/H_2O$, 2. HCl, 3. NH_4OH), Lu et al. proposed an oxide- and elemental As-free surface in a stoichiometric ratio As/Ga = 1, which also gives LEED diffraction patterns (at a relatively high electron energy of 340 eV). Hirota et al.[483] used deionized H_2O rinsing after Br_2/CH_3OH and $NH_4OH/H_2O_2/H_2O$ etch on GaAs(100) surfaces. They also proposed, based on their XPS data, that oxide-free and nearly undamaged surfaces are produced; these surfaces show different RHEED patterns at different sample temperatures (e.g., 1 × 1 at room temperature). They found that As-rich surfaces are produced after acid etches, which is also known from HF etches.[484] Similar results on the oxide removal by rinsing in deionized H_2O have been reported by Massies and Contour,[485] also for acid-etched (100) surfaces. Finally, the effects of different etches were investigated by Tufts et al.[486] by the examination of XPS data and current–voltage behavior (see also Refs. 487 and 488). Their alkaline etch (etch A) produced a surface with oxide coverages (Ga_2O_3, As_2O_3) below one monolayer and no As; their acid etch (etch B) led to less oxides but to elemental As on the surface. An H_2O_2-oxidized surface (etch C) had a Ga_2O_3-enriched surface oxide layer with smaller amounts of As_2O_3. Interestingly, when the effects of the different etches on current–voltage behavior were compared, only photoelectrochemical cells (PECs) with $FeCp_2^{0/+}$ in CH_3CN and Au/GaAs Schottky barriers showed drastic differences (PEC: A equal to B, C very poor; Schottky barrier: B better than A, C very poor). PECs with $Se_x^{-/2-}$ in solution showed no differences, and subsequent XPS analysis indicated removal of oxide and elemental As, which is further evidence of the beneficial effect of S- (Se-) containing solutions mentioned above.[478,479]

The above reports of oxide-free surfaces of polar planes showing perfect surface stoichiometries and structures do not agree with the UHV results for polar surfaces. As presented in Section VI.2(i), clean polar (100) and (111) surfaces show complex reconstructions to reduce the large polarity of these surfaces. Therefore, it seems to be evident that extra surface species must exist also on the electrochemically treated polar faces, changing their structure and stoichiometry. These, however, seem to be below the detection limit of XPS. As also shown above [Section VI.2(ii)], many of the adsorbates are not easily detected even in the more surface-sensitive SXP spectra at coverages below a monolayer. Therefore, more sophisticated experiments with optimized transfer procedures and additional surface science techniques with a sensitivity below the monolayer regime (e.g., SXPS, HREELS, LEED) are needed to get an improved insight into the surface conditions after electrode pretreatment.

Because of the facile and complex oxidation mechanism of GaAs surfaces, changes of surface composition after electrochemical reactions can only be analyzed in a reliable way when the transfer process is carried out without any exposure of the specimen to ambient conditions. Without the use of such a transfer system, reasonable results may only be expected for the investigation of anodically formed oxidation products. A study by Solomun et al.[189] on the anodic treatment of GaAs(100) surfaces using a combined electrochemical and UHV analysis system indeed indicated similar oxidation products. After anodic treatment, Ga_2O_3 and As_2O_3 and minor amounts of As were clearly identified, in good correspondence with the data reported above. Additional rinsing with the electrolyte led to As-enriched surfaces, which was attributed to the formation of As clusters. Similar surface stoichiometries after anodic photoelectrochemical etching have been reported by Woodall et al.[489] They reported the formation of an As-passivated GaAs layer, which could be heated off under mild temperature conditions (290°C), leading to nice clean XP and UP spectra and sharp LEED patterns.

Unfortunately, these surface science investigations on GaAs electrodes after etching and electrochemical treatment are only aimed at the analysis of surface composition. The possible shifts of core-level and valence-band spectra, which would give information on the electronic surface potentials and changes of energetic conditions (see Section V), have not been analyzed. Therefore, a conclusive comparison with the surface studies of UHV-prepared, clean and adsorbate-covered surfaces, described in Sections VI.2(i) and VI.2(ii), is not readily possible. In principle, the bulk core-level binding energy positions as measured by XPS can be taken as a basis for comparison, e.g., by the determination of band bending values. However, as the effects are expected in a range below 1 eV, a precise calibration of the binding energy scale and comparable measurement conditions are necessary. Therefore, the beneficial effects that may be expected from chemical and electrochemical treatments [see, e.g., the Si case, Section VI.1(iii)] are not evident for GaAs electrodes.

The electrochemical equilibrium properties of GaAs electrodes, as deduced from electrochemical measurement techniques, are strongly influenced by the surface conditions reported above. The electric responses of the (100) and (111) faces are characterized by Fermi level pinning, which has been found for aqueous as well as organic electrolytes.[490–496] The pinning position for many different electrolyte compositions is found at $E_F - E_{VB} = 0.5 \pm 0.1$ eV, which is in reasonable agreement with the position of the pinning measured for the clean polar faces and for most adsorbate interfaces [see Sections VI.2(i) and VI.2(ii)]. As the electrode treatment leads to oxidized surfaces of at least one to a few monolayers thickness, it may be concluded from a comparison with the adsorbate interfaces that a combination of surface acceptor states and broken GaAs back bonds can be considered responsible for the pinning effect. The existence of oxide coverage is evident from the shift of the flatband potential with pH, as reported in more detail below. It is interesting to note that the distribution of surface states in the band gap on GaAs(100) as measured by impedance spectra[490] (Fig. 49) shows a close resemblance to the DOS of oxygen-covered GaAs(110) surfaces measured in UHV (cf. Fig. 48). This may be taken as additional evidence that the adsorbate surfaces and the surfaces in electrolytes are very similar to each other, even for different surfaces under investigation, and supports the above discussed formation of an amorphous unordered oxide layer.

There are also some reports that do not show Fermi level pinning in the case of certain GaAs electrodes, such as electrodes having polymer-covered surfaces[497–499] and chemically modified surfaces.[487,500] As has been demonstrated in a number of systematic investigations on the elec-

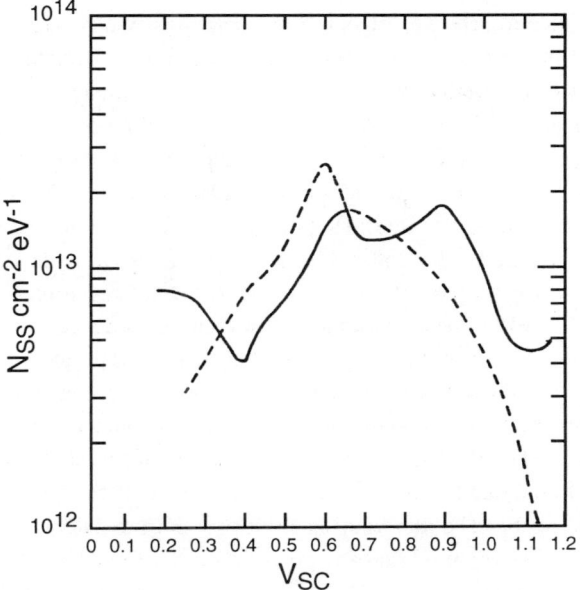

Figure 49. Distribution of surface states as determined by impedance measurements as a function of space-charge potential V_{SC}. (The dotted curve is considered to be unreliable; after Ref. 490.)

tronic properties of GaAs electrodes in organic electrolytes,[501–503] there is also no Fermi level pinning of the (110) cleavage plane[501] in contrast to the polar surfaces, which also have been investigated. In the cases of polar surfaces, the chemically modified GaAs electrode is evidently stabilized after electrode etch toward subsequent oxidation, as the overlayer may kinetically prevent the oxidative attack of H_2O. It has been shown that rinsing with deionized H_2O leads to oxide-free surfaces (see above), whereas steady exposure to H_2O leads to thin surface oxide layers.[484,485] Unfortunately, to our knowledge the conditions of chemically modified interfaces have only been analyzed for surfaces treated with Co complexes.[487] No GaAs oxide layer was found, but instead the formation of a Co(II)-oxo overlayer [Co(OH)$_2$] was detected by XPS and EXAFS after Co-complex chemisorption. After exposure to a polyselenide solution, a CoSe$_{1.8}$ surface phase was formed.

The kinetic schemes of corrosion reactions, which are, of course, important in the formation of the anodic oxides, for example, have been discussed based on models that involve charge-transfer processes via surfaces states.[485,504] However, a reasonable modeling of these processes, assuming electron-transfer mechanisms, requires a precise knowledge of the concentration, energetic distribution, and nature of these surface states, as well as the energetic interface potentials (see Fig. 9 and Section IV), which for the GaAs electrodes does not seem to be available so far. This is also true for the evolution of H_2 and the beneficial effects of metal deposits.[309] Additional information from surface science techniques is urgently needed to provide a better experimental basis for an educated speculation on the molecular mechanism of surface etching and corrosion.

The energetic position of GaAs electrodes in contact with electrolyte solutions can be inferred from the determination of flatband potentials. It has been found that for most GaAs electrodes there is the typical dependence of E_{fb} on the pH of the solution, indicative of the existence of a surface oxide layer. The reported flatband potentials differ for different electrodes and different pretreatments[490–496,501–503,505–508]; typical values are presented in Table 4. There are evidently pronounced differences due to differences in the solution pH (with a shift of E_{fb} of 59 mV/pH unit), but E_{fb} is also affected by the orientation of the GaAs electrodes and in some cases by the redox couple.

Table 4
Flatband Potentials of GaAs Electrodes

Type	pH	V_{fb}^a	Remarks	Reference
n-(111) As	0	−1.13	Aqueous solution	505
n-(111) As	13	−1.75	Aqueous solution	508
n-(111) Ga	13	−1.55	Aqueous solution	508
n-(111)		−2.0	CH_3CN, acid etch	492
n-(111)		−1.5	CH_3CN, HCl etch	492
n-(100)	14	−1.4	Aqueous solution	506
n-(100)	14	−2.1	1M Se^{2-} present	507
n-(100)	1	−0.9	Aqueous, redox couple	494
n-(100)	13	−1.8	Aqueous solution	508
n-(110)		−1.0	CH_3CN	501

^{a}In volts vs. SCE.

A precise comparison of the flatband potentials to work-function measurements in UHV is only possible when the contribution of Helmholtz potential drops due to adsorbed ionic species is considered. Cachet and co-workers[494] have investigated the flatband potentials of GaAs electrodes after different pretreatments and in different electrolytes. They also confirmed the pH-dependent E_{fb} shift and determined that the point of zero zeta potential (pzzp) of the oxide-covered GaAs electrodes is at pH = 2.8. If this value is considered, a plot of barrier height versus electrolyte potential, as determined for different redox couples, gives a straight line (Fig. 50). The slope characterizes the deviation from the Schottky limit and the amount of Fermi level pinning (see Section IV.3). The slopes determined from these plots, $S = 0.7$–0.8, are considerably larger than those for metal contacts on GaAs and indicate a considerably

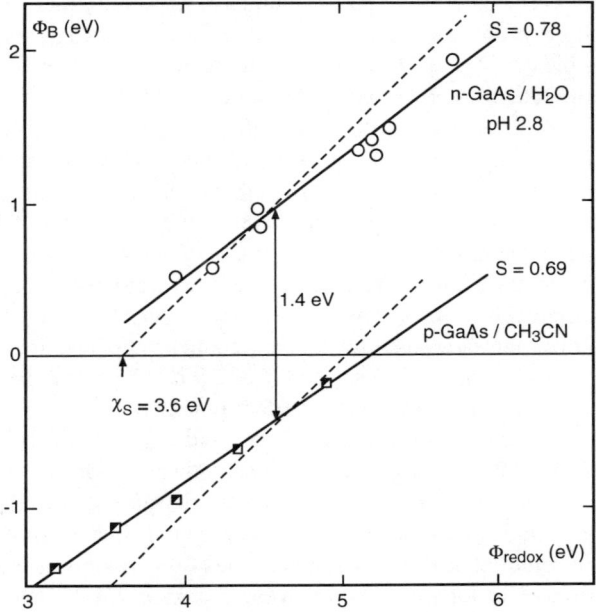

Figure 50. Variation of semiconductor/electrolyte barrier height ϕ_B for different redox couples versus their absolute potentials, after correction for the pzzp. (The dashed lines give the Schottky limit; ■, Nagasubramanian et al.; ○, this work, after Ref. 494.).

lower density of surface states. It is interesting to note that the value of the electron affinity, which is also given by these plots (by the intersection with the abscissa, $\Phi_B = 0$), is $E_A = 3$–3.3 eV, which differs considerably from the values obtained in UHV experiments presented above [clean (100) surfaces: $E_A = 3.7$–4.2 eV; disordered oxygen-covered (110) surface: $E_A = 4.0$ eV]. There are evidently additional, as yet unidentified, dipolar potential contributions at the electrolyte interface that shift the band-edge position.

This problem has been addressed in very great detail in a paper by Schlesinger and Janietz.[509] In this work, the energetic conditions and potential distributions at the GaAs electrode interface have been thoroughly revisited on the basis of a similar concept to that introduced in the introductory part of this chapter (Sections II–IV and Fig. 9). Schlesinger and Janietz's main criticism is related to the explicit and implicit assumed estimates of the different potentials that may contribute to the potential drop at the GaAs/electrolyte phase boundary: Especially the value taken for E_A and the problems in estimating the surface dipoles χ_{SC} and χ_{El} and their possible changes are discussed. Their paper gives clear evidence that many of the estimates are wrong and more precise values are needed. For example, they determined a different value of pzzp and established the point of zero charge (pzc) to be at pH = 5.9. They proposed considerably lower S values ($S = 0.14$–0.25) based on their calculation of the variation in barrier height with redox potentials (in comparison to Fig. 50). Also, the additional potential drops at the GaAs/oxide/electrolyte interface were addressed. However, unfortunately, these authors were restricted to working with improved estimates, as experimental values directly addressing the problem of interface potentials, such as could be obtained, for example, in modeling experiments as described in Section V.3, are still lacking.

In their papers investigating the surface potential distribution of GaAs electrodes in organic electrolytes, Cachet and co-workers[501–503] found many different effects competing with each other. Besides the influence of crystal orientation and the pH-dependent charging of residual oxide layers, the charging of interface states must be considered as well when the electric potential distribution is determined for variations in electrolyte composition. The clearest results were obtained for (110) cleavage planes,[501] and these can readily be compared to results from UHV studies. Here no Fermi level pinning was observed for many different redox couples. Cachet and co-workers determined the value of the flatband potential ($E_{fb} = -1.0$ V vs. SCE), which translates to an electron affinity value of 3.6–3.8 eV (for E_{abs}^0 = 4.5–4.7 eV), compared to the UHV-determined value of about 4.0 eV. Thus, an overall potential difference due to interface dipoles, $^{SC}\Delta^{El}\chi$, of at least 0.2–0.4 eV may be deduced, and this value may be even larger if the reduced electrode potential of E_{abs}^0 (reduced by the electrolyte surface dipole χ^{El} of 0.1–0.2 eV) is considered.[509,501] This surface dipole may be related to the dipole of adsorbed acetonitrile but may also reflect the removal of the GaAs(110) surface relaxation (tilt of Ga–As surface pairs), decreasing the evident electron affinity. The relative contributions of the different surface dipole layers and their changes due to contact formation [$\Delta\chi_{SC}$, $\Delta\chi_{El}$, $^{SC}\Delta^{El}g(ion)$] cannot be deduced on the basis of the data presented and therefore are simply a matter of speculation. However, the GaAs(110) electrode in H$_2$O-free organic electrolytes may indeed be a good candidate for a detailed investigation of surface potential distributions by a combination of electrochemical and surface science techniques.

3. Layered Chalcogenide Semiconductors

Whereas the structure, stoichiometry, and electronic properties of the surfaces of three-dimensional semiconductors such as Si and GaAs may differ considerably after different pretreatments as discussed above, the ideal van der Waals surfaces of two-dimensional layered chalcogenides are, in most cases, very similar to their bulk truncated surfaces. The (0001) van der Waals plane can easily be prepared by cleaving the crystal inside or outside a vacuum chamber. It contains a hexagonally close-packed layer of chalcogenide ions with chemically saturated bonds: as a consequence, the ideal surface is free of surface states and chemically rather inert (with some variations for the different layered chalcogenides[2]). For this reason, the layered chalcogenides may be considered to be ideal model systems for the investigation of fundamental aspects of semiconductor electrochemistry and especially the microscopic details of contact formation. The electrochemical and surface science properties of layered chalcogenides have been summarized in detail in Refs. 2 and 3. As is expected for perfect van der Waals surfaces, there is no Fermi level pinning, and the work function is determined by the bulk doping (shifting from values close to E_A for n-doped to values close to I_p for p-doped semiconductors). A list of experimentally determined work functions is given in Ref. 180. Within this section, we will focus on adsorption studies directly aimed at modeling electrolyte interfaces that have been performed in our lab in recent years. We will refer to electrochemical investigations and other surface science studies only in relation to our modeling experiments. In the experiments reported below, H_2O and halogens are adsorbed on the (0001) plane to investigate fundamental aspects of electrolyte interaction; these data will be presented in Section VI.3(i). We also performed a number of coadsorption studies (Na, H_2O, Br_2) to approach in our model adsorption phase the complexity of the electrolyte solution, and this work will be described in Section VI.3(ii). We will not discuss our UHV investigations on fundamental aspects of intercalation reactions, which may also be considered as model experiments of electrochemical processes (see, e.g., Refs. 510–512 and references cited therein).

(i) Monoadsorption Phases

(a) H_2O adsorption

The interaction of adsorbed H_2O with clean single-crystalline layered chalcogenide (0001) surfaces has been studied mostly by photoelectron spectroscopy. Adsorption of H_2O on UHV-cleaved van der Waals surfaces only occurs for samples cooled to liquid-nitrogen temperatures or below (LT). At room temperature (RT) the adsorbed H_2O is completely desorbed again. The desorption temperature has not been determined precisely, but it may be estimated to be in the temperature range 150–180 K, which indicates the weak interaction of H_2O with the van der Waals surfaces. As a consequence, the adsorbate tends to form three-dimensional clusters on the substrate. Therefore, we do not observe any ordered adsorbate phase: the LEED pattern is slowly attenuated in the course of H_2O adsorption, and only a diffuse background remains for large coverages.

As an example, the He I spectra of n- and p-doped WSe_2(0001) in the course of H_2O adsorption at 100 K (low temperature, LT) are shown in Fig. 51.[513,514] Due to adsorption of H_2O, extra emission features appear at around 8, 11, and 14 eV, which are assigned to the $1b_1$, $3a_1$, and $1b_2$ orbitals of molecularly adsorbed H_2O.[257] The energy differences of the orbitals as measured by SXPS indicate that H_2O is adsorbed in ice clusters. Also, the increase of the H_2O emission intensity and the attenuation of the substrate emission lines suggest adsorption of H_2O clusters, which is often found for molecular adsorption on rather weakly interacting substrates. In XPS or SXPS the substrate emission peaks do not show any chemical shifts indicative of surface reaction. Also, the O $1s$ (XPS measurements) and O $2s$ (SXPS measurements) core lines do not show extra emission lines besides those of molecularly adsorbed H_2O [E_B ($1s$) = 533 eV;

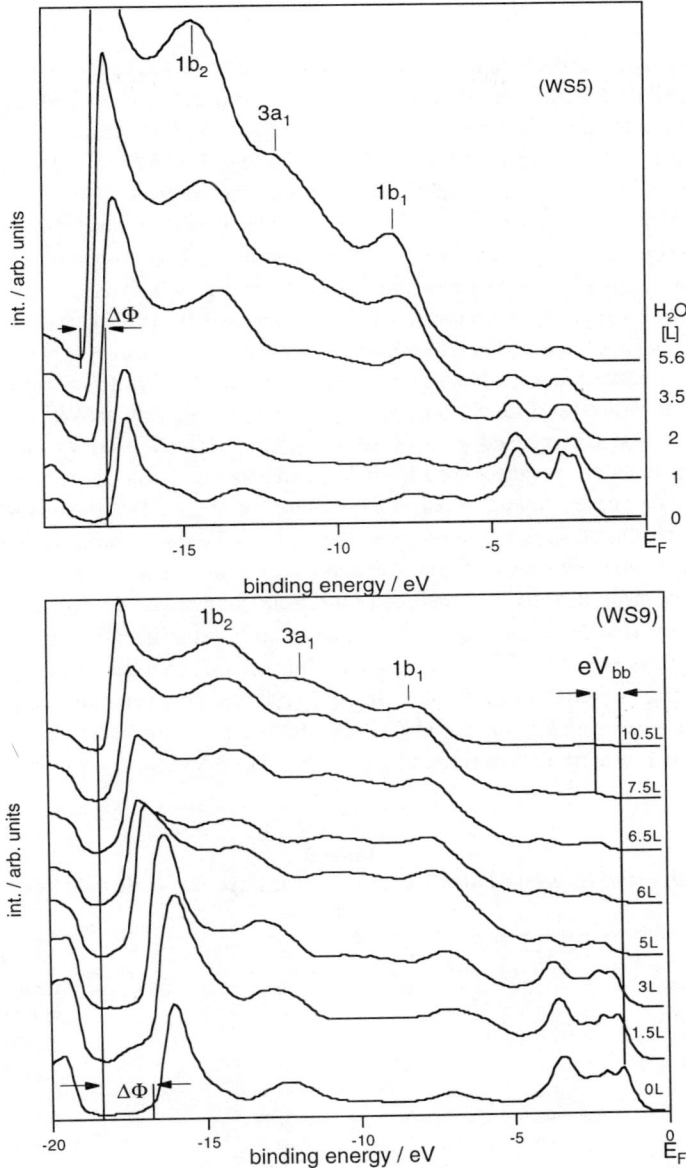

Figure 51. Valence-band spectra of the WSe$_2$/H$_2$O interface for increasing amounts of the adsorbate. *Top*: *n*-doped sample; *bottom*: *p*-doped sample. The changes of the band bending eV_{bb} and work function Φ are indicated.

E_B (2s) = 26 eV]. When the WSe$_2$ samples are annealed to room temperature (RT), the molecularly adsorbed H$_2$O is completely desorbed, and the spectra obtained are the same as those for the clean surface. Similar experiments have been performed with n- and p-doped MoS$_2$,[516,3] n-InSe, and p-GaSe[513,514] all of which show the same adsorption behavior.

The spectral series on p-WSe$_2$ (Fig. 51) shows a strong shift of the valence-band maximum (and all other substrate emission peaks) to higher binding energies with increasing H$_2$O coverage, which is due to band bending (eV_{bb}) by formation of a depletion layer (saturation value up to eV_{bb} = 1.1 eV). Band bending induced by H$_2$O can be directly proven by SPV measurements. Whereas no bias-light-induced SPV shifts are measured for the cleaved surface, illumination by white bias light (150 mW/cm^2) leads to SPV-induced reversal shifts of band bending by 0.3 eV (0.6 eV). Also, the work function, as determined from the secondary electron onset, is drastically decreased (by up to $\Delta\Phi$ = 1.6 eV) by the presence of the adsorbate. The binding energy and work function shifts $\Delta\Phi$ are summarized in Table 5. The difference between the $\Delta\Phi$ and eV_{bb} values is due to the change in the electron affinity ΔE_A induced by dipole changes $\Delta\chi$ at the semiconductor/adsorbate interface. The changes are established very slowly, which is related to the strong clustering on the van der Waals plane. Contrary to the findings for p-WSe$_2$, only a slight increase of the measured binding energies is observed (0.2 eV) for n-WSe$_2$, probably due to the formation of an accumulation layer (Table 5). In addition, we observed a

Table 5
Changes of Interface Potentials at Different Layered Semiconductor/H$_2$O Interfaces

	ΔE_B^F	$\Delta\Phi$	$\Delta\chi$	SPV[a]	References
p^+-WSe$_2$	−1.1[b]	−1.6	−0.5	+0.6	513, 517
p-WSe$_2$	−0.9	−1.4	−0.6		513, 517
i-WSe$_2$	−0.5	−1.2	−0.7	+0.3	513, 517
n-WSe$_2$	−0.2	−0.7	−0.5		513, 517
n-MoS$_2$	0.0	−0.3	−0.3		516, 3
p-MoS$_2$	−0.5	−0.5	0.0		513, 3
p-GaSe	−0.3	−1.0	−0.7		513, 517
n-InSe	−0.1	−0.9	−0.8		513, 517

[a]SPV, surface photovoltage.

[b]Negative E_B values correspond to a downward bending toward the surface.

considerable decrease in the electron affinity, comparable to that for the p-type sample.

The principal changes of surface potentials (band bending) and the electron affinity changes due to adsorbed H_2O are qualitatively comparable for all layered semiconductors studied so far (Table 5), even when the measured values are evidently different. In all cases a reversible band bending (downward to the surfaces) is only induced on p-type substrates. The effect is negligible for n-doped substrates. However, the change in electron affinity is similar for n- and p-doped substrates. Also, these changes are completely reversed after desorption and only depend on the amount of adsorbed H_2O. As soon as H_2O desorbs, the flatband potential and the original work function are restored.

Therefore, adsorbed H_2O clearly behaves as an electron donor in our adsorption experiments. A schematic model that has been proposed as an explanation for the formation of the extrinsic surface state of donor character is shown in Fig. 52. This scheme is a qualitative LCAO energy-level diagram describing the formation of surface states as discussed in Fig. 18 for Cs–GaAs interaction. For the observed band bending on p-type layered substrates, electrons have to be donated from occupied electron states above the semiconductor Fermi level into the extended semiconductor space-charge region. When alkali adsorption is considered as in Fig. 18, the adsorbate state $|A\rangle$ is well above the conduction-band edge, and electron donation is envisaged. However, the occupied orbitals of adsorbed H_2O molecules are well beyond E_F^{SC} (the positions are directly accessible in the valence-band spectra and are in good agreement with values reported in the literature).[257] Therefore, a direct electron transfer should be not possible. Since the dissociation of H_2O can be excluded on the basis of the experimental results, we suggest the formation of an extrinsic surface state by the electronic interaction of the occupied $1b_1$ and $3a_1$ H_2O MOs (O $2p$ based) with the occupied valence-band states of the semiconductor substrate (W $5d_{z^2}$ based) as shown in Fig. 52. Such interaction of occupied states does not lead to binding interaction in free molecules but may result in weak bonding on surfaces when electrons can be transferred to the solid from the antibonding states.[148,149] Interestingly, we observe a much larger effect (0.7 eV) on p-WSe_2 and an intermediate effect for p-MoS_2 (0.5 eV) as compared to the considerably reduced value of 0.3 eV for p-GaSe. The observed difference in the amount of band bending seems to be determined by electronic factors as the structure of the (0001) van der Waals plane is similar for all the investigated layered

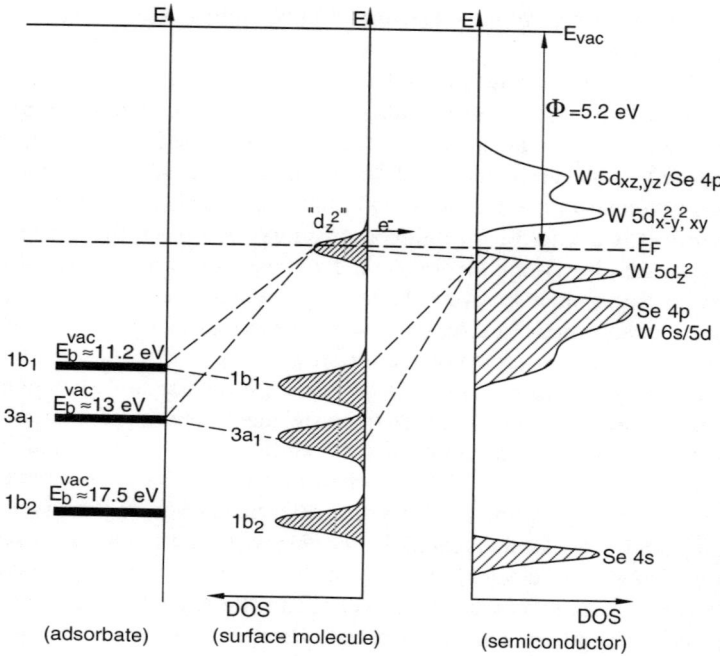

Figure 52. LCAO diagram for the formation of an extrinsic donor surface state at WSe_2/H_2O interfaces.

materials. However, the electronic structures of the compounds differ considerably. The group 6B chalcogenides are d-band semiconductors with the valence- and conduction-band edges derived from metal d-states. These are mainly involved in charge-transfer interaction with adsorbates. As a consequence, we expect a stronger electronic coupling of the H_2O molecular orbitals to the semiconductor states, as shown in Fig. 52. The valence-band edge of GaSe, on the other hand, is derived from the bonding σ (Ga $4p_z$) molecular orbital of the Ga pairs and the Se $4p$ levels.[180] The electronic coupling of these valence-band electron states to the H_2O orbitals is probably much weaker. As a consequence, we expect a less pronounced charge transfer from adsorbed H_2O to the semiconductor.

In relation to semiconductor/electrolyte interfaces, it is interesting to note that strong band bending is observed for H_2O adsorption on the inert

(0001) plane of the p-type MX_2 semiconductors. Evidently, the van der Waals plane cannot be considered to be completely inactive in electronic interactions with the solvent molecules. Our results seem to indicate that there is at least a weak electronic overlap between the adsorbate and the semiconductor band-edge states. As these are derived from the metal d orbitals, we take this as evidence for non-negligible metal electron density outside the chalcogenide plane. Such weak surface interaction was also proposed by Tributsch [518] in his early photoelectrochemical studies with MX_2 electrodes (M = Mo, W). He also proposed a weak interaction of H_2O even on the van der Waals plane, which may be enhanced in photoelectrochemical reactions, when holes arrive on metal surface atoms. Also, recent STM investigations[519] suggest the penetration of metal d-states through the van der Waals planes as they show tunneling contrast, probably due to the metal d-states inside the threefold hollow sites of the topmost chalcogenide layer.

The dipole contribution $\Delta\chi$ at the adsorbate interface will be considered in more detail. The resulting interface potential distribution of the p-WSe_2/H_2O interface is schematically shown in Fig. 53 for two possible reference levels. In Fig. 53a (reference level E_F as measured in the photoelectron spectra) the band bending eV_{bb} due to charge transfer from the H_2O-induced surface state leads to increased substrate binding energies with increased coverage Θ. The increase in E_B^F of the H_2O levels is even larger. If the measured binding energy values of the photoemission spectra (p-WSe_2/H_2O interface in Fig. 51) are plotted with reference to the position of the valence-band maximum (Fig. 53b) and thus to E_{vac}, the ionization potential of the adsorbate levels remains constant. We have observed a similar behavior for all H_2O/layered chalcogenide interfaces studied. The binding energy values of the H_2O emission lines are always the same when referred to the vacuum level. They are only shifted from the free-molecule ionization potential of H_2O vapor by a typical relaxation energy of 1.2 eV. The spatial position of the dipole change $\Delta\chi$ at the interface becomes evident. As the H_2O levels remain constant versus E_{vac}, we conclude that $\Delta\chi$ is changed at the interface between the semiconductor surface and the adsorbate and may result mostly from a changed surface dipole of the substrate, $\Delta\chi^{SC}$, induced by the adsorbed ice layer.[521,514] However, the adsorbed H_2O molecules are outside this double layer (see also Section V.3). This effect has also been observed at metal/H_2O adsorbate interfaces.[32] In the latter case also, the H_2O emission lines are coupled to the vacuum level (change their energetic position in parallel with the

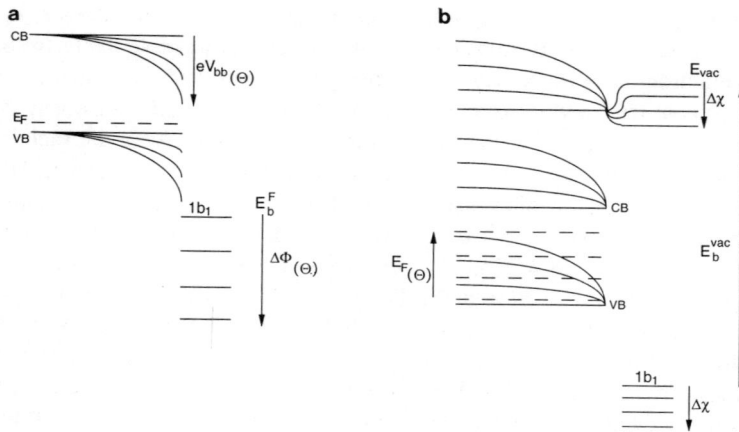

Figure 53. Schematic representation of interface potentials at the WSe_2/H_2O interface referred to E_F (a) and to E_{VB} (b).

work-function change $\Delta\Phi$) and not to the Fermi level (which at metals would lead to a constant binding energy).

A more complicated adsorption behavior is deduced for macroscopically stepped MoS_2 basal planes.[3,520] The edge planes show up as tails or steps of lower Φ in the secondary electron onset. For perfectly cleaved samples, such steps are not observed, and the emission onset is rather steep. Also, a weak contribution of surface oxides is detected. For the stepped surfaces, different ranges of adsorption can be identified. At lower dosages, adsorption evidently starts at steps, reducing their contribution in the secondary onset; the main secondary electron onset is not changed. In this coverage regime the O $1s$ peak shows up in the XP spectra at 531.6 eV; this peak is assigned to hydroxides. At higher dosage, after saturation of the edge planes a decrease of Φ is measured, which corresponds to the behavior of crystals with perfect van der Waals surfaces. For these conditions the XP spectra show the typical O $(1s)$ emission line of adsorbed H_2O around 533 eV, and the UP spectra show the emission lines of physisorbed H_2O. We also performed a few adsorption experiments with non-van der Waals planes of layered chalcogenides, which are produced by cutting crystals in UHV with a knife.[522] The experimental procedure and some properties of such cut surfaces are given in Ref. 523. In contrast to the well-known and well-documented tendency of the edge planes to be oxidized after extended exposure to air or aqueous electrolytes,[180,3] we did not find any evidence for any strong interaction of H_2O with the UHV-cut surfaces. H_2O remains molecularly adsorbed and desorbs completely at RT. Evidently, there is only the formation of a surface coordinative bond to H_2O as an extra ligand for the coordinatively unsaturated metals on the edges. A dissociation is kinetically hindered by the lack of coordination sites in close proximity.

It is interesting to speculate on the implications of Fig. 52 for elementary electron-transfer steps in (photo)electrochemical reactions. Based on this energy-level diagram, a directed hole transfer to adsorbed H_2O molecules is very improbable. The energetic distance between the valence-band edge E_{VB} and the HOMO ($1b_1$) of H_2O is about 5 eV. This seems to be too large to be overcome, even when a reorganizational broadening and an additional relaxation of the $1b_1$ level by the solvent is taken into account. Therefore, a direct hole transfer is not expected for energetic reasons. Instead, the surface-molecule surface state, dominated by the contribution of the metal d-state, will be oxidized by hole transfer. Subsequently, chemical reaction steps will follow, which lead to the oxidation of H_2O. Similar mechanisms involving hole transfer to surface-molecule states are probably of general validity in multistep electron-transfer reactions that involve strong interactions with the surface [e.g., in (photo)electrocatalysis and (photo)electrochemical etching].

(b) Halogen adsorption

Halogens (Cl_2, Br_2, I_2) are also adsorbed only at LT on (0001) planes of $MoSe_2$[229,230] and WSe_2.[524,525,514] The halogens adsorb as intact molecules without reacting with the surfaces, again proving the chemical inertness of the (0001) van der Waals plane. For Cl_2 and Br_2 the adsorption was found to be completely reversible. After annealing of the samples at RT, the adsorbed halogens desorbed and the original spectra were restored completely. The adsorption energy of I_2 seems to be a little larger than those of the other halogens as small amounts of I_2 remain on the surface of WSe_2 after annealing at RT.[525] Based on XPS and SXPS data, a reaction of the adsorbates with the substrate can also be excluded as no new chemical species were detected in the substrate core-level emission lines.[526,514] This behavior is in good agreement with electrochemical results which indicate that halogens can be evolved from group 6B layered chalcogenide electrodes without corroding the semiconductor.[527]

Typical adsorption sequences for adsorption of Br_2 onto n-$WSe_2(0001)$ in the dark and under bias illumination are shown in Fig. 54. The additional Br_2 emission lines are marked and assigned according to literature values and gas-phase photoelectron spectra (E_B^F values: $\pi_g = 3.0$ eV; $\pi_u = 5.2$ eV; $\sigma_g = 6.4$ eV). The binding-energy shifts of the substrate lines and the valence-band edge due to Br_2 adsorption in the dark indicate strong band bending, as the position of the valence-band edge is found close to the Fermi level. Under bias illumination, however, the binding energy shift due to band bending is reversed and, as a consequence, the Br_2-induced spectral shifts are drastically reduced (Fig. 54). For all nonreactive van der Waals interfaces, a similar behavior is obtained. The adsorbed halogens lead to strong band bending and thus to a large SPV-induced rebending of energy bands. These results are summarized in Table 6 and compared to results for halogen adsorption at electrochemical interfaces.

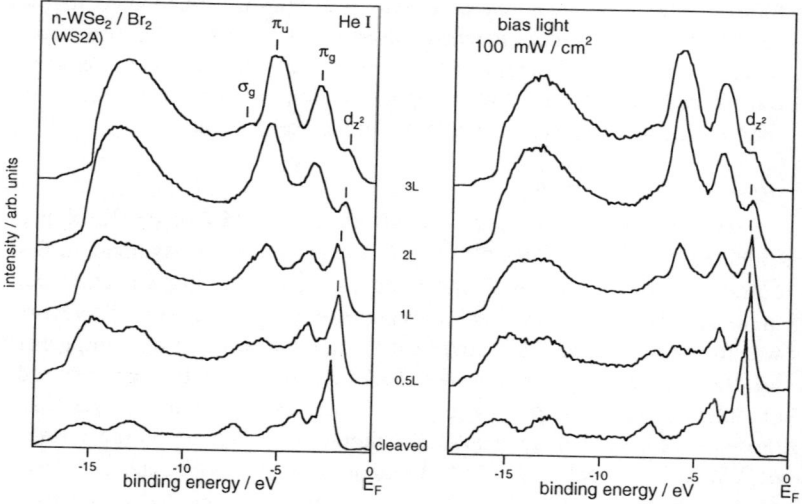

Figure 54. He I UP spectra of Br_2 adsorption onto UHV-cleaved (0001) faces of n-WSe_2 (binding energy shift due to band bending and change of Φ is indicated). The assignments of the emission features of molecularly adsorbed Br_2 are also given. The spectrum in the dark (*left*) shows band bending, whereas with bias illumination (*right*) no shift is seen due to SPV band flattening.

Table 6
Comparison of Surface Potentials in UHV Adsorption Experiments on n-Doped Layered Metal Chalcogenides of Group 6B and at Electrochemical Interfaces

Semiconductor/absorbate interface	ΔE_B^F (eV)[a]	$\Delta\Phi$ (eV)[b]	U_{ph} (eV)[c]	E_0 (V)[d]	U_{ph} (V)[e]	Reference
$MoSe_2/Cl_2$	1.4	1.8	0.7	1.4	0.7	230
$MoSe_2/Br_2$	1.2	1.8	0.7	1.1	0.5	230
$MoSe_2/I_2$	0.5	0.9	0.5	0.5	0.3	230
WSe_2/Br_2	0.9	1.8	0.6	1.1	0.6	524
WSe_2/I_2	1.0	1.4	0.6	0.5	0.6	525

[a] Binding energy shift.
[b] Work function change.
[c] UHV surface photovoltage.
[d] Standard redox potential vs. NHE.
[e] Photovoltage in electrolyte solution. Refs. 528 and 529.

For halogens, as typical electron acceptors, a reverse behavior is obtained with respect to possible charge transfer compared to that described for H_2O adsorption (see above). Owing to electron flow from the n-type semiconductor to the adsorbed halogens, E_B^F decreases, which indicates an upward bending of energy bands toward the vacuum interface. In addition, Φ increases by an even larger amount, which is due to the formation of a surface dipole with the negative end directed toward vacuum. The shift of E_B^F can partly be removed by illumination of the sample with white bias light, creating a surface photovoltage (SPV). The SPV is in most cases considerably smaller than the shift of E_B^F due to the adsorbates. As photovoltage saturation is assumed under the experimental conditions (low temperatures), the difference between the band bending eV_{bb} and the photovoltage SPV is explained by the formation of an inversion layer in the small-band-gap semiconductors, in agreement with results reported for inert organic electrolytes.[528,529] This inversion layer adds to the shift of energy bands but can be tunneled very easily due to its limited thickness. For this reason it does not contribute considerably to the surface photovoltage. A schematic energy diagram of the semiconductor/adsorbate interface for different experimental conditions is presented in Fig. 55.[230] The electric potentials at the $MoSe_2$/adsorbate interface in

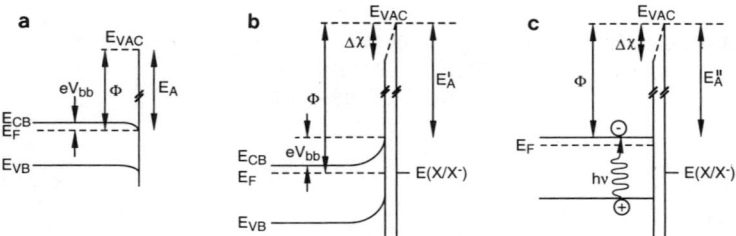

Figure 55. Schematic energy diagram of $MoSe_2/X_2$ interfaces for different experimental conditions. a, after cleavage; b, X adsorbed; c, X adsorbed, illuminated.

UHV follow the trend measured for semiconductor/electrolyte interfaces (contact potential difference eV_{bb} and obtained photovoltage U_{ph}; Table 6). The values obtained for the WSe_2 interfaces[524,525,514] are in qualitative agreement with the $MoSe_2$ results.[229,230] However, the band bending and surface photovoltage values are smaller than might be expected, considering the lower sample work function. This finding might tentatively be related to an increased shift of the band edges due to increased surface dipoles resulting from stronger surface/adsorbate interaction. The energy diagram of the WSe_2/Br_2 interface will be considered in more detail in the next section on coadsorbate systems. The flatband potentials of n-WSe_2 and n-$MoSe_2$ in organic electrolytes have been reported to be, respectively, -0.2 and $+0.25$ V vs. NHE.[528] These values may be compared to the work function Φ measured in UHV, $\Phi(WSe_2) = 4.0$ eV and $\Phi(MoSe_2) = 4.6$ eV, respectively. A reasonable correspondence for both compounds would be obtained by taking $E_{abs}^0 = 4.4$ eV; however, for $E_{abs}^0 = 4.8$ eV the positions of the flatbands differ considerably for vacuum and organic electrolyte interfaces. Therefore, it should be emphasized again that additional work is needed to precisely determine the absolute value of the redox potential scale (cf. Section II.3). The layered compounds are probably good candidates for a systematic study, as the influence of the surface dipole layer may be weak for certain electrolytes.

The detrimental effect of surface defects on the achievable SPV due to increased surface recombination has been demonstrated with intentionally damaged $MoSe_2$ (0001) cleavage planes.[230] After careful Ne^+-ion bombardment, surface states were evidently introduced which drastically reduced the photopotentials. The spectral shifts and spectral features in the UP spectra of the $MoSe_2/Br_2$ adsorption system were comparable to those obtained for perfect van der Waals surfaces. However, the surface photovoltage was considerably smaller (SPV = 0.3 eV) and indicates that defect states were introduced which led to efficient surface recombination. These defect states are situated in the forbidden energy gap and show up as additional photoemission intensity between E_F and the photoemission onset at the valence-band maximum. A very similar reduction of surface photovoltage at $MoSe_2/I_2$ interfaces was obtained after treating the surface with a low-energy electron beam ($E < 100$ eV) for extended times.[230] The undisturbed basal plane produced a surface photovoltage of 0.5 eV, which was reduced to 0.2 eV after electron bombardment. Evidently, electrons may also introduce surface defects; however, in this case they are below the detection limit of UPS. These results clearly demonstrate the sensitivity of nonequilibrium measurements, such as the SPV effect, as indicators of the optoelectronic activity of surface defects.

The experiments presented above show that certain aspects of the semiconductor/electrolyte junction can principally be simulated in UHV model experiments for monoadsorption systems. The important energy band diagram of the energy-converting junction can be determined and shows qualitative agreement with those derived from measurements in inert organic electrolytes. In addition, spectroscopic information on the species and electronic states involved in semiconductor/ adsorbate interaction is available. The similarity of the model interface to the electrolyte interface may be improved by performing coadsorption experiments (e.g., coadsorption of halogens with H_2O) or by buildup of surface redox couples. These experiments are described in the next section. Also, a more detailed comparison of surface potentials in UHV and at electrolyte contacts will be given there.

The model experiments are also very important when surface reactions must be considered, which may reduce the achievable photovoltage. For such reactive semiconductor surfaces, the surface reaction products can be identified and the initial reaction steps of the corrosion reaction may be investigated. In addition, the influence on the electronic properties may be studied. As one example of a reactive interface, the $InSe/Br_2$ adsorbate system is considered here.[524,530] A surface reaction is only derived after adsorption of Br_2 at LT and performing XPS measurements. The UP spectra show strong bromide-related emissions around 4 and 7 eV, assigned to the p_z and p_x, p_y levels of adsorbed Br^- (Fig. 56). In addition, a Br $3d$ signal at $E_B = 69.6$ eV remains in the XP spectrum after RT annealing. The reaction seems to be activated by the X-ray source

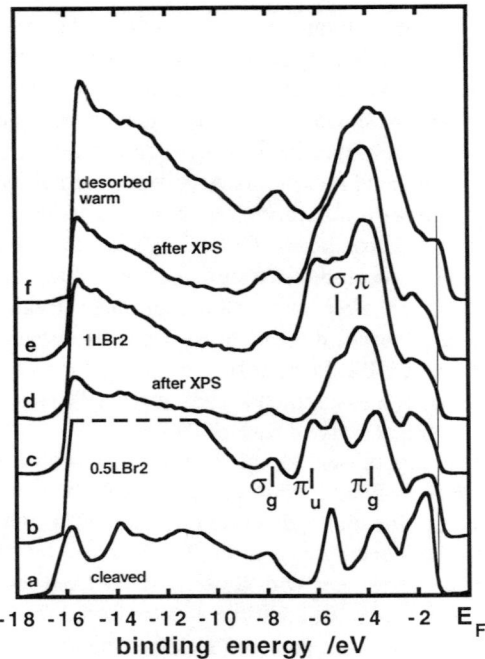

Figure 56. Irreversible adsorption of Br_2 on InSe(0001) with the formation of surface Br^- ions (no reversible surface photovoltage; He I UP spectra). T = 100K.

during the XPS measurements, as immediately after adsorption only the typical spectral features of molecularly adsorbed Br_2 are observed in the UP spectra. It is interesting to note that SPV values up to 0.6 eV can be measured before the surface reaction, but the SPV reduces to zero after the corrosion reaction. There are several layered chalcogenide semiconductors that chemically react with adsorbed halogens, such as ZrS_2/Cl_2,[531] $ZrSe_2/Cl_2$,[531] $ZrSe_2/Br_2$,[522,521] and SnS_2/Br_2.[522,521] In these cases it is evident from UP spectra that adsorption of halogens at LT leads to the formation of dissociatively adsorbed X^- ions, which remain on the surface after annealing of the sample at room temperature. Again, two strong emission lines show up at E_B^F = 4–8 eV. Also, in the core-level spectra, halogenide-induced peaks are clearly identified, and chemically shifted high-energy peaks appear in the metal emission lines. As a result of this surface reaction, a band bending is not induced by the adsorbed halogen. A reversible photovoltage is also completely absent, indicating that strong Fermi level pinning occurs for such reactive interfaces. However, an irreversible shift of the photoemission spectrum with bias light is observed which is in the opposite direction to that expected for a surface photovoltage effect.[531] It may tentatively be associated with a photochemical reaction at the interface.

Based on our present understanding, we relate the observed differences in chemical inertness between the different layered chalcogenides, which is also evident from photoelectrochemical studies, to two different factors: (i) the existence of large, chemically saturated chalcogenide ions on the van der Waals faces, which geometrically shield the metal atom, and (ii) the large contribution of metal d-states in the highest occupied and lowest unoccupied electron states of d-band semiconductors. Most probably, both factors are important. As the electronic interaction (charge transfer from and to the semiconductor) occurs through metal states (present at the band edges), the shielding of the metal sites by the chalcogenide layer prevents a reaction of the semiconductors with metal d-derived band-edge states. Hole injection into the chalcogenide-derived electronic states is immediately compensated by metal states in the band edges (relaxation of "hot" charge carriers) before a reaction can occur. In other words, the electronic interaction occurs with different atomic sites than the steric interaction. In addition, the metal d-derived electron bands are "quasi" nonbonding with respect to cation–anion interaction. For the reactive layered materials without valence bands derived from metal d-states (e.g., ZrS_2 or InSe), electronic and steric interactions occur with the same surface sites (the chalcogenide), and therefore the reactivity is increased.[531] However, also for these semiconductors the close-packed saturated chalcogenide atoms on the van der Waals surface provide a weakly interacting plane, which reduces the sticking coefficients of adsorbates and thus their reactivity compared to that on three-dimensional semiconductors.

(ii) 3.2. Coadsorption Phases

In order to approach the conditions of "real" electrolytes, coadsorption experiments were performed. One aspect to be investigated in more detail is the energy diagram of the adsorbate interface in relation to semiconductor/electrolyte interfaces containing a redox couple. As in the model experiments reported above, only one component of a redox couple was adsorbed (e.g., ox = Br_2); the approach taken was to attempt to produce both components on the surface by a proper selection of a coadsorbate (e.g., Na, which should reduce Br_2 to Br^-).

The valence-band photoelectron spectra (synchrotron induced, hv = 21 eV) of the WSe_2/Br_2, Na coadsorbate interface at different stages of the experiment are shown in Fig. 57.[532] Br_2 is initially adsorbed on the van der Waals (0001) plane and shows the typical valence-band features of the Br_2 emission lines. In good agreement with previous experiments on halogen adsorption on layered semiconductors [see Section VI.3(i)], we observed for different n-doping densities a shift of the Fermi level always

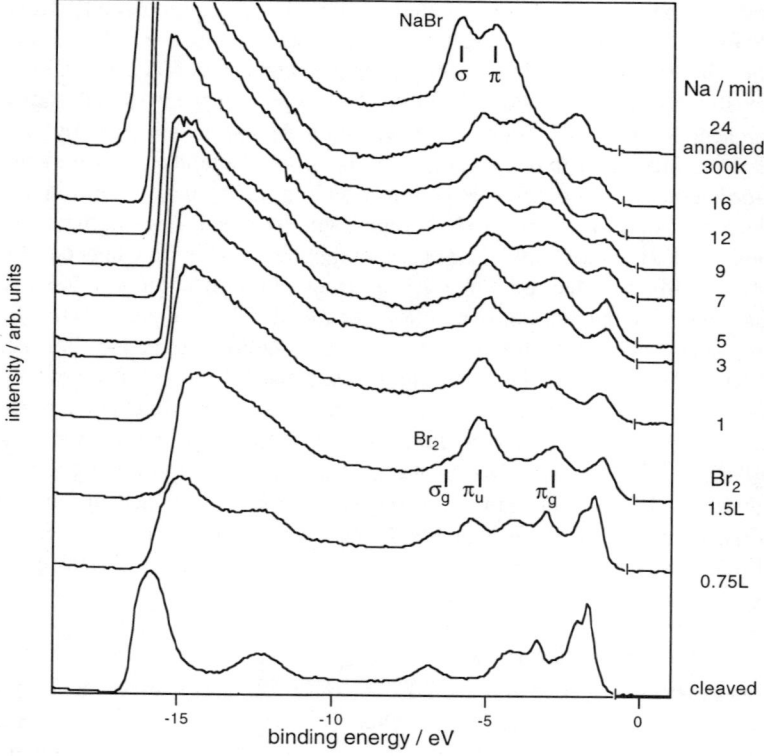

Figure 57. Valence-band spectra (hv = 21 eV) of the $WSe_2/Br_2/Na$ interface during the course of the adsorption experiment. (The adsorption conditions are marked on the figure.)

close to the valence-band edge (inversion), which is a result of band bending eV_{bb} induced by the adsorbed electron acceptor Br$_2$. E_B^F increased up to 0.9 eV for $n = 10^{17}$ cm^{-3},[524,514] whereas in the experiment presented in Fig. 57 eV_{bb} was only 0.5 eV owing to the low doping of the semiconductor ($n = 10^{15}$ cm^{-3}). The work function Φ increased by 1.7 eV and 2.0 eV, respectively. The substrate did not react with the Br$_2$ overlayer, as described above. We only observed a decrease of E_B^F due to band bending. The Br $3d$ emission lines are shown in Fig. 62. For the molecularly adsorbed Br$_2$, we observed the expected $3d_{5/2-3/2}$ doublet ($E_B^F\ 3d_{5/2}$ around 69 eV). The E_B^F values actually measured changed with substrate band bending. For some adsorption experiments, we also observed a weak shoulder at 67.4 eV due to the formation of adsorbed Br$^-$. It is formed by electron transfer from the n-type semiconductor to the adsorbed electron acceptor Br$_2$ to establish thermodynamic (electronic) equilibrium.

The surface concentration of the reduced species (Br$^-$) can be increased by coadsorbed Na. The valence-band region (Fig. 57) shows a gradual transformation of the molecularly adsorbed Br$_2$ to an adsorbed Br$^-$ species with increasing doses of Na. The E_B^F values of NaBr are marked, together with the band assignments according to the literature[258,259] (p_{xy} = 4.5 eV, p_z = 5.9 eV). As the substrate core levels still exclude any chemical reaction, the Br$^-$ formation is assigned to a reaction of Br$_2$ with the coadsorbed Na. The Na $2p$ core-level emission (E_B^F = 31.1 eV) is typical for ionic Na$^+$ on WSe$_2$ (cf. Ref. 249). For the final step of Na adsorption, the molecularly adsorbed Br$_2$ is completely transformed to NaBr, and at this point all substrate emission lines shift back to the original E_B^F values measured before the adsorption experiment was started. This back shift indicates that the electron transfer from the n-type semiconductor to the adsorbed electron acceptor Br$_2$ is reversed and the band bending is reduced again. After annealing the sample at room temperature for 12 h, we obtained the original flatband conditions with NaBr staying adsorbed on the surface. When the adsorption experiment was performed only with Br$_2$, we observed a complete desorption of the physisorbed Br$_2$ and of the ionosorbed Br$^-$, and, in this case also, the original flatband condition was restored.

The above results provide evidence that adsorption of Br$_2$ as well as Br$_2$/Na coadsorption onto n-WSe$_2$(0001) surfaces leads to energetic conditions that may be qualitatively related to those of the semiconductor/electrolyte interface for an electrolyte containing a Br$_2$/Br$^-$ redox couple. As discussed in Sections III and IV, thermodynamic equilibrium between the semiconductor and the adsorbate (electrolyte) is attained when the Fermi level within the semiconductor, $E_F^{SC} \equiv \eta_e^{SC}$, coincides with the electrochemical potential η_e^{El} of the adsorbate (electrolyte). For the adsorbate, η_e^{El} is determined by the energetic position, distribution, and occupation of the highest occupied MO (HOMO) of the reduced species and the lowest unoccupied MO (LUMO) of the oxidized species (Figs. 5 and 9). For an electron transfer from the n-type semiconductor to the adsorbate, unoccupied electron states should be available below E_F^{SC}. The LUMO of Br_2 as the electron-accepting electron state is the $2\sigma_u$ antibonding level. Its energy can be estimated to be 3.0 eV above the HOMO (π_g) based on recent electron energy loss data for molecular Br_2 adsorbed on Fe.[533] Thus, we conclude that in this case, in contrast to the H_2O adsorption experiment, a direct charge transfer from the n-type semiconductor (with a low η_e or Φ) to the adsorbed Br_2 (with a high electron affinity of the LUMO in the adsorbate stage) will be possible, as shown in Fig. 58. The energetic difference between the HOMO (π_g) and the LUMO (σ_u) of about 3 eV corresponds reasonably well to our measured E_E^F value for the π_g level of Br_2 of 3.0 eV in the adsorption experiment, which implies that E_F^{SC} is at the same energetic position as the $2\sigma_u$ level of Br_2. Therefore, it may be concluded that the electronic equilibrium of the WSe_2/Br_2 adsorption interface is determined by the energetic position of this level (see Fig. 59).

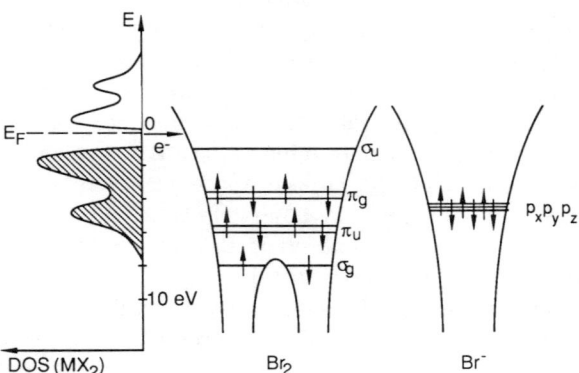

Figure 58. Schematic representation of the electron transfer to adsorbed Br_2, leading to Br^- formation.

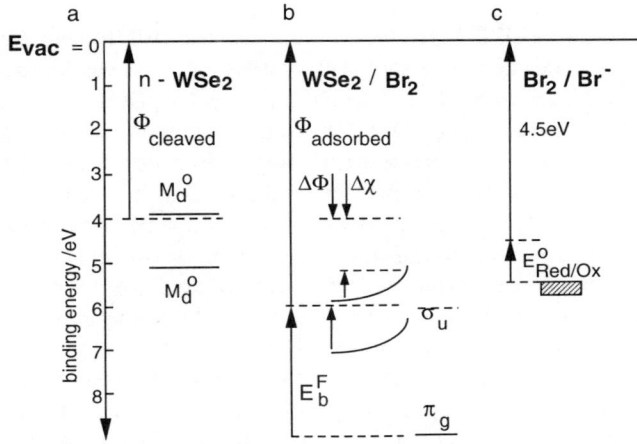

Figure 59. Schematic representation of energetic interface conditions at the $WSe_2/Br_2/Br^-$ interface in comparison to electrolyte contacts before the adsorption experiment (a) and after electronic equilibrium (b) in comparison to the redox potential of a Br_2/Br^- redox couple (c). (See text for details.)

In Fig. 59 the energy band diagrams of the clean n-WSe(0001) surface and the adsorbate-covered surface are presented. The overall shift of the work function Φ of the semiconductor, which is the sum of band bending (0.9 eV for high-doped and 0.4 eV for low-doped WSe_2) and the increased electron affinity (1.1 eV for high-doped and 0.8 eV for low-doped WSe_2), amounts to an overall change $\Delta\Phi$ of 2.0 and 1.2 eV, respectively, and to a final Φ for the adsorbate interface of around 6.0 eV (4.0 + 2.0 eV for high-doped and 4.5 + 1.2 eV for low-doped WSe_2). We may now compare this value to the redox potential of a Br_2/Br^- redox couple referred to the vacuum scale (Fig. 59). For electrolytes the standard redox potential E_0 and the concentration (activity a) of reduced (Br^-) and oxidized (Br_2) components are important for the definition of the electrochemical potential according to the Nernst equation:

$$\eta(Br_2/Br^-) \equiv E_{Red/Ox}^{(NHE)}(Br_2/Br^-) = E_0 + \frac{RT}{2 \cdot F} \ln \frac{a_{Br_2}}{a_{Br^-}^2} \quad (100)$$

The value of E_0 for liquid Br_2 is 1.1 V vs. NHE,[534] and $E_{0(Br_2/Br^-)}^{vac}$ is approximately 5.5–5.8 eV, which is in good agreement with the above determined value for E_F^{SC} in equilibrium with adsorbed Br_2 (Fig. 59). We take this as evidence that adsorbed Br_2 on $WSe_2(0001)$ can be approximated as a redox couple (Fig. 59). Evidently, $\eta_e^{El} \equiv \eta(Br_2/Br^-)$ is determined by the energetic position of $2\sigma_u$. The concentration dependence of E_F^{SC} of the adsorbate interface with changes in Br_2/Br^- concentrations may be estimated. The existence of oxidized (Br_2) and reduced (Br^-) species on the surface is formally equivalent to a surface redox couple. Accordingly, the concentration dependence of the Fermi level position in UHV adsorption experiments will be considered here in terms of a Nernst dependence. When the low temperature (100 K) of the experiment is considered, one would expect about a 10-mV change in the equilibrium potential for every decade change in the activity (concentration) ratio based on the Nernst equation. We can estimate the number of adsorbate species involved in charge transfer. The formation of the depletion layer needs about 10^{13} cm^{-2}, which is about 1% of a monolayer. The additional potential drop due to the formation of a strong inversion layer involves 10^{14} cm^{-2} (about 10% of a monolayer) owing to the smaller thickness of this double layer. The coadsorption of Na and reaction with the adsorbed Br_2 involves about one monolayer (10^{15} cm^{-2}). From a comparison of these concentration differences, it is clear that changes of energetic conditions (e.g., band bending) can hardly be measured in this type of model electrolytes formed by coadsorption as they are below 0.1 eV. Only in the extreme case, when nearly all of the adsorbed Br_2 is reduced by the coadsorbed Na, a back shift of E_{Br_2/Br^-} and E_F^{SC} will be observed experimentally. The constant level observed for E_F^{SC} over a wide Na codeposition range is taken as additional evidence that the surface coadsorption system behaves qualitatively as a redox couple according to the Nernst equation (constant eV_{bb} in Fig. 57).

The close agreement between the energy band diagram (Fig. 59) for the WSe_2/Br_2 interface (with and without coadsorbed Na) and that for the electrolyte interface proves again how closely model experiments may approach "real" electrolyte contacts. In addition, such experiments provide information on the species and electronic states involved in equilibrium formation.

Another aspect of interest that may be approached in coadsorption experiments is the solvent interaction with the redox species and/or the surface and its consequences for the contact formation. As an example, the H_2O/Br_2 interaction will be considered.[517,535] Valence-band spectra of the coadsorption system, corresponding to various stages during the adsorption of H_2O on Br_2-covered n-WSe_2, are shown in Fig. 60. The Br_2 emission features are as reported above. Three main emission features due to coadsorbed H_2O grow in; these belong to the molecular orbitals $1b_1$, $3a_1$, and $1b_2$ of adsorbed ice clusters [see Section VI.3(i)(a)]. Difference spectra obtained from UP spectra taken with different excitation energies as well as the substrate core-level spectra exclude any reaction of Br_2 with adsorbed H_2O or the substrate. From the first valence-band

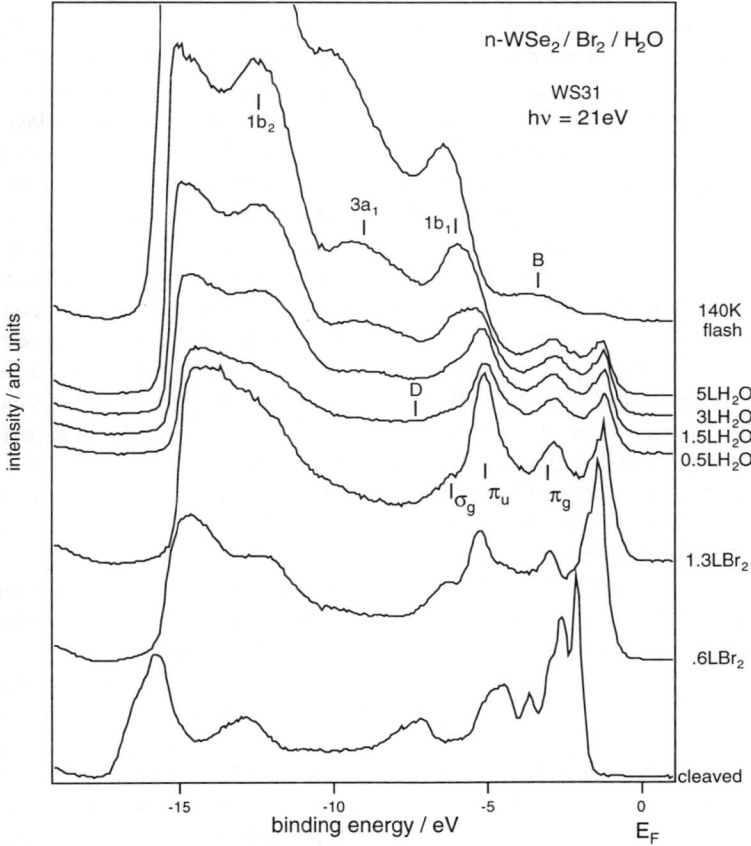

Figure 60. SXPS valence-band spectra of the coadsorption system WSe_2/Br_2, H_2O recorded at different stages of the experiment as indicated on the figure.

emission feature (W $5d_{z^2}$ level, Fig. 60) and the core-level spectra (not shown), it is evident that the band bending induced by the adsorbed Br_2 is not changed by the coadsorbed H_2O. However, the work function Φ is found to be affected. The changes of band bending are summarized in Fig. 61a. It should be noted that a temperature flash to 140 K changes the emission intensity ratios of the substrate versus the H_2O and Br_2 adsorbates (Fig. 60). Also, an additional emission line grows in, which is assigned to an enhanced contribution of adsorbed Br^-. This is more evident from the Br $3d$ core-level lines (Fig. 62), which clearly show that the Br^- contribution is increased due to coadsorbed H_2O, especially after the annealing step. Since also the binding energy of Br^- is lowered by the coadsorbed H_2O, we tentatively relate this result to the interaction (solvation) of the ionosorbed Br^- with H_2O. Corresponding spectra have also been measured for the reverse adsorption series (adsorption of Br_2 onto preadsorbed H_2O on WSe_2).[517] Again, only molecular adsorption is found, and no reaction can be identified. The Br $3d$ core-level lines are, as before, composed of contributions of molecular adsorbed Br_2 and ionosorbed Br^- (Fig. 62). The Br^- contribution is stronger compared to that in the monoadsorption case (also shown in Fig. 62), which again is attributed to the influence of interadsorbate interaction (solvation). The changes in surface potentials for this coadsorption series on p-doped substrates are summarized in Fig. 61b. The adsorption of H_2O leads to strong band bending on p-doped samples: the adsorbed H_2O behaves as an electron donor [see Section VI.3(i)(a)]. While H_2O as coadsorbate does not affect the Br_2-induced band bending (Fig. 61a), Br_2 as coadsorbate affects the H_2O-induced band bending (Fig. 61b). The H_2O-induced band bending is reversed back to flatband conditions, as is expected for a strong electron acceptor on a p-type substrate.

The coadsorption of H_2O with Br_2 may approximate the influence of the solvent. It is anticipated from our experiments that there is indeed an interaction of the coadsorbed species with each other, changing the evident (spectroscopically measured) surface composition and binding energies. We attribute this finding to a stabilization of the ionosorbed Br^- by surrounding H_2O molecules, which may be considered as solvation. However, the energetic conditions of the semiconductor as determined by adsorbed Br_2 as the redox-active species are not influenced by the coadsorbed H_2O. In contrast, Br_2 coadsorbed onto H_2O/p-WSe_2 reverses the original H_2O-induced band bending. It is evidently a stronger electron acceptor than H_2O is an electron donor as it dominates the energetic

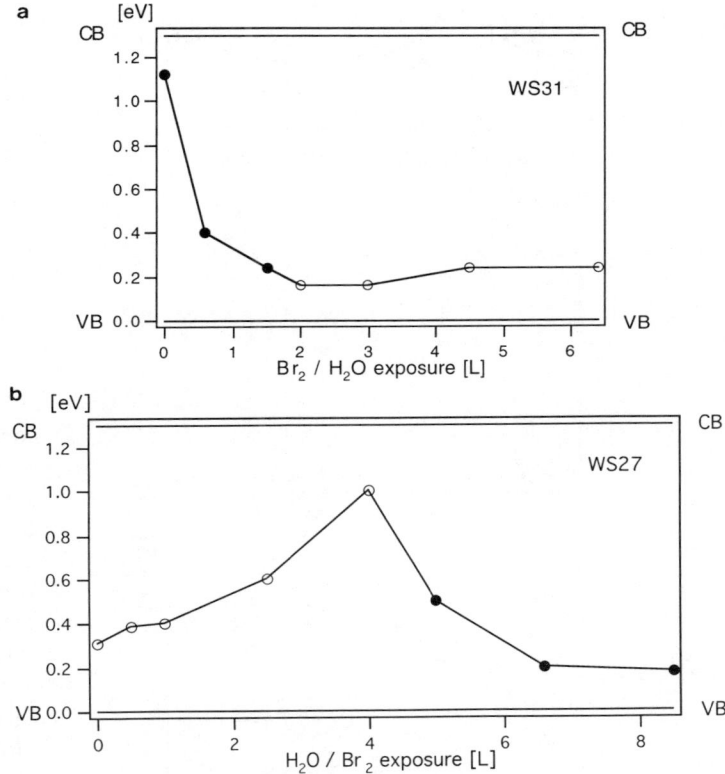

Figure 61. Changes of surface Fermi level position (band bending) for the coadsorption system $WSe_2/Br,H_2O$: (a) H_2O on Br_2-precovered n-WSe_2; (b) Br_2 on H_2O-precovered p-WSe_2. O, H_2O; ●, Br_2.

conditions of the interface in the coadsorption phase. This is attributed to the different character of the charge transfer: Br_2 may directly accept electrons into its LUMO (see Fig. 58) whereas H_2O may only act as an electron donor by the interaction with substrate electron states (Fig. 52).

The relative form and intensity of the core lines reflects the interadsorbate interaction more clearly. In Fig. 62, a comparison of the Br $3d$ emissions is shown for Br_2 monoadsorption, Br_2/H_2O coadsorption, and Br_2/Na coadsorption on $WSe_2(0001)$. It is clearly evident that for Br_2 adsorption the amount of Br^- is rather small; it is significantly increased

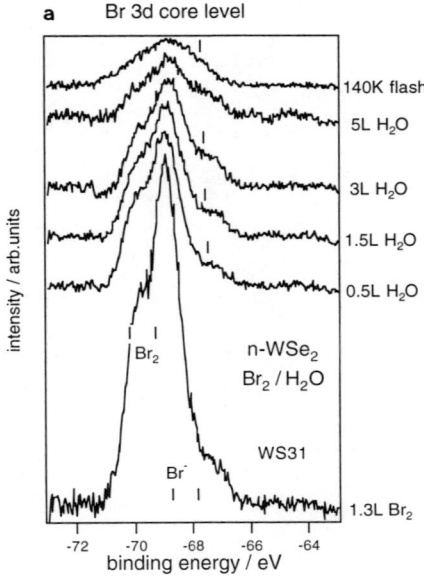
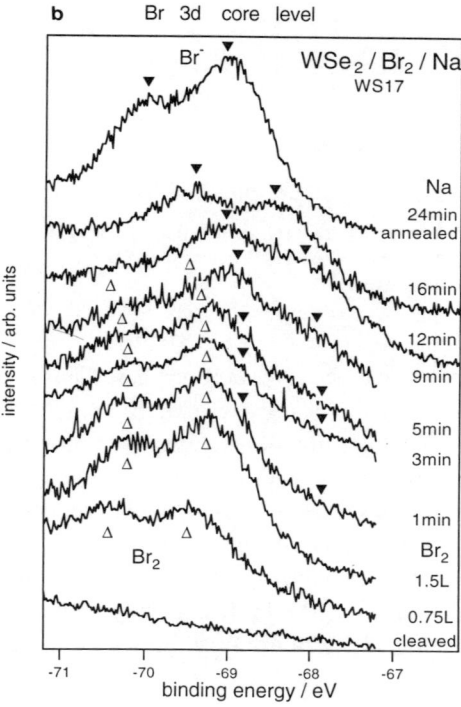

for Br_2/H_2O coadsorption and, as expected, is proportional to coadsorbed Na for the Br_2/Na system. There is also a clear indication of interadsorbate interaction in the O and Na core-level lines, showing changes in their intensity and also shifted binding energy values.[536]

Complementary coadsorption experiments for coadsorption of Na with H_2O on p-WSe_2 have been performed.[536,521] In this case the interadsorbate interaction depends on the sequence of the adsorption experiment. Adsorption of Na has to be performed at low temperatures as otherwise an intercalation reaction occurs.[249] If H_2O is coadsorbed onto preadsorbed Na (at coverages such that metallic Na layers are formed), dissociatively adsorbed H_2O is formed, in close correspondence to the behavior observed on Na substrates or Na-pretreated metallic substrates.[537] The formation of adsorbed OH is deduced from the appearance of two emission lines at E_B^F = 6 and 10 eV, due to the 1π and 3σ MOs of adsorbed OH, and from the O $2s$ core-level line measured at $E_B^F \approx 24$ eV. The band bending is dominated by the Na as is described in Ref. 249. As a very electropositive element, it donates its electron into the p-doped substrate, and the surface Fermi level moves into the conduction band ($eV_{bb} = -1.1$ eV). If, however, Na is deposited onto preadsorbed H_2O, H_2O remains molecularly adsorbed as for clean (0001) surfaces. There is, however, a strong interadsorbate interaction, which is evident from a shift of the Na core-level lines by 1 eV to lower binding energy. We attribute this shift to the increased stabilization of the Na^+ by the surrounding H_2O molecules. Further investigation is required to ascertain whether this effect is only due to an initial-state effect, such as the electrostatic stabilization of the Na^+ ions by H_2O, which would be equivalent to (a partial) solvation, or whether a final-state effect, such as screening of the core holes produced by photoemission, is the dominant contribution. Also for the latter coabsorption sequence the band bending is dominated by the adsorbed Na.

An n-WSe_2/Br_2, Na, H_2O phase was also prepared in UHV to approach the complex composition of an electrolyte even more closely. As shown in Fig. 63, Na was coadsorbed onto Br_2-precovered n-WSe_2, with equivalent results to those described above. Subsequently, H_2O was adsorbed to check whether a solvent would induce additional effects. In good

◄────────────

Figure 62. Comparison of Br $3d$ core-level lines for different adsorbate compositions on WSe_2(0001) substrates: (a) Br_2/H_2O coadsorption; (b) Br_2/Na coadsorption; $h\nu$ = 100 eV. The absorption of Br_2 alone is visible from the spectra at the bottom.

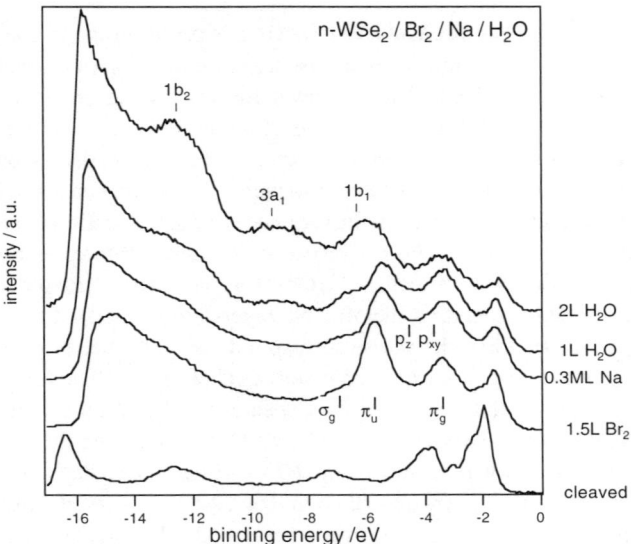

Figure 63. SXP valence-band spectra ($h\nu = 21$ eV) for the triadsorption phase n-WSe$_2$/Br$_2$, Na, H$_2$O. The sequence of adsorption is shown on the figure.

agreement with the results reported above, the band bending is determined by the strong redox couple Br$_2$/Br$^-$. Coadsorbed H$_2$O does not affect the charge distribution between the semiconductor and the model "redox couple." However, we observe again a strong influence of H$_2$O as solvent ion on the measured binding energies of all electrolyte components. For example, the Na $2p$ line of Na$^+$ is shifted by 0.5 eV to lower binding energies. The substrate core-level lines (W $4f$) remain unchanged. There are also small shifts and intensity changes observed in the Br $3d$ and O $2s$ core-level lines in these spectra as compared to the spectra for the monoadsorbate interfaces, also indicating interadsorbate interaction. This example should only be taken to indicate that, in principle, in experiments of this type a large amount of information is available about basic semiconductor/electrolyte interactions. This information can systematically be inferred from such complex UHV model experiments, which also take the interatomic interactions in the electrolyte into account. Unfortunately, the experiment described above was the first in a series that we intend to address in the near future and is still of a preliminary nature. The relative concentrations and the effect of low-temperature annealing steps are evidently of extreme importance. We attribute these effects to a kinetic activation of diffusion to reach an even distribution of adsorbate species in the model electrolyte adsorption phase, which may otherwise be still influenced by the sequence of adsorption. The changes induced by annealing steps have been noticed but have not yet been systematically explored.

In summary, the examples and results obtained with layered chalcogenides show that adsorption experiments designed to model semiconductor/electrolyte interfaces may provide detailed information on semiconductor/adsorbate interactions, which are important in electrochemistry. Surface reactions, the related formation of surface states, and their effects on interface band energy diagrams and especially band bending can easily be identified. On nonreactive interfaces, the electron states involved in charge-transfer processes, which induce band bending, can also be identified and related to electrochemical interfaces, as demonstrated, for example, with H_2O and Br_2/Na adsorption. Comparison of the results obtained in the halogen adsorption and Br_2/Na coadsorption experiments with those obtained for electrolyte solutions containing these redox couples suggests a surprisingly close correspondence between the adsorbate phase (model electrolytes) and electrolyte solutions. We feel that a more quantitative comparison may be possible for more advanced and complex adsorbate phases, which approach the electrolyte composition (solvent addition) and state variables (e.g., room temperature).

One might expect to determine the density of states distribution (DOS) of the electrolyte by using photoelectron spectroscopy. The DOS of the adsorbate electron states is directly given by the additional photoemission intensity in the valence-band region (neglecting final-state effects). In our experiments we did not obtain any clear evidence for a high concentration of occupied electron states induced by the adsorbate around the Fermi level. Owing to the high photoemission intensity of the substrate in the E_B^F range of interest, we were not able to obtain reliable difference spectra in this energy range. In addition, we still do not know whether the low sample temperature and the clustering of our adsorbates led to deviations in relation to electrolyte solutions. For a more realistic model of the electrolyte, the concentration and distribution of the solvent are also crucial parameters. Finally, and probably most importantly, the halogens are not "reversible" redox couples (there are two equivalent redox steps involved). Therefore, the electrolyte DOS cannot be described by the reorganizational broadening of only one electron state (the LUMO $2\sigma_u$ of Br_2) due to its interaction with its surroundings. It is not possible to decide on these aspects of electrochemical theory based on the simulation experiments presented here. However, as the surface spectroscopies and especially photoelectron spectroscopy may provide very interesting tools for a microscopic evaluation of interfaces, we feel it would be worthwhile to investigate these questions in more detail. The next steps should involve the improved coadsorption of the solvent and probably the investigation of "easier" (one-equivalent) redox-couple electrolyte/electrode model interfaces.

VII. SUMMARY AND CONCLUSION

The complexity of the structural and related electronic properties of semiconductor/electrolyte interfaces has been addressed in this chapter from a surface science point of view. The possible application of surface science techniques for investigating semiconductor/electrolyte interfaces was introduced, with a special emphasis on photoelectron spectroscopy. Model experiments have been suggested as a very promising surface science approach to the study of semiconductor/electrolyte interfaces that allows nearly all fundamental aspects of adsorbate interaction to be investigated. In combination with an *ex situ* surface analysis of electrochemically treated semiconductor electrodes, much valuable information can thus be deduced, including information on the structural, chemical, and electronic properties of interfaces. As the theoretical concepts used were originally developed for semiconductor/metal interfaces, the electrolyte interface is compared to related metal junctions.

After electronic equilibrium has been reached, several interfacial double layers must be distinguished which have different physical origins and also exhibit different spatial positions and dimensions. The semiconductor band bending results from charge transfer across the phase boundary and is equivalent to the Volta potential difference. For ideal semiconductor interfaces without active surface states in the band gap, this electronically induced double layer leads to an extended space-charge layer (>1000 Å) in the semiconductor. However, even for these ideal interfaces, changes of surface reconstructions and/or oriented adsorption will lead, in general, to changes of the surface dipoles and electrochemical double layers. These structural, ionic or dipolar, double layers shift the inner potential and thus the electron states of the two phases with respect to each other. The ionic or dipolar double layers are of atomic dimensions, ranging up to a few angstroms.

When electronically active surface states are involved in the charge redistribution across the phase boundary, the distribution and occupation of these interface states will determine the fraction of the Volta potential difference that will be transferred to the space-charge layer and the fraction that will remain on the semiconductor surface. In the extreme case of high surface-state density, the band bending remains constant with variation of the metal work function or electrolyte redox potential. Such interfaces show Fermi level pinning. Surface states of different character may exist at semiconductor surfaces and interfaces. Their origin as well as their position and density of states determines the electronic properties of contacts. Surface states in the band gap take part in the electronic charge redistribution across the phase boundary. Surface states outside the fundamental semiconductor band gap (surface resonances) will only affect the surface dipoles. The surface states and their electronic properties as well as the related spatial distribution of double layers can readily be analyzed by surface science techniques. The relative position of the semiconductor band edges and surfaces states may be determined versus adsorbate or electrolyte density of states distributions, and vice versa, which is crucial for an understanding of electron-transfer mechanisms across the phase boundary.

The type of information that is available from surface science investigations of semiconductors has been demonstrated for Si, GaAs, and layered chalcogenides. In the case of Si and GaAs, the structure and electronic structure of the clean surface may already be considerably different from those of ideal bulk truncated surfaces. As a consequence, the electronic properties and surface potentials may also deviate from ideal expectations. These changes are directly available from surface science studies. Adsorbates or surface layers formed in UHV as well as in electrolyte solutions may change the distribution of surface states and thus the surface Fermi level position. These shifts are directly given by parallel shifts of the measured binding energies and work functions in photoelectron spectra. Independently, the ionization potential or electron affinity may be affected, shifting the band-edge positions. These are evident from work-function changes that leave the binding energy values unaffected. The surface science experiments on clean and adsorbate-covered surfaces that have been presented in this chapter were not specifically directed toward electrochemical problems. However, the results can be compared to those obtained by an *ex situ* UHV analysis of emersed electrodes and to interface potentials determined from electrochemical experiments.

Si has a tendency to show Fermi level pinning for clean surfaces due to dangling-bond surface states. The adsorbate-covered surfaces show rather ideal behavior when the dangling bonds are perfectly saturated by the contact phase, as demonstrated for H-terminated surfaces or perfect Si/SiO_2 interfaces. Nonideal behavior and Fermi level pinning may be related to remaining defects in Si back bonds. The positions of the band edges as measured in electrolyte solutions or in UHV experiments are determined by the dipole of heteronuclear bonds to chemisorbed electronegative species.

Clean GaAs surfaces shows rather ideal behavior without surface states only for the nonpolar cleavage plane. The polar surfaces undergo complicated reconstructions to reduce the surface polarity and show Fermi level pinning. On adsorbate-covered surfaces in UHV, as well as in electrolyte solutions, surface states are introduced into the band gap for most adspecies. These surface states result from the antibonding states formed by the bonds between the As-derived valence-band states and the adatom. Rather ideal junction behavior, as compared to that of UHV-prepared surfaces, has been reported for polar surfaces after etching and (electro)chemical treatment; the origin of this behavior in some cases seems to be unclear. The nonpolar cleavage plane shows rather ideal behavior in organic electrolytes, which is in agreement with UHV results. Therefore, it may be taken as a good substrate surface for further systematic comparison of UHV and electrochemical environments.

The layered chalcogenide semiconductors are ideal substrates for a systematic comparison of UHV and electrochemical interfaces because their cleavage planes exhibit ideal structural and electronic properties. Therefore, many aspects of adsorbate interactions with semiconductor substrates can be investigated in UHV model experiments, and the findings may then be directly transferred to electrolyte interfaces. For example, the electronic states involved in contact formation as well as the electrostatic potential difference and spatial distribution of the different interfacial double layers are directly available from UHV experiments. In addition, it may be possible to deduce information about interadsorbate interaction, which may be related to solvation effects, but this will have to be further investigated.

Future UHV surface science investigations of electrochemical interfaces should compare, in a more systematic way, the results of *ex situ* experiments, which give information on a final condition of the electrode surface, and model experiments, which give information on initial states and fundamental aspects of surface interactions. Of course, these experiments should always be performed in close relation to electrochemical investigations and preferentially be complemented by *in situ* spectroscopies to get a complete set of information on interfacial processes. The *ex situ* experiments should provide perfectly clean emersed electrodes, which allow the use of more surface-sensitive techniques such as UPS, SXPS, and HREELS in addition to the more often applied ones, such as LEED and XPS. The model experiments should involve a systematic variation of experimental parameters in such a manner as to approach electrolyte conditions, e.g., systematically varied and small concentrations of redox-active species in relation to the amount of solvent molecules and the variation of sample temperature and deposition conditions. Also, volatile one-electron outer-sphere redox couples should be studied, in addition, in order to investigate the properties of more ideal electrolytes.

We expect that such a surface science approach to electrochemical problems may contribute valuable experimental information to a better understanding of solid/electrolyte interfaces at the molecular level. The influence of structural and electronic properties of the electrode surfaces on the interactions with adsorbates and electrolyte species and the related distribution of double layers across the interfaces will be clarified by such an approach. In addition, basic aspects of solvation interaction and especially of the electrolyte density of states distribution can be approached experimentally in more detail. For these type of investigations, semiconductor substrates have an advantage over metal substrates as the Volta potential contribution (variation of band bending) can easily be discriminated from the surface dipole (double) layers (variation in ionization potential). We also feel that additional work is needed for a more precise experimental determination of the value of the absolute electrode potential.

ACKNOWLEDGMENTS

First of all, I am happy to acknowledge the valuable and stimulating discussions with T. Mayer, as many of the concepts presented here were developed during his Ph.D. work. In addition, I would like to thank A. Klein, S. Rauscher, C. Pettenkofer, and F. Forstmann for their helpful comments and proofreading. I also would like to acknowledge

F. Forstmann, H. Cachet, and D. Vanmaekelbergh for providing reprints and preprints. S. Kubala created the extensive artwork by computer graphics in a very professional manner. Last but not least, I also would like to thank K. Zimba for the efficient typing of the manuscript. Our experiments presented here were supported by grants of the BMBF, Germany.

REFERENCES

[1] M. Schiavallo, ed., *Photoelectrochemistry, Photocatalysis and Photoreactors*, NATO ASI C146, Reidel, Dordrecht, 1985.
[2] A. Aruchamy, ed., *Photoelectrochemistry and Photovoltaics of Layered Semiconductors*, Kluwer, Dordrecht, 1992.
[3] W. Jaegermann and H. Tributsch, *Prog. Surf. Sci.* **29** (1988) 1.
[4] J. Janata, *Principles of Chemical Sensors*, Plenum Press, New York, 1989.
[5] M. J. Madou and S. R. Morrison, *Chemical Sensing with Solid State Devices*, Academic Press, San Diego, 1989.
[6] W. Göpel, J. Kesse, and J. N. Zemel, eds., *Sensors, Vol. 2, Chemical and Biochemical Sensors*, VCH, Weinheim, 1992.
[7] W. P. Gomes and H. H. Goossens, in *Advances in Electrochemical Science and Engineering*, Vol. 3, Ed. by H. Gerischer and C. W. Tobias, VCH, Weinheim, 1993.
[8] P. H. L. Notten, J. E. A. M. von den Meerakkes, and J. J. Kelly, *Etching of III–V Semiconductors*, Elsevier Advanced Technology, Oxford, 1991.
[9] H. Gerischer, in *Physical Chemistry, An Advanced Treatise*, Vol. 9A, Ed. by H. Eyring, D. Henderson, and W. Jost, Academic Press, New York, 1970, pp. 463–542.
[10] S. R. Morrison, *Electrochemistry at Semiconductor and Oxidized Metal Electrodes*, Plenum Press, New York, 1980.
[11] V. A. Myamlin and Y. V. Pleskov, *Electrochemistry of Semiconductors*, Plenum Press, New York, 1967.
[12] R. Memming, in *Electroanalytical Chemistry*, Vol. 11, Ed. by A. J. Bard, Marcel Dekker, New York, 1979, pp. 1–84.
[13] A. J. Bard and L. R. Faulkner, *Electrochemical Methods, Fundamentals and Applications*, John Wiley & Sons, New York, 1980.
[14] R. Greef, R. Peat, L. M. Peter, D. Pletcher, and J. Robinson, *Instrumental Methods in Electrochemistry*, Ellis Harwood, Chichester, U.K., 1985.
[15] H. D. Abruna, ed., *Electrochemical Interfaces*, VCH Publishers, New York, 1991.
[16] C. A. Melendres and A. Tadjeddine, eds., *Synchrotron Techniques in Interfacial Electrochemistry*, NATO ASI C432, Kluwer, Dordrecht, 1994.
[17] S. Trasatti and K. Wandelt, ed., *Surf. Sci.*, **335** (1995) 1.
[18] C. Gutierrez and C. Melendres, eds., *Spectroscopic and Diffraction Techniques in Interfacial Electrochemistry*, NATO ASI C320, Kluwer, Dordrecht, 1990.
[19] A. T. Hubbard, *Acc. Chem. Res.* **13** (1980) 177.
[20] P. N. Ross and F. T. Wagner, in *Advances in Electrochemistry and Electrochemical Engineering*, Vol. 13, Ed. by H. Gerischer and C. W. Tobias, John Wiley & Sons, New York, 1984.
[21] D. M. Kolb, *Z. Phys. Chem.* **154** (1987) 179.
[22] R. Kötz, in *Spectroscopic and Diffraction Techniques in Interfacial Electrochemistry*, Ed. by C. Gutierrez and C. Melendres, NATO ASI C320, Kluwer, Dordrecht, 1990, pp. 409–438.

[23] J. K. Sass, *Vacuum* **33** (1983) 741.
[24] E. M. Stuve, K. Bange, and J. K. Sass, in *Trends in Interfacial Electrochemistry*, Ed. by A. F. Silva, Reidel, Dordrecht, 1986.
[25] K. Bange, B. Straehler, J. K. Sass, and R. Parsons, *J. Electroanal. Chem.* **229** (1987) 87.
[26] F. T. Wagner and T. E. Moylan, *Surf. Sci.* **206** (1988) 187.
[27] T. Solomun, K. Christmann, and H. Baumgärtel, *J. Phys. Chem.* **93** (1989) 7199.
[28] T. Solomun, K. Christmann, A. Neumann, and H. Baumgärtel, *J. Electroanal. Chem.* **309** (1991) 95.
[29] J. K. Sass, D. Lackey, J. Schott, and B. Straehler, *Surf. Sci.* **247** (1991) 239.
[30] P. Baumann, G. Ping, D. Reuter, and H. P. Bonzel, *Surf. Sci.* **335** (1995) 186.
[31] F. T. Wagner, in *Structure of Electrified Interfaces*, Ed. by J. Lipkowski and P. N. Ross, VCH Publishers, New York, 1993, pp. 309–400.
[32] G. Pirug and H. P. Bonzel, in *Structure of Electrified Interfaces*, Ed. by J. Lipkowski and P. N. Ross, VCH Publishers, New York, 1993.
[33] E. M. Stuve, A. Krasnopoler, and D. E. Sauer, *Surf. Sci.* **335** (1995).
[34] H. Lüth, *Surfaces and Interfaces of Solids*, Springer, Berlin, 1993.
[35] W. Mönch, *Semiconductor Surfaces*, Springer, Berlin, 1993.
[36] L. J. Brillson, *Surf. Sci. Rep.* **2** (1982) 123.
[37] G. Margaritondo, ed., *Electronic Structure of Semiconductor Heterojunctions*, Kluwer, Dordrecht, 1988.
[38] D. A. King and D. P. Woodruff, eds., *The Chemical Physics of Solid Surfaces and Heterogeneous Catalysis*, Vol. 5, *Surface Properties of Electronic Materials*, Elsevier, Amsterdam, 1988.
[39] N. D. Lang, *Solid State Phys.* **28** (1973) 225.
[40] W. F. Egelhoff, Jr., *Surf. Sci. Rep.* **6** (1987) 253.
[41] S. Trasatti, in *Comprehensive Treatise of Electrochemistry*, Vol. 1, Ed. by J. O'M Bockris, B. E. Conway, and E. Yeager, Plenum Press, New York, 1980, pp. 45–81.
[42] H. Gerischer and W. Ekardt, *Appl. Phys. Lett.* **43** (1983) 393.
[43] N. W. Ashcroft and N. D. Mermin, *Solid State Physics*, Holt, Rinehart, and Winston, New York, 1976.
[44] R. Parsons, in *Modern Aspects of Electrochemistry*, No. 1, Ed. by J. O'M Bockris, Butterworths, London, 1954, pp. 3–179.
[45] E. Lange and K. Mishenkom, *Z. Phys. Chem. (Frankfurt/Main)* **149** (1930) 1.
[46] M. Lannoo and P. Friedel, *Atomic and Electronic Structure of Surfaces*, Springer-Verlag, Berlin, 1991.
[47] C. Herring and M. H. Nichols, *Rev. Mod. Phys.* **21** (1949) 185.
[48] M. Cardona and L. Ley, *Photoemission in Solids*, Vol. 1, Springer-Verlag, Berlin, 1978.
[49] N. D. Lang and W. Kohn, *Phys. Rev. B* **1** (1970) 4555; **3** (1971) 1215; **7** (1973) 2541.
[50] R. Smoluchowski, *Phys. Rev.* **60** (1941) 661.
[51] E. Wimmer, A. J. Freeman, J. R. Hiskes, and A. M. Karo, *Phys. Rev. B* **28** (1983) 3074.
[52] C. G. Scott and C. E. Reed, eds., *Surface Physics of Phosphors and Semiconductors*, Academic Press, New York, 1975.
[53] R. Paul, *Halbleiterphysik*, VEB Verlag Technik, Berlin, 1974.
[54] E. Spenke, *Elektronische Halbleiter*, Springer-Verlag, Berlin, 1965.
[55] J. Bardeen, *Phys. Rev.* **71** (1947) 717.
[56] C. Mailhiot and C. B. Duke, *Phys. Rev. B* **33** (1986) 1118; *J. Vac. Sci. Technol. B*, **3** (1985) 1170.
[57] C. B. Duke, *J. Vac. Sci. Technol.* **6** (1968) 152.
[58] M. Erbudak and T. E. Fischer, *Phys. Rev. Lett.* **29** (1972) 732.
[59] C. B. Duke, *Adv. Solid State Phys.* **33** (1994) 1.
[60] A. Kahn, *Surf. Sci. Rep.* **3** (1983) 193.
[61] W. Ranke and Y. R. Xing, *Phys. Rev. B* **31** (1985) 2246.

[62] H. J. Kuhr and W. Ranke, *Solid State Commun.* **61** (1987) 285.
[63] W. Ranke, *Phys. Rev. B* **27** (1983) 7837.
[64] H. Reiss, *J. Phys. Chem.* **89** (1985) 3783.
[65] H. Reiss, *J. Electrochem. Soc.* **135** (1988) 247.
[66] F. Forstmann, private communication.
[67] K. Uosaki and H. Kita, in *Modern Aspects of Electrochemistry*, No. 18, Ed. by R. E. White, J. O'M. Bockris, and B. E. Conway, Plenum Press, New York, 1986, pp. 1–60.
[68] J. O'M. Bockris and S. U. M. Khan, *Surface Electrochemistry*, Plenum Press, New York, 1993.
[69] F. Williams, S. P. Varma, and S. Hillenius, *J. Phys. Chem.* **64** (1976) 1549.
[70] T. Watanabe, S. Suzuki, K. Honda, and H. Gerischer, *Chem. Phys. Lett.* **105** (1984) 12.
[71] M. U. Sander, K. Luther, and J. Troe, *Ber. Bunsenges. Phys. Chem.* **97** (1993) 953.
[72] P. A. Thiel and T. E. Madey, *Surf. Sci. Rep.* **7** (1987) 211.
[73] R. A. Marcus, *J. Chem. Phys.* **24** (1956) 966.
[74] R. A. Marcus, *J. Chem. Phys.* **43** (1965) 679.
[75] H. Gerischer, *Z. Phys. Chem.* **26** (1960) 223, 325.
[76] S. U. M. Khan and J. O'M. Bockris, *J. Phys. Chem.* **87** (1983) 2599.
[77] D. Scherson, W. Ekardt, and H. Gerischer, *J. Phys. Chem.* **89** (1985) 554.
[78] S. U. M. Khan and J. O'M. Bockris, *J. Phys. Chem.* **89** (1985) 555.
[79] S. Nabayashi, A. Fujishima, and K. Honda, *J. Phys. Chem.* **87** (1983) 3487.
[80] S. Nabayashi, K. Itoh, A. Fujishima, and K. Honda, *J. Phys. Chem.* **87** (1983) 5301.
[81] F. Williams and A. Nozik, *Nature (London)* **271** (1978) 137.
[82] H. Tributsch, in *Modern Aspects of Electrochemistry*, No. 17, Ed. by J. O'M. Bockris, B. E. Conway, and R. E. White, Plenum Press, New York, 1986.
[83] H. Gerischer and W. Ekardt, *Appl. Phys. Lett.* **43** (1983) 395.
[84] S. Trasatti, *J. Electroanal. Chem.* **209** (1986) 417.
[85] R. Gomer and G. Tryson, *J. Chem. Phys.* **66** (1977) 4413.
[86] W. M. Hansen and D. M. Kolb, *J. Electroanal. Chem.* **100** (1979) 493.
[87] E. R. Kötz, H. Neff, and K. Müller, *J. Electroanal. Chem.* **215** (1986) 331.
[88] Z. Samec, B. W. Johnson, and K. Doblhofer, *Surf. Sci.* **264** (1992) 440.
[89] R. Moberg, F. Bökman, C. Bohman, and H. O. G. Siegbahn, *J. Chem. Phys.* **94** (1991) 5226; *J. Am. Chem. Soc.* **113** (1991) 3663.
[90] H. Morgner, in *Linking the Gaseous and Condensed Phases of Matter*, Ed. by L. G. Chistophorou *et al.*, Plenum Press, New York, 1994.
[91] M. Faubel and B. Steiner, in *Linking the Gaseous and Condensed Phases of Matter*, Ed. by L. G. Chistophorou *et al.*, Plenum Press, New York, 1994, pp. 517–523.
[92] W. Schottky, *Naturwissenschaften* **26** (1938) 843.
[93] N. F. Mott, *Proc. Cambridge Philos. Soc.* **34** (1938) 568.
[94] S. M. Sze, *Physics of Semiconductor Devices*, 2nd ed., John Wiley & Sons, New York, 1981.
[95] E. H. Rhoderick and R. H. Williams, *Metal–Semiconductor Contacts*, 2nd ed., Oxford Science Publ., Oxford, 1988.
[96] B. L. Sharma, ed., *Metal Semiconductor Schottky Barrier Junctions and Their Applications*, Plenum Press, New York, 1984.
[97] A. G. Milnes and D. L. Feucht, *Heterojunction and Metal–Semiconductor Junctions*, Academic Press, New York, 1972.
[98] A. M. Cowley and S. M. Sze, *J. Appl. Phys.* **36** (1965) 3212.
[99] S. Kurtin, T. C. Mc Gill, and C. A. Mead, *Phys. Rev. Lett.* **22** (1969) 1433.
[100] M. Schlüter, *Phys. Rev. B* **17** (1978) 5044.
[101] J. Bardeen, *Phys. Rev.* **71** (1947) 717.
[102] A. Many, Y. Goldstein, and N. B. Grover, *Semiconductor Surfaces*, North-Holland, Amsterdam, 1965.

[103] D. R. Frankl, *Electrical Properties of Semiconductor Surfaces*, Pergamon Press, Oxford, 1967.
[104] G. Ertl and H. Gerischer, in *Physical Chemistry*, Vol. X, Ed. by W. Jost, Academic Press, New York, 1970.
[105] W. Schottky, *Z. Phys.* **113** (1939) 367; **118** (1942) 539.
[106] P. Debeye and E. Hückel, *Phys. Z.* **24** (1923) 185, 305.
[107] R. T. Tung, *J. Vac. Sci. Technol. B* **11** (1993) 1546.
[108] G. Paasch and E. von Faber, *Prog. Surf. Sci.* **35** (1991) 19.
[109] R. W. Grant, E. A. Kraut, J. P. Waldrop, and S. P. Kowalzyk in *Heterojunction Band Discontinuities*, Ed. by F. Capasso and G. Margaritondo, Elsevier, Amsterdam, 1987, pp. 167–206.
[110] R. S. Bauer and H. W. Sang, Jr., *Surf. Sci.* **132** (1983) 479.
[111] F. Flores and R. Miranda, *Adv. Mater.* **6** (1994) 540.
[112] W. P. Gomes and F. Cardon, *Prog. Surf. Sci.* **12** (1982) 155.
[113] J. N. Chazalviel, *Electrochim. Acta* **35** (1990) 1545.
[114] D. E. Aspens and A. Frova, *Phys. Rev. B* **2** (1970) 1037.
[115] A. Hamnett, in *Comprehensive Chemical Kinetics*, Vol. 27, Ed. by R. G. Compton, Elsevier, Amsterdam, 1987, pp. 61–246.
[116] M. Cardona, *Modulation Spectroscopy*, Academic Press, New York, 1969.
[117] H. Gerischer, *Top. Appl. Phys.* **31** (1979) 115.
[118] H. Gerischer, in *Photoelectrochemistry, Photocatalysis and Photoreactors*, Ed. by M. Schiavallo, NATO ASI C146, Reidel, Dordrecht, 1985, pp. 39–106.
[119] M. A. Butler and D. S. Ginley. *J. Electrochem. Soc.* **125** (1978) 228.
[120] H. Moormann, D. Kohl, and G. Heiland, *Surf. Sci.* **80** (1979) 261.
[121] S. R. Morrison, *The Chemical Physics of Surfaces*, Plenum Press, New York, 1977.
[122] M. T. Gutierrez and J. Ortega, *Thin Solid Films* **174** (1989) 295.
[123] J. E. Tamm, *Z. Phys.* **76** (1932) 849.
[124] W. Shockley, *Phys. Rev.* **56** (1939) 367.
[125] F. J. Himpsel, *Appl. Phys. A* **38** (1985) 295.
[126] G. V. Hansson and R. I. G. Uhrberg, *Surf. Sci. Rep.* **9** (1988) 197.
[127] R. M. Feenstra, in *Scanning Tunneling Microscopy and Related Methods*, Ed. by R. J. Behm, N. Garcia, and H. Rohrer, NATO ASI E184, Reidel, Dordrecht, 1990, pp. 211–240.
[128] J. Pollmann, in *Festkörperprobleme XX*, Ed. by J. Treusch, Vieweg, Braunschweig, 1979, pp. 117–175.
[129] M. Schlüter, in *Festkörperprobleme XVIII*, Ed. by J. Treusch, Vieweg, Braunschweig, 1978, pp. 155–196.
[130] N. P. Lieske, *J. Phys. Chem. Solids* **45** (1984) 821.
[131] F. Forstmann, in *Photoemission and the Electronic Properties of Surfaces*, Ed. by B. Feuerbacher, B. Fitton, and R. F. Willis, John Wiley & Sons, Chichester, U.K., 1978, pp. 193–226.
[132] R. O. Jones, in *Surface Physics of Semiconductors and Phosphors*, Academic Press, New York, 1975, pp. 95–142.
[133] F. Garcia-Moliner and F. Flores, *Introduction to the Theory of Solid Surfaces*, Cambridge University Press, Cambridge, 1979.
[134] M. C. Desjonqueres and D. Spanjaard, *Concepts in Surface Physics*, Springer, Berlin, 1993.
[135] R. T. Tung, *J. Vac. Sci. Technol. B* **11** (1993) 1546.
[136] H. Fujitani and S. Asano, *Phys. Rev. B* **42** (1990) 1696.
[137] J. B. Goodenough, in *Photoelectrochemistry, Photocatalysis and Photoreactors*, Ed. by M. Schiavallo, NATO ASI C146, Reidel, Dordrecht, 1985, pp. 3–38.
[138] J. B. Goodenough, *Magnetism and the Chemical Bond*, Interscience, New York, 1963.
[139] J. Zaanen and G. A. Sawatzky, *Prog. Theor. Phys. Suppl.* **101** (1990) 231.

[140] J. D. Joannopoulos and G. Lucovsky, eds., *Top. Appl. Phys.* **56** (1984).
[141] J. E. Klepeis and W. A. Harrison, *J. Vac. Sci. Technol., B* **7** (1989) 964.
[142] J. Hebenstreit and M. Scheffler, *Phys. Rev. B* **46** (1992) 10134.
[143] H. Flietner, *Surf. Sci.* **200** (1988) 463.
[144] L. M. Peter, *Chem. Rev.* **90** (1990) 753.
[145] R. A. Batchelor and A. Hamnett, in *Modern Aspects of Electrochemistry*, No. 22, Ed. by J. O'M. Bockris, B. E. Conway, and R. E. White, Plenum Press, New York, 1992.
[146] G. Blyholder, in *Modern Aspects of Electrochemistry*, Vol. 8, Ed. by J. O'M. Bockris and B. E. Conway, Plenum Press, New York, 1972, pp. 1–46.
[147] W. A. Harrison, *Electronic Structure and the Properties of Solids*, Dover Publications, New York, 1989.
[148] R. Hoffmann, *Angew. Chem. Int. Ed. Engl.* **26** (1987) 846.
[149] R. Hoffmann, *Rev. Mod. Phys.* **60** (1988) 601.
[150] C. Kittel, *Introduction to Solid State Physics*, John Wiley & Sons, New York, 1976.
[151] H. Ibach and H. Lüth, *Festkörperphysik*, Springer, Berlin, 1990.
[152] A. Zangwill, *Physics at Surfaces*, Cambridge University Press, Cambridge, 1988.
[153] J. W. Gadzuk, *Surf. Sci.* **43** (1974) 44.
[154] T. B. Grimley, in *The Chemical Physics of Solid Surfaces and Heterogeneous Catalysis*, Ed. by D. A. King and D. P. Woodruff, Elsevier, Amsterdam, 1984.
[155] V. Bortolani, N. H. March, and M. P. Tosi, *Interaction of Atoms and Molecules with Solid Surfaces*, Plenum Press, New York, 1990.
[156] L. M. Peter, in *RSC Specialist Periodical Reports, Electrochemistry*, Vol. 9, Ed. by D. Pletcher, Royal Society of Chemistry, London, 1984, pp. 66–100.
[157] R. Memming, in *Photoelectrochemistry, Photocatalysis and Photoreactors*, Ed. by M. Schiavallo, NATO ASI C146, Reidel, Dordrecht, 1985, pp. 107–154.
[158] W. E. Spicer, T. Kendelewicz, N. Newman, R. Cao, C. McCants, K. Miyano, I. Lindau, Z. Liliental-Weber, and E. R. Weber, *Appl. Surf. Sci.* **33/34** (1988) 1009.
[159] W. E. Spicer, T. Kendelewicz, N. Newman, K. K. Chin, and I. Lindau, *Surf. Sci.* **168** (1986) 240.
[160] V. Heine, *Phys. Rev.* **138** (1965) 1689.
[161] S. G. Louie, J. R. Chelikowski, and M. L. Cohen, *Phys. Rev. B* **15** (1977) 2154.
[162] J. Tersoff, *Phys. Rev. Lett.* **52** (1984) 465; *Phys. Rev. B* **30** (1984) 4874.
[163] N. G. R. Daniels and G. Margaritondo, *Surf. Sci.* **132** (1982) 212.
[164] W. Mönch. *Phys. Rev. Lett.* **58** (1987) 1260.
[165] W. Mönch, *Europhys. Lett.* **7** (1988) 275.
[166] W. Mönch, *J. Vac. Sci. Technol. B* **7** (1989) 1216.
[167] R. T. Tung, *Phys. Rev. B* **45** (1992) 13509.
[168] R. Schlaf, A. Klein, C. Pettenkofer, and W. Jaegermann, *Phys. Rev. B* **48** (1993) 14242.
[169] A. Klein, C. Pettenkofer, W. Jaegermann, M. Lux-Steiner, and E. Bucher, *Surf. Sci.* **321** (1994) 19.
[170] R. Cimino, A. Giarante, K. Horn, and M. Pedio, *Europhys. Lett.* **32** (1995) 601.
[171] A. J. Bard, A. B. Bocarsly, F. R. F. Fan, E.G. Walton, and M. S. Wrighton, *J. Am. Chem. Soc.* **102** (1980) 3671.
[172] A. B. Bocarsly, D. C. Bookbinder, R. N. Dominey, N. S. Lewis, and M. S. Wrighton, *J. Am. Chem. Soc.* **102** (1980) 3683.
[173] A. Aruchamy and M. S. Wrighton, *J. Phys. Chem.* **84** (1980) 2848.
[174] M. L. Rosenbluth and N. S. Lewis, *J. Phys. Chem.* **93** (1989) 3753.
[175] R. L. van Meirhaeghe, R. Cardon, and W. P. Gomes, *J. Electroanal. Chem.* **188** (1985) 287.
[176] H. J. Lewerenz, *J. Electroanal. Chem.* **356** (1993) 121.
[177] P. A. M. Sherwood, *Chem. Soc. Rev.* **14** (1985) 1.
[178] A. T. Hubbard, *Chem. Rev.* **88** (1988) 633.

[179] M. P. Soriaga, ed., *Electrochemical Surface Science*, American Chemical Society, Washington, D.C., 1988.
[180] W. Jaegermann, in *Photoelectrochemistry and Photovoltaics of Layered Semiconductors*, Ed. by A. Aruchamy, Kluwer, Dordrecht, 1992.
[181] H. J. Lewerenz and T. Bitzer, *J. Electrochem. Soc.* **139** (1992) L21; T. Bitzer and H. J. Lewerenz, *Surf. Sci.* **269/270** (1992) 886.
[182] T. Bitzer, H. J. Lewerenz, M. Gruyters, and K. Jacobi, *J. Electroanal. Chem.* **359** (1993) 287; *J. Electrochem. Soc.* **140** (1993) L44.
[183] K. Jacobi, M. Gruyters, P. Geng, T. Bitzer, M. Aggour, S. Rauscher, and H. J. Lewerenz, *Phys. Rev. B* **51** (1995) 5432.
[184] W. Hirschwald et al., *Curr. Top. Mater. Sci.* **7** (1981) 143.
[185] W. Göpel, *Prog. Surf. Sci.* **20** (1985) 9.
[186] G. Heiland and H. Lüth, in *The Chemical Physics of Solid Surfaces and Heterogeneous Catalysis*, Vol. 3B, Ed. by D. A. King and D. P. Woodruff, Elsevier, Amsterdam, 1984.
[187] R. H. Williams and I. T. McGovern, in *The Chemical Physics of Solid Surfaces and Heterogeneous Catalysis*, Vol. 3B, Ed. by D. A. King and D. P. Woodruff, Elsevier, Amsterdam, 1984, pp. 267–309.
[188] A. Khan, *Surf. Sci. Rep.* **3** (1983) 193.
[189] T. Solomun, W. Richtering, and H. Gerischer, *Ber. Bunsenges. Phys. Chem.* **91** (1987) 412; T. Solomun, R. McIntyre, W. Richtering, and H. Gerischer, *Surf. Sci.* **169** (1986) 414.
[190] H. M. Kühne and H. Tributsch, *Ber. Bunsenges. Phys. Chem.* **88** (1984) 10; *J. Electroanal. Chem.* **201** (1986) 263.
[191] K. Büker, N. Alonso-Vante, R. Scheer, and H. Tributsch, *Ber. Bunsenges. Phys. Chem.* **98** (1994) 674.
[192] G. Ertl and J. Küppers, *Low Energy Electrons and Surface Chemistry*, Verlag Chemie, Weinheim, 1985.
[193] D. P. Woodruff and T. A. Delchar, *Modern Techniques of Surface Science*, Cambridge University Press, Cambridge, 1986.
[194] H. Cardona and L. Ley, eds., *Photoemission in Solids*, Vols. 1 and 2, Springer-Verlag, Berlin, 1978 and 1979.
[195] B. Feuerbach, B. Fitton, and R. F. Willis, eds., *Photoemission and the Electronic Properties of Surfaces*, John Wiley & Sons, New York, 1978.
[196] A. M. Bradshaw and K. Scheffler, in *The Chemical Physics of Solid Surfaces and Heterogeneous Catalysis*, Vol. 2, Ed. by D. A. King and D. P. Woodruff, Elsevier, Amsterdam, 1983, pp. 165–257.
[197] E. E. Koch, ed., *Handbook of Synchrotron Radiation*, Vol. 1a, 1b, North-Holland, Amsterdam, 1983.
[198] D. Briggs, ed., *Handbook of X-Ray and UV Photoelectron Spectroscopy*, Heyden, London, 1977.
[199] D. Briggs and M. P. Seah, eds., *Practical Surface Analysis by Auger and X-Ray Photoelectron Spectroscopy*, John Wiley & Sons, New York, 1983.
[200] C. R. Brundle and A. D. Baker, eds., *Electron Spectroscopies*, Vols. 1–4, Academic Press, New York, 1977.
[201] G. E. Muilenberg, ed., *Handbook of X-Ray Photoelectron Spectroscopy*, Perkin-Elmer, Eden Prairie, Minnesota, 1978.
[202] V. Dose, *Prog. Surf. Sci.* **13** (1983) 225.
[203] T. Fauster and V. Dose, in *Chemistry and Physics of Solid Surfaces VI*, Ed. by R. Vanselow and R. Howe, Springer-Verlag, Berlin, 1986, pp. 483–507.
[204] H. Raether, *Springer Tracts in Modern Physics*, Vol. 88, Springer, Berlin, 1980.
[205] H. Lüth, *Festkörperprobleme XXI*, Vieweg, Braunschweig, 1981.
[206] H. Ibach and D. L. Mills, *Electron Energy Loss Spectroscopy and Surface Vibrations*, Academic Press, New York, 1982.

[207] J. T. Grant, *Appl. Surf. Sci.* **13** (1982) 35.
[208] L. E. Davies et al., *Handbook of Auger Electron Spectroscopy*, Physical Electronics, Eden Prairie, Minnesota, 1976.
[209] D. E. Ramaker, R. Vanselow, and R. Howe, eds., *Chemistry and Physics of Solid Surfaces IV*, Springer, Berlin, 1982.
[210] M. A. van Hove, W. H. Weinberg, and C. M. Chan, *Low Energy Electron Diffraction*, Springer, Berlin, 1979.
[211] J. B. Pendry, *Low Energy Electron Diffraction*, Academic Press, New York, 1974.
[212] A. Benninghoven, F. G. Rüdenauer, and H. W. Herner, *Secondary Ion Mass Spectrometry*, Wiley-Interscience, New York, 1987.
[213] E. Taglauer and W. Heiland, *Appl. Phys.* **9** (1976) 261.
[214] E. Taglauer, *Appl. Phys. A* **38** (1985) 161.
[215] V. M. Agranovich and A. A. Maradudin, in *Modern Problems in Condensed Matter Science*, Vol 11, Ed. by V. M. Agranovich and A. A. Maradudin, North-Holland, Amsterdam, 1985.
[216] T. E. Madey and R. Stockbauer, in *Methods of Experimental Physics*, Ed. by R. L. Park, Academic Press, New York, 1985.
[217] T. E. Madey, in *Springer Series in Chemical Physics*, Vol 17, Springer, Berlin, 1983.
[218] T. L. Einstein, *Appl. Surf. Sci.* **11/12** (1982) 42.
[219] J. Stöhr, in *Chemistry and Physics of Solid Surfaces V*, Ed. by R. Vanselow and R. Howe, Springer, Berlin, 1984.
[220] R. J. Behm and W. Hösler, in *Chemistry and Physics of Solid Surfaces VI*, Ed. by R. Vanselow and R. Howe, Springer, Berlin, 1986, pp. 361–407.
[221] R. Wiesendanger and H. J. Güntherodt, eds., *Scanning Tunneling Microscopy*, Springer, Berlin, 1992.
[222] L. Reimer, *Scanning Electron Microscopy*, Springer, Berlin, 1985.
[223] C. Kunz, ed., *Synchrotron Radiation, Techniques and Applications*, Springer, Berlin, 1979.
[224] H. Lüth and G. Heiland, *Nuovo Cimento* **39** (1977) 748.
[225] H. C. Gatos and J. Lagowski, *J. Vac. Sci. Technol.* **10** (1973) 130.
[226] J. E. Demuth, W. J. Thompson, N. J. DiNardo, and R. Imbihl, *Phys. Rev. Lett.* **56** (1986) 1408.
[227] P. John, T. Miller, T. C. Hsieh, A. P. Shapiro, A. L. Wachs, and T. C. Chiang, *Phys. Rev. B* **34** (1986) 6704.
[228] G. Margaritondo, L. J. Brillson, and N. G. Stoffel, *Solid State Commun.* **35** (1980) 277.
[229] W. Jaegermann, *Chem. Phys. Lett.* **126** (1986) 301.
[230] W. Jaegermann, *Ber. Bunsenges. Phys. Chem.* **92** (1988) 537.
[231] A. L. Fahrenbruch and R. H. Bube, *Fundamentals of Solar Cells*, Academic Press, New York, 1983.
[232] H. J. Hovel, in *Semiconductors and Semimetals*, Vol 11, Ed. by R. R. Willardson and A. C. Beer, Academic Press, New York, 1975.
[233] W. W. Gärtner, *Phys. Rev.* **116** (1959) 84.
[234] J. I. Hanoka and R. O. Bell, *Annu. Rev. Mater. Sci.* **11** (1981) 353.
[235] M. Krumrey, Dissertation, TU Berlin, 1990.
[236] A. Many, Y. Goldstein, and N. B. Grover, *Semiconductor Surfaces*, North-Holland, Amsterdam, 1965.
[237] D. R. Frankl, *Electrical Properties of Semiconductor Surfaces*, Pergamon Press, Oxford, 1967.
[238] G. G. B. Garrett and W. H. Brattain, *Phys. Rev.* **99** (1955) 376.
[239] E. O. Johnson, *Phys. Rev.* **111** (1958) 153.
[240] E. O. Johnson, *RCA Rev.* **18** (1957) 556.
[241] A. L. Goodman, *J. Appl. Phys.* **32** (1961) 2550.

[242] M. Alonso, R. Cimino, and K. Horn, *Phys. Rev. Lett.* **64** (1990) 1947.
[243] D. M. Aldao, G. D. Wadill, P. J. Benning, C. Capasso, and J. H. Weaver, *Phys. Rev. B* **41** (1990) 6092.
[244] M. Hecht, *Phys. Rev. B* **41** (1990) 7918.
[245] K. Jacobi, U. Myler, and P. Althainz, *Phys. Rev. B* **41** (1990) 10721.
[246] S. Chang, I. M. Vitomirov, L. J. Brillson, D. F. Rioux, P. D. Kirchner, G. D Pettit, J. M. Woodall, and M. H. Hecht, *Phys. Rev. B* **41** (1990) 12299.
[247] D. Mao, A. Khan, M. Marsi, and M. Margaritondo, *Phys. Rev. B* **42** (1990) 3228; *Appl. Surf. Sci.* **48/49** (1991) 324.
[248] O. Lang, T. Mayer, C. Pettenkofer, and W. Jaegermann, unpublished results.
[249] A. Schellenberger, R. Schlaf, C. Pettenkofer, and W. Jaegermann, *Phys. Rev. B* **45** (1992) 3538.
[250] M. Sander, W. Jaegermann, and H. J. Lewerenz, *J. Phys. Chem.* **96** (1992) 782.
[251] M. Bronold, C. Pettenkofer, and W. Jaegermann, *J. Appl. Phys.* **76** (1994) 5800.
[252] T. Mayer, C. Pettenkofer, W. Jaegermann, and C. Levy-Clement, *Surf. Sci. Lett.* **254** (1991) L423.
[253] M. Henzler, *Appl. Phys. A* **34** (1984) 205.
[254] K. Bange, T. E. Madey, J. K. Sass, and E. M. Stuve, *Surf. Sci.* **183** (1987) 334.
[255] D. Lackey, J. Schott, B. Straehler, and J. K. Sass, *J. Chem. Phys.* **91** (1989) 1365.
[256] J. K. Sass, J. Schott, and J. Lackey, *J. Electroanal. Chem.* **283** (1990) 441.
[257] P.A. Thiel and T. E. Madey, *Surf. Sci. Rep.* **7** (1987) 211.
[258] P. A. Dowben, *CRC Crit. Rev. Solid State Mater. Sci.* **13** (1987) 191.
[259] M. Grunze and P. A. Dowben, *Appl. Surf. Sci.* **10** (1982) 209.
[260] H. P. Bonzel, *Surf. Sci. Rep.* **8** (1988) 43.
[261] H. P. Bonzel, A. M. Bradshaw, and G. Ertl, eds., *Physics and Chemistry of Alkali Metal Adsorption*, Elsevier, Amsterdam, 1989.
[262] F. P. Coenen, M. Kästner, G. Pirug, H. P. Bonzel, and U. Stimming, *J. Phys. Chem.* **98** (1994) 7885.
[263] E. M. Stuve and N. Kizhakevariam, *J. Vac. Sci. Technol. A* **11** (1993) 2217.
[264] F. T. Wagner and T. E. Moylan, *ACS Symp. Ser.* **378** (1988) 166.
[265] J. K. Sass and K. Bange, *ACS Symp. Ser.* **378** (1988) 54.
[266] T. Solomun, H. Baumgärtel, and K. Christmann, *J. Phys. Chem.* **95** (1991) 10041.
[267] A. Neumann, H. Rabus, D. Arvanitis, T. Solomun, K. Christmann, and K. Baberschke, *Chem. Phys. Lett.* **201** (1993) 108.
[268] D. A. King and D. P. Woodruff, eds., *The Chemical Physics of Solid Surfaces and Heterogeneous Catalysis, Vol 3B, Chemisorption Systems*, Elsevier, Amsterdam, 1984.
[269] J. W. Gadzuk, in *Photoemission and the Electronic Properties of Surfaces*, Ed. by B. Feuerbach, B. Fitton, and R. F. Willis, John Wiley & Sons, New York, 1978, pp. 111–136.
[270] J. Hölzl and F. K. Schulte, *Springer Tracts in Modern Physics* **85** (1979) 1.
[271] G. A. Somorjai, *Chemistry in Two Dimensions*, Cornell University Press, Ithaca, New York, 1981.
[272] J. Topping, *Proc. Roy. Soc. (London) Ser. A* **114** (1927) 67.
[273] H. O. Finklea, ed., *Semiconductor Electrodes*, Studies in Physical and Theoretical Chemistry, Vol. 55, Elsevier, New York, 1988.
[274] Y. V. Pleskov, *Solar Energy Conversion, A Photoelectrochemical Approach*, Springer-Verlag, Berlin, 1990.
[275] R. Memming, in *Photochemistry and Photophysics*, Vol. 2, Ed. by J. F. Rabek, CRC Press, Boca Raton, Florida, 1990, pp. 143–189.
[276] Ref. 68, Chapter 5.
[277] N. Alonso Vante and H. Tributsch, in *The Electrochemistry of Novel Materials*, Ed. by L. Lipkowski and P. N. Ross, VCH Publishers, New York, 1994.
[278] R. I. G. Uhrberg and G. V. Hansson, *Crit. Rev. Solid State Mater. Sci.* **17** (1991) 133.

[279] C. Sebenne, in *Handbook on Semiconductors*, Vol. 2, Ed. by M. Balkanski, Elsevier, Amsterdam, 1994, pp. 33–160.
[280] F. J. Himpsel, in *Handbook on Semiconductors*, Vol. 2, Ed. by M. Balkanski, Elsevier, Amsterdam, 1994; *Surf. Sci. Rep.* **12** (1990) 1.
[281] J. Pollmann, R. Kall, P. Krüger, A. Mazur, and G. Wolfgarten, *Appl. Phys. A* **41** (1986) 21.
[282] G. K. Wertheim, J. M. Riffe, J. E. Rowe, and P. H. Citrin, *Phys. Rev. Lett.* **67** (1991) 120.
[283] W. Mönch, P. Koke, and S. Krüger, *J. Vac. Sci. Technol.* **19** (1981) 313.
[284] C. A. Sebenne, J.-P. Lacharme, I. Andriamanantenasoa, and M. Khial, *Appl. Surf. Sci.* **41–42** (1989) 352.
[285] K. C. Pandey, *Phys. Rev. Lett.* **47** (1981) 1913; **49** (1982) 223.
[286] R. E. Schlier and H. E. Farnsworth, *J. Chem. Phys.* **30** (1959) 917.
[287] K. Takyanagi, Y. Tanishiro, M. Takahishi, and S. Takahashi, *J. Vac. Sci. Technol., A* **3** (1985) 1502; *Surf. Sci.* **164** (1985) 367.
[288] K. Hricovini, R. Günther, P. Thiry, A. Taleb-Ibrahimi, G. Indlekofer, J. E. Bonnet, P. Dumas, Y. Petroff, X. Blase, Xuejun Zhu, Steven G. Louie, Y. J. Chabal, and P. A. Thiry, *Phys. Rev. Lett.* **70** (1993) 1992.
[289] J. J. Paggel, W. Theis, K. Horn, Ch. Jung, C. Hellwig, and H. Petersen, *Phys. Rev. B* **50** (1994) 18686.
[290] J. E. Demuth, W. J. Thompson, N. J. DiNardo, and R. Imbihl, *Phys. Rev. Lett.* **56** (1986) 1408.
[291] R. C. Helms and B. E. Deal, eds., *The Physics and Chemistry of SiO_2 and the $Si-SiO_2$ Interface 2*, Plenum Press, New York, 1993.
[292] P. Morgen, U. Höfer, W. Wurth, and E. Umbach, *Phys. Rev. B* **39** (1989) 3720.
[293] U. Höfer, P. Morgen, W. Wurth, and E. Umbach, *Phys. Rev. B* **40** (1989) 1130.
[294] W. Ranke and Y. R. Xing, *Surf. Sci.* **157** (1985) 353.
[295] A. Stockhausen, T. U. Kampen, and W. Mönch, *Appl. Surf. Sci.* **56–58** (1992) 795.
[296] J. M. Seo, K. J. Kim, H. W. Yeom, and C. Park, *J. Vac. Sci. Technol., A* **12** (1994) 2255.
[297] H. Flietner, *Phys. Halbleiteroberfläche*, **21** (1990) 1.
[298] G. Hollinger and F. J. Himpsel, *J. Vac. Sci. Technol., A* **1** (1983) 640.
[299] G. Hollinger and F. J. Himpsel, *Appl. Phys. Lett.* **44** (1984) 93.
[300] F. J. Himpsel, F. R. McFeely, A. Taleb-Ibrahimi, J. A. Yarmoff, and G. Hollinger, *Phys. Rev. B* **38** (1988) 6084.
[301] M. M. Banarszak Holl, S. Lee, and F. R. McFeely, *Appl. Phys. Lett.* **65** (1994) 1097.
[302] W. Ranke, *Phys. Rev. B* **39** (1989) 1595.
[303] G. Hollinger, S. J. Sferzo, and M. Lannoo, *Phys. Rev. B* **37** (1988) 7149.
[304] S. Ciraci, E. Ellialtioglu, and S. Erkoc, *Phys. Rev. B* **26** (1982) 5716.
[305] Y. Miyamoto and A. Oshiyama, *Phys. Rev. B* **41** (1990) 12680; Y. Miyamoto, A. Oshiyama, and A. Ishitani, *Solid State Commun.* **74** (1990) 343.
[306] C. Silvestre and M. Shayegan, *Phys. Rev. B* **37** (1988) 10432.
[307] T. Engel, *Surf. Sci. Rep.* **18** (1993) 91.
[308] H. Froitzheim, in *The Chemical Physics of Solid Surfaces and Heterogeneous Catalysis*, Vol. 5, *Surface Properties of Electronic Materials*, Ed. by D. A. King and D. P. Woodruff, Elsevier, Amsterdam, 1988, pp. 183–234.
[309] P. Allongue, in *Modern Aspects of Electrochemistry*, No. 23, Ed. by B. E. Conway, J. O'M. Bockris, and R. E. White, Plenum Press, New York, 1992, pp. 239–314.
[310] G. Schulze and M. Henzler, *Surf. Sci.* **124** (1983) 336.
[311] H. Froitzheim, U. Köhler, and H. Lammering, *Surf. Sci.* **149** (1985) 537.
[312] A. Akremi, J.-P. Lacharme, and C.A. Sebenne, *Physica B* **170** (1991) 503.
[313] T. Sakurai and H. D. Hagstrum, *Phys. Rev. B* **14** (1976) 1593.
[314] S. Ciraki, R. Butz, E. M. Oellig, and H. Wagner, *Phys. Rev. B* **30** (1984) 74.
[315] J. E. Northrup, *Phys. Rev. B* **44** (1991) 1429.

[316] E. G. McRae and C. W. Caldwell, *Phys. Rev. Lett.* **46** (1981) 1632.
[317] R. Butz, E. M. Oellig, H. Ibach, and H. Wagner, *Surf. Sci.* **147** (1984) 343.
[318] K. Mortensen, D. M. Chen, P. J. Bedrossian, J. A. Golovchenko, and F. Besenbacher, *Phys. Rev. B* **43** (1991) 1816.
[319] F. Owman and P. Martensson, *Surf. Sci. Lett.* **303** (1994) L367.
[320] L. S. O. Johansson, R. I. G. Uhrberg, and G. V. Hansson, *Phys. Rev. B* **38** (1988) 13490.
[321] C. J. Karlsson, E. Landemark, L. S. O. Johansson, U. O. Karlsson, and R. I. G. Uhrberg, *Phys. Rev. B* **41** (1990) 1521.
[322] P. Koke and W. Mönch, *Solid State Commun.* **36** (1980) 1007.
[323] N. Safta, J.-P. Lacharme, C. A. Sebenne, and A. Akremi, *J. Phys.: Condens. Matter* **5** (1993) 6623.
[324] A. Stockhausen, T. U. Kampen, H. Nienhaus, and W. Mönch, in *Proceedings of the 4th International Conference on the Formation of Semiconductor Interfaces*, Ed. by B. Lengeler, H. Lüth, and W. Mönch, World Scientific, Singapore, 1993.
[325] J. R. Engstrom, M. M. Nelson, and T. Engel, *Surf. Sci.* **215** (1989) 437.
[326] F. R. McFeely, J. F. Morar, N. D. Shinn, G. Landgren, and F. J. Himpsel, *Phys. Rev. B* **30** (1984) 764.
[327] F. R. McFeely, J. F. Morar, and F. J. Himpsel, *Surf. Sci.* **165** (1986) 277.
[328] M. J. Bozack, M. J. Dresser, W. J. Choyke, P. A. Taylor, and J. T. Yates, Jr., *Surf. Sci.* **184** (1987) L332.
[329] H. H. Farrell, in *The Chemical Physics of Solid Surfaces and Heterogeneous Catalysis*, Vol. 3B, Ed. by D. A. King and D. P. Woodruff, Elsevier, Amsterdam, 1984, pp. 225-266.
[330] J. E. Rowe, *Phys. Rev. Lett.* **34** (1975) 398; *Surf. Sci.* **48** (1975) 461.
[331] J. E. Rowe, G. Margaritondo, and S. B. Christmann, *Phys. Rev. B* **16** (1977) 1581.
[332] P. H. Citrin, J. E. Rowe, and P. Eisenberger, *Phys. Rev. B* **28** (1983) 2299.
[333] G. Thornton, P. L. Wincott, R. McGrath, I. T. McGovern, F. M. Quinn, D. Norman, and D. D. Wedensky, *Surf. Sci.* **211/212** (1989) 959.
[334] L. S. O. Johansson, R. I. G. Uhrberg, R. Lindsay, P. L. Wincott, and G. Thornton, *Phys. Rev. B* **42** (1990) 9534.
[335] P. K. Larsen, N. V. Smith, M. Schlüter, H. H. Farrell, K. M. Ho, and M. L. Cohen, *Phys. Rev. B* **17** (1978) 2612.
[336] R. D. Schnell, D. Rieger, A. Bogen, F. J. Himpsel, K. Wandelt, and W. Steinmann, *Phys. Rev. B* **32** (1985) 8057.
[337] L. J. Whitman, S. A. Joyce, S. A. Yarmoff, F. R. McFeely, and L. J. Teminello, *Surf. Sci.* **232** (1990) 297.
[338] D. J. Elliot, *Integrated Circuit Fabrication Technology*, McGraw-Hill, New York, 1982.
[339] G. Ghidini and F. W. Smith, *J. Electrochem. Soc.* **131** (1984) 2924.
[340] E. Schröder-Bergen and W. Ranke, *Surf. Sci.* **236** (1990) 103.
[341] J. A. Schaefer, F. Stucki, D. J. Frankel, W. Goepel, and G. J. Lapeyre, *J. Vac. Sci. Technol., B* **2** (1984) 359.
[342] J. A. Schaefer, J. Anderson, and G. J. Lapeyre, *J. Vac. Sci. Technol., A* **3** (1985) 1443.
[343] K. Fives, R. McGrath, C. Stephens, I. T. McGovern, R. Cimino, D. S.-L. Law, A. L. Johnson, and G. Thornton, *J. Phys.: Condens. Matter* **1** (1989) SB105.
[344] W. Ranke and Y. R. Xing, *Surf. Sci.* **157** (1985) 339.
[345] H. Ibach, H. Wagner, and D. Bruchmann, *Solid State Commun.* **42** (1982) 457.
[346] D. Schmeisser and J. E. Demuth, *Phys. Rev. B* **33** (1986) 4233.
[347] D. Schmeisser, F. J. Himpsel, and G. Hollinger, *Phys. Rev. B* **27** (1983) 7813; D. Schmeisser, *Surf. Sci.* **137** (1984) 197; W. Ranke, D. Schmeisser, and Y. R. Xing, *Surf. Sci.* **152/153** (1985) 1103.
[348] M. Nishijima, K. Edamoto, Y. Kubota, S. Tanaka, and M. Onchi, *J. Chem. Phys.* **84** (1986) 6458.
[349] D.-X. Dai, F.-R. Zhu, Y.-S. Luo, and I. Davoli, *J. Phys.: Condens. Matter* **4** (1992) 5855.

[350] S. Ciraki and H. Wagner, *Phys. Rev. B* **27** (1983) 5180.
[351] N. S. Lewis, *Annu. Rev. Phys. Chem.* **42** (1991) 543.
[352] N. S. Lewis and A. B. Bocarsly, in *Semiconductor Electrodes*, Studies in Physical and Theoretical Chemistry, Vol. 55, Ed. by H. O. Finklea, Elsevier, New York, 1988, pp. 241–276.
[353] H. J. Lewerenz, *Electrochim. Acta* **37** (1992) 847.
[354] C. Levy-Clement, A. Lagoubi, R. Tenne, and M. Neumann-Spallart, *Electrochim. Acta* **37** (1992) 877.
[355] W. Kern, *J. Electrochem. Soc.* **137** (1990) 1887.
[356] D. B. Fenner, D. K. Biegelsen, and R. D. Bringans, *J. Appl. Phys.* **66** (1989) 419.
[357] E. Yablonovitch, D. L. Allara, C. C. Chang, T. Gmitter, and T. B. Bright, *Phys. Rev. Lett.* **57** (1986) 249.
[358] G. S. Higashi, Y. J. Chabal, G. W. Trucks, and K. Raghavachari, *Appl. Phys. Lett.* **56** (1990) 656.
[359] J. M. C. Thornton and R. H. Williams, *Phys. Scr.* **41** (1990) 1047.
[360] T. Dittrich, H. Angermann, H. Flietner, T. Bitzer, and H. J. Lewerenz, *J. Electrochem. Soc.* **141** (1994) 3595.
[361] S. Rauscher, T. Dittrich, M. Aggour, J. Rappich, H. Flietner, and H. J. Lewerenz, *Appl. Phys. Lett.* **66** (1995) 3018.
[362] T. Sunada, T. Yasaka, M. Takakura, T. Sugiyama, S. Miyazaki, and M. Hirose, *Jpn. J. Appl. Phys.* **29** (1990) L2408.
[363] D. Gräf, S. Bauer-Mayer, and A. Schnegg, *J. Vac. Sci. Technol., A* **11** (1993) 940.
[364] T. Tagahagi, A. Ishitani, H. Kuroda, and Y. Nagasawa, *J. Appl. Phys.* **69** (1992) 803.
[365] M. Niwano, Y. Takeda, K. Kurita, and M. Miyamoto, *J. Appl. Phys.* **72** (1992) 2488.
[366] H. Gerischer, P. Allongue, and V. Costa-Kieling, *Ber. Bunsenges. Phys. Chem.* **97** (1993) 753.
[367] F. Ozanam and J.-N. Chalzalviel, *J. Electron Spectrosc. Rel. Phenom.* **64/65** (1993) 395.
[368] G. W. Trucks, K. Raghavachari, G. S. Higashi, and Y. J. Chabal, *Phys. Rev. Lett.* **65** (1990) 504.
[369] P. Dumas, Y. J. Chabal, and G. S. Higashi, *Phys. Rev. Lett.* **65** (1990) 1124.
[370] P. Dumas and Y. J. Chabal, *Chem. Phys. Lett.* **181** (1991) 537.
[371] J. Rappich, H. Jungblut, M. Aggour, and H. J. Lewerenz, *J. Electrochem. Soc.* **141** (1994) L99.
[372] S. Rauscher and H. J. Lewerenz, unpublished results.
[373] D. Gräf, M. Grundner, R. Schulz, and L. Mühlhoff, *J. Appl. Phys.* **68** (1990) 5155.
[374] D. Gräf, M. Grundner, and R. Schulz, *J. Vac. Sci. Technol., A* **7** (1989) 808.
[375] M. Egawa and H. Ikoma, *Jpn. J. Appl. Phys.* **33** (1994) 943.
[376] B. J. Tufts, A. Kumar, A. Bansal, and N. S. Lewis, *J. Phys. Chem.* **96** (1992) 4581.
[377] G. Schlichthörl and L. M. Peter, *J. Electrochem. Soc.*, in press.
[378] M. J. Madou, B. H. Loo, K. W. Frese, and S. R. Morrison, *Surf. Sci.* **108** (1981) 135.
[379] H. Gerischer, *Electrochim. Acta* **34** (1989) 1005.
[380] A. P. Abbott, D. J. Schiffrin, and S. A. Campbell, *J. Electroanal. Chem.* **328** (1992) 355.
[381] J. Ronga, A. Bsiesy, F. Gaspard, R. Herino, M. Ligeon, F. Muller, and A. Halimaoui, *J. Electrochem. Soc.* **138** (1991) 1403.
[382] P. Allongue, V. Costa-Kieling, and H. Gerischer, *J. Electrochem. Soc.* **140** (1993) 1009.
[383] S.-L. Yau, F. R. F. Fan, and A. J. Bard, *J. Electrochem. Soc.* **139** (1992) 2825.
[384] G. Schlichthörl and L. M. Peter, *J. Electroanal. Chem.* **381** (1995) 55.
[385] C. B. Duke, in *The Chemical Physics of Solid Surfaces and Heterogeneous Catalysis*, Vol. 5, *Surface Properties of Electronic Materials*, Ed. by D. A. King and D. P. Woodruff, Elsevier, Amsterdam, 1988, pp. 69–118.
[386] C. B. Duke, *Adv. Solid State Phys.* **33** (1994) 1.
[387] C. Mailhiot, C. B. Duke, and Y. C. Chang, *Phys. Rev. B* **30** (1984) 1109.

[388] D. V. Froelich, M. E. Lapeyre, J. D. Dow, and R. E. Allen, *Superlattices & Microstruct.* **1** (1985) 87.
[389] G. W. Gobeli and F. G. Allen, *Phys. Rev.* **137** (1965) A245.
[390] G. M. Guichar, C. A. Sebenne, and G. A. Jarry, *Phys. Rev. Lett.* **37** (1976) 1158.
[391] J. van Laar, A. Huijser, and T. L. Rooy, *J. Vac. Sci. Technol.* **14** (1977) 894.
[392] D. E. Eastman, T.-C. Chiang, P. Hermann, and F. J. Himpsel, *Phys. Rev. Lett.* **45** (1980) 656.
[393] W. Ranke and K. Jacobi, *Prog. Surf. Sci.* **10** (1981) 1.
[394] R. Z. Bachrach, R. S. Bauer, P. Chiriada, and G. W. Hansson, *J. Vac. Sci. Technol.* **19** (1981) 335.
[395] H. H. Farrel and C. J. Palmstrøm, *J. Vac. Sci. Technol.*, *B* **8** (1990) 903.
[396] D. K. Biegelsen, R. D. Bringans, J. E. Northrup, and L. E. Swartz, *Phys. Rev. B* **41** (1990) 5701.
[397] K. Jacobi, *Surf. Sci.* **132** (1983) 1.
[398] M. Alonso, F. Soria, and J. L. Sacedón, *J. Vac. Sci. Technol., A* **3** (1985) 1598.
[399] T. C. Chiang, R. Ludeke, M. Aono, G. Landgren, F. J. Himpsel, and D. E. Eastman, *Phys. Rev. B* **27** (1983) 4770.
[400] W. Chen, M. Dumas, D. Mao, and A. Kahn, *J. Vac. Sci. Technol., B* **10** (1992) 1886.
[401] K. Hirose, A. Uchiyama, T. Noguchi, and M. Uda, *Appl. Surf. Sci.* **56** (1992) 11; K. Hirose, E. Foxman, T. Noguchi, and M. Uda, *Phys. Rev. B* **41** (1990) 6076.
[402] I. M. Vitomirov, A. Raisanen, A. C. Finnefrock, R. E. Viturro, L. J. Brillson, P. D. Kirchner, G. D. Pettit, and J. W. Woodall, *Phys. Rev. B* **46** (1992) 13293.
[403] S. Y. Tong, G. Xu, and W. N. Mei, *Phys. Rev. Lett.* **52** (1984) 1693.
[404] D. J. Chadi, *Phys. Rev. Lett.* **52** (1984) 1911.
[405] K. W. Haberern and M. D. Pashley, *Phys. Rev. B* **41** (1990) 3226.
[406] D. K. Biegelsen, R. D. Bringans, J. E. Northrup, and L. E. Swartz, *Phys. Rev. Lett.* **65** (1990) 452.
[407] K. Cierocki, D. Troost, L. Koenders, and W. Mönch, *Surf. Sci.* **264** (1992) 23.
[408] F. A. Cotton and G. Wilkinson, *Advanced Inorganic Chemistry*, 3rd ed., Interscience, New York, 1972.
[409] R. H. Williams and I. T. McGovern, in *The Chemical Physics of Solid Surfaces and Heterogeneous Catalysis, Vol. 3B, Chemisorption Systems*, Ed. by D. A. King and D. P. Woodruff, Elsevier, Amsterdam, 1984, pp. 267–309.
[410] L. J. Brillson and G. Margaritondo, in *The Chemical Physics of Solid Surfaces and Heterogeneous Catalysis, Vol. 5, Surface Properties of Electronic Materials*, Ed. by D. A. King and D. P. Woodruff, Elsevier, Amsterdam, 1988, pp. 119–182.
[411] K. A. Bertness, J. J. Yeh, D. J. Friedman, P. M. Mahowald, A. K. Wahi, T. Kendelewicz, I. Lindau, and W. E. Spicer, *Phys. Rev. B* **38** (1988) 5406.
[412] G. Landgren, R. Ludeke, Y. Jugnet, J. F. Morar, and F. J. Himpsel, *J. Vac. Sci. Technol., B* **2** (1984) 351; *Phys. Rev. B* **30** (1984) 4839.
[413] S. G. Anderson, T. Komeda, J. M. Seo, C. Capasso, G. D. Waddill, P. J. Benning, and J. H. Weaver, *Phys. Rev. B* **42** (1990) 5082.
[414] F. Bartels and W. Mönch, *Surf. Sci.* **143** (1984) 315.
[415] G. Hughes and R. Ludeke, *J. Vac. Sci. Technol., B* **4** (1986) 1109.
[416] C. Y. Su, I. Lindau, P. W. Chye, P. R. Skeath, and W. E. Spicer, *Phys. Rev. B* **25** (1982) 4054.
[417] K. D. Childs and M. G. Lagally, *Phys. Rev. B* **30** (1984) 5724.
[418] C. F. Yu, T. Schmidt, D. W. Podlesnik, E. S. Yang, and R. M. Osgood, Jr., *J. Vac. Sci. Technol., A* **6** (1988) 754.
[419] C. D. Thurmond, G. P. Schwartz, G. W. Kammlott, and B. Schwartz, *J. Electrochem. Soc.* **127** (1980) 1366.
[420] R. Ludeke, *Solid State Commun.* **21** (1977) 815.

[421] D. Flamm, A. Meisel, and E.-H. Weber, *J. Electron Spectrosc. Relat. Phenom.* **42** (1987) 73.
[422] W. Mönch, *Surf. Sci.* **168** (1986) 577.
[423] A. Kahn, D. Kanani, P. Mark, C. Y. Su, I. Lindau, and W. E. Spicer, *Surf. Sci.* **87** (1979) 325.
[424] R. Dorn, H. Lüth, and G. J. Russel, *Phys. Rev. B* **10** (1974) 5049.
[425] W. E. Spicer, P. W. Chye, C. M. Garner, I. Lindau, and P. Pianetta, *Surf. Sci.* **86** (1979) 763.
[426] K. Stiles, D. Mao, and A. Kahn, *J. Vac. Sci. Technol., B* **6** (1988) 1170.
[427] W. Mönch, *Appl. Surf. Sci.* **22/23** (1985) 705.
[428] H. Nienhaus and W. Mönch, *Appl. Surf. Sci.* **65** (1993) 632.
[429] C. D. Thuault, G. M. Guichar, and C. A. Sebenne, *Surf. Sci.* **80** (1979) 273.
[430] J. Szuber, *Appl. Surf. Sci.* **55** (1992) 143.
[431] W. Ranke and K. Jacobi, *Surf. Sci.* **81** (1979) 504.
[432] M. Alonso and F. Soria, *Surf. Sci.* **182** (1987) 530.
[433] R. Haight and J. Bokor, *Phys. Rev. Lett.* **26** (1986) 2846.
[434] R. D. Bringans and R. Z. Bachrach, *Solid State Commun.* **45** (1983) 83.
[435] P. K. Larsen and J. Pollmann, *Solid State Commun.* **53** (1985) 277.
[436] H. Qi, P. E. Gee, T. Nguyen, and R. F. Hicks, *Surf. Sci.* **323** (1995) 6; H. Qi, P. E. Gee, and R. F. Hicks, *Phys. Rev. Lett.* **72** (1994) 250.
[437] X.-Y. Hou, S. Yang, G.-S. Dong, X.-M. Ding, and X. Wang, *Phys. Rev. B* **35** (1987) 8015.
[438] K. Jacobi, G. Steinert, and W. Ranke, *Surf. Sci.* **57** (1976) 571.
[439] K. Mochiji, J. Ochiai, S. Yamamoto, and S. Takatani, *Surf. Sci.* **311** (1994) L677.
[440] H. Watanabe and S. Matsui, *Jpn. J. Appl. Phys.* **32** (1993) 6158.
[441] S. M. Mokler, P. R. Watson, L. Ungier, and J. R. Arthur, *J. Vac. Sci. Technol., B* **10** (1992) 2371.
[442] T. Onno, *Phys. Rev. B* **44** (1991) 8387.
[443] P. E. Gregory and W. E. Spicer, *Surf. Sci.* **54** (1976) 229.
[444] O. M'Hamedi, F. Proix, and C. Sebenne, *Semicond. Sci. Technol.* **2** (1987) 418.
[445] H. Lüth and R. Matz, *Phys. Rev. Lett.* **46** (1991) 1652.
[446] L. H. Dubois and G. P. Schwartz, *Phys. Rev. B* **26** (1982) 794.
[447] F. Antonangeli, C. Calandra, E. Colavita, S. Nannarone, C. Rinaldi, and L. Sorba, *Phys. Rev. B* **29** (1984) 8.
[448] C. Astaldi, L. Sorba, C. Rinaldi, R. Mercuri, S. Nannarone, and C. Calandra, *Surf. Sci.* **162** (1985) 39.
[449] L. Sorba, M. Pedio, S. Nannarone, S. Chang, A. Raisanen, A. Wall, P. Philip, and S. Franciosi, *Phys. Rev. B* **41** (1990) 1100.
[450] A. Santoni, L. Sorba, D. S. Shuh, L. J. Terminello, A. Franciosi, and S. Nannarone, *Surf. Sci.* **269/270** (1992) 893.
[451] A. Plesanovas, A. Castellani Tarabini, I. Abbati, S. Kaciulis, G. Paolicelli, L. Pasquali, A. Ruocco, and S. Nannarone, *Surf. Sci.* **307–309** (1994) 890.
[452] F. Manghi, C. M. Bertoni, C. Calandra, and E. Molinari, *J. Vac. Sci. Technol.* **21** (1982) 371.
[453] A. F. Wright, C. Y. Fong, and I. P. Patra, *Surf. Sci.* **244** (1991) 51.
[454] F. Bartels, L. Surkamp, H. J. Clemens, and W. Mönch, *J. Vac. Sci. Technol., B* **1** (1983) 756.
[455] T. U. Kampen, L. Koenders, K. Smit, M. Rueckschloss, and W. Mönch, *Surf. Sci.* **242** (1991) 314.
[456] T. U. Kampen, D. Troost, X.-Y. Hou, L. Koenders, and W. Mönch, *J. Vac. Sci. Technol., B* **9** (1991) 2095.
[457] G. Margaritondo, R. E. Rowe, C. M. Bertoni, C. Calandra, and F. Mangli, *Phys. Rev. B* **20** (1979) 1538; *Phys. Rev. B* **23** (1981) 509.

[458] P. D. Schnell, D. Rieger, A. Bogen, K. Wandelt, and W. Steinmann, *Solid State Commun.* **53** (1985) 205.
[459] F. Stepniak, D. Rioux, and J. H. Weaver, *Phys. Rev. B* **50** (1994) 1929.
[460] J. C. Patrin and J. H. Weaver, *Phys. Rev. B* **48** (1993) 17913.
[461] D. Troost, L. Koenders, L. Y. Fan, and W. Mönch, *J. Vac. Sci. Technol., B* **5** (1987) 1119.
[462] J. Pankratz, H. Nienhaus, and W. Mönch, *Surf. Sci.* **307–309** (1994) 211.
[463] D. Troost, H. J. Clemens, L. Koenders, and W. Mönch, *Surf. Sci.* **286** (1993) 97.
[464] A. B. McLean, L. J. Terminello, and F. R. McFeely, *Phys. Rev. B* **40** (1989) 11778.
[465] C. Ju, Y. Chen, T. R. Ohno, and J. H. Weaver, *Phys. Rev. B* **46** (1992) 10197.
[466] D. Troost, L. Koenders, and W. Mönch, *Appl. Surf. Sci.* **65/66** (1993) 619.
[467] M. Büchel and H. Lüth, *Surf. Sci.* **87** (1979) 285.
[468] M. Liehr and H. Lüth, *J. Vac. Sci. Technol.* **16** (1979) 1200.
[469] D. Mokwa, D. Kohl, and G. Heiland, *Surf. Sci.* **139** (1984) 98.
[470] C. Webb and M. Lichtensteiner, *J. Vac. Sci. Technol.* **21** (1982) 659.
[471] K. D. Childs, W.-A.- Lup, and M. G. Lagally, *J. Vac. Sci. Technol., A* **2** (1984) 593.
[472] C. R. Carlson, W.F. Buechter, F. Che-Ibrahim, and E. G. Seebauer, *J. Chem. Phys.* **99** (1993) 7190.
[473] C. W. Wilmsen, *J. Vac. Sci. Technol.* **19** (1981) 279.
[474] C. W. Wilmsen, in *Physics and Chemistry of III–V Compound Semiconductor Interfaces*, Ed. by C. W. Wilmsen, Plenum Press, New York, 1985.
[475] T. Ishikawa and H. Ikoma, *Jpn. J. Appl. Phys.* **31** (1992) 3981.
[476] G. Hollinger, R. Skheyta-Kabbani, and M. Gendry, *Phys. Rev. B* **49** (1994) 11159.
[477] N. T. Barrett, G. N. Greaves, S. Pizzini, and K. J. Roberts, *Surf. Sci.* **227** (1990) 337.
[478] P. Moriarty, B. Murphy, L. Roberts, A. A. Cafolla, G. Hughes, L. Koenders, and P. Bailey, *Phys. Rev. B* **50** (1994) 14237.
[479] X. Y. Hou, W. Z. Cai, Z. Q. He, P. H. Hao, Z. S. Li, X. M. Ding, and X. Wang, *Appl. Phys. Lett.* **60** (1992) 2252.
[480] K. W. Frese, Jr., in *Semiconductor Electrodes*, Studies in Physical and Theoretical Chemistry, Vol. 55, Ed. by H. O. Finklea, Elsevier, New York, 1988, pp. 373–410.
[481] R. E. Williams, Gallium Arsenide Processing Techniques, Artech, Dedham, Massachusetts, 1984.
[482] Z. H. Lu, C. Lagarde, E. Sacher, J. F. Currie, and A. Yelon, *J. Vac. Sci. Technol., A* **7** (1989) 646.
[483] Y. Hirota, K. Sugii, and Y. Homma, *J. Electrochem. Soc.* **138** (1991) 799; *Appl. Phys. Lett.* **58** (1991) 2794; *Appl. Phys. Lett.* **59** (1991) 3412.
[484] D. Gräf, M. Grundner, D. Lüdecke, and R. Schulz, *J. Vac. Sci. Technol., A* **8** (1990) 1955.
[485] M. Massies and J. P. Contour, *J. Appl. Phys.* **58** (1985) 806; *Appl. Phys. Lett.* **46** (1985) 1152.
[486] B. J. Tufts, L. G. Casagrande, N. S. Lewis, and F. J. Grunthaner, *Appl. Phys. Lett.* **57** (1990) 1242.
[487] B. J. Tufts, I. L. Abrahams, C. E. Caley, S. R. Lunt, G. M. Miskelly, M. J. Sailor, P. G. Santangelo, N. S. Lewis, A. L. Roe, and K. O. Hodgson, *J. Am. Chem. Soc.* **112** (1990) 5123.
[488] R. P. Vasquez, B. F. Lewis, and F. J. Grunthaner, *J. Vac. Sci. Technol., B* **1** (1983) 791.
[489] J. M. Woodall, P. Oelhafen, T. N. Jackson, J. L. Freeouf, and G. D. Pettit, *J. Vac. Sci. Technol., B* **1** (1983) 795.
[490] K. W. Frese, Jr. and S. R. Morrison, *J. Electrochem. Soc.* **126** (1979) 1235.
[491] F. R. F. Fan and A. J. Bard, *J. Am. Chem. Soc.* **102** (1980) 3677.
[492] P. A. Kohl and A. J. Bard, *J. Electrochem. Soc.* **126** (1979) 59.
[493] G. Nagasubramanium, B. L. Wheeler, and A. J. Bard, *J. Electrochem. Soc.* **130** (1983) 1680.
[494] G. Horowitz, P. Allongue, and H. Cachet, *J. Electrochem. Soc.* **131** (1984) 2563.

[495] P. Allongue and H. Cachet, *Surf. Sci.* **168** (1986) 356.
[496] O. Savadogo, *Can. J. Chem.* **67** (1989) 382.
[497] R. Noufi, D. Tench, and L. F. Warren, *J. Electrochem. Soc.* **127** (1980) 2310.
[498] K. Rajeshwar, M. Kaneko, and. A. Yamada, *J. Electrochem. Soc.* **130** (1983) 38.
[499] G. Horowitz, G. Tourillon, and F. Garnier, *J. Electrochem. Soc.* **131** (1984) 151.
[500] I. L. Abrahams, B. J. Tufts, and N. S. Lewis, *J. Am. Chem. Soc.* **109** (1987) 3742.
[501] B. Ba, H. Cachet, B. Fotouhi, and A. Gorochov, *Semicond. Sci. Technol.* **9** (1994) 1529; H. Cachet, J. Bruneaux, B. Ba, B. Fotouhi, and A. Gorochov, in *Proceedings of the 4th International Conference on the Formation of Semiconductor Interfaces*, Ed. by B. Lengeler, H. Lüth, and W. Mönch, World Scientific, Singapore, 1994.
[502] B. Ba, H. Cachet, B. Fotouhi, and O. Gorochov, *J. Electroanal. Chem.* **351** (1993) 337; *Electrochim. Acta* **37** (1992) 309.
[503] B. Ba, B. Fotouhi, N. Gabouze, O. Gorochov, and H. Cachet, *J. Electroanal. Chem.* **334** (1992) 263.
[504] J. J. Kelly and D. Vanmaekelbergh, to be published.
[505] W. H. Laflere, F. Cardon, and W. P. Gomes, *Surf. Sci.* **44** (1974) 541; W. H. Laflere, R. L. Van Meirhaeghe, and F. Cardon, *Surf. Sci.* **59** (1976) 401.
[506] K. Rajeshwar and T. Mraz, *J. Phys. Chem.* **87** (1983) 742.
[507] P. Allongue, H. Cachet, and G. Horowitz, *J. Electrochem. Soc.* **130** (1983) 2352.
[508] J. J. Kelly and P. H. L. Notten, *J. Electrochem. Soc.* **134** (1987) 444.
[509] R. Schlesinger and P. J. Janietz, *J. Electrochem. Soc.* **139** (1992) 1936.
[510] A. Schellenberger, W. Jaegermann, C. Pettenkofer, and Y. Tomm, *Ionics* **1** (1995) 115.
[511] W. Jaegermann, C. Pettenkofer, A. Schellenberger, C. A. Papageorgopoulos, and M. Kamaratos, *Chem. Phys. Lett.* **221** (1994) 441.
[512] C. Pettenkofer and W. Jaegermann, *Phys. Rev. B* **50** (1994) 8816.
[513] T. Mayer, A. Klein, O. Lang, C. Pettenkofer, and W. Jaegermann, *Surf. Sci.* **269/270** (1992) 909.
[514] T. Mayer and W. Jaegermann, in *Synchrotron Techniques in Interfacial Electrochemistry*, Ed. by C. A. Melendres and A. Tadjeddine, NATO ASI C432, Kluwer, Dordrecht, 1994, pp. 451–468.
[515] D. Schmeisser, F. J. Himpsel, G. Hollinger, B. Reihl, and K. Jacobi, *Phys. Rev. B* **27** (1983) 3279.
[516] W. Jaegermann and D. Schmeißer, *J. Vac. Sci. Technol., A* **5** (1987) 627.
[517] T. Mayer, C. Pettenkofer, and W. Jaegermann, *J. Phys. Chem.*, in press.
[518] H. Tributsch, *J. Electrochem. Soc.* **125** (1978) 1086.
[519] D. Hanemann and H. Tributsch, *Chem. Phys. Lett.* **216** (1993) 81.
[520] W. Jaegermann, D. Schmeißer, J. Lilie, and H. Tributsch, unpublished results.
[521] T. Mayer, Dissertation, TU Berlin, 1993.
[522] T. Mayer, A. Klein, C. Pettenkofer, and W. Jaegermann, unpublished results.
[523] A. Klein, J. Lehmann, C. Pettenkofer, W. Jaegermann, M. Lux-Steiner, and E. Bucher, *Appl. Surf. Sci.* **70/71** (1993) 470.
[524] T. Mayer, C. Pettenkofer, and W. Jaegermann, *J. Phys.: Condens. Matter* **3** (1991) 1.
[525] O. Henrion and W. Jaegermann, submitted.
[526] T. Mayer, C. Pettenkofer, and W. Jaegermann, to be published.
[527] C. P. Kubiak, L. F. Scheemeyer, and M. S. Wrighton, *J. Am. Chem. Soc.* **102** (1980) 6898.
[528] W. Kautek and H. Gerischer, *Ber. Bunsenges. Phys. Chem.* **84** (1980) 645.
[529] L. F. Schneemeyer and M. S. Wrighton, *J. Am. Chem. Soc.* **102** (1980) 6964.
[530] T. Mayer, C. Pettenkofer, and W. Jaegermann, *Surf. Sci. Lett.* **254** (1991) L423.
[531] W. Jaegermann and C. Pettenkofer, *Ber. Bunsenges. Phys. Chem.* **92** (1988) 1354.
[532] T. Mayer, J. Lehmann, C. Pettenkofer, and W. Jaegermann, *Chem. Phys. Lett.* **198** (1992) 621.
[533] P. A. Dowben, M. Grunze, and S. Varma, *Solid State Commun.* **57** (1986) 631.

[534] R. C. Weast, ed., *CRC Handbook of Chemistry and Physics*, 65th ed., CRC Press, Boca Raton, Florida, 1984.
[535] W. Jaegermann and T. Mayer, *Surf. Sci.* **335** (1995) 343.
[536] T. Mayer, C. Pettenkofer, and W. Jaegermann, to be published.
[537] H. P. Bonzel, *Surf. Sci. Rep.* **8** (1987) 43.

2

Photovoltaic and Photoelectrochemical Cells Based on Schottky Barrier Heterojunctions

Waheed A. Badawy

Department of Chemistry, Faculty of Science, University of Kuwait, Safat, 13060 Kuwait.
Permanent address: Department of Chemistry, Faculty of Science, University of Cairo, Giza, Egypt

I. INTRODUCTION

The considerable increase in recent years of interest in solar energy, the main source of all kinds of energy in our universe, and its direct conversion to work has increased the importance of photovoltaic and photoelectrochemical systems. In order to understand the direct conversion of solar energy to work, we may consider the experiment performed in what is called a vacuum bulb radiometer. The device consists of four vanes, each having one side silvered and the other blackened, enclosed in an evacuated glass bulb. A beam of light is capable of exerting a pressure on a surface on which it is incident. The pressure exerted on the silvered surface is almost double that exerted on the blackened one, owing to reflection. When one pair of opposite vanes is oriented toward a beam of light, a vane whose silvered surface is exposed to the beam experiences a greater force than an opposite vane whose blackened surface faces the beam. A net torque produced by these forces causes the vanes to rotate about a perpendicular axis (Fig. 1). The mechanical power developed is equal to the torque multiplied by the angular velocity of the vanes. This experiment illustrates that the energy of solar beams can be converted directly to work without passing through an intermediate step, unlike the familiar process whereby solar energy is used to produce heat, which is then converted to

Modern Aspects of Electrochemistry, *Number 30*, edited by Ralph E. White *et al.* Plenum Press, New York, 1996.

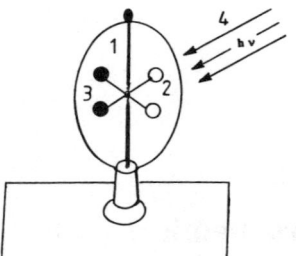

Figure 1. Vacuum bulb radiometer. 1, Evacuated glass bulb; 2, silvered plates; 3, blackened plates; 4, incident solar beam.

work. In this case the direct conversion efficiency is always not limited by the temperature that is produced during the solar energy conversion. The important point in this experiment is that solar energy can be converted directly to work without involving a thermodynamic cycle, and thus no heat reservoirs will be required.[1]

A number of approaches to the direct conversion of solar energy have been investigated.[2,3] One important process is the direct conversion of solar energy to chemical energy as happens in nature. Green plants absorb a very limited fraction of the solar spectrum to convert water and carbon dioxide to carbohydrates.[4] This is a very slow process, and many trials have been carried out to simulate it. There have been many investigations involving photoelectrolytic cells in which the absorbed solar energy is used in an electrolytic nonspontaneous reaction[6] to produce valuable chemicals, either in an anodic process (e.g., the production of Cl_2 from brine solutions) or a cathodic process (e.g., the production of hydrogen as fuel).[5-11]

The most investigated process in the last decade has been the direct conversion of solar energy to electricity using photovoltaics.[12] A photovoltaic is a device that generates voltage when it is subjected to sunlight. Most of these devices are $p–n$ junction devices that are made of silicon and are capable of producing 0.5 V per cell with a typical efficiency of 10–12%.[13] The high manufacturing costs of these devices and the technological and theoretical complications associated with these devices are the underlying reasons for their limited use as solar energy converters. In all solar energy converters, semiconductors play an important role in the energy absorption and conversion process, and therefore a brief discussion of semiconductors will be presented in the next section.

II. INTRINSIC AND EXTRINSIC SEMICONDUCTORS

For any pure crystalline solid, the outer energy levels broaden into energy bands containing many closely spaced states. States within these bands are accessible to electrons, whereas energy levels between these bands are forbidden. Near the absolute zero of temperature, the electrons tend to occupy the lowest energy states, and each material reaches only a certain energy level.[14]

Depending on the distribution of electrons in these energy levels near absolute zero, materials are classified as conductors or insulators. If the outermost band is completely filled with electrons, the material is said to be an insulator. The completely filled band is the valence band, and the following empty band is the conduction band. If the crystalline solid has a sufficient number of electrons to partially fill the conduction band, these electrons are easily excitable in the energy states of this band and acquire the necessary drift velocity for conduction. In this case the solid behaves as a metallic conductor. When the temperature of the insulator solid is raised to the normal working temperature of 25°C, electrons near the top of the valence band are excited into the conduction band, and a small number of electrons will acquire drift velocity and hence exhibit a small conductivity. This occurs for materials that have a small energy gap between the valence band and conduction band; such materials are termed semiconductors (cf. Fig. 2). If the material is highly pure, it is termed an intrinsic semiconductor, and its conductivity is dependent only on the temperature.[14] The conductivity increases drastically with increasing temperature of the solid. The carrier density of an intrinsic semiconductor is given by

$$n_i = p_i = A\ T^{3/2}\ e^{-E_g/2kT} \qquad (1)$$

where n_i and p_i are the number of excited electrons in the conduction band and the number of corresponding positive holes in the valence band, respectively, A is an empirical parameter of the material, k is the Boltzmann constant, and E_g is the energy gap. The value of A is typically $\sim 10^{16}$ cm^{-3} K$^{-3/2}$, E_g is 1.1 eV for silicon, and k is equal to 8.625×10^{-5} eV/K.

The conductivity of an intrinsic semiconductor is zero at absolute zero [cf. Eq. (1)] and the material behaves as a perfect insulator. This conductivity can be artificially made independent of temperature by doping and transforming the intrinsic semiconductor to an extrinsic or doped material.

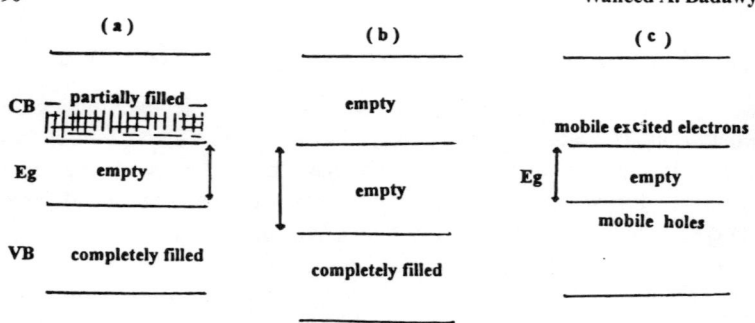

Figure 2. Band structure of a metallic conductor (a), an insulator (b), and a semiconductor (c) at room temperature (~298 K). CB, Conduction band; VB, valence band; Eg, energy gap (band gap).

In this context, doping means the addition of small amounts of foreign atoms of higher or lower valency to the intrinsic material, e.g., the addition of phosphorus with its +5 valence to silicon with its +4 valence. Such addition will create an allowed energy level at the top of the forbidden region, i.e., just below the conduction band, and the fifth electron of the phosphorus will fill this level. These electrons are always excited in the conduction band, and a large number of electrons will contribute to the electrical conductivity. The electron carrier density will be supplied mainly by the doped atoms (phosphorus in this example), which are termed donor atoms, and hence

$$n \approx N_d \quad (n\text{-type semiconductor}) \qquad (2)$$

where N_d is the donor atom concentration. The material is said to be an n-type material. If the intrinsic material is doped with foreign atoms with lower valency, e.g., if silicon is doped with arsenic atoms whose valence is +3, the allowed energy level will be created just above the valence band. This level will contribute to the conductivity by a number of vacancies equivalent to the number of dopant atoms, which are termed acceptors. The large number of holes left in the valence band are the majority carriers, and the material is designated p-type semiconductor. The majority carrier concentration (p) is given by

$$p \approx N_a \quad (p\text{-type semiconductor}) \qquad (3)$$

where N_a is the acceptor (dopant) atom concentration.

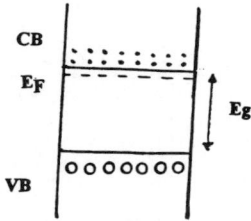

Figure 3. Band structure of n-type semiconductor. CB, Conduction band; VB, valence band; ● mobile electrons in the conduction band (majority carriers); ○ mobile holes in the valence band (minority carriers); - - - E_F, Fermi level, vacant donor level; E_g, band-gap energy.

To summarize, an n-type semiconductor consists of a large number of temperature-independent free electrons and a small temperature-dependent concentration of free holes. In a p-type semiconductor the reverse occurs; i.e., the semiconductor consists of a large number of temperature-independent free holes and a small temperature-dependent concentration of free electrons. In both materials there is a large number of immobile ions (donor and acceptor ions, respectively). The band structure and distributions of charge carriers and ions in an n-type semiconductor is presented in Fig. 3.

III. THE PHOTOVOLTAIC JUNCTION

In most applications, the junction photovoltaic consists of an isolated p–n junction fabricated of a wafer of p-type semiconductor (mainly silicon) on the top of which a thin film of n-type material has been deposited. The wafer is the absorber base, and the film is called the surface layer. Electrodes are affixed to the outer surfaces of the device. The front contact consists of an extremely thin metallic grid of transparent nature that allows the solar spectrum to reach the base material with minimum losses (Fig. 4). When the solar spectrum reaches the absorber, some photons create hole-electron pairs, which generate photocurrent that flows from the n-type to the p-type material. The photocurrent is proportional to the rate of creation of hole–electron pairs per unit area n', i.e.,

$$J_{\text{ph}} = e\, n' \qquad (4)$$

If the device is electrically isolated, there is no steady-state current, and hence the junction current (Jj) must be equal to the photocurrent (J_{ph}):

$$J_{\text{ph}} = J_j = J_0\, (e^{V_{\infty}/kT} - 1) \qquad (5)$$

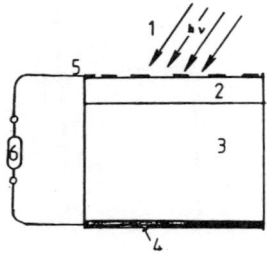

Figure 4. Typical p–n photovoltaic. 1, Incident solar spectrum, $h\nu$; 2, p-type surface film; 3, n-type base material (absorber); 4, back contact layer (n-electrode); 5, p-type front contact (p-electrode, metal grid); 6, load.

where V_{oc} is the open-circuit voltage of the device, and J_0 is the reverse saturation current or the dark current, which is the current created by minority carriers that are created at the junction as hole–electron pairs by thermal excitation,

$$J_0 = DT^3 \, e^{-E_g/kT} \tag{6}$$

where D is a characteristic constant of the junction. The open-circuit voltage (V_{oc}) that appears across the terminals of the device is an important factor in determining the efficiency of the junction. It is given by

$$V_{oc} = kT \ln[(J_{ph}/J_0) + 1] \tag{7}$$

It is clear that the value of V_{oc} is dependent on the temperature and both the photocurrent and the reverse saturation current.

During operation through any load placed across the terminals of the photovoltaic device, the current passed through the load will be given by

$$J = J_{ph} - J_j \tag{8}$$

where J_j represents the fraction of the photocurrent shunted from the diode; i.e.,

$$J = J_{ph} - J_0 \, (e^{V/kT} - 1) \tag{9}$$

where V is the working voltage. Thus, the working voltage is given by

$$V = kT \ln\{[(J_{ph} - J)/J_0] + 1\} \tag{10}$$

from which it is clear that the output or terminal voltage increases as the photocurrent increases but decreases when a current is driven off from the junction. The power characteristic (P) of a photovoltaic device at any point of operation is given by

$$P = V \times J \quad (11)$$

It vanishes under both open-circuit ($J_{oc} = 0$) and short-circuit ($V_{sc} = 0$) conditions. Its maximum is reached at certain intermediate values of current and voltage, termed the maximum power point. The values of the current and voltage at the maximum power point are designated as J_m and V_m, respectively.

The values of J_m and V_m are very important in obtaining the efficiency of the device. The solar conversion efficiency of a photovoltaic device is given by the ratio between the output usable electrical energy ($J_m \cdot V_m$) and the efficient absorbed input solar energy ($V_{oc} \cdot J_{ph}$), i.e.,

$$\eta = J_m \cdot V_m / V_{oc} \cdot J_{sc} \quad (12)$$

It differs from the maximum theoretical efficiency obtained from a certain device, which is given by

$$\eta_{max} = V_{oc} \cdot J_{sc} / F \quad (13)$$

where F is the incident solar flux in the wavelength range of the solar spectrum and is related to the spectral responsivity of the photocurrent. In the case of a silicon photovoltaic, the average conversion efficiency ranges between 10 and 14% whereas the calculated maximum theoretical efficiency of silicon reaches 24%.[13] The maximum theoretical efficiency, which represents the absolute upper limit of efficiency, is highly dependent on the energy gap of the semiconductor. It is low for materials of either very high or very low band-gap energies. The highest theoretical limit occurs for materials whose band gap is in the range of 1.5 eV.

IV. PHOTOELECTROCHEMICAL DEVICES

Although the basis of energy conversion in photoelectrochemical devices may be considered to be the same as in photovoltaic devices, photoelectrochemical devices are distinguished by the use of a semiconductor/electrolyte interface which replaces the solid state junction of a photovoltaic device. In that sense, photoelectrochemical cells have been treated in the same way as photovoltaic devices.[5-11,15-24] The development of the field of photoelectrochemistry and especially semiconductor electrochemistry and its application to photovoltaic systems has been described extensively from a historical perspective in the literature.[24–37] The concept of semiconductor liquid junctions has its origins in the work of Brattain and Garrett,[38] followed by the studies of Williams[39] and the proposal of a solar cell based on a semiconductor/redox couple junction by Gerischer.[40] The work of Fujishima and Honda,[6] in which a photoanode was shown to assist in the photoelectrolysis of water, attracted many groups in the world to follow up this subject. The intensive work in this area has led to photoelectrochemical devices of greater than 10% solar conversion efficiency,[40] and the understanding of the theoretical aspects of these systems has increased the number of applications based on photoelectrochemical processes.[27,33,42]

The semiconductor–liquid-junction solar cell may be viewed as a battery that operates on light illuminating its photosensitive electrode (the semiconductor electrode) and consumes only photons coming from the solar spectrum.[38,39] No fuels or cell materials are consumed during cell operation.

In order to describe the interface at a semiconductor/electrolyte junction, it is essential to understand the distribution of the electronic energy levels that govern the charge transfer and current flow at the electrode/electrolyte interface and to compare them with the energy levels of a semiconductor/metal junction. The simplest way of representing these energy levels can be exemplified by a Schottky barrier model, as presented in Fig. 5. In semiconductor/metal junctions, the relative positions of the Fermi levels before contact determine the nature of the contact. The potential difference extends very deep into the bulk of the semiconductor (cf. Fig. 5a). This is because of the limited number of free carriers in the semiconductor by comparison with the number in the metal.[43] In the case of a semiconductor/electrolyte interface, the solid-state Fermi statistics are not applicable to the electrolyte and the energy level of the electrolyte, V_{redox}, should be considered (cf. Fig. 5b).

In electrochemical systems the energy level of an electrolyte is represented by the redox potential of the solution, which can be measured easily using an inert metal electrode (e.g., Pt) immersed in the electrolyte containing the redox couple and a reference electrode, usually a saturated calomel electrode (SCE) or a silver/silver chloride reference electrode.[44] When electronic equilibrium is attained, the measured potential, E, between the two electrodes will be given by the Galvani potential difference[44]; i.e.,

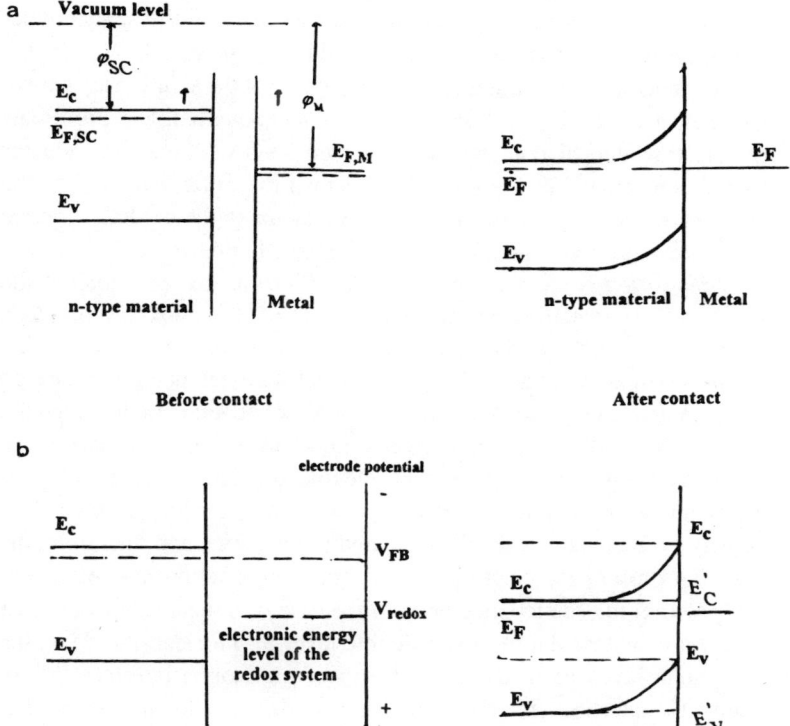

Figure 5. Energy diagrams for n-type semiconductor/metal (a) and n-type semiconductor/electrolyte (b) before and after contact. E_c and E_v are the energy levels of the conduction and valence band, respectively, when all the potential drop occurs within the semiconductor. E'_C and E'_V are the corresponding energy levels when all the potential drop occurs within the electrolyte.

$$eE = e\,(\phi_{M_{ref}} - \phi_{M_{redox}}) \tag{14}$$

where $\phi_{M_{ref}}$ and $\phi_{M_{redox}}$ are the Galvani potentials of the reference electrode and the inert electrode, respectively, in the redox system, and e is the electronic charge. E represents the position of the energy level of the redox electrolyte with respect to that of a reference electrode. It is conventionally referred to the potential of the standard hydrogen electrode (SHE) as the arbitrary zero of potential. There is no need to determine the absolute energy level of the redox couple in order to describe the semiconductor/electrolyte interface as long as the flatband potential, V_{FB}, of the semiconductor electrode and the redox potential of the solution are known with respect to a certain reference. The flatband potential, V_{FB}, represents the electrochemical potential or the Fermi level of the semiconductor before contact. If V_{FB} is more negative than the redox potential of the solution, electrons flow from the semiconductor to the oxidized species of the redox system, and a potential difference builds up at the interface in order to attain an equilibrium. At equilibrium, the potential of the semiconductor electrode is the same as that of the redox system. At a metal/metal interface, the excess charge builds up just within a few angstroms of the interface, whereas at a metal/semiconductor interface the charge extends very deep into the bulk of the semiconductor, in which the whole potential drop occurs, if no electronic states exist at its surface. In comparison with this picture, the potential distribution at the electrolyte/semiconductor interface is complicated. The electrolyte has a low density of free charge carriers compared to the metal, and, therefore, the potential drop of the electrolyte/semiconductor junction may occur entirely within the semiconductor as in the case of a metal/semiconductor junction or within the electrical double layer in the electrolyte, i.e., the Helmholtz layer, as in the case of a metal/electrolyte interface.[37,46] In reality, the question of whether the potential drop occurs in the semiconductor or the double layer in the solution depends on the carrier density, dielectric constant, and concentration of surface states of the semiconductor and, of course, on both the dielectric constant and concentration of the redox solution.[44-47]

In their book *Surface Electrochemistry*,[11] Bockris and Khan give a detailed description of photoelectrochemical processes and discuss the determination of the flatband potential of a semiconductor, the concept of surface states and their determination, potential scales, distribution of electronic states in solution, potential dependence of electronic states at the interface, and the theories of photocurrent. These authors describe precisely the effect of surface states on the photoelectrochemical kinetics at different electrodes. They give examples of photoelectrocatalysis and photoelectrolysis, and discuss models for photoelectrochemical kinetics and the application of these models to photoelectrochemical and photovoltaic cells.

V. SCHOTTKY BARRIER SOLAR CELLS (SBSC)

The Schottky barrier type of solar cells is currently being used in technical devices for conversion of solar energy to electrical energy. Such electronic devices are also used for optoelectronic applications.[31] As in the case of p–n junctions (Fig. 4), the cell consists of a bulk semiconductor, as the base or absorber, covered with a thin semitransparent film of a metal or an oxide (photovoltaic cells) or in contact with a redox electrolyte (photoelectrochemical SBSC).

The difference between the work functions of the metal (or oxide layer) and the semiconductor causes a depletion region, and an electric field exists across the interface and extends considerably into the semiconductor owing to the relatively low carrier density in the semiconductor compared to that in the metal, the oxide layer, or even the electrolyte. In the case of a semiconductor/oxide junction, the oxide film should be highly conducting in order to ensure the formation of the depletion layer. Most usable metal oxides are semiconductors, which should be highly doped so as to reach the properties of degenerate semiconductors and approach metallic conductivity. In order to obtain a suitable SBSC, the electrolyte also should be highly conducting; this criterion is easily met with most aqueous solutions and also with some nonaqueous electrolytes. The oxide film or the electrolyte should be highly transparent to ensure that the whole solar spectrum is transmitted to the absorber layer.

The absorption of photons by the base material will create electron–hole pairs. These charges will be separated spontaneously under the action of the junction electric field and will flow to the corresponding contacts of the junction. The separated charges tend to forward-bias the junction, causing the photoelectric current to flow in any external load connected between the terminals of the cell, i.e., the metal/oxide or metal/electrolyte contact electrode and the semiconductor contact electrode. Such a charge separation is presented in Fig. 6. The Schottky barrier height, φ_B, is extremely sensitive to the surface properties (i.e., surface states, surface charges, etc.) of the semiconductor. It bears little relation to the difference between the work function of the surface layer (metal, metal oxide, or even electrolyte) and that of the semiconductor. The separated holes and electrons lead to a decrease of the electric field under open-circuit conditions, and a corresponding photovoltage (V_{oc}) is developed. If the junction is short-circuited, the band bending will not be changed by illumination; the electrons excited into the conduction band will move into the bulk of the

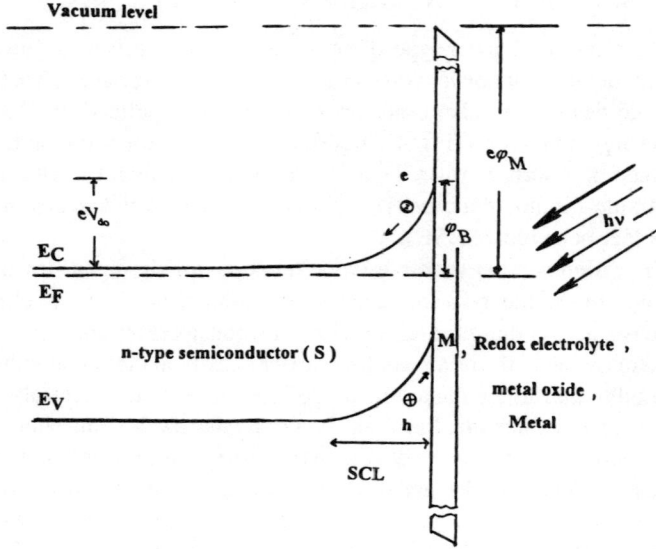

Figure 6. Principle of operation of a Schottky barrier solar cell (SBSC). M, metal or metal oxide with a much greater work function, φ_M, than the semiconductor, S (or an electrolyte with a suitable redox potential); SCL, space-charge layer or depletion layer; φ_B, Schottky barrier height given by the difference between the Fermi level, E_F, and the edge of the conduction band, E_C; E_V, valence band; $h\nu$, incident photons.

n-type material, via the external circuit, toward the surface layer, where they recombine with the holes coming from the valence band, leading to a corresponding photocurrent, J_{ph}, which is dependent on the light intensity. In order to reach a large photovoltage, the semiconductor should be heavily doped so that the Fermi level is close to the conduction band in the bulk in the case of an n-type material or to the valence band in the case of a p-type material. The maximum theoretically obtainable photovoltage is given by the band-gap energy, i.e.,

$$V_{ph,max} = E_g/e \tag{15}$$

The solar conversion efficiency of any threshold device, represented by Eq. (12), can be calculated using an equation of the form[26,48]

$$\eta = \frac{E^* \int_{E_{th}}^{E} A(E)N(E)\, dE}{\int_{0}^{\infty} EN(E)\, dE} \quad (16)$$

where E^* is the energy equivalent to the cell voltage, E_{th} is the optical energy gap of the semiconductor, $A(E)$ represents the absorbance of the semiconductor in a certain energy region E, and $N(E)$ represents the flux density of incident photons in the same region. According to theoretical calculations of maximum efficiency, GaAs represents one of the best candidates.[49] Efficiencies up to 23% have been reported with GaAs single crystals. However, high costs and technological difficulties together with environmental aspects limit the use of this material for terrestrial applications. Absorption of parts of the incident solar spectrum in the surface layer of the Schottky barrier type junction accounts for the main losses that affect the solar conversion efficiency.

1. Current Transport Mechanism in SBSC

The photovoltaic or photoelectrochemical junction in SBSC consists of a Schottky diode as an essential constituent and a source of photocurrent (cf. Fig. 7). The current flow across the depletion layer is normally rectified; i.e., a voltage applied in one direction (the forward direction) will induce a large current whereas in the reverse direction the current saturates at a very low value. The band model of an n-type semiconductor during biasing is compared to that under an open-circuit condition in Fig. 8. The applied voltage, appearing across the space-charge region of the semiconductor, separates the Fermi energies; the situation in the case of a positively and a negatively biased junction is shown in Fig. 8a and 8c, respectively, and may be compared with the equilibrium situation of Fig. 8b. In the dark, where it is assumed that there is no minority carrier injection, the current flow is associated with the flow of the majority carriers, i.e., electrons in the case of an n-type material.[50] The electron current to the surface has to overcome the potential barrier, V_s. This potential barrier is given by

$$V_s = V_B + V_a \quad (17)$$

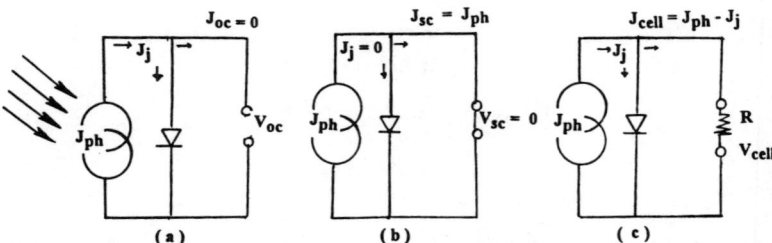

Figure 7. Electronic model of Schottky barrier solar cell. (a) Open-circuit condition; (b) short-circuit condition; (c) operation condition with a load R. ⊖, photocurrent source; ▽, Schottky diode.

where V_B is the Schottky barrier height at equilibrium (Fig. 8b) and V_a is the applied potential. With a negative bias on the semiconductor, the electron current to the surface is very high, and this is the forward direction of the Schottky diode. The current density is given by the thermionic emission as in the case of Fig. 8b[50,51]:

$$J = AT^2 e^{-V_s/kT} \qquad (18)$$

where J is the thermionic emission current and A is a constant.

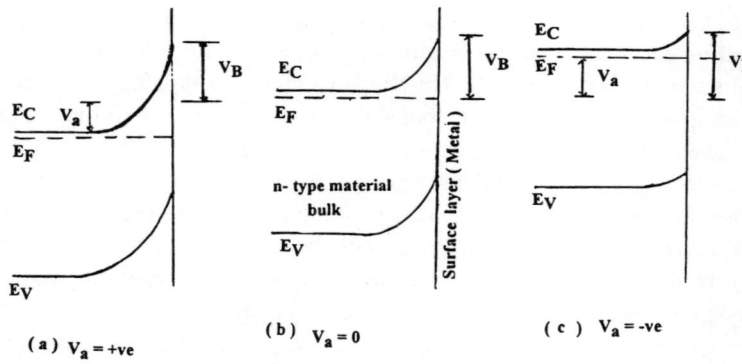

Figure 8. Band configuration of a Schottky barrier diode type junction with an n-type semiconductor under different conditions: (a) positive applied potential ($V_a = +$ve); (b) equilibrium condition ($V_a = 0$); (c) negatively biased junction ($V_a = -$ve).

Injection of electrons from the surface layer into the semiconductor requires them to overcome the potential barrier V_B, i.e., the energy difference between the Fermi energy of the surface layer and the Fermi energy of the semiconductor (the bottom of the conduction band at the surface, whether the junction is biased or not). This barrier is independent of the applied voltage, and therefore the injection current will never be high. In the configuration of Fig. 8a, neither electron current to the surface nor injection from the surface can take place. The junction is said to be reverse-biased since this is the reverse direction of the diode, where the current is very small and independent of the potential. The reverse current under illumination is actually the current of minority carriers produced within a diffusion length of the barrier region. Once a minority carrier reaches this region, it is moved to the surface by the electric field and reaches it. The value of the reverse saturation current is given by an equation identical to Eq. (6). The basis of operation of this Schottky diode is the basis of operation of SBSC. The most important factor is to ensure that the configuration holds. Such a configuration holds when the semiconductor surface is covered by a very thin insulating film that separates the semiconductor from the conducting surface layer (metal, metal oxide, or electrolyte), as in what are termed metal/insulator/semiconductor (MIS or MOS) junctions. In these configurations, as was discussed above, the absorption of photons by the solar cell will create electron–hole pairs, minority carriers produced within a diffusion length of the surface are forced to move to the Schottky barrier, and the majority carriers are forced to the bulk and hence to the semiconductor contact electrode. The ideal band model and solar cell characteristics of such a cell are presented in Fig. 9a and 9b, respectively. As the minority carriers reach the surface, they develop current or voltage or both in the external circuit. The equations for short-circuit conditions, open-circuit conditions, and operation conditions are identical to those given above for the junction photovoltaic [Eqs. (5), (7), and (10)].

2. Role of Surface States

The calculations of photocurrent and the development of a theory of the photocurrent based on the Schottky barrier model have their foundations in the pioneering work of Gaertner.[52] In this work, the photocurrent at an illuminated semiconductor was calculated for a semiconductor/metal interface in contact with an inert gas, neglecting the influence of the

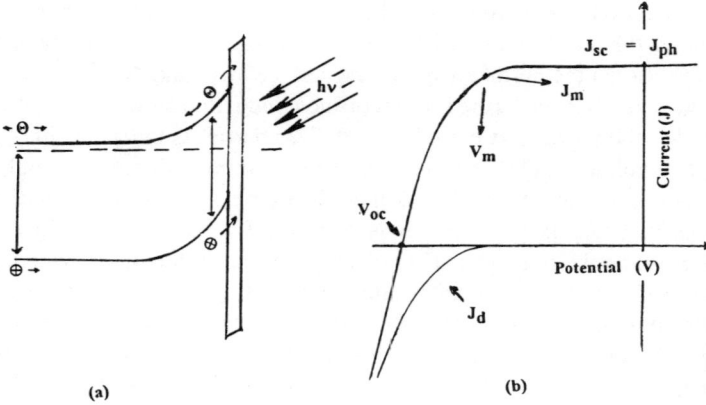

Figure 9. Band model (a) and current–voltage characteristics (b) of an illuminated ideal Schottky barrier solar cell. J_{sc}, short-circuit current or photocurrent (J_{ph}); J_m, maximum-power-point current; V_m, maximum-power-point voltage; V_{oc}, open-circuit voltage; J_d, dark or reverse saturation current.

semiconductor surface and the properties of the electrical double layer at the semiconductor/electrolyte interface. The acceptance of the Schottky barrier model by Gerischer[53] and Bard[54] led to an emphasis on this model in theoretical work on photoelectrochemical systems. Although the Schottky barrier model does apply in many photoelectrochemical systems,[55] it proved essential to consider surface properties (surface states) and their effect on the Helmholtz layer and to recognize the corresponding Helmholtz barrier model in photoelectrochemical kinetics.[37] In the study of photoelectrochemical systems, it is beneficial to use models of photoelectrochemical kinetics that are based on the surface properties or, better yet, the properties of the solution side of the interface, while the Schottky barrier model stresses what is happening inside the semiconductor itself. In this chapter we will focus only on the Schottky barrier type solar cells either in the photovoltaic or in the photoelectrochemical forms. However, it should be emphasized that in dealing with photoelectrochemical reactions and photoelectrochemical kinetics the effects of surface states should be considered. The concept of surface states in semiconductors originated with the work of Bardeen in 1947.[56] In this work, surface states associated with the unpaired valencies (dangling bonds) set up by interruption of the crystal to form a surface were considered. In solutions, such bands are removed because they tend to react with water. A detailed description and discussion of the models and the effects of surface states at a semiconductor/electrolyte interface is given by Bockris and Khan.[11] The subject has also been discussed by Batchelor and Hamnett,[57] who have focused on the aspects of semiconductor behavior that have been identified to originate with surface states and vary with their concentration and have shown how deviations from the ideal model occur. This is a subject that deals with the understanding of surface electrochemistry on a molecular level.

VI. PHOTOELECTROCHEMICAL PROCESSES IN SOLAR ENERGY CONVERSION

In the preceding sections we have treated the solid-state junction as the main part of the energy converter as is always done in the Schottky barrier model. In the metal/insulator (oxide)/semiconductor (MIS) or semiconductor/insulator (oxide)/semiconductor (SIS) cells based on this model, limitations on energy conversion arise from the decrease of the usable voltage, i.e., the junction potential, V_j, by the ohmic drop of the series resistance arising from the resistance, R_s, of the surface collector electrode and the limited conductivity of the surface layer. (Limitations on the percent transmittance of the surface layer also play an important role and will be discussed later.) The usable voltage in this case will be given by

$$V_{cell} = V_j - JR_s \qquad (19)$$

In photoelectrochemical cells, the processes occurring at the solid/solution interface play an important role in the limitation of the cell voltage. When a highly conducting electrolyte is used, as is usually the case, the limitations arise from the overvoltage η accompanying the electrochemical processes occurring at the electrode/electrolyte interface. The cell voltage will then be given by[58,59]

$$V_{cell} = V_j - \eta \qquad (20)$$

Such limitations should lead to the consideration of the processes occurring at the solid/solution interface in a photoelectrochemical cell and the classification of these cells according to the nature of the processes taking place.

1. Regenerative Photoelectrochemical Cells

In regenerative photoelectrochemical cells, solar energy is converted into electrical energy. The energy scheme for an n-type semiconductor/electrolyte interface is very similar to that shown in Figs. 5, 6 and 9, and the nature of the photoeffects is identical. The only difference is that the Fermi level of the metal or oxide surface layer is replaced by the redox potential of the electrolyte, taking into consideration the filled and empty states (cf. Fig. 10.). The value of the redox potential affects the possible photopotential, which depends on the band bending below the semiconductor surface occurring at equilibrium in the dark. Therefore, an important criterion in the selection of the semiconductor is that a large band bending should be obtained. For n-type semiconductors, the selection of a redox electrolyte having a redox potential, corresponding to $E_{F,el}$, that is very close to the valence band of the semiconductor is also important. An appropriate selection can be made by consideration of the energy levels of semiconductors and the redox potentials of electrolytes given in Tables 1 and 2, respectively. Based on the values in these tables, an ideal combination can be made by selecting n-Cd Se as an n-type material and Ce^{4+}/Ce^{3+} to give a good photoelectrochemical device.

The cell operation occurs in the same way as discussed for photovoltaic systems in Section V and illustrated graphically in Fig. 8. Under illumination, the band bending decreases by the amount of photovoltage

Table 1
Energy-Level Positions and Band-Gap Energies of Some Semiconductors at pH1

Semiconductor	E_c (V)	E_v (V)	E_g (eV)
Si (n and p)	−0.50	+0.60	1.10
GaAs (n and p)	−0.40	+1.00	1.40
GaP (n and p)	−1.00	+1.25	2.25
CdSe (n)	−0.05	+1.65	1.70
CdS (n)	−0.45	+2.05	2.50
ZnO (n)	−0.15	+3.10	3.25
WO_3 (n)	+0.25	+3.45	3.20
SnO_2 (n)	+0.40	+4.20	3.80
TiO_2 (n)	−0.05	+3.15	3.20

Table 2
Redox Potentials of Some Redox Systems at pH 1 vs. the Normal Hydrogen Electrode (NHE)

Redox system	$E°$ (V)
$Eu^{2+/3+}$	−0.60
H_2/H^+	−0.06
$[Fe(CN)_6]^{3-/4-}$	+0.40
$Fe^{2+/3+}$	+0.70
$Ce^{3+/4+}$	+1.60

obtained. The value of the photovoltage can reach E_g/e, as is given by Eq. (15).

In our case, which involves the use of an n-type semiconductor, the holes created by photon excitation will be transferred across the interface to the reduced species, Red, of the redox system. The excited electrons in the conduction band reach the oxidized species, Ox, through the counter electrode. The charge-transfer processes can be summarized in the following way:

(a) Anodic process at the n-type surface:

$$Red + h^+ \rightarrow Ox \qquad (21)$$

(b) Cathodic process at the counter electrode:

$$Ox + e^- \rightarrow Red \qquad (22)$$

These reactions proceed in the regenerative cell without any irreversible consumption of the components of the redox system.[43]

As was discussed above, the photocurrent J_{ph} depends linearly on the light intensity whereas the photovoltage is an exponential function of the light intensity.[24,25,60] The most serious problem facing these ideal photoelectrochemical cells, such as n-CdSe/$Ce^{4+/3+}$/counter electrodes, is the dissolution of the semiconductor (anodic dissolution of CdSe). The anodic dissolution reaction of the semiconductor (see Table 3) competes with the anodic process of the redox reaction (Eq. 21). The stabilization of the semiconductor in the solution and the factors affecting its stability represent a major field of investigation. If one considers the thermodynamic

Table 3
Anodic Dissolution Reactions of Some Semiconductors

$Si + 2H_2O + 4h^+ \rightarrow SiO_2 + 4H^+$
$GaAs + 6H_2O + 6h^+ \rightarrow Ga(OH)_3 + H_3AsO_3 + 6H^+$
$GaP + 6H_2O + 6h^+ \rightarrow Ga(OH)_3 + H_3PO_3 + 6H^+$
$CdS + 2HCl + 2h^+ \rightarrow Cd^{2+} + S + 2H^+ + 2Cl^-$
$TiO_2 + 4HCl + 4h^+ \rightarrow TiCl_4 + O_2 + 4H^+$
$SnO_2 + 4HCl + 4h^+ \rightarrow SnCl_4 + O_2 + 4H^+$

approach, a semiconductor decomposes in aqueous solutions if its calculated thermodynamic decomposition potential is more negative than the O_2/H_2O potential. Consideration of the thermodynamic decomposition potentials of some semiconductors given in Table 4 indicates that TiO_2 and SnO_2 are the most stable toward photodecomposition in aqueous solutions.[53,61,62] If we consider the large band-gap energy of these materials (>3 eV), then we cannot use them as absorbers of solar energy. The oxidized species (Ox) of a redox system oxidizes the semiconductor also if the redox potential V_{redox} is more positive than the thermodynamic decomposition potential of the semiconductor (V_{decomp}).[53] These considerations limit the use of familiar semiconductors like GaAs or Si as effective absorber electrodes in photoelectrochemical cells. Therefore, it is important to prevent the semiconductor from direct contact with the electrolyte (or even with the environment in the case of photovoltaic cells). Transparent and semitransparent metal oxide films can help to a great extent in this respect.[63,64]

Table 4
Thermodynamic Decomposition Potentials of Some Semiconductors in Aqueous Solutions

Semiconductor	E_g (eV)	$E°_{decomp}$
GaAs	1.40	−0.38
GaP	2.25	−0.71
CdS	2.50	−0.09
TiO_2	3.20	+1.40
SnO_2	3.80	+1.05

Calculations based only on thermodynamic data are of limited value. Kinetic limitations and the relative positions of the energy states on both sides of the interface may change the picture completely. It has been shown, by some ring-disk experiments,[65] that the redox process competes with the anodic dissolution when energy states of the redox system are not located near the valence band but above it. This implies that the energy levels at the semiconductor surface are involved. Electrons are first transferred from the redox system into these states and recombine with the photo-created holes. These energy levels, or surface states, may occur as intermediate states in the dissolution process. Holes move toward the surface and cause the breaking of a surface bond. When another hole is trapped, the next bond is destroyed and dissolution occurs. This step can be prevented by the transfer of an electron from the redox system to the surface group after only one bond has been broken. In this way, the surface bonds are restored, and stabilization of the semiconductor takes place.

The photoelectrochemical regenerative cells have some advantages over solid-state devices. These include the ease of preparation of the junction, minimum losses due to absorption or reflection by the surface layers because of the use of highly conducting and transparent electrolytes, and also the availability of many n-type semiconductors that are suitable for use in these types of cells. A serious problem is the instability of the photoanodes, but this can be solved to large extent.[22] Photoelectrochemical regenerative cells are essentially used for the conversion of solar energy into electricity as in the case of photovoltaic cells. The limitations on semiconductor stability and the competition of the photodecomposition reaction with the reaction of the redox species at the electrode/electrolyte interface lead to smaller photovoltage, which leads in turn to lower conversion efficiencies. There have also been attempts to store the produced electrical energy in the cells.[66]

2. Photoelectrolytic Cells

The most important application of photoelectrochemical systems is their use for the direct conversion of solar energy into chemical energy. One of the most important electrolysis reactions is the production of hydrogen as a clean fuel from the photoelectrolysis of water. In this respect, the stability of the absorber and the performance of the electrode during the charge-transfer reaction represent the major problems.

For the electrolysis of water, the cell consists of a suitable combination of a semiconductor/electrolyte electrode and a metal counter electrode or two semiconductor/electrolyte electrodes. The electrolyte is an aqueous solution of an inert but highly conducting salt. The principle of operation of an n-type semiconductor/electrolyte/metal counter electrode cell in a naturally aerated solution is presented in Fig. 10. The electrolyte for the water electrolysis can be represented by two redox potentials; the first corresponds to the system H_2/H_2O, and the second corresponds to the system O_2/H_2O. In the case of naturally aerated or oxygen-saturated electrolytes, the Fermi level will be close to $E^\circ_{O_2/H_2O}$ and to the valence band at the surface as in the case shown in Fig. 10. Under illumination and short-circuit conditions, the photo-created holes move toward the surface and recombine with electrons transferred from occupied levels of the O_2/H_2O system, leading to anodic oxygen evolution. The excited electrons in the conduction band move through the external circuit and reach the counter electrode, where they react with water, leading to the cathodic evolution of hydrogen. This process does not occur at or near equilibrium because the occupied levels below the Fermi level of the counter electrode are far below $E^\circ_{H_2/H_2O}$, and, therefore, at low light intensities no electron transfer occurs although the system is under short-circuit conditions. In

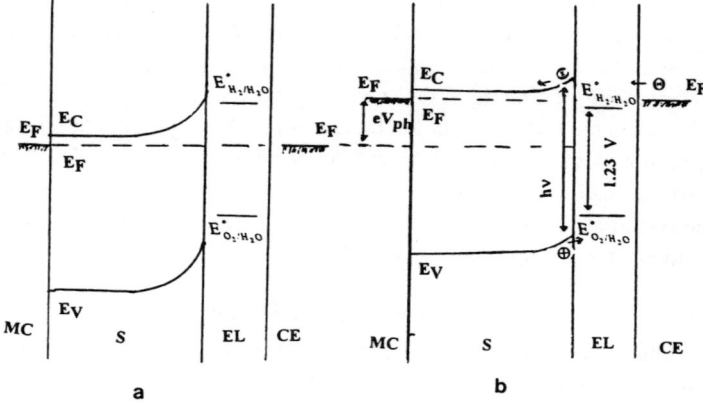

Figure 10. Energy diagram of photoelectrolytic cell at equilibrium (a) and under a short-circuit condition (b) in a naturally aerated aqueous solution containing inert ionic species. S, n-Type semiconductor (absorber); MC, metallic contact; CE, counter electrode; EL, aqueous electrolyte; V_{ph}, photovoltage.

Schottky Barrier Photoanodes

this case, the band bending is decreased, and a photovoltage V_{ph} is developed. The photovoltage increases with increasing light intensity until the Fermi level of the counter electrode is moved above the $E^\circ_{H_2/H_2O}$ level so that the electron-transfer process can occur, leading to the hydrogen evolution reaction at the counter electrode and, of course, the oxygen evolution reaction at the semiconductor electrode. The difference between this photoelectrolysis cell and a regenerative photoelectrochemical cell is that in the regenerative cell no photovoltage occurs under short-circuit conditions, because the Fermi level of the counter electrode is fixed by the Fermi level of the redox system.

In the above described photoelectrolytic cell, the Fermi level can only be moved sufficiently upward if the conduction band at the surface occurs above $E^\circ_{H_2/H_2O}$. If this condition is not fulfilled, the Fermi level can never come above the $E^\circ_{H_2/H_2O}$ level, and hence no electron transfer can occur. In this case an external applied voltage (V_{appl}) can be used to shift the cathode potential above $E^\circ_{H_2/H_2O}$ to make the electron-transfer reaction possible, leading to the cathodic hydrogen evolution.[67] As can be seen in Fig. 11, there is still an energy conversion provided that $eV_{appl} < 1.23$ eV ($E^\circ_{O_2/H_2O} - E^\circ_{H_2/H_2O} = 1.23$ eV). The oxygen level ($E^\circ_{O_2/H_2O}$) should be above the valence band at the interface to permit electron transfer to that

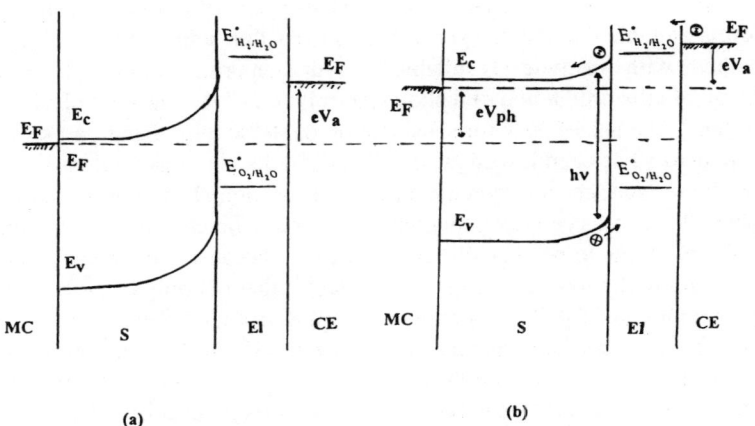

Figure 11. Energy diagram of photoelectrolytic cell under applied potential at equilibrium (a) and under a short-circuit condition (b) in a naturally aerated aqueous solution containing inert ionic species. S, n-Type semiconductor; MC, metallic contact; EL, aqueous electrolyte; CE, counter electrode; V_{ph}, photovoltage; V_a, applied voltage.

band. Thus, the following are the proper conditions for operation of a photoelectrolytic cell: the energy gap of the semiconductor should be greater than the energy of the O_2–H_2 cell, i.e., $E_g > 1.23$ eV; the valence-band energy should be above the energy of the O_2/H_2O system, i.e., $E_V > E°_{O_2/H_2O}$; and the conduction-band energy should be below the energy of the H_2/H_2O system, i.e., $E_C < E°_{H_2/H_2O}$. Consideration of the different materials and the energy of the H_2O/H_2 system shows that only a few materials meet the requirement of having their conduction-band energy above $E°_{H_2/H_2O}$, such as, $SrTiO_3$[68] and $KTaO_3$.[69,70] These materials have a very large band gap ($E_g > 3.2$ eV) and do not absorb in the visible region of the solar spectrum but absorb only in the UV region. TiO_2 was thought to be a suitable material[6] and was subjected to intensive investigations.[71-75] This material can only work at extremely high light intensities[76]; it has also an energy gap of 3.2 eV, and its absorption is mainly in the UV region.[64,77,78] Many other materials, especially semiconducting oxides like SnO_2, WO_3, and Fe_2O_3, have also been investigated.[63,64,79-85] Other materials investigated as absorbers for photoelectrolytic devices include some transition-metal chalcogenides such as MoS_2, which has a band gap of 1.75 eV.[86-89] Unfortunately, all compounds that fulfill the requirements discussed above are unstable. The main anodic process is the electrochemical dissolution of the semiconductor.

If it is possible, the combination of an n-type photoanode with energy levels suitable for the oxygen evolution reaction and a p-type photocathode with energy levels suitable for hydrogen production may be very helpful in the production of a photoelectrolytic cell for the electrolysis of water.[90] The use of dyes for sensitization of stable large-band-gap semiconductors has been investigated.[62,91] Dye molecules adsorbed on semiconductor surfaces are effective in light absorption. The realization of a photoelectrolysis cell operating with a suitable efficiency and acceptable stability seems to be very difficult to achieve because the technological fabrication of a low-band-gap semiconductor that not only can fulfill the requirements of the photoelectrolysis of water from the energetic point of view but also has the necessary stability against electrolytic dissolution is a very hard task.[92] The problem can be solved if conventional semiconductors like Si or GaAs can be covered with a stable material that fulfills the requirements of transparency and conductivity and at the same time can form a junction that is similar to a semiconductor/metal junction, i.e., a Schottky barrier photoelectrochemical cell. Materials that are suitable for this purpose are transition-metal oxides.

VII. METAL OXIDE FILMS

Metal oxide films represent an important category of materials. They are used extensively in spectroscopic devices for a wide range of applications. They are currently being used in solar cell fabrication. SnO_2 and TiO_2 thin films are the most widely used materials. They have excellent stability against chemical attack and serve as antireflectors. Therefore, they are widely used as protective windows for reactive substrates. The high transmission properties of these materials in the visible and near-infrared regions of the solar spectrum together with their high stability are the reasons for the growing interest in their application as essential parts in the fabrication of solar cells for terrestrial and extra terrestrial applications. The use of these materials in the preparation of stable low-cost solar cells and photoanodes for photoelectrolytic devices has been investigated,[6,73,93-98] and trials are ongoing. In the following section we will examine the properties of SnO_2 and TiO_2 thin films in more detail.

1. Preparation and Electrode Properties of SnO_2

Tin dioxide is the most frequently investigated semiconducting film on glass or quartz and has been used as an indicator electrode in various electrochemical applications.[99-102] The high conductivity and good transparency of SnO_2 on glass have led to its use as an electrode for simultaneous electrochemical and optical studies by internal reflection spectroscopy.[103] The electrical and optical bulk properties of SnO_2 have been investigated in detail.[104] The charge-transfer process at the SnO_2/electrolyte interface has been investigated by studying the potential distribution at the interface.[79,105-107] The charge transfer between the redox system and the n-type SnO_2 electrodes was found to occur mainly across the conduction band. Interfacial capacity measurements have shown that variations of the applied voltage appear across the space-charge region of the semiconducting SnO_2, i.e., the band bending of the material changes with any potential change whereas the potential across the Helmholtz layer remains unchanged. Electron transfer from a redox system into the conduction band of the material would not be influenced by an applied anodic potential, and electron tunneling though the space-charge region is important in the electrode process.

SnO_2 films has been used to prepare the well-known n-Si/SnO_2 photoanode, which has been extensively studied.[108-112] The heterojunctions produced by depositing such films onto n-Si were equivalent to the best junctions investigated.[109] The photocurrent–voltage characteristics of the system n-Si/SnO$_2$/metal or n-Si/SnO$_2$/electrolyte are consistent with a Schottky-barrier-like model, which allows the calculation of the current–voltage curves for such systems under different conditions of illumination and electrochemical charge transfer.[58,111]

Several methods have been used for the preparation of SnO$_2$ thin films. Among these, the spray–pyrolysis technique[85,111] has a particular advantage in that compounds that have no significant vapor pressure can be included in the spray and thus incorporation of foreign atoms in the SnO$_2$ matrix can be conveniently achieved. Polycrystalline SnO$_2$ films containing up to 10% Sb or Ru oxides have been prepared by this technique, and their electrode properties investigated.[113] The distribution of electronic states in the band gap of the semiconducting material was determined by ultraviolet photoelectron spectroscopy (UPS). The electrode kinetics of the redox systems $Fe^{2+/3+}$ and $Ce^{3+/4+}$ for the Sb- and Ru-containing SnO$_2$ films were correlated, and a model for the charge-transfer mechanism was suggested. According to this model, electrons can pass the narrow semiconductor space-charge barrier by mediation of the deep donor states in the band gap that are introduced by the incorporated atoms.

The spraying system (Fig. 12) consists of an electrically heated stainless-steel oven containing a very narrow channel through its center, which can be connected to a vacuum line, to hold the sample[114] and an atomizer containing a nozzle mounted in a 250-ml glass round-bottom flask containing the spraying solution, which consists mainly of 0.7M SnCl$_4$ in ethyl acetate. The water necessary for the hydrolysis reaction was supplied as vapor saturated in an inert gas or air. Based on the high-temperature hydrolysis of SnCl$_4$,[115] very pure, homogeneous, and reproducible SnO$_2$ films can be obtained according to

$$SnCl_{4(g)} + 2H_2O_{(g)} \rightarrow SnO_{2(s)} + 4HCl_{(g)} \qquad (23)$$

Solutions of SnCl$_4$ in ethyl acetate gave clear glasslike films, whereas other solvents such as water, water–HCl, or ethanol yielded nonhomogeneous films that were cloudy in appearance.[111,116] The thickness of the deposited films can be determined either spectrophotometrically[117] or by a tally-step instrument after the SnO$_2$ film has been partially etched off with a Zn/HCl mixture.

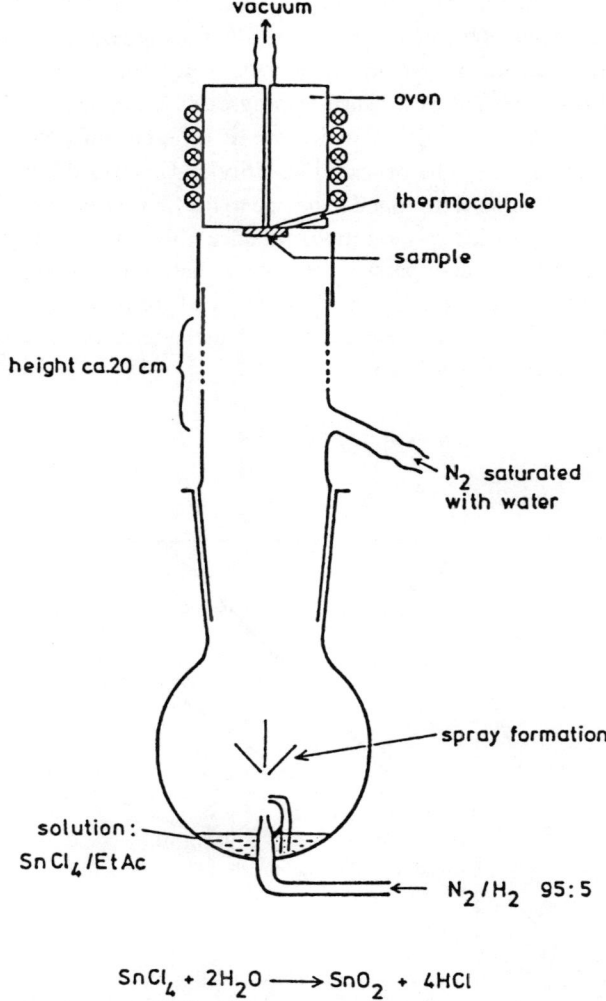

$$SnCl_4 + 2H_2O \longrightarrow SnO_2 + 4HCl$$

Figure 12. Schematic representation of the spray / chemical vapor deposition (CVD) apparatus used to prepare the oxide (SnO_2) films.

2. Homogeneity of the SnO$_2$ Films

The films prepared by the spray–pyrolysis technique were found to be homogeneous. The rate of growth of the films at any temperature is strictly independent of the film thickness, which means that the film growth process on the SnO$_2$ surface progresses unchanged with time. Examples of the film growth with time at different temperatures are presented in Fig. 13. The observed stability and reproducibility of the oxide film formation were found to be due to the limited amount of water supplied to the hydrolysis reaction. Identical films were obtained by reducing the partial vapor pressure of water to half of its initial value and doubling the reaction time.[111] The dependence of the film growth rate on the substrate temperature was found to obey the familiar Arrhenuis equation,[118] which can be written as

$$\text{Growth rate} = \frac{d\delta}{dt} = A e^{-G^*/RT} \qquad (24)$$

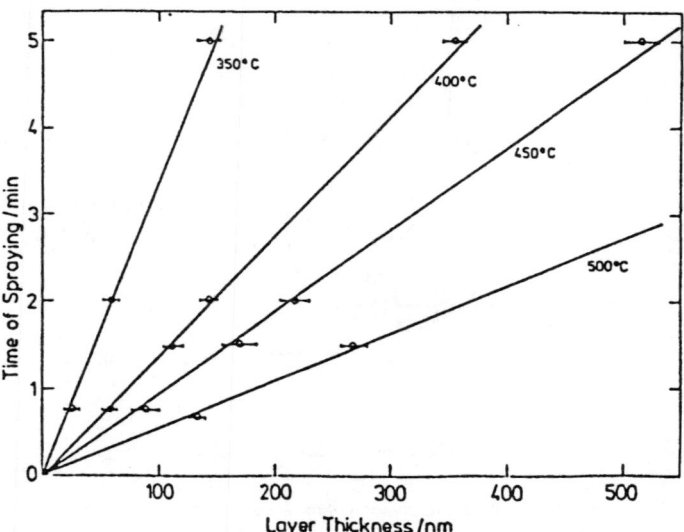

Figure 13. Growth of SnO$_2$ films with time at the indicated substrate temperatures. Spraying mixture: 1.7×10^{-4} mol SnCl$_4$ and 1.7×10^{-4} mol H$_2$O per cubic decimeter of spray. Flow rate: 6 dm^3/min.

where δ is the film thickness, G^* is the activation free energy of film deposition, A is the frequency factor, and R and T have their usual meanings. A value of 18 kJ/mol was calculated for the activation free energy, ΔG^*, from the slope of the linear Arrhenius plots obtained according to Eq. (24).

3. Conductivity of the SnO$_2$ Films

For a given temperature the specific conductance of the SnO$_2$ films prepared by the spray–pyrolysis technique described above is independent of film thickness. Figure 14 illustrates the variation of the film resistance and the calculated specific conductance with the film thickness. The resistance of the film decreases as the thickness increases. The specific conductance of the film is about 150 Ω^{-1} cm^{-1}. The lack of a dependence of this value on the film thickness reflects the high homogeneity of the films prepared by this method. The effect of substrate temperature on the film conductivity for films of the same thickness prepared at different

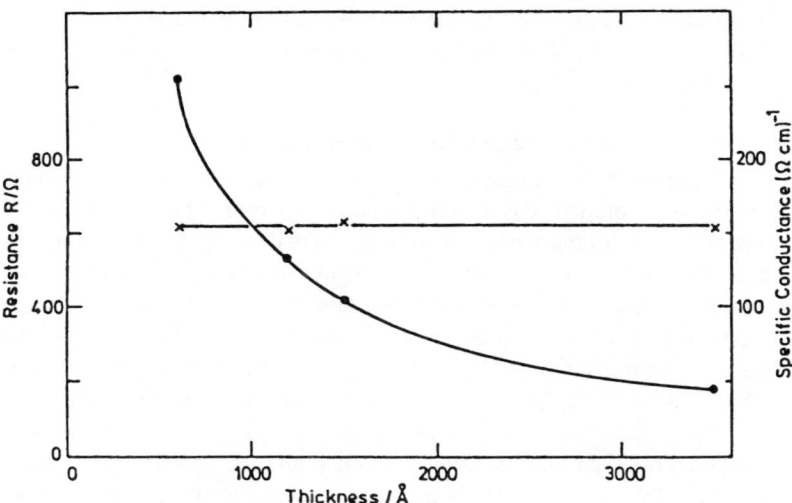

Figure 14. Film resistance measured between two gold contacts deposited on SnO$_2$ film and the calculated specific conductance of the SnO$_2$. Temperature of film formation: 380°C. ●, resistance; X, specific conductance.

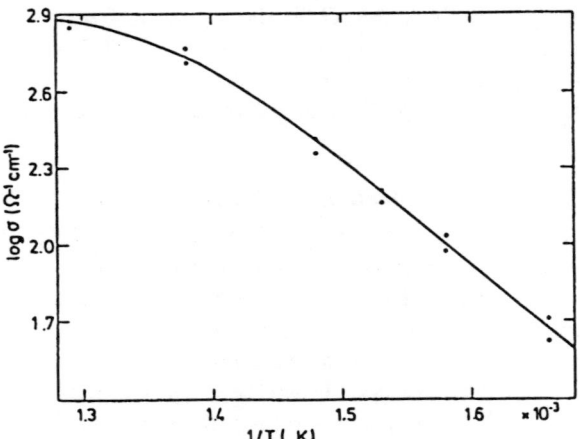

Figure 15. Dependance of the room-temperature conductivity of SnO_2 films on the temperature of film formation.

temperatures is presented in Fig. 15. The reason for the increased room-temperature conductivity of the samples (n-type material) prepared at higher temperatures is the higher concentration and mobility of the charge carrier in these samples.[119]

4. Junction Characteristics of n-Si/SnO$_2$

Transparent SnO_2 films, 90 ± 10 nm in thickness, were deposited on Si wafers to prepare the n-Si/SnO$_2$ heterojunction. The power characteristics of one of these devices under simulated sunlight illumination are presented in Fig. 16. The front contacts of these devices were not optimized to collect the photogenerated current.[122] They consist of two rectangular gold contacts deposited onto the SnO_2 surface in order to measure the film conductivity.[111] This fact is responsible for the relatively low fill factor (0.42) and a cell efficiency of ~5% for the cell presented in Fig. 16. Under 90-mW/cm^2 simulated solar light intensity (AM1), the measured open-circuit potential is 470 mV and the short-circuit current density is 18.6 mA/cm^2. Taking into consideration that the area of the cells investigated was, on average, 0.15–0.20 cm^2 and that the average value of the room-temperature conductivity of SnO_2 on the Si substrates is 160 Ω^{-1} cm^{-1} for a deposition temperature of 380°C, analysis of the system was

Schottky Barrier Photoanodes

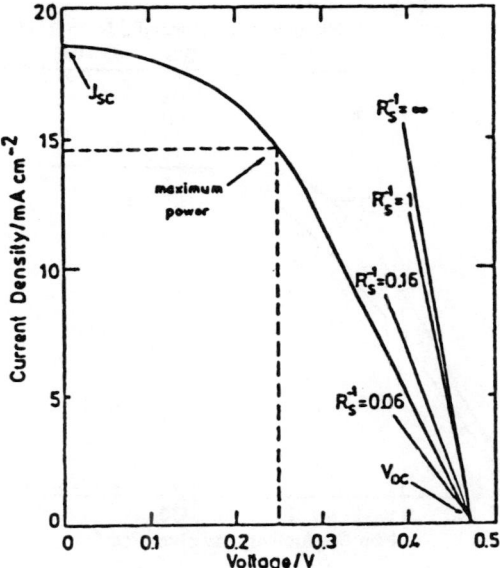

Figure 16. Typical power characteristics of an SnO_2/Si photocell prepared at 380°C. Illumination: Simulated solar spectrum, 90 mW/cm² (AM1). V_{oc} is the open-circuit potential, and J_{sc} is the short-circuit current. The straight lines starting at V_{oc} indicate the slopes calculated with Eq. (27) for selected values of the recorded series resistance (R_s^{-1}, cf. Fig. 17).

carried out by taking the current–voltage relationship of the photovoltaic device as a function of the effective series resistance[120]:

$$J = J_{ph} - J_0(e^{\frac{e}{nkT}(V+JR_s)} - 1) \quad (25)$$

The saturation current density is 10^{-7} A/cm², and n (the diode quality factor) is 1.6. From this equation one obtains

$$\frac{1}{R_s} = J_{SC}\frac{e}{nKT}\left(\ln\frac{J_{ph} - J_{SC} + J_0}{J_0}\right)^{-1} \quad (26)$$

For the same thickness of SnO_2, $1/R_s$ is proportional to the film conductivity. An increase in $1/R_s$ will increase the short-circuit current and also

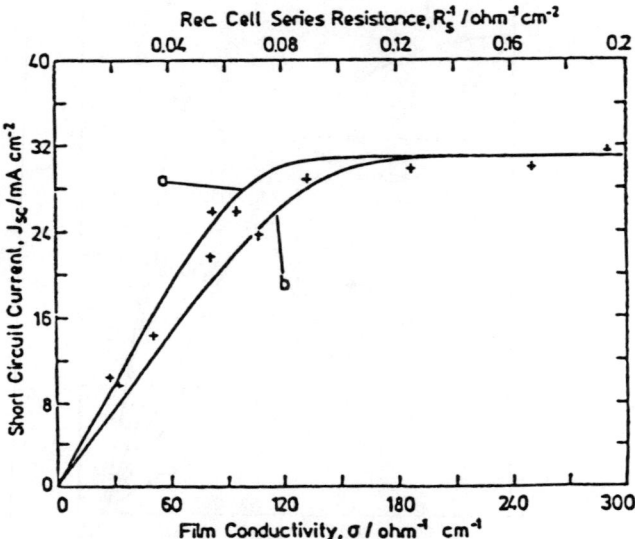

Figure 17. Short-circuit currents, J_{sc}, obtained with SnO_2/Si photocells prepared at different temperatures, plotted versus the determined SnO_2 conductivity (+). The solid lines represent J_{sc}–R_s^{-1} relations calculated with Eq. (26); curve a corresponds to the range $0 \leq R_s^{-1}\ 0.2\Omega^{-1}$ cm^{-2} plotted on the graph; curve b corresponds to the range $0 \leq R_s^{-1} \leq 0.15\Omega^{-1}$ cm^{-2}. Illumination: Tungsten lamp, unfiltered.

improves the fill factor of the photovoltaic (cf. Figs. 16 and 17). Figure 17 shows the effect of film conductivity and the $1/R_s$ values on the short-circuit current of the n-Si/SnO$_2$ photovoltaic cell. Experimental values were in excellent agreement with Eq. (26) when the film conductivity range was made to correspond to a $1/R_s$ range of 0–0.2 Ω^{-1}. A film conductivity of at least 150 Ω^{-1} cm^{-1} is necessary to eliminate the limitation of short-circuit photocurrent flow by R_s. Films of such conductivities are obtained by the spray–pyrolysis technique at substrate temperatures of $\geq 360°C$. The effect of the film conductivity on the fill factor is analyzed by considering the slope of the photocurrent–voltage curve according to Eq. (25) at the open-circuit potential, i.e.,

$$\left(\frac{dJ}{dV}\right)_{V=V_{oc}} = -\left(R_s + \frac{nKT}{eJ_0} e^{-\frac{eV_{oc}}{nKT}}\right)^{-1} \qquad (27)$$

with $V_{oc} = 0.47$ V and at 300 K a slope of -0.206 A cm^{-2} V^{-1} for the ideal situation of $R_s = 0$. This line ($R_s^{-1} = \infty$) is presented in Fig. 16, together with lines corresponding to other selected values of R_s^{-1}. A comparison of the slopes of these lines with the conductivity results presented in Fig. 17 shows that, within a certain range of R_s values, the value of R_s can be considered to have no effect on the short-circuit current but strongly affects the shape of the photocurrent–voltage curve, i.e., the fill factor of the cell. Therefore, it is very important to control the conductivity of the surface film (SnO$_2$ film) so as to attain the most suitable fill factor of the prepared heterojunction. A film conductivity of 1.5×10^3 Ω^{-1} cm^{-1}, which corresponds to an R_s value of 1 Ω, will be essential to give a good fill factor and hence better solar conversion efficiency. Such high conductivity was not obtainable with pure SnO$_2$ films (cf. Fig. 15). In order to obtain good performance from these heterojunctions, two aspects should be considered. The first is the improvement of the solid-state characteristics of the surface layer, i.e., the oxide film, by increasing its conductivity, which can be done by doping or, to use a better term, foreign atom incorporation. The second is the modification of the cell geometry in such a way as to decrease the series resistance of the current collector.

5. Improvement of the Characteristics of SnO$_2$

The SnO$_2$ employed as a thin-film-electrode material contains oxygen vacancies in the lattice. The n-type carrier concentration can be raised by the presence of donor atoms such as Cl and Sb to values above 10^{20} cm^{-3}, and the conductivity can be well above 100 Ω^{-1} cm^{-1}.[113] Many attempts have been made to improve the characteristics of SnO$_2$ films in order to fulfill the requirements of a wide range of applications, especially in photovoltaic and photoelectrochemical devices. Incorporating foreign atoms in the SnO$_2$ matrix affects the position of the Fermi level of the prepared oxide.[122] In photoelectrochemical devices it was found feasible and beneficial to separate the solid-state junction (photovoltaic) and the electrochemical performance of the device (n-Si/SnO$_2$/electrolyte).[5,59] The catalytic improvement of the SnO$_2$ films was clearly reflected in the electrochemical performance. Such improvement is achieved by incorporating foreign atoms in the oxide-film matrix. The spray–pyrolysis technique described in Section VII.1 enables the convenient incorporation of foreign atoms such as F, In, Ru, and Sb, etc.[5,58,63,64,77,98,119] Incorporating Ru or Sb into the SnO$_2$ films leads to the deterioration of the photovoltaic characteristics of the n-Si/SnO$_2$ junction, as reflected by a decrease in the open-circuit potential. On the other hand, such foreign atoms improve the charge-transfer kinetics at the SnO_2/electrolyte interface. The enhancement of the charge-transfer kinetics by foreign atom incorporation or surface deposition, such as the deposition of RuO_2 on top of the SnO_2 surface,[5] should be very carefully controlled; otherwise, a decrease in the short-circuit current of the device will occur, which will decrease the solar conversion efficiency.[5]

The modification of the properties of the oxide film should be controlled to be suitable for optimum application (i.e., which parameter should be increased or decreased to match the requirements of the device). Increasing the conductivity of the films by foreign atom incorporation satisfies the need for conducting and transparent electrodes for spectroscopic applications.[122–124] High-conducting SnO_2 films are suitable for Schottky barrier solar cells of high performance.[122,126]

(i) Effect of Foreign Atoms on the Conductivity of the SnO_2 Films

As was discussed in Section VII.3, the specific conductance of the SnO_2 films prepared by the spray–pyrolysis technique at a given temperature is independent of the film thickness. The film conductivity ($1/R$), on the other hand, increases as the film thickness increases. For the same thickness, the presence of foreign atoms incorporated in the SnO_2 matrix (e.g., In) increases the film conductivity by one to three orders of magnitude. The specific conductance of SnO_2 films prepared from solutions in ethyl acetate is always higher than that of films prepared from alcoholic solutions.[64,86] For SnO_2 films of 100-nm thickness, the specific conductance increases from 100–200 Ω^{-1} cm^{-1} for pure SnO_2 films to 3×10^3–5×10^3 Ω^{-1} cm^{-1} for SnO_2 films containing 1% In.[63,78]

(ii) Effect of Foreign Atoms on the Optical Properties and Band-Gap Energy of the SnO_2-Films

For the optical and band-gap energy measurement described below, boron glass and fused silica substrates were used.

(a) Film transmittance

The transmission spectra of SnO_2 films were recorded over the region covering the near-infrared, visible, and UV parts of the solar spectrum (2500–185 nm) using a Zeiss PMQ3 single-beam spectrophotometer. The resolving power was ca. 0.02 nm in the UV and 2.0 nm in the visible and near-infrared regions. The light sources were a tungsten lamp for the

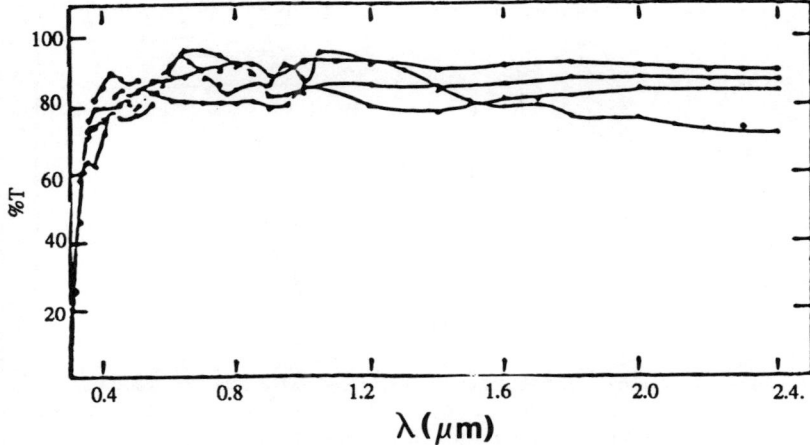

Figure 18. Transmittance (%T) of pure SnO_2 layers of different thicknesses (t) as a function of the wavelength (λ) of the incident radiation in the range 300–2500 nm. o, $t \approx$ 80 nm; Δ, $t \approx$ 240 nm; ●, $t \approx$ 160 nm; ▲, $t >$ 250 nm.

2500–325 nm region and a deuterium lamp for the 325–185 nm range. The detectors were a lead sulfide photocell for the 2500–800 nm range and an R446 photomultiplier for the 800–185 nm region. The transmission data for pure and In-incorporated SnO_2 films on borosilicate glass were obtained in the 2500–300 nm range using borosilicate glass as reference. In the 325–185 nm range, the transmittance ratio for the pure and foreign-atom-incorporated SnO_2 films on fused silica plates was obtained by placing a very thin layer sample as reference, while the thicker layer is the test sample; this technique minimizes the reflection effects.[123] Typical results for SnO_2 films with/and without fluorine doping are presented in Figs. 18 and 19, respectively. In these figures data for pure and foreign-atom-incorporated films of different thicknesses are summarized. The general features of these curves show that the pure films and the In-incorporated SnO_2 films containing up to 1% In are almost perfectly transparent in the visible and near-infrared regions. With this concentration of indium in the SnO_2 film, a slight decrease in the transmittance (ca. 3.5%) was observed,[64] but the transmittance was always over 90%. The same results were obtained for SnO_2 films with incorporated Ru and Sb.[5,59,113] Incorporation of fluorine in the SnO_2 matrix up to a concentration of 10% had

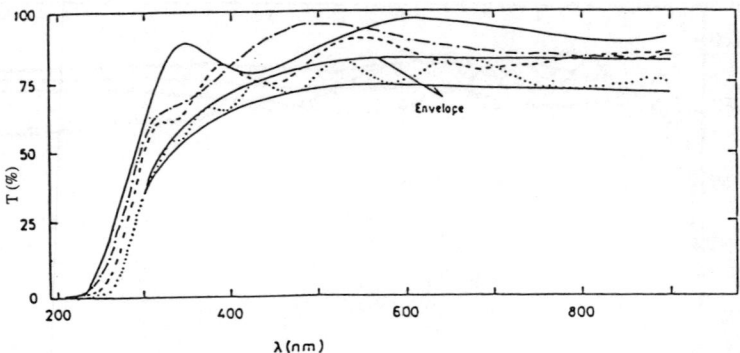

Figure 19. Transmittance (%T) of undoped SnO_2 and 10% F-doped SnO_2 thin films of different thicknesses (t) as a function of wavelength (λ), in the range 200–900 nm. —·—·, Undoped SnO_2 film, $t = 85$ nm. For the F-doped SnO_2 films, $t = 85$ nm (—), 172 nm (- - -), and 350 nm (······).

no effect on the high transparency of the prepared films (cf. Fig. 19).[126] The transmittance (T)–wavelength (λ) relations show an increasing number of interference extrema with increasing film thickness and increased incorporation of fluorine in the SnO_2 matrix. Film thickness and refractive index were calculated from the interference extrema of the T–λ spectra. The method is based on analysis of the transmittance spectrum of a weakly absorbing film/nonabsorbing substrate system.[127] Inverse transmittance can be decomposed into two components,

$$\frac{1}{T_\lambda} = \mu_\lambda + C_\lambda \cdot V_\lambda \tag{28}$$

where $\mu(\lambda)$ and $V(\lambda)$ consist of exponential and sinusoidal terms, respectively. $\mu(\lambda)$ and $C(\lambda)$ can be calculated from the experimentally traced transmittance envelope curves, T^+ and T^-, of the spectrum at a wavelength λ (cf. Fig. 19):

$$\mu_\lambda = \frac{T_\lambda^+ + T_\lambda^-}{2T_\lambda^+ \cdot T_\lambda^-} \tag{29}$$

$$C_\lambda = \frac{T_\lambda^+ - T_\lambda^-}{2T_\lambda^+ \cdot T_\lambda^-} \tag{30}$$

The optical parameters of the film, i.e., the refractive index n, the thickness t, and the extinction coefficient k, are given by

$$n_\lambda = \frac{1}{2}\{[8n_S C_\lambda + (n_S + 1)^2]^{1/2} + [8n_S C_\lambda + (n_S - 1)^2]^{1/2}\} \quad (31)$$

$$t = \left\{4\left[\frac{n_{\lambda_i}}{\lambda_i} - \frac{n_{\lambda_{i+1}}}{\lambda_{i+1}}\right]\right\}^{-1} \quad (32)$$

$$k_\lambda = \frac{\lambda}{4\pi \cdot t} \ln \frac{\mu + (\mu^2 - C^2 + \sigma)^{1/2}}{2a_+} \quad (33)$$

where

$$\sigma = \left[\frac{(n_S^2 - 1)}{8n_S}\right]^2 \left(n - \frac{1}{n}\right)^2 \quad (34)$$

$$a_+ = \frac{(n_+ + 1)^3(n + n_S^2)}{16 n_S n^2} \quad (35)$$

For the first-order approximation, n_s is the index of refraction of the substrate, and λ_i and λ_{i+1} are the wavelengths corresponding to two successive extrema. For higher order approximation,

$$n_\lambda^2 = n_\lambda - v k_\lambda^2 \quad v = f(n, n_S) \quad (36)$$

The corrections to the refractive index and the film thickness, however, are only necessary for highly absorbing films, which is not our case.

Application of this method of calculation to the transmission spectra of Fig. 19 gives the variation of the refractive index of the pure and F-incorporated SnO_2 films as a function of film thickness and incident radiation. The results of these calculations are summarized in Fig. 20. The mean feature of the plots in Fig. 20 is the presence of two regions. The first is a nondispersive region (nearly steady) at $\lambda \geq 440$ nm for F-incorporated SnO_2 and $\lambda \geq 520$ nm for pure SnO_2 of the same thickness (85 nm). In the second region, which is near the UV region, the refractive index fluctuates, indicating increased absorption of the films in this region. The refractive index of the samples decreases with increasing film thickness. A considerable decrease of n was measured for layers having a thickness

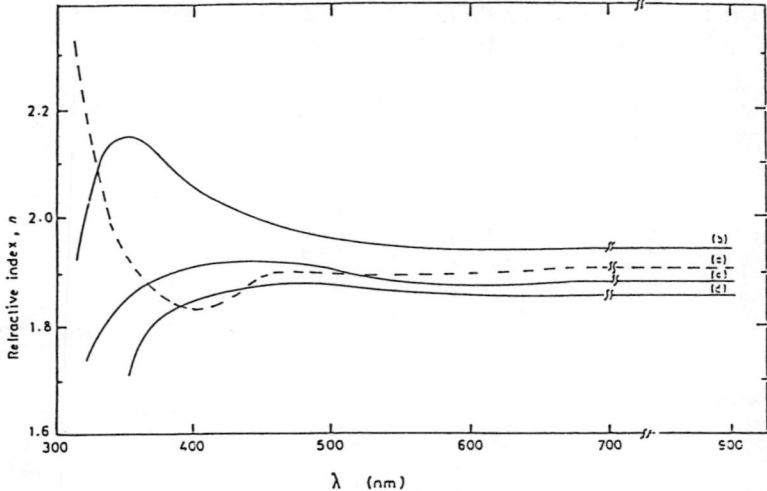

Figure 20. Refractive index (n), as a function of wavelength (λ) in the range 200–900 nm, calculated from the spectra in Fig. 19. (a) Undoped SnO_2 film, $t = 85$ nm; (b–d) F-doped SnO_2 films with $t = 85$ nm (b), 172 nm (c), and 350 nm (d).

of 350 nm and containing 10% F. Some optical data for films of different thickness are presented in Table 5. X-ray diffraction investigations have shown that the crystallinity of the samples increases with increasing F content and increasing film thickness. An improvement of the grain size occurs. Such results have increased the importance of structural investigations of SnO_2 films.

Table 5
Effect of F Incorporation on Solid-State Characteristics of SnO_2

Material	Thickness (nm)	% T at 500 nm	R_s ($\Omega/\square$)[a]	ϕ ($\times 10^{-3}$)
Undoped SnO_2	85	97	509	1.45
SnO_2–10% F	85	94	58	9.3
SnO_2–10% F	172	90	20	17.4
SnO_2–10% F	350	89	13	23.9

[a]At room temperature.

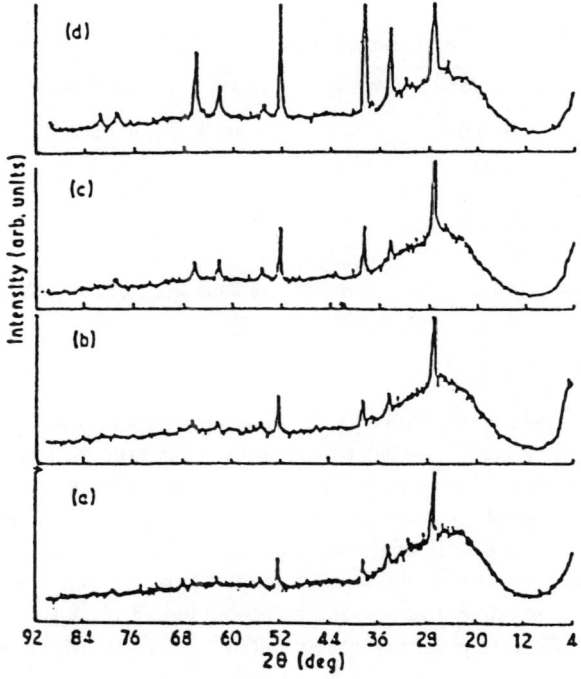

Figure 21. X-ray diffraction patterns of undoped and 10% F- doped SnO_2 films of different thicknesses (t). (a) Undoped SnO_2 film, $t = 85$ nm; (b–d) F-doped SnO_2 films with $t = 85$ nm (b), 172 nm (c), and 350 nm (d).

(b) Structural investigation of the SnO_2 films

Figure 21 shows X-ray diffraction patterns of pure and F-incorporated SnO_2 films of different thicknesses. The d spacing is calculated from 2θ, and the relative intensities (I/I_0) for the various planes are estimated from the recorded counting rate. The values of d agree with ASTM data for SnO_2 powder.[128,129] The addition of F up to 10% produces no additional lines representing the foreign atoms (F) or any other undesired phases. The lattice constants a and c were determined using high-angle reflections.[130] A comparison of the calculated values with the literature data for SnO_2 is presented in Table 6. The values of a, c, and the unit-cell volume,

Table 6
Structural Parameters of SnO$_2$

Material	Lattice constant, a (nm)	Lattice constant, c (nm)	Unit cell volume, V (nm^3)	Crystallite size, D (nm)	Film thickness (nm)
Spec. pure SnO$_2$ powder	0.4738^a	0.3188^a	—	—	—
Undoped SnO$_2$ film	0.4715	0.3154	0.70117	17	85
SnO$_2$–10% F	0.4708	0.3151	0.69843	21	85
SnO$_2$–10% F	0.4710	0.3163	0.70168	24	172
SnO$_2$–10% F	0.4731	0.3161	0.70751	30	350

aASTM data.

V, for the pure SnO$_2$ films agree with those reported by ASTM for polycrystalline SnO$_2$ powder of tetragonal cassiterite structure.[129] The intensity of diffraction planes of SnO$_2$ films shows a definite increase with fluorine incorporation and increasing film thickness, as illustrated in Fig. 22. Some of the planes of F-containing films orient themselves to give maximum reflection and hence maximum intensity. The crystallinity of the films is improved by fluorine incorporation. Both doping and film thickness affect the preferred orientation of microcrystallites in polycrystalline thin films.[131] An estimation of the preferred orientation planes of the prepared films was carried out by studying the peak profiles of the (110), (200), and (301) planes because these planes are the most sensitive to microcrystalline orientation in SnO$_2$.[132,133] The (200) plane was found to be the most sensitive for describing the preferred orientation of the prepared films. This agrees with the result reported by Agashe et al.[134] for SnO$_2$ thin films prepared at higher substrate temperatures (>375°C). (The films reported here were prepared at a substrate temperature of 450 ± 2 °C.) From Fig. 22, it can be seen that the intensity of the (200) diffraction plane increases with increasing thickness and fluorine incorporation. This is associated with an increased extent of preferred orientation of microcrystallites along the (200) plane, which is reflected in an increased crystallite size along this plane. The average crystallite size, D, was checked using the Scherrer formula[134–136]:

$$D = \frac{0.94\lambda}{\Delta(2\theta) \cos \theta} \tag{37}$$

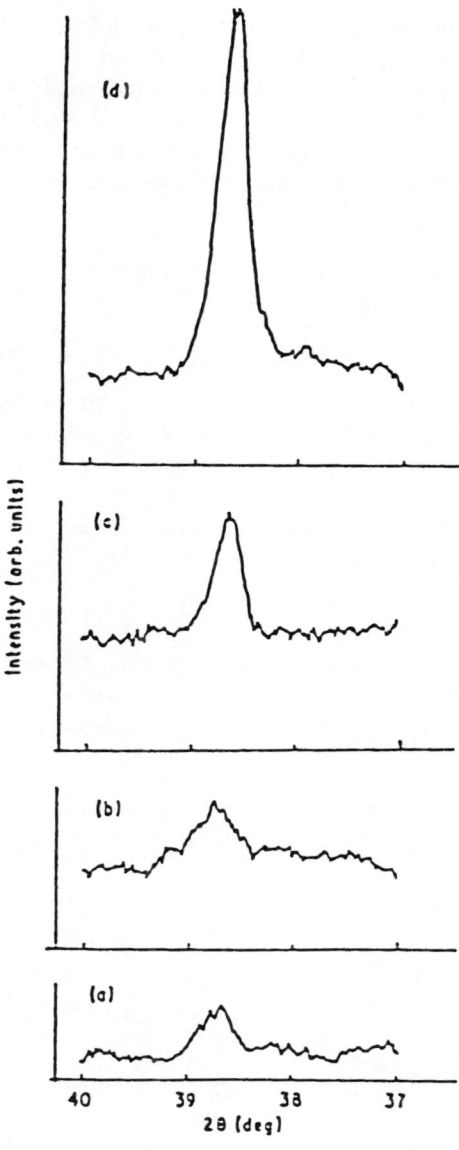

Figure 22. Effect of 10% F doping on the crystallinity of SnO_2 films. (a) Undoped SnO_2 film, $t = 85$ nm; (b–d) F-doped SnO_2 films with $t = 85$ nm (b), 172 nm (c), and 350 nm (d).

where $\Delta(2\theta)$ is the half-peak width of the (200) diffraction line in radians, θ is the diffraction angle for the (200) plane, and λ is the wavelength of the incident X-ray beam. The crystalline size increased from 17 to 30 nm with increasing thickness of the F-incorporated SnO_2 film,[127] which corresponds to 77% of the crystallite size that can be attributed to the suitable condition for crystallization in the presence of F-incorporation.

(c) Energy-gap calculations

The direct energy gap of a semiconducting film can be obtained from transmittance data using the equation[137,138]

$$\alpha = \alpha_0(h\nu - E_g)^{1/2} \qquad (38)$$

where α is the absorption-coefficient, α_0 is a constant independent of the photon energy, and $h\nu$ and E_g are the photon energy and band-gap energy, respectively. Photons having energy less than E_g will not be absorbed; i.e., for $h\nu < E_g$, $\alpha = 0$.

Calculations of the absorption coefficient, α, are based on the transmission ratio for two layers of different thickness,[124,137]

$$T_{1-2} = e^{-\alpha \Delta t} \qquad (39)$$

where T_{1-2} is the transmission ratio and Δt is the difference between the thicknesses of the two films. The value of $\alpha \Delta t$ can be calculated accurately without calculation of Δt, by placing the thinner sample in the reference

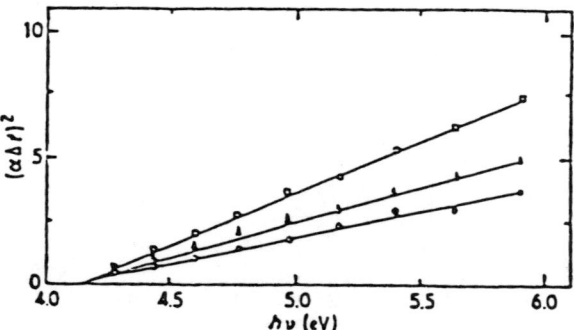

Figure 23. Variation of the square of the absorption coefficient as a function of the photon energy for undoped SnO_2 thin films of different thicknesses: o, 25 nm; Δ, 40 nm; □, 85 nm.

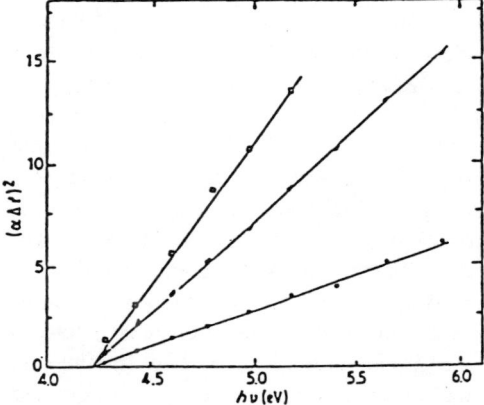

Figure 24. Variation of the square of the absorption coefficient as a function of the photon energy for F-doped SnO$_2$ thin films of different thicknesses: o, 85 nm; Δ, 172 nm; □ 350 nm.

beam and the thicker one in the sample beam of the spectrophotometer.[139] The direct band-gap energy can be obtained by plotting $(\alpha\Delta t)^2$ as a function of photon energy $(h\nu)$. The extrapolation of the linear portion of the plot to $(\alpha\Delta t)^2 = 0$ gives the band-gap energy of the film. Examples of these plots are presented in Fig. 23 for pure SnO$_2$ films[63] and in Fig. 24 for F-incorporated SnO$_2$ films[127] of different thicknesses. The band-gap energy is independent of the film thickness. For pure SnO$_2$ films, values of E_g ranging between 3.6 and 4.1 eV were obtained.[63,127] SnO$_2$ films in which foreign atoms are incorporated have higher E_g than the pure films (4.21 eV for fluorine-containing SnO$_2$ films). The increase in the band-gap energy or the band gap widening may be explained by assuming that the allowed states near the bottom of the conduction band of the n-type material are occupied to rather high levels and that the allowed transitions from the valence band would have correspondingly higher energies than the forbidden gap. Such energy gap widening has also been observed with TiO$_2$ films[64,77,78,98] and with In$_2$O$_3$, and In$_2$O$_3$–Sn.[140]

6. The Electrochemical Behavior of the Oxide Film

The resistance of SnO$_2$ against chemical attack is high. For thin films of this material containing donors like Cl or Sb, the n-type carrier concentration may be increased to values above 10^{20} cm^{-3}, and the conductivity can be well above 150 Ω^{-1} cm^{-1}, as was discussed above. The electrodes retain their semiconducting properties depending on the potential of the redox electrolyte. The exchange currents tend to be low, and the Tafel plots are unsymmetrical.[79,141-144] Electrochemical reactions on SnO$_2$ electrodes containing Ru (during chlorine evolution) proceed effectively in the anodic direction.[143,144] The behavior was attributed to electronic states in the SnO$_2$ band gap.[145] The presence of such surface states affects the interfacial distribution and the charge-transfer barrier. The surface states effectively mediate electron transfer through the barrier, depending on their energetic position with respect to the redox system in the electrolyte and on their accessibility to the conduction-band electrons and the depolarizer.[146,147] The distribution of such states in the band gap and its correlation to the electrode properties can be determined by the use of ultraviolet photoelectron spectroscopy (UPS). In the case of highly doped material, many more electronic states in the band gap are occupied and thus accessible to UPS. The sampling depth of the method is less than 1 nm; i.e., only the states at or near the surface, which are the most effective in electrode processes, can be analyzed.

The SnO$_2$ films used in the investigations that will be discussed in this section were prepared by the spray–pyrolysis technique at a substrate temperature of 400–450°C, as described previously.[111] SbCl$_5$ and RuO$_4$ were used for incorporation of Sb and Ru into the SnO$_2$ matrix. The n-type carrier density was determined by capacitance measurements in nonaqueous solvents to be $(7 \pm 2) \times 10^{19}$ cm^{-3}. Additions of more than 10% Sb produced films having a milky appearance and poor electrode properties. The substrates for the films for the electrochemical measurements were polished glassy carbon disks, 8 mm in diameter and 3 mm in height, mounted in stainless-steel holders used for the rotating disk system and insulated with Teflon tape so that only the SnO$_2$-coated circular front face (area = 0.5 cm^2) was in contact with the electrolyte. Standard potentiostatic techniques were employed for the electrochemical measurements. A Vacuum Generators combined ESCA/UPS spectrometer was used to record the UPS spectra of the samples. At least three spectra were recorded for each sample. The first spectrum was taken without any sample pretreatment. The second was taken after 10-min Ar$^+$ bombardment (500 V, 3 μA), which was expected to remove contaminants without changing the SnO$_2$ surface significantly. Further sputtering increased the photoelectron emission in the energy range of the band gap. Prolonged sputtering did not lead to a detectable signal at the Fermi energy, indicating that a metallic phase was not formed. The Ru and Sb concentrations in the SnO_2 film were determined by the same system in the ESCA mode.

(i) Electrode Kinetics of $Fe^{2+/3+}$ on SnO_2 Films

The electrochemical behavior of the different SnO_2 rotating disks in the redox system $Fe^{2+/3+}$ (0.0025M each) in 0.1M H_2SO_4 is represented by the Tafel plots in Fig. 25. The exchange current densities at $\eta = 0$ were found to have values between 2 and 20 $\mu A/cm^2$ for all samples, even those containing up to 10% Sb and 6% Ru. These values are not too different from each other from the standpoint of the experimental reproducibility, which leads to the conclusion that the exchange current densities for the system $Fe^{2+/3+}$ on the different SnO_2 disks do not differ considerably. The cathodic Tafel slopes, b_c, are similar and fall in the range $180 < b_c < 190$ mV/decade. This corresponds to a value of $\alpha_c \approx 0.35$, where α_c is the cathodic charge-transfer coefficient. In the anodic direction, remarkably linear Tafel lines of slope $b_a > 500$ mV/decade were obtained with pure SnO_2 films. This value corresponds to an anodic charge-transfer coefficient, α_a, of <0.1. Incorporation of Sb in the SnO_2

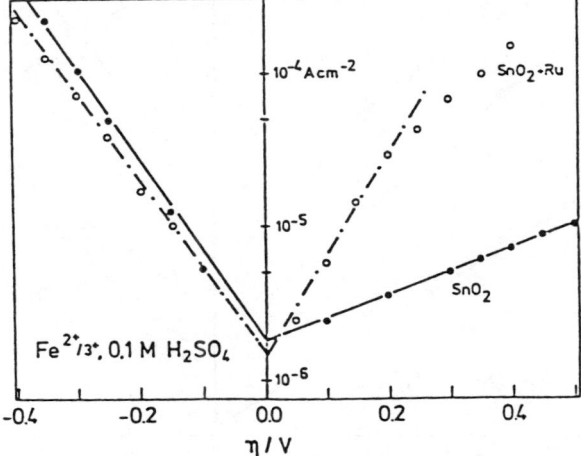

Figure 25. Tafel plots obtained for SnO_2 thin film electrodes with the redox system $Fe^{2+/3+}$ (each 0.0025M) in 0.1M H_2SO_4. Electrodes were rotated at 3300 min^{-1}. ●, SnO_2 (identical, within the limits of reproducibility, to results obtained with SnO_2 + 5% Sb); o, SnO_2 + 1.8% Ru.

matrix at concentrations up to 10% did not have a significant effect on the values of α_a or α_c. A concentration of 5% Sb produced films of particularly high stability. Incorporation of Ru, on the other hand, caused remarkable changes in the electrochemical behavior of the SnO_2 film. A dramatic change in the anodic Tafel slope was recorded. At a concentration of 0.5% Ru in the SnO_2 film, a b_a value of ~350 mV/decade was recorded. Increasing the concentration of Ru to 1.5% led to a b_a value of ~150 mV/decade. With further increases of the Ru concentration in the SnO_2 film, the values of b_a and b_c remained unchanged at values corresponding to $\alpha_a \approx \alpha_c \approx 0.38 \pm 0.05$. The exchange current density was further increased to ca. 30 $\mu A/cm^2$ at a concentration of 10% Ru.

(ii) Electrode Kinetics of $Ce^{3+/4+}$ on SnO_2 Films

The Tafel plots for the different SnO_2 rotating disks in the $Ce^{3+/4+}$ redox system (0.05M each in 1M H_2SO_4) are presented in Fig. 26. The

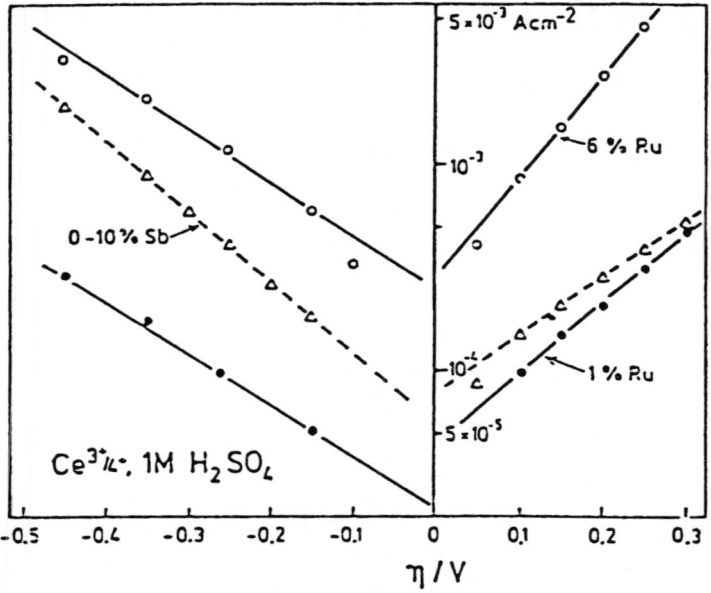

Figure 26. Tafel plots obtained for SnO_2 thin film electrodes with the redox system $Ce^{3+/4+}$ (each 0.05M) in 1M H_2SO_4. Electrodes were rotated at 3300 min^{-1}. The SnO_2 electrodes contained the indicated atom % of Sb or Ru.

presence of Sb in this case also has little effect on the charge-transfer rates. An increase in the concentration of Ru, on the other hand, increases the exchange current density, such that it reaches values comparable to those obtained with Pt electrodes in the same redox system.[106,148] The anodic branch is steeper than the cathodic branch. The anodic and cathodic branches do not extrapolate to the same exchange currents. Even at the higher concentrations of Ru, the sum of the anodic and cathodic charge-transfer coefficients ($\alpha_a + \alpha_c$) was less than 0.5.

(iii) Ultraviolet Photoelectron Spectroscopy of the SnO_2 Films

The UPS spectrum of a pure SnO_2 film is presented in Fig. 27. The sample was biased to −10 V; the secondary electron threshold was taken as the true sample threshold and thus as the zero point of the electron energy scale, E_a. The peak at ~12 eV corresponds to the SnO_2 valence band. Extrapolation of the electron density function to high energies leads to the valence-band edge, E_{VB}. Using the literature value of the band-gap energy, $E_g = 3.7$ eV,[63,149–151] the energy of electrons emitted from the conduction-band edge, E_{CB}, was calculated to be 17.5 eV. For a highly doped semiconductor, the Fermi level, E_F, is probably not more than

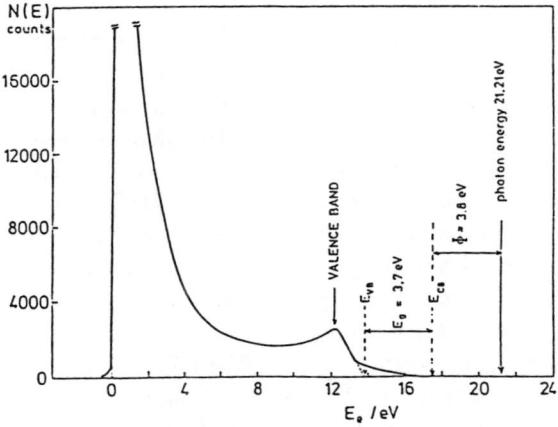

Figure 27. UPS spectrum of a pure SnO_2 film. E_g is the literature value of the SnO_2 band gap and was used to obtain the conduction-band edge, E_{CB}, from the derived valence-band edge, E_{VB}. Φ is the SnO_2 work function.

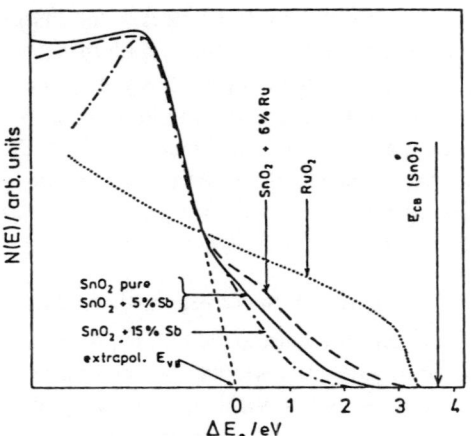

Figure 28. UPS spectra of the prepared SnO_2 electrodes. The percentages of Sb and Ru indicated on the figure are atom % values, determined by ESCA. For comparison, the UPS spectrum of pure RuO_2 is also shown. The energies of electrons from the samples' valence-band edge (Fig. 27) were taken as the zero of the ΔE_e scale.

0.1 eV below the conduction-band edge.[152] E_{VB} corresponds to the photon energy (21.21 eV) diminished by the sum of the work function plus the gap energy. Considering that $E_F = 17.4$ eV on the energy scale of Fig. 27, a work function $\varphi = 3.8$ eV is obtained. This value is consistent with the value obtained with a Kelvin probe ($\varphi = 3.8 \pm 0.1$ eV).[153] The UPS spectra of SnO_2 films into which foreign atoms had been incorporated are presented in Fig. 28. The valence-band edges derived by extrapolation were taken as the zero of the energy scale (ΔE_e), in order to facilitate comparison of the density of states in the gap. Ruthenium increased the density of electronic states in the band gap whereas Sb did not show remarkable effects up to a concentration of 10%. The UPS spectrum obtained with pure RuO_2 is included in Fig. 28 for comparison. The sharp cutoff indicates metallic character and a work function φ of 4.1 eV.

With n-type SnO_2, charge-transfer reactions with electroactive species in aqueous electrolytes are expected to proceed via conduction-band electrons. The flatband potential in $0.1M$ H_2SO_4 solution is ca. +0.1 V

(NHE).[106] The standard potentials of the redox systems considered are at more positive values. These values are not positive enough for the possibility of hole injection into the valence band, which is located at +3.8 V (NHE). If the SnO_2 electrode is polarized to the standard potentials, a depletion layer will be formed underneath the semiconductor surface, and this depletion layer constitutes a barrier for electron transfer across the interface. The dark currents on semiconductor electrodes are normally determined by the rate at which electrons can pass over or across the space-charge barrier.[32,154] The semiconductor model for the interfacial charge-transfer processes would involve only the allowed conduction-band states in the semiconductor. For lower values of donor density and dielectric constant ($N_D < 10^{18}$ cm^{-3} and $\varepsilon < 10$), the dark current density (j) / overvoltage (η) relation would be given by[159]

$$j = j_0 \left(1 - e^{\frac{-e}{kT}\eta} \right) \quad (40)$$

where j_0 is the exchange current density. The obtained Tafel slopes would correspond to $\alpha_a = 0$ and $\alpha_c = 1$. With a donor density of $N_D = 7 \times 10^{19}$ cm^{-3}, a significant fraction of the applied overvoltage (η) will appear across the Helmholtz double layer, reducing the effective space-charge barrier. This leads to values of $\alpha_a > 0$ and $\alpha_c < 1$. The sum of $\alpha_a + \alpha_c$ should remain unity.[155,156] This semiconductor model does not apply to the SnO_2 electrodes, since in no case is the condition $\alpha_a + \alpha_c = 1$ fulfilled. The $Ce^{3+\backslash 4+}$ reaction would have to proceed at a much slower rate than the $Fe^{2+\backslash 3+}$ reaction because the space-charge barrier, $e\Delta E_{sp}$, is much higher due to the difference in the standard potentials of ca. 0.8 V, as illustrated in Fig. 29.

The more reasonable possibility is the assumption that the electrons can pass the space-charge barrier by hopping from the conduction band to surface states via electronic states of donors deeper in the band gap. These states do not contribute to the bulk conductivity but can exchange electrons with the conduction band and the surface under the influence of the high electric field in the space-charge layer. The distribution of electronic states in the band gap can be obtained from the UPS spectra. The results of Fig. 28 show a signal starting 1.1 eV below the conduction-band edge with SnO_2 or Sb-incorporated SnO_2 electrodes and 0.5 eV below the band edge in the presence of 6% Ru. At these energies the density of states must be on the order of 10^{21} cm^{-3} (eV)$^{-1}$. The relevant information available from

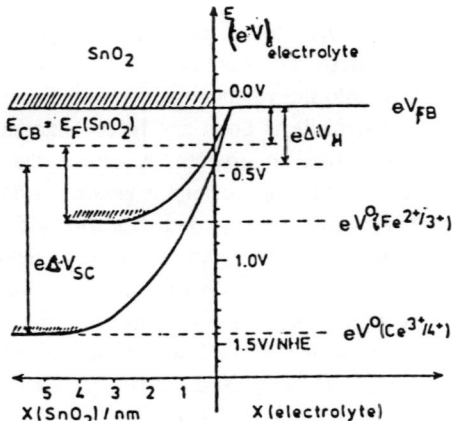

Figure 29. Schematic energy diagram of the SnO_2/aqueous redox electrolyte interface obtained by assuming the absence of electronic states in the gap except those of ionized donors of constant density $N_D = 7 \times 10^{19}$ cm^{-3} and a double-layer capacitance, C_H of 15 μF/cm^2. The potential distributions between the space-charge and the Helmholtz layers (ΔV_{SC}, ΔV_H) are shown at the standard potentials ($V°$) of the redox systems considered.

capacitance measurements concerning the distribution of states is presented in Fig. 30.

At anodic potentials at which a depletion layer is formed, deeper-lying donor states will be ionized and contribute to the space charge. The existence of such deeper donor levels has two consequences for the charge-transfer reaction:

1. The space-charge layer will be thinner than would be calculated according to a model assuming constant donor density, since the density of positively charged states increases from the bulk to the surface, and hence tunneling will be enhanced.

2. The donor states in the space-charge layer, being close enough to each other and to the surface, can mediate electron transfer by a hopping mechanism.

The second consequence will offer locally different pathways for electron exchange between the surface and the conduction band because the spatial

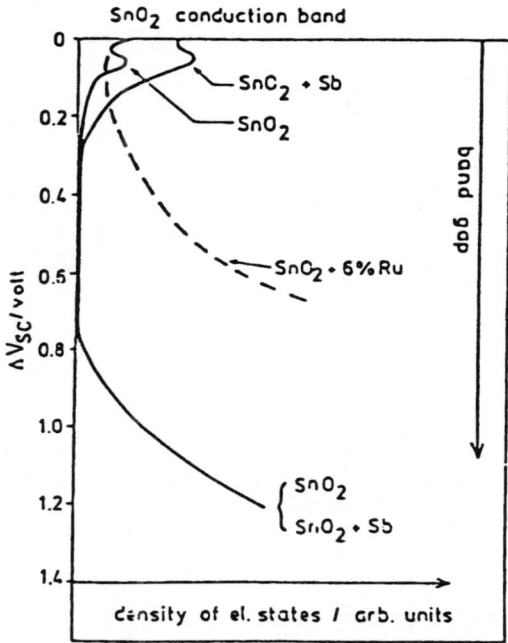

Figure 30. Estimate of the distribution of electronic states in the band gap of the indicated electrode materials, based on UPS and capacitance measurements. ΔV_{SC} is the potential drop across the space-charge layer.

distribution of donor states will be quite inhomogeneous. For such polycrystalline material, grain boundaries might be particularly preferred pathways for electron transfer. On the basis of these assumptions, the space-charge layer can be better represented by the model shown in Fig. 31. This model permits electron passage through the space-charge barrier by electron exchange between occupied and vacant donor states. If there are enough electronic states on the surface that can exchange electrons with the redox system, the rate-determining step will be electron passage through the barrier, and the occupation of surface states will be in equilibrium with the redox system. This means that a variation of the density of occupied and vacant surface states with applied potential occurs, since a part of this potential charges the voltage drop in the Helmholtz double

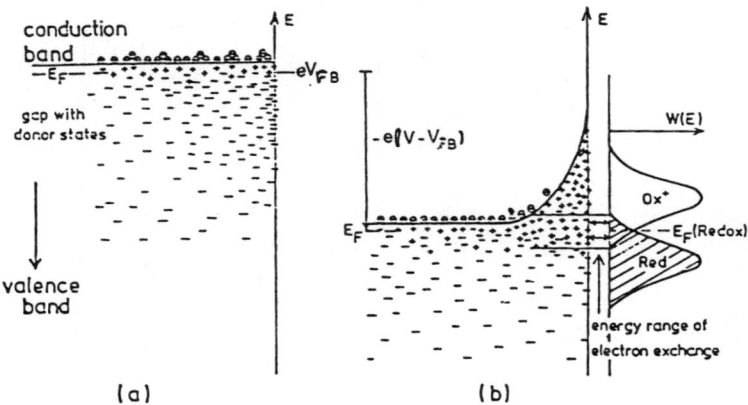

Figure 31. Semiconductor electrode containing electronic states deep in the band gap, (a) at the flatband potential, V_{FB}, and (b) at the standard potential of a redox system $[V^\circ = E_F (\text{redox})/e]$. Electrons can penetrate the space-charge barrier by a hopping mechanism. The distribution functions for electron energy states of the redox components (Ox^+, Red) in the electrolyte are also schematically shown $[W(E)]$.

layer. As long as the passage of electrons through the space-charge barrier remains rate-determining, the measurable reaction rate will not be affected, and the sum of α_a and α_c will be less than 1. The situation that would predominate during anodic and cathodic polarization as compared to the equilibrium situation is illustrated in Fig. 32. If the electrons on their way through the barrier have to pass several donor states in single steps, only the shift of these states relative to each other with a change of the voltage will affect the electron-transfer rate between two states in the space-charge barrier or these donor states and the surface or the conduction band, respectively; i.e., the single charge-transfer step has to be divided into several smaller steps, and a shift of the passage of electrons to different donor and surface states with variation of potential takes place. The apparent charge-transfer coefficients will vary considerably with the material and potential range of the redox reaction, depending on the slowest step among these several steps. In any case, smaller values of α_a and α_c than obtained for metal electrodes are to be expected.[146,147]

The model described by Fig. 32 applies well to the $Ce^{3+/4+}$ reaction on pure SnO_2 electrodes and on SnO_2 electrodes in which foreign atoms are incorporated. It applies also to the $Fe^{2+/3+}$ redox reaction on the SnO_2 (Ru) electrodes. According to Fig. 30, at the respective standard potentials

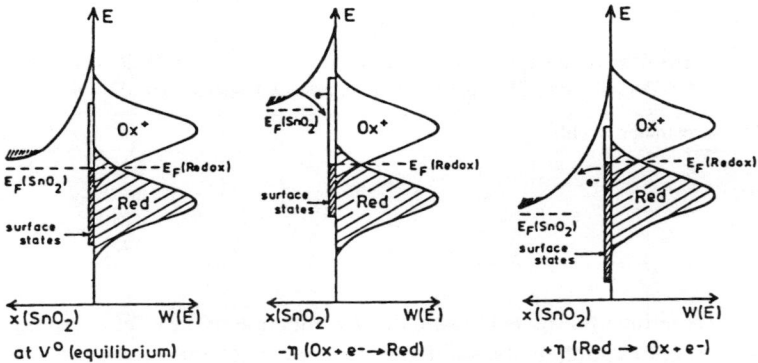

Figure 32. Model of electrochemical charge transfer involving surface states in equilibrium with a redox system. An applied overvoltage, η, is divided up between the space-charge and Helmholtz layers. The equilibrium situation corresponds to Fig. 31b.

electronic states are available in the band gap. Charging and discharging of these electronic states leads to a change of potential in the Helmholtz layer depending on the polarization, η (cf. Fig. 32). These electronic states are available for a hopping type of transfer across the space-charge barrier.

7. Titanium Oxide Films

Titanium oxide represents one of the most familiar and attractive metal oxides suitable for solar cell fabrication.[97] The growing interest in the utilization of this material for solar energy conversion purposes in terrestrial and extraterrestrial applications derives from its excellent properties as an antireflector and its stability. The high resistance of TiO_2 to chemical attack and corrosion extends its applications as a photoanode in photoelectrolytic cells.[6,94,96,157,158] Incorporation of foreign materials into the TiO_2 matrix improves the characteristics of the material in different ways. Attempts have been made to shift the response of the material as a wide-band-gap semiconductor into the visible region of the solar spectrum.[6,64,94,159]

Pure and foreign-atom-incorporated TiO_2 thin films have been prepared on glass, fused silica, glassy carbon, platinum, or silicon substrates conveniently and reproducibly using the spray–pyrolysis technique described in Section VII.1,[85,111] according to the pyrolysis reaction

$$TiCl_4 + 2H_2O \xrightarrow{723\ K} TiO_{2(S)} + 4HCl_{(g)} \quad (41)$$

Table 7
Values of the Refractive Index (n) and the Extinction Coefficient (κ) for Sb-Doped TiO_2 Thin Films (Thickness ≈ 80 nm) at λ = 550 nm

Sb concentration (mol/l)	n	κ
0.03	2.275	0.525
0.09	2.150	0.425
0.12	2.135	0.390
0.15	2.175	0.425

Experiments directed toward the investigation of the effect of foreign atom incorporation on the solid-state and optical characteristics of TiO_2, similar to those performed with SnO_2, have been carried out. It turned out that the presence of foreign atoms such as antimony[65] or indium[159] at concentrations of up to 1% in the TiO_2 matrix improves the optical characteristics. Highly transmitting films (>90% transmittance) were obtained. The conductivity of the films increased by at least one order of magnitude in the presence of Sb or In. However, the TiO_2 films are still less conducting than the SnO_2 films. Other optical parameters are very sensitive to the concentration of the foreign atom. As an example, the variation of the extinction coefficient (κ) and refractive index (n) of the prepared films with the concentration of Sb in the spraying mixture is summarized in Table 7. Incorporation of Sb did not have any significant effect on the band-gap energy.[65] Incorporation of In, on the other hand, increased the band-gap energy by ca. 6%. The value of E_g calculated for pure TiO_2 was 3.6 eV while that for films with incorporated In was 3.85 eV. The large band-gap energy calculated for pure TiO_2 was attributed to the mixed structure of the film, which depends on the preparation conditions. The electrochemical properties of TiO_2 either as a pure polycrystalline material or as an anodic film on the metal or different substrates have been extensively studied.[160–162] In the next section we will concentrate on the photovoltaic and photoelectrochemical characteristics of these oxide films coupled with n-Si for solar energy conversion.

VIII. n-Si/SnO_2 AND n-Si/TiO_2 PHOTOVOLTAIC AND PHOTOELECTROCHEMICAL SYSTEMS

Before dealing with the photovoltaic and photoelectrochemical characteristics of n-Si/SnO_2, and n-Si/TiO_2 systems, we will discuss the effect of the thin insulating layer of SiO_x which is responsible for the Schottky barrier structure (SIM or SIS) as was described before. The thickness of this layer should be controlled; otherwise, deterioration of the prepared cells or photoanodes will occur.[112]

1. Effect of the Interfacial SiO_x Layer on the n-Si/Oxide Heterojunction

The Schottky barrier solar cells based on the n-Si/oxide system have shown high conversion efficiencies, especially in photovoltaic applications. Conversion efficiencies of more than 14% have been reported.[77,78,111,163–166] The n-Si/oxide systems, especially n-Si/SnO_2, were found to deteriorate with aging.[111,165] The instability of the n-Si/SnO_2 system was attributed to irreversible changes occurring at the thin insulating SiO_x layer at the Si/SnO_2 interface.[166,167] The cell deterioration after aging or heat treatment is reflected in decreases of the open-circuit potential V_{oc}, and the short-circuit current J_{ph}, and hence a reduction in the solar conversion efficiency. This is illustrated in Fig. 33. Such changes are mainly due to interaction with the atmospheric O_2 and/or H_2O (cf. the results obtained in an argon atmosphere). The dark current–voltage curves

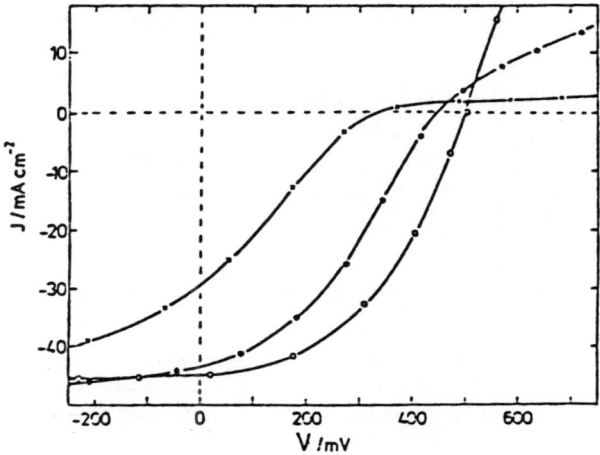

Figure 33. Photocurrent (J)–cell voltage (V) characteristics of freshly prepared (○) and heat-treated (200°C, 1 h; ●) SnO_2/n-Si solar cells in ambient atmosphere. Illumination: 100 mW/cm^2.

corresponding to the photocurrent–voltage curves of Fig. 33 are presented in Fig. 34. For silicon Schottky barrier cells of highest efficiency, a thin interfacial insulating SiO_x layer is essential. The theory of metal/insulator/semiconductor (MIS) devices is useful in evaluating the performance of these systems.[163,164,168] The superposition principle is found to be valid:

$$J = J_{ph} + J_d \qquad (42)$$

where J is the cell current, J_{ph} is the photocurrent, and J_d is the dark current, which is assumed to be limited by the rate of thermionic emission of electrons from the conduction band to the surface layer (SnO_2). The value of J_d is given by

$$J_d = J_0 e^{\frac{eV}{nKT}} \qquad (43)$$

where J_0 is the reverse saturation current, and n is the diode quality factor, which characterizes the distribution of the potential, V, between the semiconductor space-charge layer and the insulating film.[164] The conduction-band electrons can penetrate the barrier by tunneling at a high rate

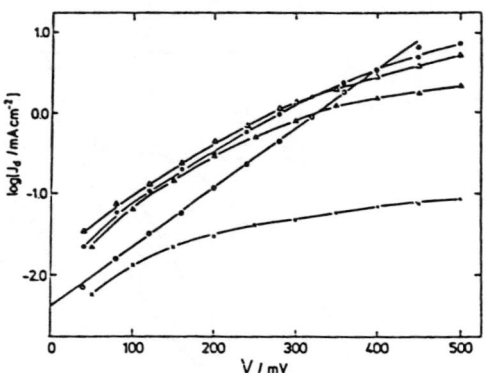

Figure 34. Dark current (J_d) flow upon forward biasing (V) of SnO_2/n-Si heterojunctions. The linear plot (o) was obtained with a freshly prepared cell (diode quality factor $n = 2.25$). The lower curve (×) was measured after heat treatment in air. The other curves were measured after storage of the cells in air at ambient temperature for the indicated periods of time. ●, 15 days; ▲, 35 days; △, 120 days.

without affecting J_d of the n-Si. The insulator film introduces charged states that increase the semiconductor space-charge barrier,[164] causing a reduction of the dark current. The deterioration of the cell upon heating in air (cf. Fig. 33) is due to the diffusion of H_2O and/or O_2 along the grain boundaries of the polycrystalline SnO_2 to react with SiO_x, reducing the density of charged states. This leads to a reduction in the barrier height for the thermionic emission process[165] and hence causes the increased dark current at low potentials observed with cells stored at room temperature (cf. Fig. 34). After heat treatment, the dark currents were decreased by two orders of magnitude (cf. Fig. 34), and the diode quality factor increased with increasing potential. These changes are due to the poor electronic conduction properties of the SiO_x layer.[169] The asymmetry of the barrier to electrons and holes[170] can be explained by a hopping model.[171,172] The understanding of this model is based on the consideration of the nature of the interface between the Si substrate and the thermally grown amorphous SiO_2.[173] There exists an oxygen-deficient transition region, 0.4 to 4.0 nm thick.[173,174] The SiO_2 structure is not obtained until a film thickness of 5 nm.[175] The insulating layer of the Si/oxide Schottky barrier cells is on the order of 2 nm,[111] and therefore the nonstoichiometry prevails over the larger part of the insulating layer. UPS experiments can give an estimate of the energy distribution of the associated electronic states in SiO_x.[176] As presented in Fig. 35, a large concentration of states extends from the SiO_2 valence-band edge ca. 4 eV into the band gap. Comparison of the distribution of states with the Si band diagrams (Fig. 35) shows that $N(SiO_x)$ at E_{VB} (Si) $\geq N(SiO_x)$ at $E_{CB}(Si)$. Since states of $E_{VB}(Si)$ are effective in mediating hole transport, it follows that this process is favored in relation to the electron hopping, which require states at E_{CB}. On the basis of the UPS spectra, the results of Figs. 33 and 34 can be interpreted in terms of two mechanisms:

1. The freshly prepared insulating films are thin enough to permit direct tunneling of electrons and holes. Heating in air causes film growth by additionally formed SiO_x, and hence interfacial electron transfer is blocked while hole transport proceeds via SiO_x states.

2. Transport of electrons and holes across the insulating film occurs via SiO_x states in the freshly prepared cells.[173,177] During heat treatment, SiO_x transforms into SiO_2 and the concentration of SiO_x electronic states is reduced, leading to deterioration of the photovoltaic characteristics.

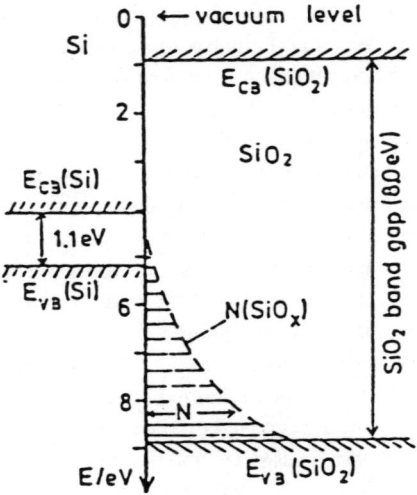

Figure 35. Schematic representation of the distribution of SiO_x electronic states, $N(SiO_x)$, as determined by UPS. The energetic positions of the Si and SiO_2 valence-band and conduction-band edges (E_{VB}, E_{CB}) are included.

The latter mechanism, represented in Fig. 36, is supported by the fact that it is difficult for SiO_x to grow at 200 °C, although its preparation under an open atmosphere at $T > 400$ °C has been reported.[111] The beneficial effect of the interfacial SiO_x layer results from the particular distribution of hopping states. It leads to reduction of the majority carrier thermionic emission current while the transport of the photogenerated minority carriers is not hindered, as shown in Fig. 36. Changes in the stoichiometry of the SiO_x to SiO_2 as explained by mechanism 2 or even an increase in the thickness of the SiO_x layer during cell preparation, aging, or thermal stress leads to the deterioration of the cell quality.

2. Photovoltaic Characteristics

In this section we will demonstrate the effect of foreign atom incorporation on the photovoltaic characteristics of the n-Si/oxide system. Foreign atom incorporation in the oxide matrix (either SnO_2 or TiO_2)

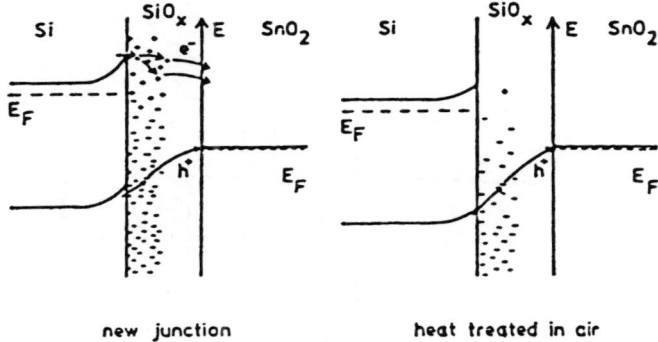

Figure 36. Schematic energy diagram of the Si/SiO$_x$/SnO$_2$ interface during photocurrent flow. The symbols + and − in the SiO$_x$ layer indicate unoccupied and occupied electronic states. E_F is the Fermi energy level. The arrows symbolize the flow of majority carriers (e^-) and photogenerated holes (h^+) across the heterojunction.

affects the position of the Fermi level. The solar conversion efficiency of the n-Si/oxide junction will increase when the Fermi level assumes lower values, and vice versa, as illustrated in Fig. 37. Typical power characteristics of n-Si/SnO$_2$ under illumination [100-mW/cm^2 simulated solar spectrum (AM1)] are presented in Fig. 38. The figure presents the photocurrent–voltage curves for junctions with pure SnO$_2$ (curve 1), SnO$_2$– 1% Ru (curve 2), and SnO$_2$ with RuO$_2$ deposited at a concentration of 20 μg/cm^2 (curve 3) and 100 μg/cm^2 (curve 4). The results show that the presence of Ru in the SnO$_2$ matrix leads to a decrease in the open-circuit potential from ca. 600 mV to ca. 450 mV. With the same parameters of preparation and measurements adjusted, the open-circuit potential could be reproduced within ca. 30 mV. The variation is attributed to the SiO$_x$ barrier layer formed on Si during the preparation of the heterojunction, as discussed before. Deposition of RuO$_2$ onto the SnO$_2$ surface up to a concentration of 20 μg/cm^2 did not affect either the open-circuit potential or the short-circuit current significantly (curve 3). Further increase of the concentration of the RuO$_2$ deposit (>20 μg/cm^2) is accompanied by a decrease in the short-circuit current obtained using the same illumination intensity. The short-circuit current of a cell containing 100 μg of RuO$_2$ per square centimeter of SnO$_2$ surface is approximately half that of a cell with pure SnO$_2$ or RuO$_2$ deposited at a concentration of less than 20 μg/cm^2.

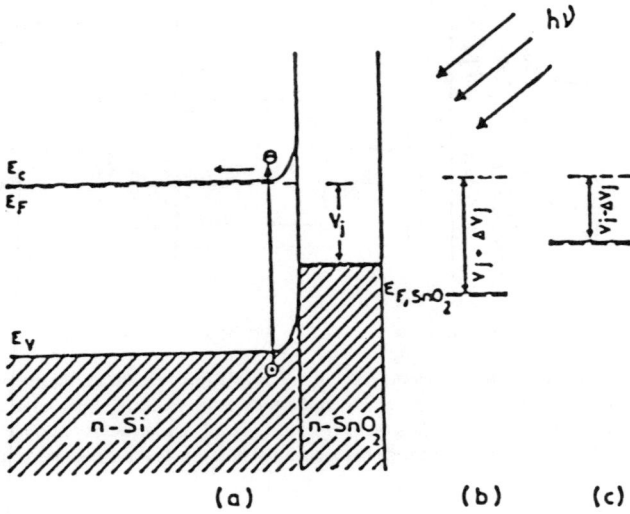

Figure 37. Effect of doping of an SnO_2 film with foreign materials on the junction potential V_j. (a) Junction potential of pure SnO_2; (b) increased junction potential due to lowering of the Fermi level of SnO_2; (c) decreased junction potential (effect of Ru).

The open-circuit potential of all these cells remains about the same (compare curves 1, 3, and 4 in Fig. 38). The important effect of foreign atom incorporation is the increase of the fill factor of the cells—from 0.615 for the pure SnO_2 film to 0.723 for the SnO_2–1% Ru film. The solar conversion efficiency remains in the same range, about 11%. Controlled deposition of RuO_2 on the SnO_2 surface to a concentration of $\leq 20\,\mu g/cm^2$ does not affect the open-circuit potential, while an increase in the cell fill factor takes place. This is reflected in an increase in the solar conversion efficiency, which reaches a value of 12.5%. The incorporation of Sb or In in the SnO_2 matrix did not have significant effects on the cell efficiency compared to the effect of Ru. The important effect is the increase of the fill factor by more than 25%.[5,59,63,78] The effect of foreign atom incorporation on the photovoltaic characteristics of n-Si/TiO_2 is illustrated in Fig. 39. In this figure typical power characteristics of both n-Si/TiO_2 and n-Si/TiO_2–1% Sb solar cells under [illumination 100-mW/cm² simulated solar spectrum (AM1)] are presented. From these results, it is clear that the incorporation of Sb in the TiO_2 material even at a low concentration,

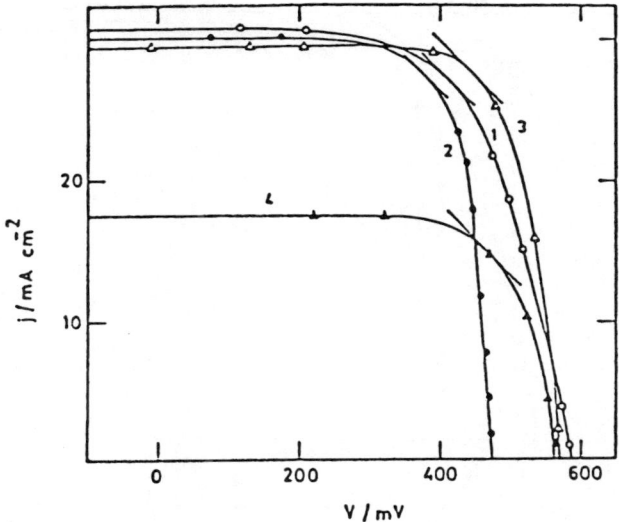

Figure 38. Typical power characteristics of n-Si/SnO$_2$ and n-Si/SnO$_2$–Ru photovoltaic cells (solid-state devices). Illumination: 100 mW/cm^2 simulated solar spectrum (AM1), ○, Pure SnO$_2$; ●, SnO$_2$ with Ru incorporated; △, SnO$_2$ with RuO$_2$ deposited on the surface (20 μg/cm^2); ▲, SnO$_2$ with RuO$_2$ deposited on the surface (ca. 100 μg/cm^2).

determined by ESCA to be 1%, significantly improves the power characteristics of the solar cells. The fill factor increased from ca. 0.6 (pure TiO$_2$) to ca. 0.8 (TiO$_2$–Sb). This improvement manifests itself in the solar conversion efficiency of the prepared cells: An increase from ca. 10% (pure TiO$_2$) to ca. 14% (TiO$_2$–Sb) was obtained.[77] Similar results were obtained with n-Si/TiO$_2$–1% In cells.[78,98] The presence of In or Sb in the TiO$_2$ surface film has its main effect on the open-circuit potential of the photovoltaic device; an increase of over 100 mV was recorded. The presence of such foreign atoms at concentrations of up to 1% has no significant effect on the measured photocurrent. Under these illumination conditions, a value between 27 and 30 mA/cm^2 was recorded.[69,98]

3. Photoelectrochemical Characteristics

In one of our publications,[58] a mathematical description of the heterojunction photocells in contact with an electrochemical system was

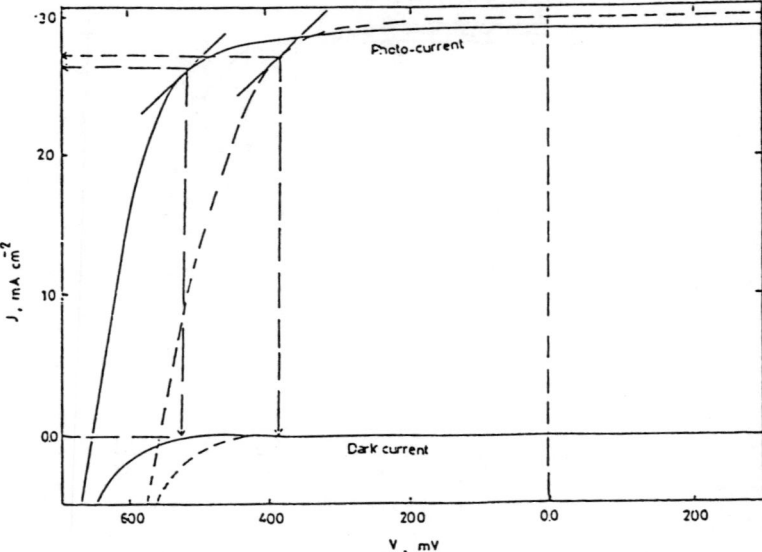

Figure 39. Typical power characteristics of n-Si/TiO$_2$ (- - -) and n-Si/TiO$_2$–Sb (——) photovoltaic cells. Illumination intensity: 100 mW/cm^2, simulated solar spectrum (AM1).

derived. Thereby, the photovoltaic characteristics as well as the mass-transport and charge-transfer overvoltage of the coupled electrochemical reaction are considered. The current limitation by either the photon flux or the rate of redox ion diffusion is well characterized. A low charge-transfer rate was demonstrated to have a deteriorating effect on the fill factor. The redox system Fe(CN)$_6^{3-/4-}$ was shown to have a rather low electrochemical rate constant on the SnO$_2$ electrode, which leads to a very low fill factor (0.28 under AM1 illumination), compared to the value of 0.62 that would be obtained under the same conditions if the electrode reactions were electrochemically reversible. The solid-state characteristics of the Si/oxide heterojunction meet the requirements of an efficient photovoltaic cell. It is necessary to increase the rate of the electrochemical reaction at the oxide/electrolyte interface in order to improve the performance of photoelectrochemical devices based on these systems. A slow electrochemical charge-transfer step leads to deterioration of the cell quality. The photoelectrochemical behavior of n-Si/SnO$_2$ with different concentrations

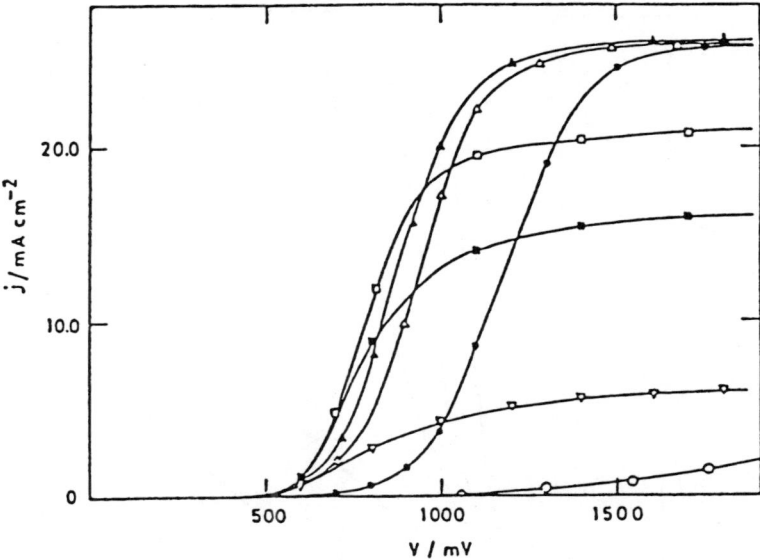

Figure 40. Current–voltage curves of Si/SnO$_2$/3.5M NaCl photoelectrochemical cells using pure, doped, and surface-improved SnO$_2$ films at 298 K. Light intensity: 100 mW/cm^2. Scan rate: 50 mV/s; speed of rotation: 1500 rpm. o, Pure SnO$_2$; ●, 1% incorporation of Ru; Δ, 5 μg RuO$_2$/cm^2 on SnO$_2$ surface; ▲, 10 μg RuO$_2$/cm^2; □, 25 μg RuO$_2$/cm^2; ■, 100 μg RuO$_2$/cm^2; ▽, 400 μg RuO$_2$/cm^2.

of Ru either incorporated in the SnO$_2$ film or deposited as RuO$_2$ on the SnO$_2$ surface as well as with no Ru present is shown in Fig. 40. The results show that the photocurrent is very sensitive to the concentration of RuO$_2$ on the SnO$_2$ surface. The photoelectrochemical performance of the cell is at a maximum when the concentration of RuO$_2$ is in the range of 5–20 μg per square centimeter of SnO$_2$ surface. At higher concentrations of RuO$_2$, drastic decreases in the photocurrent and also in the fill factor are obtained. The open-circuit potential remains approximately unchanged. The electrochemical performance of the SnO$_2$ surface improves as the amount of deposited RuO$_2$ increases. This effect is clear if we consider the behavior of films deposited on glassy carbon, as presented in Fig. 41. The electrochemical behavior of SnO$_2$ gradually approaches the behavior of pure RuO$_2$ as the amount of deposited RuO$_2$ increases. The improvement is clearly reflected in the charge-transfer coefficients obtained from the

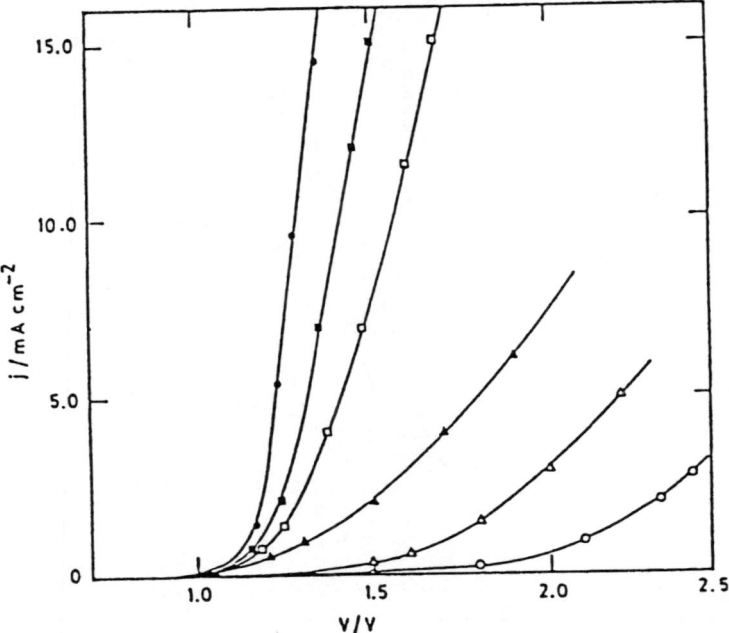

Figure 41. Current–voltage curves obtained with pure and modified SnO_2 films deposited on glassy carbon in 3.5M NaCl at 298 K. Scan rate: 50 mV/s; speed of rotation: 1500 rpm. o, Pure SnO_2; △, 5 μg RuO_2/cm^2 on SnO_2 surface; ▲, 10 μg RuO_2/cm^2; □, 25 μg RuO_2/cm^2; ■, > 100 μg RuO_2/cm^2; ●, pure RuO_2 electrode.

corresponding Tafel plots. The Tafel slopes and charge-transfer coefficients for the Cl_2 evolution reaction on the different electrodes are summarized in Table 8. The electrodes of high electrochemical performance ($RuO_2 \geq 100$ μg/cm^2) have poor photovoltaic performance (cf. Fig. 39). The analysis of the current–voltage characteristics of heterojunction photoelectrochemical cells is based on the concept of separating the photovoltaic junction from the electrochemical properties[59,60] using the following equation:

$$V_j = V_{cell} + \eta + JR_s \qquad (44)$$

where V_j is the junction potential, V_{cell} is the cell potential, η is the overpotential of the electrochemical reaction at the oxide/electrolyte in-

Table 8
Kinetic Parameters for the Anodic Oxidation of Cl^- at Pure and Surface Improved SnO_2 and Pure RuO_2 Rotating Disk Electrodes in $3.5M$ NaCl at 298 K [a]

Electrode	a (V)	b (mV)
Pure SnO_2	−2.82	339
SnO_2 with RuO_2 deposited on surface		
5 μg RuO_2/cm^2	−2.04	274
10 μg RuO_2/cm^2	−0.58	270
>20 μg RuO_2/cm^2	−0.36	160
Pure RuO_2	−0.14	123

[a]Scan rate, 50 mV/s; rotation speed, 1500 rpm.

terface, and RJ_s is the ohmic drop due to any ohmic series resistance. In this equation the junction potential V_j is given by the sum of the voltage drops in the whole system, including ohmic drops (JR_s) and the overvoltage of the electrochemical reaction. The typical current–voltage curves of the illuminated junction obtained from solid-state systems demonstrate how the Si/oxide junction may be protected from deterioration due to foreign atom incorporation. In the example given, although RuO_2 is essential for catalyzing the chlorine evolution reaction, its presence throughout the oxide film matrix decreases the open-circuit potential of the solar cell. The presence of RuO_2 as a deposit on the SnO_2 surface provides the specific catalytically active Ru sites required for the oxidation of Cl^-. Such deposition must be controlled, or otherwise the photocurrent will decrease.

The exchange current density is determined from the current density–potential plots. Near the equilibrium potential, the linearized form of the Butler–Volmer equation is conveniently used:

$$j = j_0 \frac{zF}{RT} \eta_{ct} \tag{45}$$

The exchange current density values for the $Fe(CN)_6^{3-/4-}$ redox system are very close to those obtained earlier for pure SnO_2 films and SnO_2 films with Ru or Sb incorporated. The value of the rate constant, k_0, in an electrolyte containing $0.05M$ K_3 $Fe(CN)_6$ + $0.05M$ K_4 $Fe(CN)_6$ + $0.5M$ KNO_3 was 1.5×10^{-2} cm/s compared to a value of 2×10^{-1} cm/s for the same system on Pt.[178] The increased values of k_0 and hence the exchange current density reflect the improvement in the performance of the SnO_2 electrodes afforded by Ru deposits or even by Ru incorporation. For the chlorine evolution reaction, Tafel slopes for pure SnO_2 electrodes range between 300 and 400 mV per decade, owing to the charge-transfer limitations across the electrode/electrolyte interface since the electrode (SnO_2 or even TiO_2) represents an n-type semiconductor. The presence of incorporated Ru or RuO_2 deposits increases the rate of charge transfer, and hence lower values of the Tafel slopes are obtained (cf. Table 8).

Foreign atom incorporation in TiO_2 improves the electrochemical properties of the oxide. Thus improvement is demonstrated by the increased Tafel slopes obtained with TiO_2 containing 1% Sb or In. As an example, the Tafel plots obtained with pure TiO_2 and TiO_2–1% Sb are presented in Fig. 42. The improvement in the electrochemical characteristics of TiO_2 film afforded by foreign atom incorporation is reflected in the photoelectrochemical behavior of the photoanodes. Figure 43 presents the photocurrent–potential behavior of n-Si/TiO_2/electrolyte and

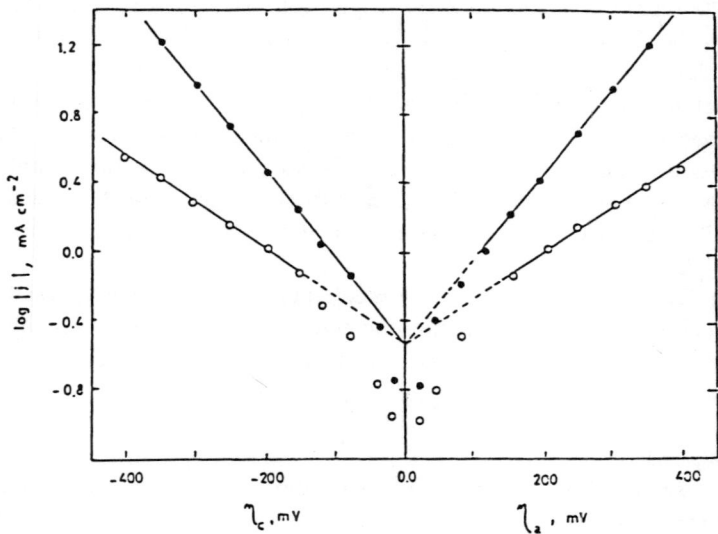

Figure 42. Tafel plots obtained with pure (○———○) and 1% Sb-doped (●- - - -●) TiO_2 electrodes in $0.05M$ $K_3Fe(CN)_6$ + $0.05M$ $K_4Fe(CN)_6$ + $0.5M$ KNO_3 at 298K. Scan rate: 50 mV s^{-1}; speed of rotation: 2500 rpm.

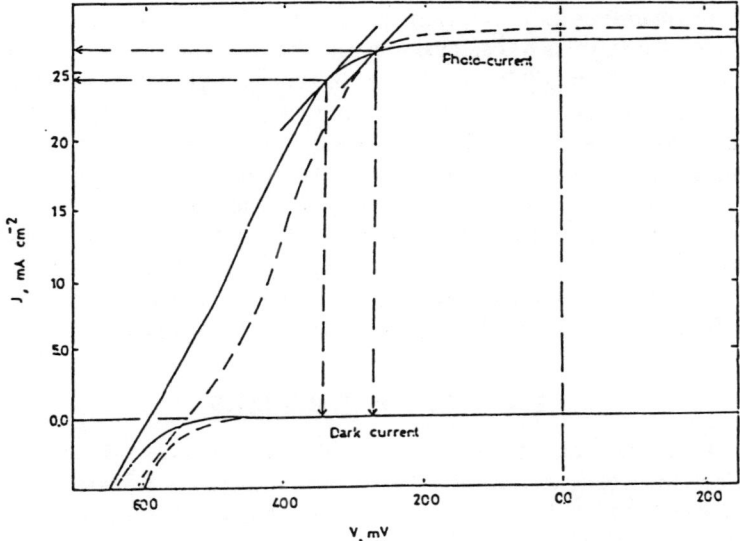

Figure 43. Current density–potential curves of n-Si/TiO$_2$ (- - -) and n-Si/TiO$_2$–Sb (——) photoanodes in $0.05M$ K$_3$Fe(CN)$_6$ + $0.05M$ K$_4$Fe(CN)$_6$ + $0.5M$ KNO$_3$ at 198 K. Scan rate: 50 mV/s; rotation speed: 2500 rpm. Illumination intensity: 100 mW/cm^2 (AM1).

n-Si/TiO$_2$–1% Sb/electrolyte systems. As in the case of photovoltaic systems, foreign atom incorporation has no effect on the photocurrent. The value of the saturation current density under AM1 illumination lies between 26 and 29 mA/cm^2 for these systems. The improved charge-transfer

Table 9
Photovoltaic Parameters of n-Si/Oxide/Metal and n-Si/Oxide/Electrolyte Solar Cells Under 100-mW/cm^2 Illumination (AM1)

Solar cell	J_{sc} (mA/cm^2)	V_{oc} (mV)	Fill factor	η (%)
n-Si/SnO$_2$/M	27	560	0.45	6.70
n-Si/SnO$_2$–(1%)In/M	26	530	0.62	8.51
n-Si/SnO$_2$/electrolyte	27	490	0.34	4.50
n-Si/SnO$_2$–In/electrolyte	26	440	0.49	5.60
n-Si/TiO$_2$/M	30	540	0.61	9.90
n-Si/TiO$_2$–(1%)In/M	28.5	620	0.81	14.0
n-Si/TiO$_2$/electrolyte	27.5	530	0.43	6.25
n-Si/TiO$_2$–In/electrolyte	27	580	0.54	8.58

kinetics manifests itself mainly in the fill factor of the photoelectrochemical cell. An increase of ca. 25% was calculated,[77] which is reflected, in turn, in the efficiency of the photoelectrochemical cells: The solar conversion efficiency increased from 6.7% (pure TiO_2) to 8.1% (TiO_2–1% Sb). The open-circuit potential did not show a significant increase with foreign atom incorporation, as was observed with the corresponding photovoltaic cells. A comparison between the photovoltaic and photoelectrochemical conversion efficiencies of the n-Si/SnO_2 and n-Si/TiO_2 systems, either pure or with foreign atoms incorporated, is presented in Table 9. The values of the short-circuit current, J_{sc}, the open-circuit potential, V_{oc}, and the fill factor, FF, for the different cells are also summarized.

IX. CONCLUDING REMARKS

Heterojunctions of the type n-Si/oxide represent a class of Schottky barrier solar cells that could be very useful for practical applications. Modified SnO_2 or TiO_2 films ~100 nm in thickness are very useful for this purpose. Modification of the oxide film can be achieved by incorporation of small amounts of foreign atoms such as In, Ru, Sb, etc. It is always beneficial to modify the oxide film without affecting the n-Si/oxide heterojunction (i.e., the solid state heterojunction). Improvement of the charge-transfer kinetics at the oxide/electrolyte interface in photoelectrochemical cells can be carried out by depositing controlled amounts of a catalyst like RuO_2 on top of the oxide film. Foreign atom incorporation in the oxide film at levels of up to 1% improves the solid-state characteristics, especially in the case of TiO_2, without affecting the film transparency. Solar conversion efficiencies amounting to 14% can be obtained with the oxide-modified photovoltaic cells. Photoelectrochemical cells having efficiencies of up to 9 or 10% can be obtained by modifying the oxide film. The control of the insulating SiO_x film at the n-Si/oxide interface is very important; the thickness of this insulating film should not exceed 1–2 nm. Highly stoichiometric SiO_x or thicker films lead to deterioration of the cell quality.

ACKNOWLEDGMENTS

The author is very grateful to the Max-Planck Society and colleagues at the Fritz-Haber Institute in Berlin, Germany, for their helpful cooperation in establishing the basic material for this work. The author will always remember Professor Heinz Gerischer, who gave him the opportunity to learn a lot in this field.

Thanks are also due to the Alexander von Humboldt Foundation and colleagues at the Free University of Berlin in Germany. To my students and colleagues at the Cairo University and National Research Center in Egypt, I wish to express my heartfelt thanks and gratitude for their patience and hard work in establishing our measuring systems at the Cairo University. I would also like to express my appreciation and gratitude to Ms. Amany S. Al-Azab for her assistance in the typing and final preparation of the manuscript.

REFERENCES

[1] J. Millman, *Vacuum Tube and Semiconductor Electronics*, McGraw-Hill, New York, 1985, Chapters 3 and 5.
[2] S. W. Angrist, *Direct Energy Conversion*, 3rd ed., Allyn & Bacon, Boston, 1976, Chapters 3 and 5.
[3] A. B. Meinel and M. P. Meinel, *Applied Solar Energy*, Addison-Wesley, Reading, Massachusetts, 1976, Chapter 15.
[4] B. J. Brinkworth, *Solar Energy for Man*, John Wiley & Sons, New York, 1972, Chapter 8.
[5] W. A. Badawy, *J. Electroanal. Chem.* **281** (1990) 85.
[6] A. Fujishima and K. Honda, *Nature (London)* **238** (1972) 37.
[7] A. Nozik, *Nature, (London)* **257** (1975) 383.
[8] M. S. Wrighton, D. S. Ginley, P. T. Tolczanaki, A. S. Ellis, A. S. Morse, and A. Linz, *Proc. Natl. Acad. Sci. U.S.A.* **72** (1975) 1518.
[9] J. G. Mavroides, J. A. Kefalas, and D. F. Kolesar, *Appl. Phys. Lett.* **28** (1976) 241.
[10] K. C. Hardee and A. J. Bard, *J. Electrochem. Soc.* **123** (1976) 1024.
[11] J. O'M. Bockris and S. U. M. Khan, *Surface Electrochemistry*, Plenum Press, New York, 1993, Chapter 5.
[12] J. A. Merrigan, *Sunlight to Electricity*, MIT Press, Cambridge, Massachusetts, 1975.
[13] D. L. Pulfrey, *Photovoltaic Power Generation*, Van Nostrand Reinhold, New York, 1978.
[14] C. Kittel, *Elementary Solid State Physics*, John Wiley and Sons, New York, 1962, Chapters 6 and 7.
[15] Sol. Wieder, *An Introduction to Solar Energy for Scientists and Engineers*, Krier Publisher, Florida, 1992, Chapter 5.
[16] J. Manassen, D. Calven, and G. Hodes, *Nature (London)* **263** (1976) 97.
[17] H. Gerischer and J. Gobrecht, *Ber. Bunsenges. Phys. Chem.* **80** (1976) 327.
[18] A. J. Bard, *J. Photochem.* **10** (1979) 59.
[19] H. Ehrenreich and J. Martin, *Phys. Today* **32** (1979) 25.
[20] H. Gerischer, *Pure Appl. Chem.* **52** (1980) 2649.
[21] A. J. Bard, *Science* **207** (1980) 139.
[22] A. Heller and B. Miller, *Electrochim. Acta* **25** (1980) 29.
[23] R. Memming, *Electrochim. Acta* **25** (1980) 77.
[24] A. Heller, *Sol. Energy* **29** (1982).
[25] H. Gerischer, in *Advances in Electrochemistry and Electrochemical Engineering*, Vol. 1, Ed. by P. Delahay, Interscience, New York, 1961, Chapter 4.

[26] M. D. Arsher, *J. Appl. Electrochem.* **5** (1975) 17.
[27] K. Rajieshwar, P. Singh, and J. Dubow, *Electrochim. Acta* **20** (1975) 1117.
[28] L. Harris and R. H. Wilson, *Annu. Rev. Mater. Sci.* **8** (1978) 99.
[29] A. J. Nozik, *Annu. Rev. Phys. Chem.* **29** (1978) 189.
[30] M. Tomkiewicz and H. Fay, *Appl. Phys.* **18** (1979) 1.
[31] M. Green, in *Modern Aspects of Electrochemistry*, No.2, Ed. by J. O'M. Bockris, Academic Press, New York, 1980, Chapter 5.
[32] S. R. Morrison, *Electrochemistry at Semiconductor and Oxidized Metal Electrodes*, Plenum Press, New York, 1980.
[33] R. H. Wilson, *CRC Crit. Rev. Solid State Mater. Sci.* **10** (1980) 1.
[34] R. Memming, in *Electroanalytical Chemistry*, Vol. 11, Ed. by A. J. Bard, Marcel Dekker, New York, 1981.
[35] M. M. Khan and J. O'M. Bockris, in *Modern Aspects of Electrochemistry*, No. 14, Ed. by J. O'M. Bockris, B. E. Conway, and R. E. White, Plenum Press, New York, 1982, Chapter 3.
[36] K. Usaki and H. Kita, in *Modern Aspects of Electrochemistry*, No. 18, Ed. by R. E. White, J. O'M. Bockris, and B. E. Conway, Plenum Press, New York, 1986, Chapter 1.
[37] M. E. Orazem and J. Newman, in *Modern Aspects of Electrochemistry*, No. 18, Ed. by R. E. White, J. O'M. Bockris, and B. E. Conway, Plenum Press, New York, 1986, Chapter 2.
[38] W. H. Brattain and G. G. B. Garrett, *Bell Syst. Tech. J.* **34** (1955) 128.
[39] R. Williams, *J. Chem. Phys.* **32** (1960) 1505.
[40] H. Gerischer, *J. Electroanal. Chem.* **58** (1975) 263.
[41] B. Parkinson, *Acc. Chem. Res.* **17** (1984) 431 and references therein.
[42] M. Graetzel, in *Modern Aspects of Electrochemistry*, No. 15, Ed. by R. White, J. O'M. Bockris, and B. E. Conway, Plenum Press, New York, 1983, pp. 83–165.
[43] A. G. Milnes and D. L. Feucht, *Heterojunctions and Metal/Semiconductor Junctions*, Academic Press, New York, 1972.
[44] J. O'M. Bockris and A. K. N. Reddy, *Modern Electrochemistry*, Vol. 2, Plenum Press, New York, 1970, Chapter 7.
[45] S. Trasatti, *J. Electroanal. Chem.* **139** (1982) 1.
[46] H. Gerischer, in *Photovoltaic and Photoelectrochemical Solar Energy Conversion*, Ed. by F. Cardon, W. P. Gomes, and W. Dekeyser, NATO ASI Summer School, August 1980, Gent, Belgium, Vol. B69, Plenum Press, New York, 1981, pp. 129–262.
[47] S. R. Morrison, *Electrochemistry at Semiconductor and Oxidized Metal Electrodes*, Plenum Press, New York, 1980, Chapter 1.
[48] J. J. Loferski, *J. Appl. Phys.* **27** (1956) 777.
[49] C. F. Fisher, *Festkörperprobleme*, Vol. 14, Pergamon Press, Braunschweig, 1974.
[50] G. F. Neumark and K. Kosai, in *Semiconductors & Semimetals*, Vol. 19, Academic Press, New York, 1983.
[51] S. R. Morrison, in *Photoelectrochemical Solar Cells*, Ed. by K. S. V. Santhanan and M. Sharon, Elsevier, New York, 1988, Chapter 2.
[52] W. W. Gaertner, *Phys. Rev.* **116** (1959) 84.
[53] H. Gerischer, in *Semiconductor-Liquid-Junction Solar Cells*, Ed. by A. Heller, The Electrochemical Society, Princeton, New Jersey, 1977, p. 1.
[54] D. Laser and A. J. Bard, *J. Electrochem. Soc.* **123** (1976) 1828.
[55] L. A. Abrantes, R. Peat, L. M. Peter, and A. Hamnett, *Ber. Bunsenges. Phys. Chem.* **91** (1987) 369.
[56] J. Bardeen, *Phys. Rev.* **71** (1947) 71.
[57] R. A. Batchelor and A. Hamnett, in *Modern Aspects of Electrochemistry*, No. 22, Ed. by J. O'M. Bockris, B. E. Conway, and R. E. White, Plenum Press, New York, 1992, p. 265.

[58] F. Decker, H. Fracastero-Decker, W. A. Badawy, K. Doblhofer, and H. Gerischer, *J. Electrochem. Soc.* **130** (1983) 2173.
[59] W. A. Badawy, *Ind. J. Technol.* **24** (1986) 118.
[60] R. Memming, in *Comprehensive Treatise of Electrochemistry*, Vol. 7, Ed. by B. E. Conway, J.O' M. Bockris, E. Yeager, S. U. M. Khan, and R. E. White, Plenum Press, New York, 1983, Chapter 8.
[61] A. J. Bard and M. S. Wrighton, *J. Electrochem. Soc.* **124** (1977) 1706.
[62] R. Memming, *J. Electrochem. Soc.* **125** (1978) 117.
[63] W. A. Badawy, H. H. Afify, and E. M. El-Giar, *J. Electrochem. Soc.* **137** (1990) 1592.
[64] W. A. Badawy, R. S. Momtaze, H. H. Afify, and E. M. El-Giar, *J.Mater. Sci. Materials in Electronics* **2** (1992) 11 2.
[65] R. Memming, *Ber. Bunsenges. Phys. Chem.* **81** (1977) 732.
[66] J. Manassen, G. Hodes, and D. Cahen, *J. Electrochem. Soc.* **124** (1977) 532.
[67] A. J. Nozik, *Semiconductor Liquid Junction Solar Cells*, Ed. by A. Heller, Proceedings Vol. 77-3, The Electrochemical Society, Princeton, New Jersey, (1977).
[68] M. S. Wrighton, A. B. Ellis, P. T. Wolczanski, D. L. Morse, and H. B. Abrahamson, *J. Am. Chem. Soc.* **98** (1976) 44.
[69] M. S. Wrighton, A. B. Ellis, P. T. Wolczanski, D. L. Morse, and D. S. Ginby, *J. Am. Chem. Soc.* **98** (1976) 2774.
[70] J. M. Bolts and M. S. Wrighton, *J. Phys. Chem.* **80** (1976) 2641.
[71] T. Ohmiski, Y. Nakato, and H. Tsubomura, *Ber. Bunsenges. Phys. Chem.* **79** (1975) 523.
[72] K. L. Hardee and A. J. Bard, *J. Electrochem. Soc.* **122** (1975) 739.
[73] L. A. Harris and R. H. Wilson, *J. Electrochem. Soc.* **123** (1976) 1010.
[74] D. Laser and A. J. Bard, *J. Electrochem. Soc.* **123** (1976) 1027.
[75] A. K. Ghosh and H. P. Maruska, *J. Electrochem. Soc.* **124** (1977) 1516.
[76] W. Gissler, P. L. Lensi, and S. Pizzini, *J. Appl. Electrochem.* **6** (1976) 9.
[77] W. A. Badawy, *Sol. Energy Mater. Sol. Cells* **28** (1993) 293.
[78] W. A. Badawy, 186th Electrochemical Society Meeting, Miami, Florida, 1994; Corrosion and Dielectric Science and Technology Divisions Proceedings Vol. 94–25, p. 292.
[79] F. Moellers and R. Memming, *Ber. Bunsenges. Phys. Chem.* **76** (1972) 469.
[80] M. A. Butler, R. D. Nashby, and R. K. Quin, *Solid State Commun.* **19** (1976) 1011.
[81] W. Gissler and R. Memming, *J. Electrochem. Soc.* **124** (1977) 1710.
[82] K. L. Hardee and A. J. Bard, *J. Electrochem. Soc.* **123** (1976) 1024.
[83] K. L. Hardee and A. J. Bard, *J. Electrochem. Soc.* **124** (1977) 215.
[84] J. H. Kennedy and K. W. Frese, *J. Electrochem. Soc.* **125** (1978) 709; **125** (1978) 723.
[85] W. A. Badawy and E. A. El-Taher, *Thin Solid Films* **158** (1988) 277.
[86] H. Tributsch and J. C. Bennett, *J. Electroanal. Chem.* **81** (1977) 97.
[87] M. Kunst, G. Beck, and H. Tributsch, *J. Electrochem. Soc.* **131** (1984) 954.
[88] L. Borrell, S. Cervera-March, J. Gimenez, R. Simmaro, and J. M. Andujar, *Sol. Energy Mater. Sol. Cells* **25** (1992) 25.
[89] H. J. Lewerenz and J. Stumper, *Ber. Bunsenges. Phys. Chem.* **92** (1988) 1350.
[90] A. J. Nozik, *Appl. Phys. Lett.* **29** (1976) 150.
[91] H. Tsubomura, M. Matsumura, K. Nakatani, and Y. Nomura, in *Semiconductor Liquid Junction Solar Cells*, Ed. by A. Heller, Proceedings Vol. 77-3, The Electrochemical Society, Princeton, New Jersey, 1977.
[92] M. A. Butler and D. S. Ginley, *J. Electrochem. Soc.* **125** (1978) 223.
[93] L. A. Harris and R. H. Wilson, *Am. Rev. Mater. Sci.* **8** (1978) 99.
[94] H. P. Maruska and A. K. Ghosh, *Sol. Energy Mater.* **1** (1979) 237.
[95] H. P. Maruska and A. K. Ghosh, *Sol. Energy* **20** (1979) 433.
[96] Y. Matsumoto, J. Kurimoto, Y. Amagasaki, and E. Sats, *J. Electrochem. Soc.* **127** (1980) 2360.
[97] J. H. Wohlgemuth, D. B. Warfield, and G. A. Johnson, IEEE (1982) 809.

[98] W. A. Badawy, International Symposium on Electrochemical Science & Technology (ISEST), Hong Kong, August 24–26, 1995, pL13, pp. 1–8.
[99] T. Kuwana, R. K. Dovington, and D. W. Leedy, *Anal Chem.* **36** (1964) 2023.
[100] W. N. Hansen, T. Kuwana, and R. A. Osteryoung, *Anal. Chem.* **38** (1966) 1810.
[101] J. W. Strojik and T. Kuwana, *J. Electroanal. Chem.* **16** (1960) 471.
[102] J. W. Strojik, *Chem. Anal. (Warsaw)* **17** (1972) 1023.
[103] H. N. Blount, N. Winograd, and T. Kuwana, *J. Phys. Chem.* **72** (1970) 3231.
[104] M. Nagasawa and S. Shionoya, *Jpn. J. Appl. Phys.* **10** (1971) 472, 727.
[105] H. A. Laitinen, C. A. Vincent, and T. M. Bednarstki, *J. Electrochem. Soc.* **115** (1968) 1024.
[106] O. Elliot, D. L. Zellmer, and H. A. Laitinen, *J. Electrochem. Soc.* **117** (1970) 1343.
[107] R. Memming and F. Moellers, *Ber. Bunsenges. Phys. Chem.* **76** (1972) 475.
[108] T. Nishino and Y. Hamakawa, *Jpn. J. Appl. Phys.* **9** (1970) 299.
[109] T. Feng, A. Ghosh, and C. Fishman, *Appl. Phys. Lett.* **35** (1979) 266.
[110] I. Chambouleyron and E. Saucedo, *Sol. Energy Mater.* **1** (1979) 299.
[111] W. A. Badawy, F. Decker, and K. Doblhofer, *Sol. Energy Mater.* **8** (1983) 363.
[112] W. A. Badawy and K. Doblhofer, *Appl. Phys. A* **35** (1984) 189.
[113] W. A. Badawy, K. Doblhofer, I. Eiselt, H. Gerischer, S. Krause, and J. Melsheimer, *Electrochim. Acta* **29** (1984) 1617.
[114] H. Kato, A. Yoshido, and T. Arizume, *Jpn. J. Appl. Phys.* **15** (1976) 1819.
[115] R. N. Ghoshtagor, in *International Conference on Chemical Vapor Deposition (CVD) VI*, Proceedings Vol. 77-5, The Electrochemical Society, Princeton, New Jersey, 1977, p. 433.
[116] W. A. Badawy and M. S. Morsi, *B. Electrochem.* **5** (1989) 276.
[117] J. C. Manifacier, M. de Muricia, and J. P. Fillard, *Thin Solid Films* **41** (1977) 127.
[118] P. W. Atkins, *Physical Chemistry*, 5th ed., Oxford University Press, Oxford, 1994, pp. 877–899.
[119] J. A. Aboaf, V. C. Murocotte, and N. J. Chou, *J. Electrochem. Soc.* **120** (1973) 701.
[120] T. Feng, C. Fishman, and A. K. Ghoash, *Proceedings of the 13th IEEE Photovoltaic Specialist Conference*, IEEE, New York, 1978, p. 519.
[121] H. J. Hovel, *Semiconductors and Semimetals*, Vol. 11, Ed. by R. K. Willardson, Academic Press, New York, 1975.
[122] W. Townsend, in *Photovoltaic and Photoelectrochemical Energy Conversion*, Ed. by F. Cardon, W. P. Gomes, and W. Dekeyser, Plenum Press, New York, 1981, pp. 66–155.
[123] O. P. Agnihotri, M. T. Mohammed, A. K. Abbass, and K. I. Arshak, *Solid State Commun.* **47** (1983) 195.
[124] A. C. Abbass and M. T. Mohammed, *J. Appl. Phys.* **59** (1986) 1641.
[125] A. C. Abbass, *Solid State Commun.* **61** (1987) 507.
[126] A. K. Ghosh, C. Fishman, and T. Feng, *J. Appl. Phys.* **49** (1978) 3490.
[127] H. H. Afify, R. S. Momtaz, W. A. Badawy, and S. A. Nasser, *J. Mater. Sci. Materials in Electronics* **2** (1991) 40.
[128] H. Demiryont, J. R. Sites, and K. Geib, *J. Appl. Opt.* **24** (1985) 490.
[129] Inorganic Index to Powder Diffraction File, ASTM, Powder Diffraction File, Card No. 5-0467, American Society for Testing and Materials, Philadelphia.
[130] M. U. Cohen, *Rev. Sci. Instrum.* **6** (1935) 68.
[131] A. F. Corroland and L. H. Stack, *J. Electrochem. Soc.* **123** (1976) 1889.
[132] N. S. Murty and S. R. Jawalekar, *Thin Solid Films* **100** (1983) 219.
[133] S. R. Vishwakarma, J. P. Upadhyay, and H. C. Prasad, *Thin Solid Films* **176** (1989) 99.
[134] C. Agashe, B. R. Marathe, M. G. Takawale, and V. G. Bhide, *Thin Solid Films* **164** (1988) 261.
[135] K. Tominaga, T. Yuasa, M. Kume, and O. Tada, *J. Appl. Phys.* **24** (1985) 944.
[136] H. Gzternastek, A. Brundik, and M. Jachimowski, *Solid State Commun.* **65** (1988) 1025.
[137] J. Bardeen, F. J. Slatt, and L. J. Hall, in *Photoconductivity Conference, 1954*, John Wiley & Sons, New York, 1956, p. 146.

[138] A. K. Abbass, H. Bakr, S. A. Jassim, and T. A. Fahad, *Sol. Energy Mater.* **17** (1988) 425.
[139] O. P. Agnihotri, B. K. Gupta, and A. K. Sharma, *J. Appl. Phys.* **49** (1978) 425.
[140] I. Hamberg, C. G. Granquist, K. F. Berggren, B. E. Sernelives, and L. Engstrom, *Sol. Energy Mater.* **12** (1985) 479.
[141] H. A. Leitinen, *Denki Kagaku*, **44** (1976) 626.
[142] N. R. Armstrong, A. W. C. Lin, M. Fujihira, and T. Kuwana, *Anal. Chem.* **48** (1976) 741.
[143] T. A. Chertykovtseva, Z. D. Skuridina, D. M. Shub, and V. A. Veselovskii, *Elektrokhimiya*, **14** (1978) 1412.
[144] C. Iwakura, M. Inai, T. Uemura, and H. Tamura, *Electrochim. Acta* **26** (1981) 579.
[145] J. B. Goodenough and A. Hamnet, 34th ISE Meeting, Erlangen, September, 1983, Lecture 111.3.
[146] W. Schmickler, *Ber. Bunsenges. Phys. Chem.* **82** (1976) 477.
[147] J. Ulstrup, *Charge Transfer Processes in Condensed Media*, Lecture Notes in Chemistry, No. 10, Springer, Berlin, 1979.
[148] Z. M. Jarzebski and J. P. Marten, *J. Electrochem. Soc.* **123** (1976) 299C.
[149] T. Arai, *J. Phys. Soc. Jpn.* **15** (1960) 916.
[150] H. Kanko and K. Miyake, *J. Appl.Phys.* **53** (1982) 3629.
[151] H. Kim and H. Laitinen, *J. Electrochem. Soc.* **122** (1975) 53.
[152] G. Mierdel, *Elektrophysik*, VEB Verlag Technik, Berlin, 1972, Chapter 3.4.
[153] K. Bange and J. Sass, private communication. K. Bange, Dissertation, TU Berlin, 1983.
[154] H. Gerischer, *Semiconductor Electrode Reactions*, in *Advances in Electrochemistry and Electrochemical Engineering*, Vol. 1, Ed. by P. Delahay and C. W. Tobias, Interscience, New York, 1961, p. 139.
[155] H. Gerischer, *Z. Phys. Chem.* **27** (1961) 48.
[156] K. Usaki and H. Kita, *J. Electrochem. Soc.* **130** (1983) 985; discussion by H. Gerischer, discussion section of *J. Electrochem. Soc.*, December 1983.
[157] A. Monnier and J. Augustinski, *J. Electrochem. Soc.* **127** (1980) 1576.
[158] A. K. Ghosh and H. P. Maruska, *J. Electrochem. Soc.* **124** (1977) 1516.
[159] W. A. Badawy, R. S. Momtaz, and E. M. El-Giar, *Phys. Status Solidi A* **118** (1990) 197.
[160] E. J. Kelly, in *Modern Aspects of Electrochemistry* No. 14, Ed. by J. O'M. Bockris, B. E. Conway, and R. E. White, Plenum Press, New York, 1982, pp. 319–420.
[161] A. Felske, W. A. Badawy, and W. J. Plieth, *J. Electrochem. Soc.* **137** (1990) 1804.
[162] W. A. Badawy, S. S. El-Egamy, and Kh. M. Ismail, *Br. Corros. J.* **28** (1993) 133.
[163] D. L. Pulfrey, *IEEE Trans. Electron Devices* **ED-25** (1978) 1308.
[164] S. L. Fonash, *Solar Cell Device Physics*, Academic Press, New York, 1981.
[165] H. P. Maruska, A. K. Ghosh, D. J. Eustace, and T. Feng, *J. Appl. Phys.* **54** (1983) 2489.
[166] S. T. Nash and R. L. Anderson, *IEEE Trans. Electron Devices* **ED-24** (1977) 468.
[167] S. M. Goodnick, J. F. Wagner, and C. W. Wilmsen, *J. Appl. Phys.* **51** (1980) 527.
[168] O. P. Agnihotri and B. K. Gupta, *Solar Selective Surfaces*, John Wiley & Sons, New York, 1981, Chapter 3.
[169] J. Sewchun, J. Du Bow, A. Myszkowski, and R. Singh, *J. Appl. Phys.* **49** (1980) 955.
[170] K. K. Ng and H. C. Card, *J. Appl. Phys.* **51** (1980) 2153.
[171] H. Overhof, in *Festkörperprobleme* XVI, Vieweg, Braunschweig, 1976, p. 239.
[172] S. Kar, S. Ashok, and S. J. Fonash, *J. Appl. Phys.* **51** (1980) 3417.
[173] R. Williams, *J. Vac. Sci. Technol.* **14** (1977) 1106.
[174] E. E. Aspens and J. B. Theeten, *J. Electrochem. Soc.* **127** (1980) 1359.
[175] J. P. Ponpon, R. Stuck, and P. Siffert, in *Proceedings of the 12th IEEE Photo. Spec. Conference*, 1976, p. 900.
[176] M. Schultz, *Insulating Films on Semiconductors*, Institute of Physics, Bristol, 1980, pp. 75–80.
[177] M. A. Green, *Appl. Phys. Lett.* **33** (1979) 178.
[178] M. Morita, C. Iwakura, and H. Tamura, *Electrochim. Acta* **24** (1973) 639.

3

The Mechanism of Formation of Coarse and Disperse Electrodeposits

K. I. Popov, N. V. Krstajić, and M. I. Čekerevac

*Faculty of Technology and Metallurgy, University of Belgrade,
11000 Belgrade, Yugoslavia*

I. INTRODUCTION

Morphology is probably the most important property of electrodeposited metals. It depends mainly on the kinetic parameters of the deposition process and on the deposition overpotential or current density. The mechanisms of formation of different growth forms as a function of the above variables have been recently elucidated. Hence, it is now possible to correlate the morphology of metal electrodeposits and deposition process conditions.

In this chapter, the initiation and growth of coarse (cauliflower-like and carrot-like), dendritic, spongy, and dendritic spongy deposits growth will be considered, first through the elaboration step was the explication of corresponding mathematical models, followed by physical simulation and finally comparison with real growth forms. It will be shown that current theories based mainly on the Barton–Bockris and Diggle–Despić–Bockris approaches account well for all basic aspects of the morphology of metal electrodeposits in a quantitative way.

There is some minimal overlap of the material presented in this chapter with previously published theoretical treatments in chapters and books related to this field, but completely new experimental evidence is presented.

Modern Aspects of Electrochemistry, Number 30, edited by Ralph E. White *et al.* Plenum Press, New York, 1996.

II. COARSE DEPOSITS

1. Mathematical Model

Any solid metal surface that serves as a substrate for metal electrodeposition possesses a certain coarseness. It is convenient to define the surface coarseness as the difference in thickness of the metal at the highest and lowest points above an arbitrary reference plane facing the solution. In early models in which the surface was described by a periodic function, this is equal to twice the amplitude of the function.[1]

Historically, it was first established that under certain conditions of dissolution the surface coarseness tends to decrease.[2] Krichmar[3] was the first to point out that in some cases of deposition, under conditions somewhat analogous to those in which the coarseness decreases, the opposite effect occurs; i.e., in prolonged cathodic reduction, under conditions at which the process is close to being under complete diffusion control, amplification of the surface coarseness takes place.

Taking a sinusoidal profile for the electrode

$$H = H_0 \sin\left(\frac{2\pi x}{a}\right) \tag{1}$$

Krichmar[3] obtained the relationship

$$H_0(t) = H_0 \exp\left(\frac{j}{j_L} \frac{Q}{Q_{0,K}}\right) \tag{2}$$

for $H_0(t) \ll a$, where

$$Q = jt \tag{3}$$

and

$$Q_{0,K} = \frac{\rho n F a}{2\pi M} \tanh\left(\frac{2\pi \delta}{a}\right) \tag{4}$$

In the above equations a is the wavelength of the sinusoidal profile, F is the Faraday constant, H is the local elongation, H_0 is the initial amplitude of the sinusoidal profile, $H_0(t)$ is the amplitude of the sinusoidal profile at time t, j is the current density, j_L is the limiting diffusion current density, M is the molar mass, n is the number of electrons, Q is the quantity of electricity, t is the time, x is the coordinate normal to the plane of the electrode, δ is the thickness of the diffusion layer, and ρ is the density of the metal.

Simpler mathematics were used in another, independently derived theory of the same phenomenon put forward by Despić, Diggle, and Bockris.[1,4,5] A somewhat simplified treatment will be given here.

Consider the model of a surface irregularity shown in Fig. 1. The surface irregularity is buried deep in the diffusion layer, which is characterized by a steady linear diffusion to the flat portion of the surface. The current densities at the various parts of the surface are as follows:

(a) At the flat part of the surface, the limiting diffusion current density is that for steady-state linear diffusion, i.e.,

$$j_L = \frac{nFDC_0}{\delta} \quad (5)$$

where D is the diffusion coefficient and C_0 is the bulk concentration of the depositing ions.

(b) At the side of an irregularity, even when a possible lateral diffusion flux supplying the depositing ions is neglected, the current density, j_S,

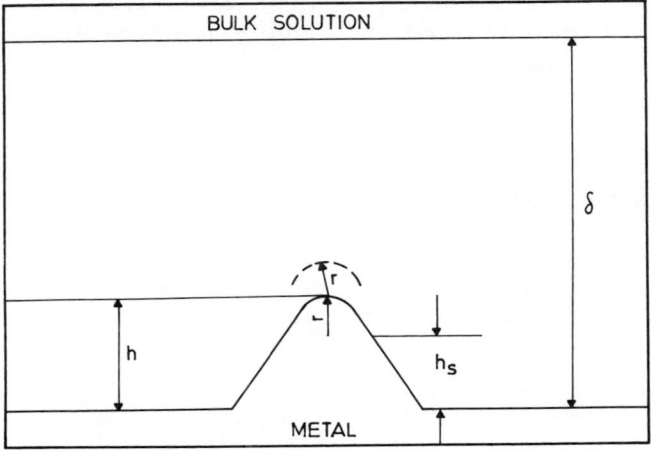

Figure 1. Model of a surface irregularity; h is the height of the protrusion relative to the flat portion of the surface, h_s is the corresponding local side elongation, r is the radius of the protrusion tip, and δ is the thickness of the diffusion layer.

at any point of height h_S must be larger than that at the flat part of surface, j. This is because the point is closer to the diffusion-layer boundary; i.e., the effective diffusion layer is thinner, and hence the diffusion flux and resulting current density are larger. Obviously, this is valid if the protrusion height does not affect the outer limit of the diffusion layer, i.e., if $\delta \gg h_S$. The limiting diffusion current density, $j_{L,S}$, is given as

$$j_{L,S} = \frac{nFDC_0}{\delta - h_S} = j_L \frac{\delta}{\delta - h_S} \qquad (6)$$

(c) At the tip of an irregularity, there is an additional reason for the increased current density. The lateral flux cannot be neglected, and the situation can be approximated by assuming a spherical diffusion current density, $j_{L,\text{tip}}$, given by

$$j_{L,\text{tip}} = \frac{nFDC^*}{r} \qquad (7)$$

where C^* is the concentration of the diffusing species at a distance r from the tip, assuming that around the tip a spherical diffusion layer having a thickness equal to the radius of the protrusion tip is formed.[6,7] If deposition to the macroelectrode is under full diffusion control, the distribution of the concentration C inside the linear diffusion layer is given by[5]

$$C = C_0 \frac{h}{\delta} \qquad (8)$$

where $0 \leq h \leq \delta$. Hence,

$$C^* = C_0 \frac{h+r}{\delta} \qquad (9)$$

and

$$j_{L,\text{tip}} = j_L \left(1 + \frac{h}{r}\right) \qquad (10)$$

because of Eqs. (5), (7), and (9).[8]

The general polarization curve equation for the flat surface is given by[9]

$$j = \frac{j_0(f_c - f_a)}{1 + \frac{j_0 f_c}{j_L}} \quad (11)$$

where j_0 is the exchange current density and

$$f_c = \exp\left(\frac{\alpha_c F \eta}{RT}\right); \quad f_a = \exp\left(\frac{-\alpha_a F \eta}{RT}\right) \quad (12)$$

where α_c and α_a are the cathodic and anodic transfer coefficients, respectively, R is the gas constant, T is the temperature, and η is the overpotential. The current densities j_S and j_{tip} to the different points of the electrode surface can then be obtained by substitution of j_L in Eq. (11) by appropriate values from Eqs. (6) and (10) as

$$j_S = \frac{j_0(f_c - f_a)}{1 + \left(\frac{j_0 f_c}{j_L}\right)\left(\frac{\delta - h_S}{\delta}\right)} \quad (13)$$

and

$$j_{tip} = \frac{j_0(f_c - f_a)}{1 + \left(\frac{j_0 f_c}{j_L}\right)\left(\frac{r}{r + h}\right)} \quad (14)$$

for the side and the tip of the protrusion, respectively. It should be noted that Eq. (14) is valid only if the radius of the protrusion tip is sufficiently large to make the surface energy term negligible.[6] Obviously, the diffusion layer around the tip of the electrode surface protrusion can be formed only if the condition

$$r < \delta - h \quad (15)$$

is satisfied.

The effective rate of growth of the side elevation is equal to the rate of motion of the side elevation relative to the rate of motion of the flat surface[1]:

$$\frac{dh_S}{dt} = \frac{M}{\rho n F}(j_S - j) \quad (16)$$

Substitution of j_S from Eq. (13) and j from Eq. (11) in Eq. (16) and further rearrangement gives[8]

$$\frac{dh_S}{dt} = \frac{j^2 M h_S}{j_L \rho n F \delta} \quad (17)$$

if $\delta \gg h_S$ and $f_c \gg f_a$, or in the integral form

$$h_S = h_{0,S} \exp\left(\frac{jQ}{j_L Q_{0,D}}\right) \quad (18)$$

just as in the previous case (Eq. 2), where $h_{0,S}$ is the initial height of the local side elevation, Q is given by Eq. (3), and

$$Q_{0,D} = \frac{nF\delta\rho}{M} \quad (19)$$

According to both mechanisms [Eqs. (2) and (18)], an increase of the surface coarseness can be expected with increasing quantity of deposited metal for the same deposition current density as well as with increasing current density for the same quantity of electrodeposited metal.

In the same way, the protrusion tip propagation rate can be obtained by substituting j_{tip} from Eq. (14) and j from Eq. (11) in Eq. (16) and further rearranging to the form

$$\frac{dh}{dt} = \frac{Mjj_{tip}h}{\rho n F j_L (r + h)} \quad (20)$$

It is obvious from Eqs. (17) and (20) that

$$\frac{dh}{dt} > \frac{dh_S}{dt} \quad (21)$$

because $j_{tip} > j$ and $h/(r+h) > h_S/\delta$, meaning that the protrusion tip propagation will be larger under spherical diffusion control.

2. Physical Simulation

To test the validity of the above equations, Despić and Popov[10,11] carried out experiments on diffusion-controlled metal electrodeposition on a well-defined, triangularly shaped surface profile, through a diffusion layer of well-defined thickness $\delta \gg h$. A phonograph disk negative was used as a substrate upon which a layer of an agar-containing copper

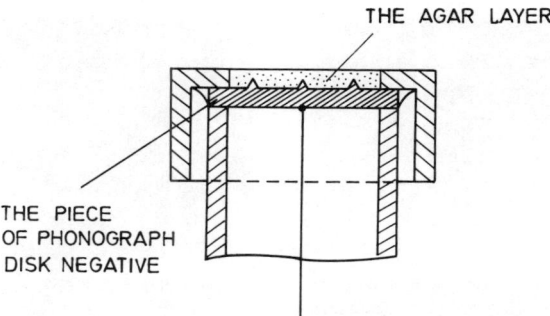

Figure 2. The model electrode for diffusion-controlled electrodeposition of metals. (From Ref. 11.)

sulfate–sulfuric acid solution was placed and left to solidify, as illustrated in Fig. 2. As current was passed and the layer was depleted of copper ions, an increase in the height of the triangular ridges was observed. Metallographic samples were made in wax, and cross sections of the deposit were photographed under the microscope. Figure 3 was thus obtained.[11]

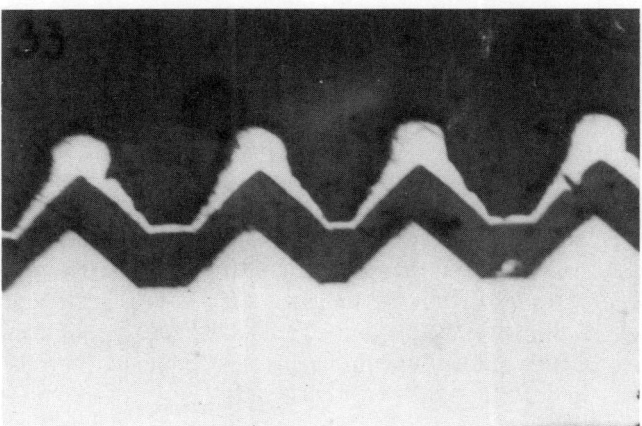

Figure 3. Cross section of a copper electrodeposit obtained by deposition in the system from Fig. 2 from 0.5 mol dm^{-3} CuSO$_4$ in 0.5 mol dm^{-3} H$_2$SO$_4$. The thickness of the agar diffusion layer was 1.0 mm. Deposition overpotential, 300 mV; deposition time, 120 min. Magnification, 450×. The substrate is a piece of phonograph disk negative. (From Ref. 11.)

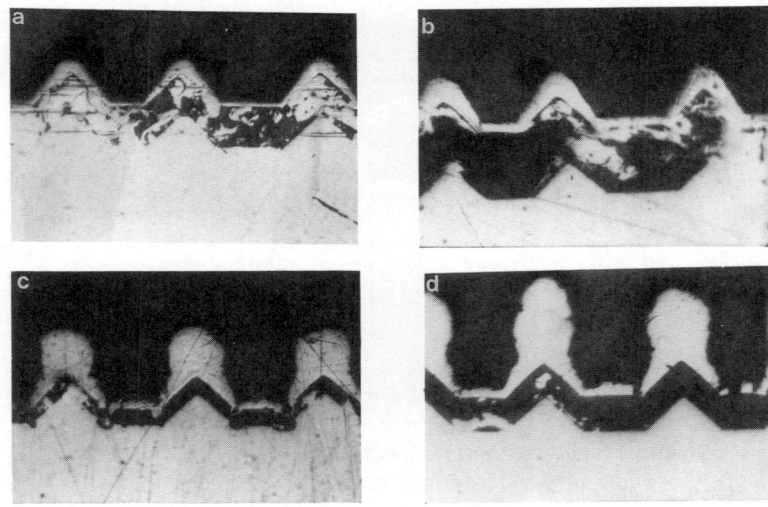

Figure 4. Cross sections of copper deposits obtained as in Fig. 3 for different deposition times: (a) 60 min; (b) 120 min; (c) 180 min; (d) 240 min. (From Ref. 11.)

In accordance with the discussion presented in Section I.1, three parts of the surface can be seen in Fig. 3: the flat part of the electrode and the sides and the tips of irregularities, providing an excellent physical illustration of the mathematical model.

The effect of deposition time at a given current density, i.e., the effect of the quantity of electrodeposited metal, on the protrusion height is illustrated in Fig. 4. The electrodeposits shown in Fig. 4 were obtained in the same manner as the electrodeposit shown in Fig. 3 but on a substrate with more widely spaced ridges. As a consequence, a difference in the shape of the growing protrusions in Figs. 3 and 4 is seen. In the latter case, the spherical diffusion layer is formed around a larger part of the protrusion and the effect around the tip itself is less pronounced in the early stages of deposition.

3. Real Systems

The effect of deposition current density on the increase in surface coarseness for a fixed quantity of electrodeposited metal is illustrated by Fig. 5. As expected, the surface coarseness increases strongly with increas-

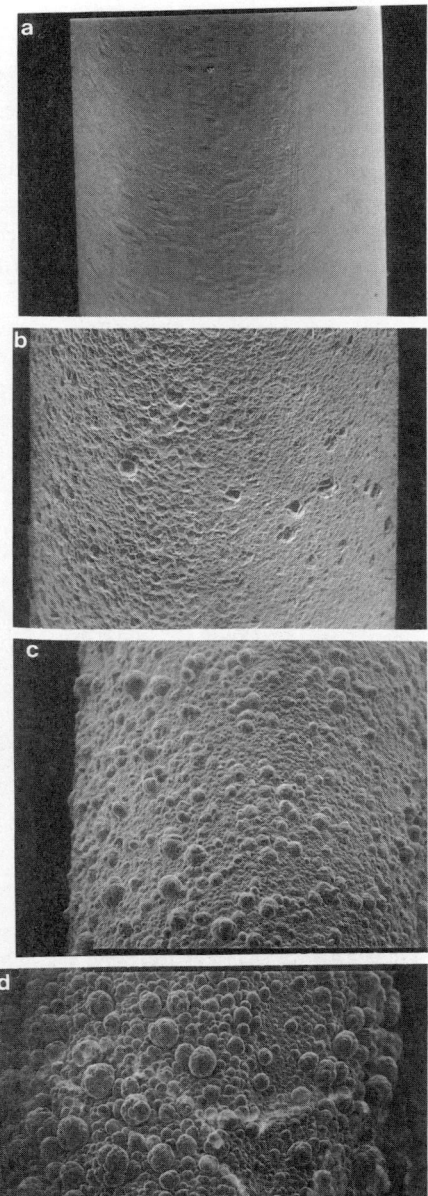

Figure 5. Copper deposits obtained from 0.5 mol dm^{-3} CuSO$_4$ in 0.5 mol dm^{-3} H$_2$SO$_4$ by electrodeposition under mixed activation–diffusion control. Quantity of electricity, 48 mAh cm^{-2}. The substrate is a copper wire electrode. (a) Initial surface; (b) deposition overpotential, 180 mV; (c) deposition overpotential, 230 mV; (d) deposition overpotential, 300 mV. Magnification, 100×. (From Ref. 12, with kind permission from Elsevier Science S.A., Lausanne, Switzerland.)

ing current density.[12,13] It should be noted that the theories describing the increase of electrode surface coarseness are valid for $H_0(t) \ll \delta$ or $h \ll \delta$, i.e., for short deposition times. For qualitative investigations the effect of current density on the increase of surface coarseness, large quantities of metal were used, as it was assumed that the qualitative picture of the phenomenon would not be changed. In quantitative investigations, the deposition times in which the approximation $\delta \gg h$ is valid[14] must be determined and taken into account.

All deposition overpotentials corresponded to mixed activation–diffusion control, and the deposits obtained were polycrystalline. In the region of activation-controlled electrodeposition, at the same initial surface, the situation becomes quite different as can be seen from Fig. 6. Activation-controlled deposition of copper produces large grains with relatively well-defined crystal shapes. This can be explained by the fact

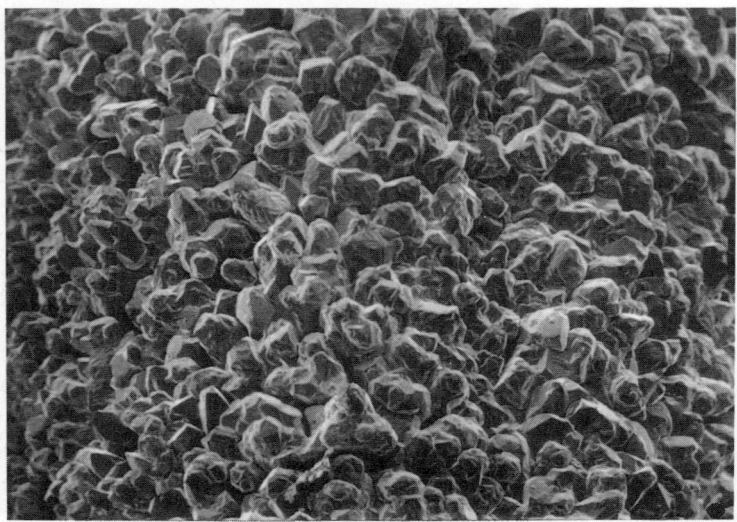

Figure 6. Copper deposits obtained from 0.5 mol dm^{-3} CuSO$_4$ in 0.5 mol dm^{-3} H$_2$SO$_4$ by activation-controlled electrodeposition. Quantity of electricity, 48 mAh cm^{-2}. Deposition overpotential, 120 mV. Magnification, 100×. The substrate is a copper wire electrode. (From Ref. 12, with kind permission from Elsevier Science S.A., Lausanne, Switzerland.)

Electrodeposit Formation

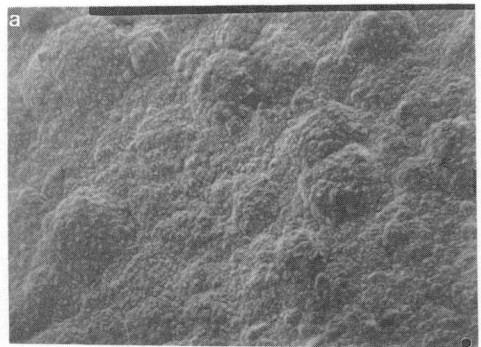
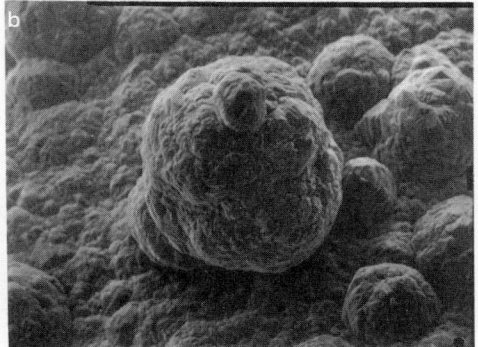

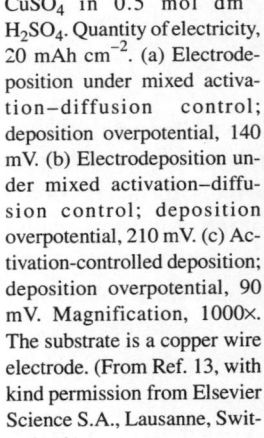

Figure 7. Copper deposits obtained from 0.1 mol dm^{-3} CuSO$_4$ in 0.5 mol dm^{-3} H$_2$SO$_4$. Quantity of electricity, 20 mAh cm^{-2}. (a) Electrodeposition under mixed activation–diffusion control; deposition overpotential, 140 mV. (b) Electrodeposition under mixed activation–diffusion control; deposition overpotential, 210 mV. (c) Activation-controlled deposition; deposition overpotential, 90 mV. Magnification, 1000×. The substrate is a copper wire electrode. (From Ref. 13, with kind permission from Elsevier Science S.A., Lausanne, Switzerland.)

that the values of the exchange current densities on different crystal planes are quite different, whereas the reversible potential is approximately the same for all planes.[15] This can lead to preferential growth of some crystal planes, because the rate of deposition depends only on orientation, leading

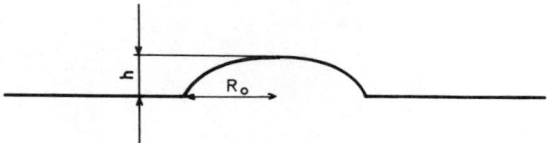

Figure 8. Cross section of an ellipsoidal electrode surface protrusion. (From Ref. 8, with kind permission from the Serbian Chemical Society, Belgrade, Yugoslavia.)

to the formation of a large-grained rough deposit. However, even at low degrees of diffusion control, the formation of large, well-defined grains is not expected, because of irregular growth caused by mass-transport limitations.

All the above facts are once again illustrated by Fig. 7.[13] It is seen from Figs. 5 and 7 that the surface protrusions are globular and mainly cauliflower-like. If the initial electrode surface protrusions are ellipsoidal, they can be characterized by the base radius R_0 and the height h as shown in Fig. 8. The tip radius is then given by

$$r = \frac{R_0^2}{h} \tag{22}$$

The initial electrode surface protrusion is characterized by $h \to 0$ and $r \to \infty$ if $R_0 \neq 0$. In this situation, a spherical diffusion layer cannot be formed around the tip of the protrusion [see Eq. (15)], and linear diffusion control takes place, leading to an increase in the height of the protrusion relative to the flat surface according to Eq. (18). When h increases, r decreases, and a spherical diffusion control can be operative around the whole surface protrusion, if it is sufficiently far from the other ones, as illustrated by Fig. 9. In this situation, second-generation protrusions can grow inside the diffusion layer of first-generation protrusions in the same way as the first-generation protrusions grow inside the diffusion layer of the macroelectrode, and so on.

A cauliflower deposit is formed under such conditions, as shown in Fig. 10. It is seen from Fig. 10a that the distance between the cauliflower grains is sufficiently large to permit the formation of spherical diffusion zones around each of them. At the same time, second-generation protru-

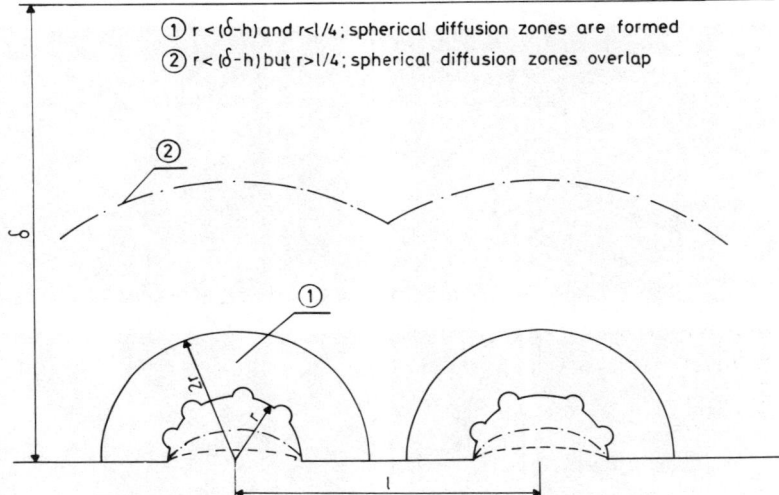

Figure 9. Schematic representation of the establishment of spherical diffusion layers around independently growing protrusions. (From Ref. 8, with kind permission from the Serbian Chemical Society, Belgrade, Yugoslavia.)

sions grow in all directions, as shown in Fig. 10b,c. This confirms the assumption that deposition takes place in spherically symmetric fashion.

On the other hand, a spherical diffusion layer cannot be formed around the closely packed protrusions because of overlapping of their diffusion fields, and they grow in the diffusion layer of the macroelectrode.

The propagation rate of the hemispherical protrusion tip can be obtained by substitution of r by h in Eq. (20) because in this situation

$$h = r \tag{23}$$

The resulting equation may be written in the form

$$\frac{dh}{dt} = \frac{Mjj_{\text{tip}}}{\rho nF \cdot 2j_L} \tag{24}$$

or

$$h = h_0 + \frac{Mj_{\text{tip}}}{\rho nF \cdot 2j_L} Q \tag{25}$$

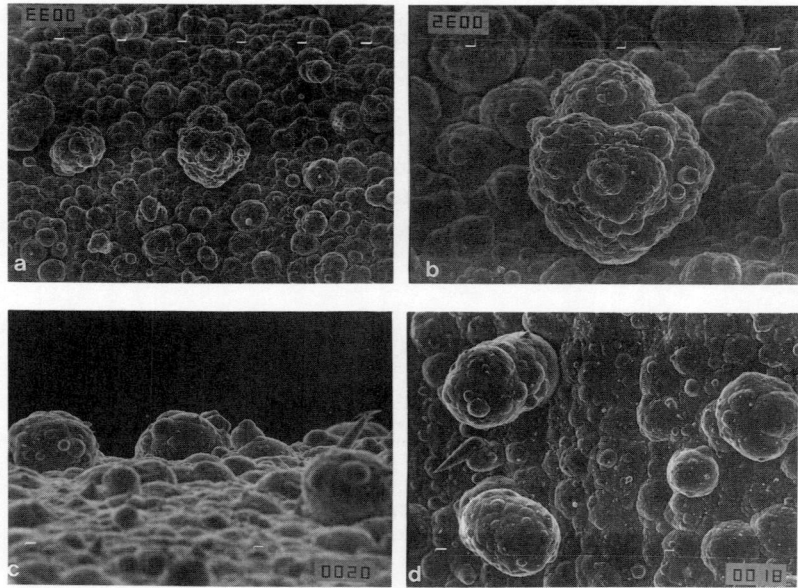

Figure 10. Copper deposits obtained from 0.3 mol dm^{-3} CuSO$_4$ in 0.5 mol dm^{-3} H$_2$SO$_4$ by electrodeposition under mixed activation–diffusion control. Deposition overpotential, 220 mV. (a) Quantity of electricity, 40 mAh cm^{-2}; magnification, 200×. (b) Same as in (a) but at 500× magnification. (c) and (d) Quantity of electricity, 20 mAh cm^{-2}; magnification, 750×. The substrate is a copper wire electrode. (From Ref. 8, with kind permission from the Serbian Chemical Society, Belgrade, Yugoslavia.)

where Q is given by Eq. (3).

To a first approximation, the propagation rate can be taken to be practically the same in all directions, meaning that the cauliflower-type deposit formed by spherically symmetric growth inside the diffusion layer of the macroelectrode will be hemispherical, as illustrated by Fig. 10a–c. This type of protrusion is much larger than that formed by linearly symmetric growth inside the diffusion layer of the macroelectrode (Fig. 10a,b), as is predicted by Eq. (21).

At the same time it is possible for a protrusion to grow in the direction of the bulk of the solution, thus maintaining a cylindrical shape, as seen in Fig. 10d. This probably happens when the lateral flux becomes lower because of the protrusion density.

It is also seen from Figs. 10c,d and 11 that the growth of some protrusions produces carrot-like forms, another typical form obtained in copper deposition under mixed activation–diffusion control. This happens under the condition $r/h \ll 1$, when spherical diffusion control takes place only around the tip of the protrusion, as illustrated by Figs. 1 and 11. In this case Eq. (20) can be rewritten in the form

$$\frac{dh}{dt} = \frac{Mjj_{\text{tip}}}{\rho n F j_L} \tag{26}$$

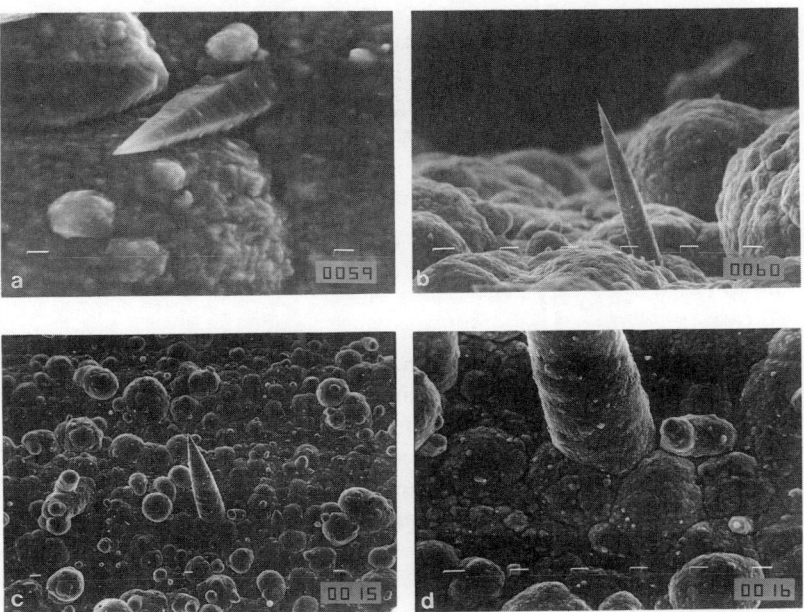

Figure 11. Copper deposits obtained from 0.3 mol dm^{-3} CuSO$_4$ in 0.5 mol dm^{-3} H$_2$SO$_4$ by electrodeposition under mixed activation–diffusion control. Deposition overpotential, 220 mV. (a) Quantity of electricity, 10 mAh cm^{-2}; magnification, 10000×. (b) Quantity of electricity, 40 mAh cm^{-2}; magnification, 2000×. (c) Quantity of electricity, 20 mAh cm^{-2}; magnification, 500×. (d) The root of the carrot from (c). Magnification, 2000×. The substrate is a copper wire electrode. (From Ref. 16, with kind permission from the Serbian Chemical Society, Belgrade, Yugoslavia.)

and

$$h = h_0 + \frac{M j_{tip}}{\rho n F j_L} Q \qquad (27)$$

where Q is given by Eq. (3).

On the other hand, if $r \ll h$, Eq. (14) can be rewritten in the form

$$j_{tip} = j_0(f_c - f_a) \qquad (28)$$

meaning that deposition to the protrusion tip can be under pure activation control. This happens if the nuclei have a shape like that in Fig. 11a. The assumption that the protrusion tip grows under activation control is confirmed by the regular crystallographic shape of the tip,[16] just as in the case of grains growing on the macroelectrode under activation control (see Figs. 6 and 7c). For sufficiently high cathodic overpotential, such that $f_c \gg f_a$, Eq. (27) can be written in the form

$$h = h_0 + \frac{M j_0 f_c}{\rho n F j_L} Q \qquad (29)$$

where Q is given by Eq. (3).

The maximum growth rate at a given overpotential corresponds to activation-controlled deposition. Because of this, the propagation rate at the tip will be many times larger than that in other directions, resulting in protrusions like that in Fig. 11b. The final form of the carrot-like protrusion is shown in Fig. 11c. It can be concluded from the parabolic shape that such protrusions grow as moving paraboloids in accordance with the Barton–Bockris theory,[6] the tip radius remaining constant because of the surface energy effect. It can be concluded from Fig. 11d that thickening of such a protrusion is under mixed activation–diffusion control because the deposit is seen to be of the same quality as that on the surrounding macroelectrode surface.

It is seen from Figs. 10–12 that new nuclei are mainly formed at the bottom of the already growing protrusions or between them. If the diffusion zones of two or more hemispherical protrusions partially overlap, as illustrated by Fig. 13, the current densities j at each point outside the overlapping field will be those required by the equation

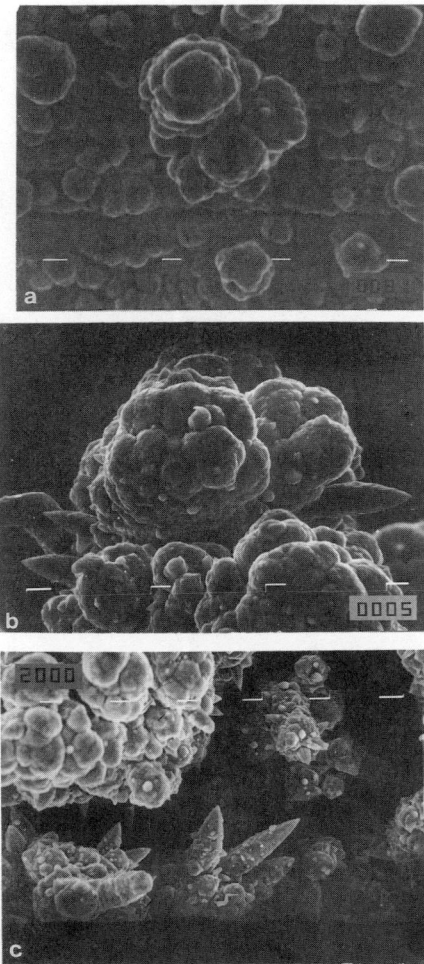

Figure 12. Copper deposits obtained from 0.3 mol dm^{-3} CuSO$_4$ in 0.5 mol dm^{-3} H$_2$SO$_4$ by diffusion-controlled deposition. Deposition overpotential, 400 mV. (a) Quantity of electricity, 2 mAh cm^{-2}; magnification, 3500×. (b) Quantity of electricity, 20 mAh cm^{-2}; magnification, 3500×. (c) Quantity of electricity, 20 mAh cm^{-2}; magnification, 2000×. (d) Quantity of electricity, 20 mAh cm^{-2}; magnification, 7500×. The substrate is a copper wire electrode. (From Ref. 17, with kind permission from the Serbian Chemical Society, Belgrade, Yugoslavia.)

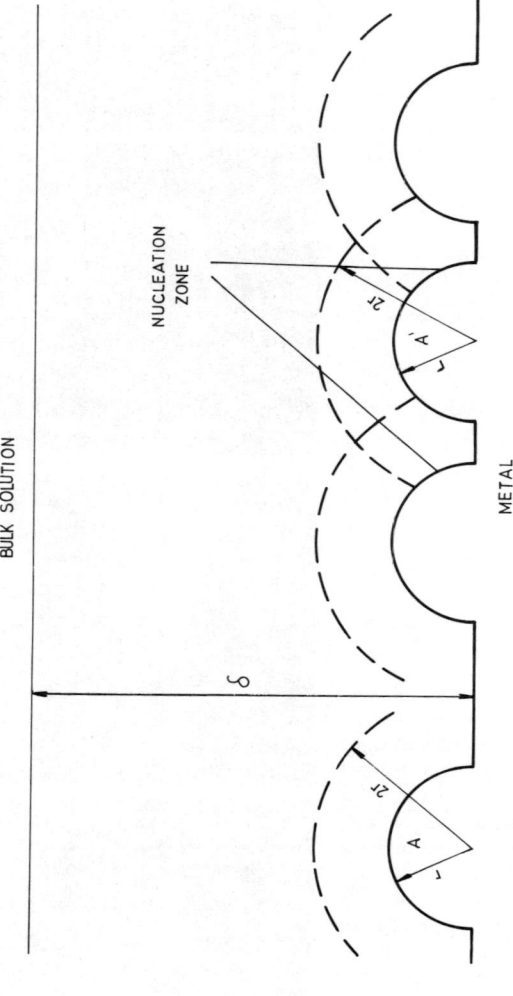

Figure 13. Schematic representation of spherical diffusion layers and their overlap around growing protrusions. (From Ref. 17, with kind permission from the Serbian Chemical Society, Belgrade, Yugoslavia.)

Electrodeposit Formation

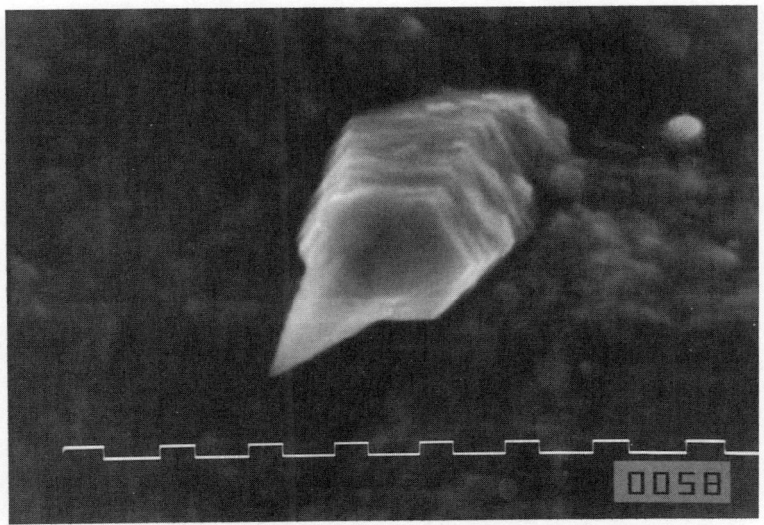

Figure 14. Copper deposit obtained from 0.3 mol dm^{-3} CuSO$_4$ in 0.5 mol dm^{-3} H$_2$SO$_4$ by electrodeposition under mixed activation–diffusion control. Quantity of electricity, 10 mAh cm^{-2}; deposition overpotential, 220 mV; magnification, 15,000×. The substrate is a copper wire electrode. (From Ref. 16, with kind permission from the Serbian Chemical Society, Belgrade, Yugoslavia.)

$$\eta = \eta_{0,c} \ln\left(\frac{j}{j_0}\right) + \eta_{0,c} \ln\left(\frac{1}{1 - \frac{j}{j_L}}\right) \quad (30)$$

if $f_c \gg f_a$, where j_L is the limiting diffusion current (linear or spherical) and $\eta_{0,c}$ is the slope of the cathodic Tafel line divided by 2.3.

In the overlapping field the current density j' at the points where the maximum possible current density is j will be lower than j. In this way the substrate becomes partially inert, and the formation of new nuclei is enhanced at these points if the entire electrode surface is at the same potential because

$$\eta = \eta_{0,c} \ln\left(\frac{j'}{j_0}\right) + \eta_{0,c} \ln\left(\frac{1}{1 - \frac{j'}{j_L}}\right) + \Delta\eta \quad (31)$$

where $\Delta\eta$ is the nucleation overpotential.[17]

Some of the new nuclei are the precursors of carrot-like protrusions, depending on their crystal orientation. In this case, they are in the form of small hexagonal pyramids, as shown in Fig. 14. Based on their morphology and because copper has a face-centered cubic crystal structure, it is reasonable to assume that they are truncated by a high-Miller-index plane. According to Pangarov and Vitkova,[18,19] the orientation of nuclei is related to the overvoltage used. It is reasonable to expect that the appearance of precursors of carrot protrusions has its own overvoltage range.

The shape of the protrusions depends on the positions of the precursors relative to other protrusions. For example, the carrot-like protrusion in Fig. 10d is probably produced by further growth of a nucleus at a position like that of the nucleus seen in Fig. 15.

It is seen from Fig. 16 that the tips of growing carrot-like protrusions can change the direction of protrusions. This is mainly observed in a later

Figure 15. Copper deposit obtained from 0.3 mol dm^{-3} CuSO$_4$ in 0.5 mol dm^{-3} H$_2$SO$_4$ by deposition under mixed activation–diffusion control. Quantity of electricity, 10 mAh cm^{-2}; deposition overpotential, 220 mV; magnification, 5000×. The substrate is a copper wire electrode. (From Ref. 16, with kind permission from the Serbian Chemical Society, Belgrade, Yugoslavia.)

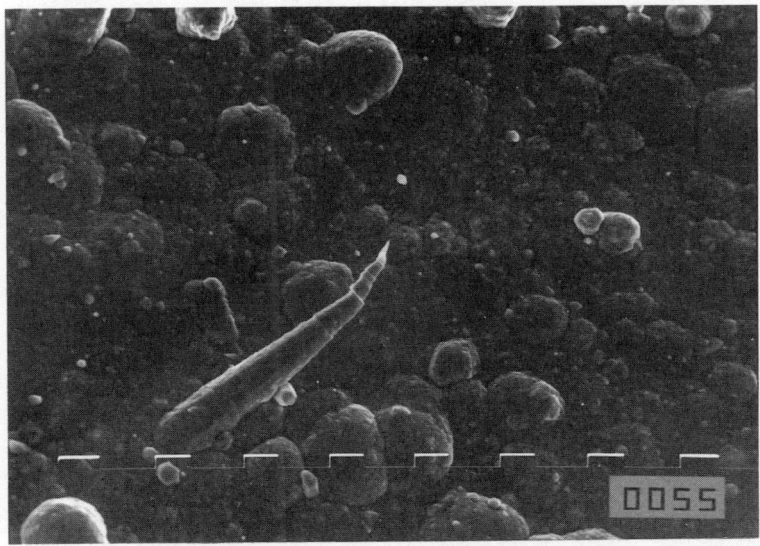

Figure 16. Copper deposit obtained from 0.3 mol dm^{-3} CuSO$_4$ in 0.5 mol dm^{-3} H$_2$SO$_4$ by deposition under mixed activation–diffusion control. Deposition overpotential, 220 mV; quantity of electricity, 10 mAh cm^{-2}; magnification, 1500×. The substrate is a copper wire electrode. (From Ref. 16, with kind permission from the Serbian Chemical Society, Belgrade, Yugoslavia.)

stage of protrusion growth. Thus, the smaller the protrusion height, the lower is the probability of a change in the growth direction. The reason for this behavior is not yet clear. However, it is probably caused by local perturbations on the growth front on the protrusion tip. This effect and the further nucleation on already growing protrusions cause the growth of different forms, even forms like those in Fig. 17.

The growth of surface protrusions in galvanostatic deposition has been considered only for deposition in the limiting diffusion current density range.[20,21]

In summary, deposition under mixed activation–diffusion control causes the formation of a number of growth forms and the increase of surface coarseness, this increase being more pronounced at higher current densities. It should be noted that electrodeposition at a periodically changing rate can change considerably the morphology of the deposit.[22]

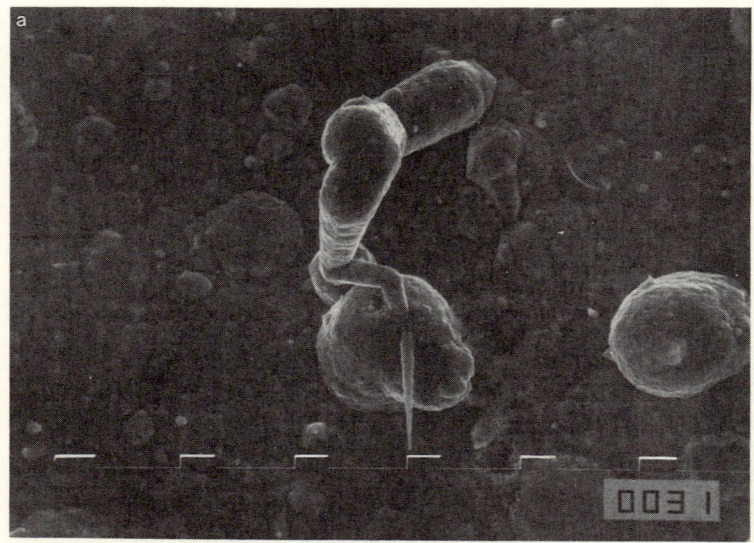

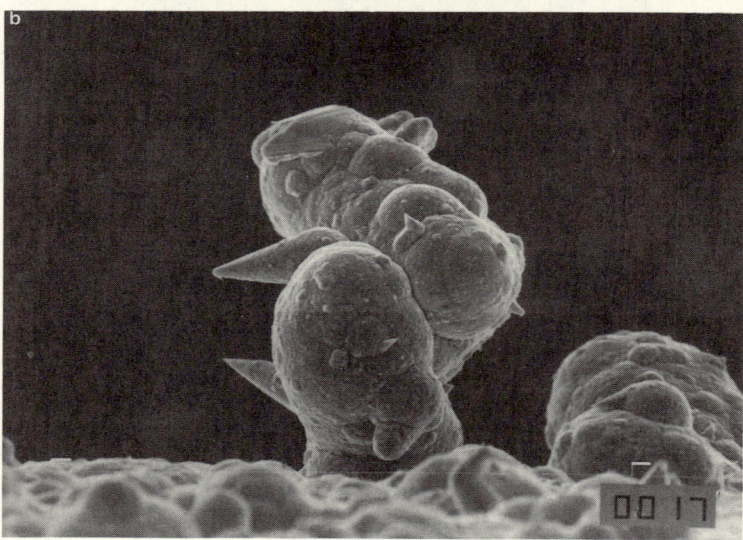

Figure 17. Copper deposits obtained from 0.3 mol dm^{-3} CuSO$_4$ in 0.5 mol dm^{-3} H$_2$SO$_4$ by deposition under mixed activation–diffusion control. Deposition overpotential, 220 mV. (a) Quantity of electricity, 10 mAh cm^{-2}; magnification, 2000×. (b) Quantity of electricity, 20 mAh cm^{-2}; magnification, 1000×. The substrate is a copper wire electrode. (From Ref. 16, with kind permission from the Serbian Chemical Society, Belgrade, Yugoslavia.)

II. SPONGY DEPOSITS

1. Mathematical Model

It follows from Eq. (11) that deposition in systems with low exchange current densities comes under full diffusion control at sufficiently large overpotentials. On the other hand, if

$$\frac{j_0}{j_L} \gg 1 \tag{32}$$

deposition will be under complete diffusion control at all overpotentials if some other kind of control does not take place (e.g., for silver deposition at low overpotentials two-dimensional nucleation is the rate-determining step[23,24]). At low overpotentials a small number of nuclei are formed, and they can grow independently. The limiting diffusion current density to the growing nucleus $j_{L,N}$ is given by

$$j_{L,N} = \frac{nFDC_0}{r_N} \tag{33}$$

or

$$j_{L,N} = \frac{j_L \delta}{r_N} \tag{34}$$

where r_N is the tip radius of the nucleus. Hence, if $r_N \to 0$, the condition given by Eq. (32) is not satisfied and deposition is under activation control. Activation-controlled deposition is thus possible even at $j_0 \gg j_L$ on very small electrodes such as nuclei on an inert substrate.

An increase in r_N leads to a decrease of $j_{L,N}$, and at, sufficiently large r_N, deposition comes under mixed activation–diffusion control. It can be assumed that this occurs at

$$j \geq k j_{L,N} \tag{35}$$

where $0 < k < 1$. By combining Eqs. (11), (34), and (35), one obtains

$$r_c = \frac{j_L k \delta}{j_0[(1-k)f_c - f_a]} \tag{36}$$

where r_c is the radius of the growing nucleus when the process comes under mixed control.[25] Obviously, Eq. (36) is valid if

$$0 < k < 1 - \frac{f_a}{f_c} \tag{37}$$

The radius of the initial stable nucleus, r_0, at overpotential η is given by

$$r_0 = \frac{2\sigma M}{\rho n F} \tag{38}$$

where σ is the surface energy, and is extremely low even at low overpotentials.[26] The radius of the growing nucleus will vary with time according to[27]

$$r_N = r_0 + \frac{1}{2} \frac{M j_0 f t}{\rho n F} \approx \frac{M j_0 f t}{2 \rho n F} \tag{39}$$

where $f = f_c - f_a$. Obviously, Eqs. (36) and (39) are only an approximation at short deposition times, because the effect of surface energy has not been taken into consideration. At longer deposition times they are valid, because the surface energy term at higher r values can be neglected.[6]

The combination of Eqs. (36) and (39) gives

$$t_i = \frac{j_L}{j_0^2} \frac{2 k n F \delta \rho}{M f [(1-k) f_c - f_a]} \tag{40}$$

where t_i is the induction period in which r_c is reached.

In deposition under mixed control, amplification of the surface irregularities on the growing nucleus takes place, leading to the formation of a spherical agglomerate of filaments, which form a spongy deposit.[28] The above reasoning is valid if spherical diffusion control can take place around growing grains. Assuming that around each grain with radius r_N, growing under spherical diffusion control, a diffusion layer of the same thickness is formed, it is obvious that initiation of spongy growth is possible if the number of nuclei per square centimeter, N, satisfies the condition

$$N \leq \frac{1}{(4 r_c)^2} \tag{41}$$

On the basis of all the above facts, it can be concluded that the formation of a spongy deposit on an inert substrate should be caused by mass-transport limitations under conditions of low nucleation rate. Hence, suitable conditions for the formation of spongy deposits arise at low overpotentials in systems where $j_L < j_0$.

2. Physical Model

According to the mathematical model presented, at a fixed value of the overpotential the growth of spongy deposits is possible if

$$r_N > r_c \tag{42}$$

where r_c is given by Eq. (36). Obviously, r_c decreases with increasing overpotential.

The situation in which spongy deposits can start to grow can easily be demonstrated.[29] Grains of desired size and distribution can be grown at low overpotentials under conditions of activation-controlled deposition. This corresponds to growth of grains at $r < r_c$. The situation in which $r > r_c$ can be simulated by increasing the overpotential to a sufficiently high value to cause diffusion control around the growing grains and the amplification of surface irregularities. This permits the simulation of the initial stage of spongy growth, as illustrated in Fig. 18. The growth of protrusions in all directions is a good proof that the deposition on the grain is under spherical diffusion control. At longer deposition times, the protrusions branch and interweave, as shown in Fig. 19, causing the macroelectrode to have a spongy appearance.

At $r_N < r_c$ the conditions of spherical diffusion control can be established only around the tip of growing grains inside the diffusion layer of the macroelectrode, leading to the growth of protrusions toward the bulk of the solution. This is illustrated in Fig. 20a. Deposition at the same overpotential as in the previous case, but on the considerably smaller grain, produces a needle growing toward the bulk solution, while at higher overpotential spongy growth on the grain of the same size takes place again, meaning $r > r_c$ (Fig. 20b). This is in good agreement with Eq. (36). Deposition at the same overpotential on grains similar to those seen in Fig. 18, but closer packed, produces a deposit similar to that in Fig. 20a, with the protrusions growing toward the bulk of the solution, governed by the diffusional flux to the macroelectrode, as illustrated in Fig. 20c. This corresponds to the situation in which grains with $r_N > r_c$ are so close to each other that the formation of a spherical diffusion layer is prevented, i.e., the condition given by Eq. (41) is not satisfied. Regardless of this, in prolonged deposition a spongy deposit is obtained because of branching and interweaving of growing protrusions as in the previous case.

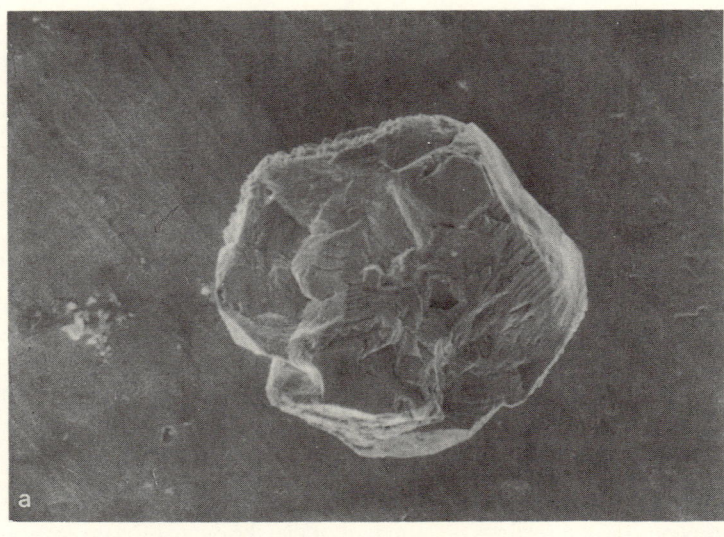

Figure 18. Cadmium deposits: (a) from 1.0 mol dm^{-3} CdSO$_4$ in 0.5 mol dm^{-3} H$_2$SO$_4$ solution at 12 mV on the copper plane electrode; deposition time, 15 min; (b) from 0.1 mol dm^{-3} CdSO$_4$ in 0.5 mol dm^{-3} H$_2$SO$_4$ solution at 120 mV on the substrate from (a); deposition time, 45 s. Magnification, 1000×. (From Ref. 29, with kind permission from Chapman and Hall, London, United Kingdom.)

Figure 19. Cadmium deposits obtained under the same deposition conditions as in Fig. 18b but after a deposition time of 120 s. Magnification, 100×. (From Ref. 29, with kind permission from Chapman and Hall, London, United Kingdom.)

3. Real Systems

Typical spongy electrodeposits are formed during zinc and cadmium electrodeposition at low overpotentials.[25] Scanning electron microscopy (SEM) images of zinc deposited at an overpotential of 20 mV onto a copper electrode from a zincate alkaline solution are shown in Fig. 21. The number of nuclei formed at constant overpotential increases with pulse duration after an induction period,[30] tending to a saturation value at higher durations.[31,32] The saturation value, N_0, i.e., the maximum number of nucleation sites that can be occupied, increases with increasing overpotential.[32] The nucleation law can be written in the form[33]

$$N = N_0[1 - \exp(-At)] \qquad (43)$$

where

$$A = K_1 j_0 \exp(-\frac{K_2}{\eta^2}) \qquad (44)$$

The increase in the number of nuclei formed with increasing deposition time can be seen in Fig. 21a and b, and a spongy deposit is formed as seen in Fig. 21b. The spongy growth takes place on a relatively small number of nuclei, as shown by Fig. 21b and c. The initiation of spongy

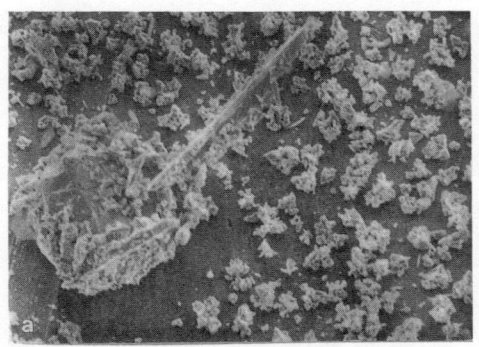

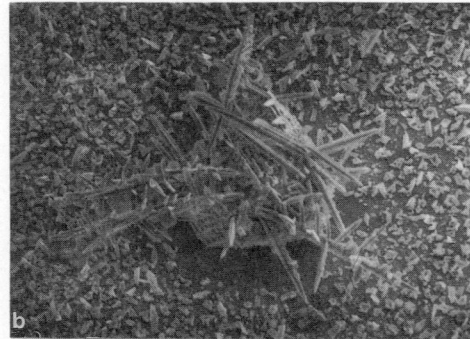

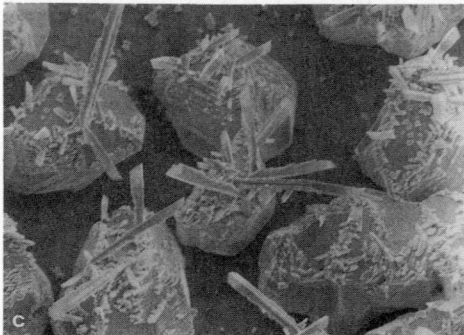

Figure 20. Cadmium deposits. (a) After deposition from 0.1 mol dm^{-3} CdSO$_4$ in 0.5 mol dm^{-3} H$_2$SO$_4$ solution at 120 mV (deposition time, 70 s) on the substrate obtained by deposition from 1.0 mol dm^{-3} CdSO$_4$ in 0.5 mol dm^{-3} H$_2$SO$_4$ solution at 12 mV and a deposition time of 6 min. Magnification, 1000×. (b) Same conditions as in (a) but at 170 mV and with a deposition time of 30 s. Magnification, 1000×. (c) After deposition from 0.1 mol dm^{-3} CdSO$_4$ in 0.5 mol dm^{-3} H$_2$SO$_4$ solution at 120 mV (deposition time, 30 s) on the substrate obtained by deposition from 1.0 mol dm^{-3} CdSO$_4$ in 0.5 mol dm^{-3} H$_2$SO$_4$ solution at 20 mV and a deposition time of 15 min. Magnification, 1000×. The substrate is a copper plane electrode. (From Ref. 29, with kind permission from Chapman and Hall, London, United Kingdom.)

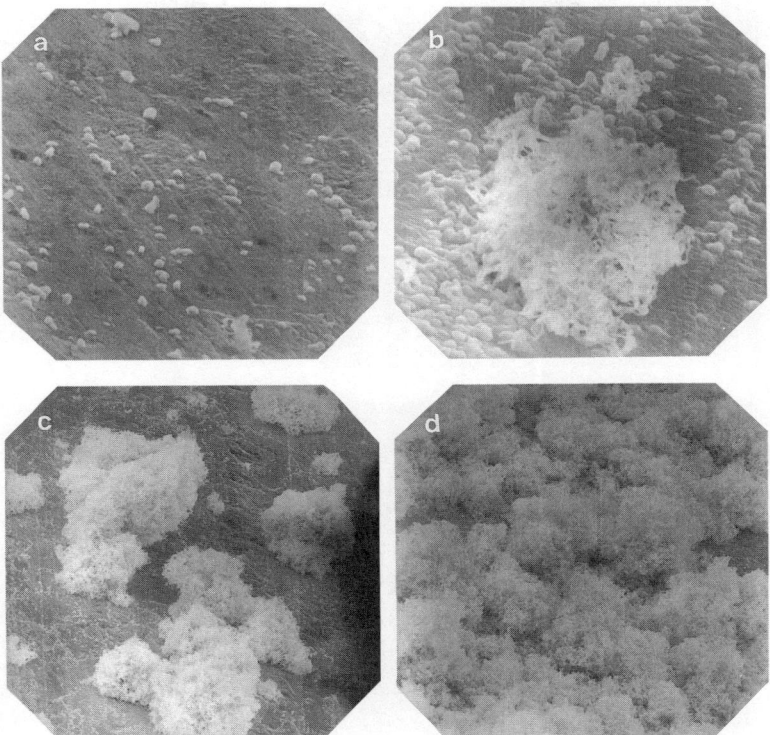

Figure 21. Zinc deposits obtained by deposition at 20 mV from 0.1 mol dm^{-3} zincate and 1.0 mol dm^{-3} KOH solution. (a) Deposition time, 10 min; magnification, 7500×. (b) Deposition time, 20 min; magnification, 7500×. (c) Deposition time, 30 min; magnification, 1000×. (d) Deposition time, 60 min; magnification, 1000×. The substrate is a copper plane electrode. (From Ref. 25, with kind permission from Chapman and Hall, London, United Kingdom.)

growth at a fixed overpotential is possible if the condition $r_N > r_c$ (Eq. 42) is satisfied, which happens after some time (Eq. 40). On the other hand, increasing deposition time leads to the formation of a larger number of nuclei, and the condition given by Eq. (42) is not satisfied over a large part of the electrode surface. Regardless of this, the coverage of the electrode surface by spongy deposits increases with increasing deposition time up to full coverage, as seen from Fig. 21d, in the same way as was illustrated

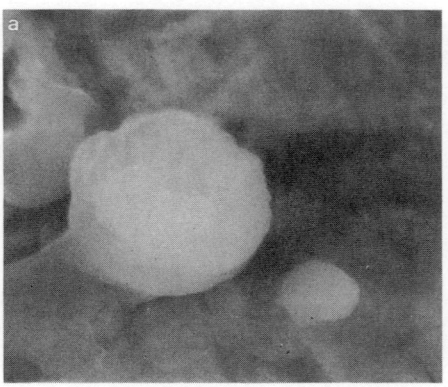

Figure 22. Zinc deposits obtained by deposition at 35 mV from 0.1 mol dm^{-3} zincate at 1.0 mol dm^{-3} KOH solution. (a) Deposition time, 7 min; magnification, 13,000×. (b) Deposition time, 15 min; magnification, 8000×. The substrate is a copper plane electrode. (From Ref. 36, with kind permission from the Serbian Chemical Society, Belgrade, Yugoslavia.)

earlier by Fig. 19. Spongy growth can start on the growing nucleus if the conditions given by Eqs. (41) and (42) are both satisfied at the same time. Obviously, substitution of t_i for t in Eq. (43) and further substitution in Eq. (41) gives

$$N_0[1 - \exp(-At_i)] < \frac{1}{(4r_c)^2} \tag{45}$$

as the condition for initiation of spongy growth. It can be satisfied at sufficiently low overpotentials where $K_2/\eta^2 \gg 1$, $A \to 0$, and, regardless of t_i being greater than 0,

$$At_i \approx 0 \tag{46}$$

i.e., when a small number of nuclei are formed during the induction period of spongy growth initiation.[34]

In the first stage of deposition, the formation of nuclei having a regular crystal shape can be expected[35] because of activation-controlled deposition. After r_c is reached, the system comes under mixed control, producing polycrystalline grains like those shown in Fig. 22a, just as in the case of mixed control of copper deposition.[12,13] In this situation, amplification of the surface irregularities on the growing grains takes place, and spongy growth is initiated.

The ideal spongy nucleus obtained in a real system is shown in Fig. 22b and illustrates the above discussion well.[36] The agglomerate of filaments in Fig. 21b is obviously formed by further growth of a nucleus like that in Fig. 22b.

Hence, it can be concluded that at low overpotentials the initiation of spongy growth is due to the amplification of surface protrusions directly inside the spherical diffusion layer formed around each independently growing grain, as in the case of the formation of cauliflower deposits. The growth of protrusions in all directions is good proof that the initial stage of deposition on the grain is under spherical diffusion control, while further growth takes place in the diffusion layer of the macroelectrode, just as shown by physical simulation (Figs. 18 and 20). In less ideal situations, the spongy nuclei are like those in Fig. 23, resulting in the same appearance of the macroelectrode in further deposition.

The validity of Eq. (36) can be easily tested using $CdSO_4$ solutions of widely varying concentrations. According to Lorenz,[37] the following relationship holds for cadmium deposition from sulfate solutions:

$$j_0 \sim C_0^{0.5} \tag{47}$$

On the other hand, it is known that in metal electrodeposition under natural convection[38] j_L varies with concentration according to

$$j_L \sim C_0^{1.25} \tag{48}$$

It follows from Eqs. (47) and (48) that

$$j_L/j_0 \sim C^{0.75} \tag{49}$$

and from Eqs. (36) and (49) it follows that

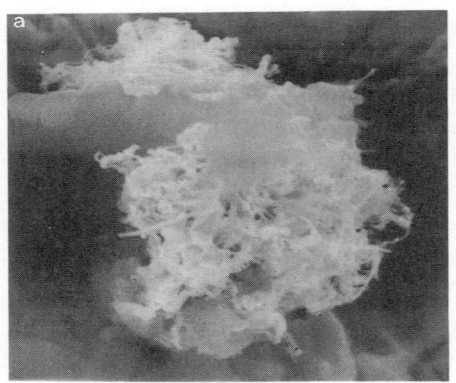

Figure 23. Zinc deposits obtained by deposition at 35 mV from 0.1 mol dm^{-3} zincate and 1.0 mol dm^{-3} KOH solution. (a) Deposition time, 20 min; magnification, 4800×. (b) Deposition time, 15 min; magnification, 10,000×. The substrate is a copper plane electrode. (From Ref. 36, with kind permission from the Serbian Chemical Society, Belgrade, Yugoslavia.)

$$r_c \sim \frac{j_L}{j_0} \sim C^{0.75} \quad (50)$$

assuming to a first approximation the same k and deposition overpotential.

The j_0 values for cadmium deposition from sulfate solution are estimated as 1 mA cm^{-2} for 0.005 mol dm^{-3} CdSO$_4$ (solution 1) and 10 mA cm^{-2} for 1.0 mol dm^{-3} CdSO$_4$ (solution 2), respectively. The corresponding limiting diffusion currents can be assumed to be 0.5 mA cm^{-2} and 100 mA cm^{-2}.[23] Hence, $r_{c2}/r_{c1} \approx 20$ and, from Eq. (41), $N_2/N_1 \approx 1/400$, where the subscripts 1 and 2 correspond to solutions 1 and 2. The cadmium deposit obtained from solution 1 is spongy, as seen in Fig. 24a, meaning that the conditions given by Eqs. (41) and (42) are

Figure 24. (a) Cadmium deposits obtained by deposition from 0.005 mol dm^{-3} CdSO$_4$ and 0.4 mol dm^{-3} K$_2$SO$_4$ in 0.1 mol dm^{-3} H$_2$SO$_4$ solution. Deposition overpotential, 10 mV; deposition time, 10 h; magnification, 1000×. (b) and (c) Cadmium deposits obtained by deposition from 1.0 mol dm^{-3} CdSO$_4$ in 0.5 mol dm^{-3} H$_2$SO$_4$ solution. Deposition overpotential, 10 mV; deposition time: (b) 24 min, (c) 10 h; magnification: (b) 200×, (c) 100×. The substrate is a copper plane electrode. (From Ref. 25, with kind permission from Chapman and Hall, London, United Kingdom.)

both satisfied. In the case of solution 2, r_c is much larger and N much lower, and suitable conditions for spongy deposit formation are not reached. Hence, the grains will grow under pure activation control until a complete surface film is formed, as illustrated by Fig. 24b and c.

It should be noted that some other possible mechanisms of spongy deposit formation have been considered in a qualitative way, as reviewed

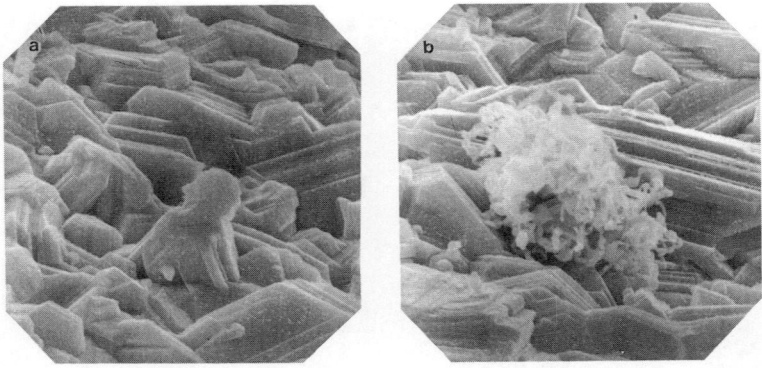

Figure 25. Zinc deposits obtained by deposition at 60 mV from 0.1 mol dm^{-3} zincate and 1.0 mol dm^{-3} KOH solution. Deposition time, 260 min; magnification: (a) 5000×, (b) 7500×. The substrate is a copper plane electrode. (K. I. Popov and N. V. Krstajić, unpublished data.)

in Refs. 39 and 40, but the mechanism presented above seems to be the most probable.[40] However, the mechanism of formation of a spongy deposit over an initial coating, which is not seen in the case of cadmium but occurs in zinc deposition,[25,39] requires clarification. For instance, the mechanism of spongy growth initiation in this case has not been elucidated. It may be due to the further growth of nuclei formed at the bottom of growing protrusions (in a similar way as in Figs. 10–12), as shown in Fig. 25. This phenomenon requires further investigation.

III. DENDRITIC GROWTH INITIATION

1. Mathematical Model

Two phenomena seem to distinguish dendritic growth from simple amplification of surface coarseness[1]:

1. There seems to exist a certain well-defined critical overpotential value below which dendrites do not grow.
2. The dendrites exhibit a highly ordered structure and grow and branch in well-defined directions. According to Wranglen,[41] a dendrite is a skeleton of a monocrystal and consists of a stalk and branches, thereby resembling a tree.

It is known that dendritic growth occurs selectively at three types of sites[1,42]:

1. Dendritic growth occurs at screw dislocations. For the screw dislocation to turn, separated one-dimensional nucleation is necessary, as for any layer growth. The dendrite stem is likely to be oriented in the natural direction of development of a spiral growth site. Swordlike dendrites with pyramidal tips are formed by this process.[1,5]

2. The regularities in the appearance of side branches and, in particular, the well-defined angle between branches of different order and the stem in most dendrites seem to have their origin in the phenomenon of twinning. Many investigations of the crystallographic properties of dendrites have reported the existence of twin structure.[43–47] In the twinning process the so-called indestructible re-entrant groove is formed.[48,49] Repeated one-dimensional nucleation in the groove is sufficient to provide for growth extending in the direction defined by the bisector of the angle between the twin planes. The direction of motion of the re-entrant edge with respect to the crystal plane is well defined by the basic twinning properties and is always the same for a given system. As a result, there is a defined angle between the direction of growth of the dendrite stem and all its branches.[1]

3. It is a particular feature of the hexagonal close-packed lattice that growth along a high-index axis does not lead to formation of planes of low indices. Grooves containing planes are perpetuated, and so is the chance for extended growth by the one-dimensional nucleation mechanism. In the case under consideration, this is a probable explanation for the development of dendrites in the high-index direction only. Further confirmation is found in the fact that the dendrites are of two-dimensional character.[42]

In all the above cases, the adatoms are incorporated into the lattice by repeated one-dimensional nucleation. On the other hand, deposition to the tip of screw dislocations can theoretically be considered as deposition to a point; in the other two cases, the deposition is to a line.

From the electrochemical point of view, a dendrite can be defined as an electrode surface protrusion that grows under activation control, while deposition to the macroelectrode is predominantly under diffusion control.[1,5] The polarization curve for the flat macroelectrode is given by Eq. (11), and that for the tip of a protrusion growing inside the macroelectrode diffusion layer is given by Eq. (14). It follows from Eq. (11) that $j \approx j_L$ if $j_0 f_c / j \gg 1$ and $f_c \gg f_a$. On the other hand, if $h \gg r$ and $h/r \gg j_0 f_c / j_L$

activation-controlled deposition to the tip of the protrusion takes place, and Eq. (14) can be rewritten in the form

$$j = j_{0,\text{tip}} f_c \tag{51}$$

where $j_{0,\text{tip}}$ is the corrected value of the exchange current density. This is so because if $r/h \ll 1$, Eq. (9) can be rewritten in the form

$$C_{\text{tip}} = C_0 \frac{h}{\delta} \tag{52}$$

where C_{tip} is the concentration at the tip of the protrusion growing under activation control inside the diffusion layer of the macroelectrode if deposition to it is under full diffusion control.[5] A similar situation can arise if cylindrical diffusion around the tip of the growing protrusion takes place.[50]

The exchange current density $j_{0,\text{tip}}$ at the tip of such a protrusion is given by[51]

$$j_{0,\text{tip}} = \left(\frac{C_{\text{tip}}}{C_0}\right)^\gamma j_0 \tag{53}$$

or, taking into account Eq. (52),

$$j_{0,\text{tip}} = \left(\frac{h}{\delta}\right)^\gamma j_0 \tag{54}$$

where γ is a function of the symmetry factor β, and j_0 is the exchange current density corresponding to the depositing ion bulk concentration. Equation (51) can then be rewritten in the form

$$j = j_0 \left(\frac{h}{\delta}\right)^\gamma f_c \tag{55}$$

Dendrites grow faster than the flat electrode surface, and the condition for initiation of dendritic growth is

$$j_L \leq j_0 \left(\frac{h}{\delta}\right)^\gamma f_{c,c} \tag{56}$$

where $f_{c,c}$ corresponds to η_c, the critical overpotential at which initiation of dendritic growth from the tip of the growing protrusion, whose height is h, is possible instantaneously after the steady-state concentration distribution inside the diffusion layer of the macroelectrode has been reached.[52,53] Taking into account Eq. (12) and the relationship

$$\eta_{0,c} = \frac{RT}{a_c F} \tag{57}$$

Eq. (56) can be rewritten in the form

$$\eta_c = \eta_{0,c} \ln\left[\frac{j_L}{j_0}\left(\frac{\delta}{h}\right)^\gamma\right] \tag{58}$$

The minimum overpotential at which dendritic growth can be initiated, η_i, corresponds to $h = \delta$, and Eq. (58) becomes

$$\eta_i = \eta_{0,c} \ln \frac{j_L}{j_0} \tag{59}$$

A different situation arises if $j_0 f_c / j_L \gg 1$ and $h/\gamma \gg j_0 f_c / j_L$ but $f_c > f_a$. The current density to the tip of the protrusion is then given by

$$j = j_{0,\text{tip}}(f_c - f_a) \tag{60}$$

For $h = \delta$, $j_{0,\text{tip}} = j_0$, while the diffusion current density to the macroelectrode is given by

$$j_d = j_L\left(1 - \frac{f_c}{f_a}\right) < j_L \tag{61}$$

Assuming that j_L can be used instead of j_d and following the same reasoning as in the derivation of Eqs. (56) and (59), one obtains

$$\eta_i = \frac{RT}{nF}\frac{j_L}{j_0} \tag{62}$$

The above relationships are the same as or very similar to those derived in earlier, somewhat cruder treatments.[1,52,54] It follows from Eq. (62) that for systems with $j_0 \to \infty$, dendritic growth is possible at all overpotentials. Experimentally, in all cases some critical overpotential of dendritic growth initiation exists, being on the order of 10 mV.[6,55,56] Assuming that under complete diffusion and surface energy control ($j_0 \to \infty$) the current density to the macroelectrode is given by[6]

$$j = \eta \frac{(nF)^2 DC_0}{\delta RT} \tag{63}$$

and, assuming that Eq. (8) is valid, the current density on the tip of the dendrite growing inside the diffusion layer of the macroelectrode is given by

$$j_d = \eta^2 \frac{(nF)^3 DC_0}{8\sigma VRT} \frac{h}{\delta} \tag{64}$$

then, it is possible to derive the relationships

$$\eta_c = \frac{8\sigma V}{nFh} \tag{65}$$

and

$$\eta_i = \frac{8\sigma V}{nF\delta} \tag{66}$$

using the same procedure as in the derivation of Eqs. (58) and (59).

It should be noted that all above derivations are valid, whether $r/h \ll 1$ or not, if the protrusion tip radius is sufficiently large to make the effect of the surface energy term[6] negligible.

2. Physical Simulation

The cross sections of copper deposits obtained in a model system are shown in Fig. 26. Deposits at 300 mV are compact; at 600 mV they are dendritic. This means that dendrites are formed at overpotentials larger than a certain critical value, as required by Eq. (59), because both overpotentials correspond to the limiting diffusion current plateau. It is seen that the current density to the tips of dendrites depends on the h/δ ratio [see Eq. (55)], so that larger dendrites are produced at more elevated points of the electrode surface. This is because the effective height of the dendrite precursor in the model diffusion layer is equal to the sum of the height of the precursor and the height of the point at which nucleation took place relative to the flat part of the electrode surface. In the same way, for the nuclei formed on the tip of the protrusion (Fig. 27), η_c [see Eq. (58)] is lower than for those formed on the flat surface, and a dendrite is formed at the tip of the protrusion while at the same overpotential dendrites are

Electrodeposit Formation

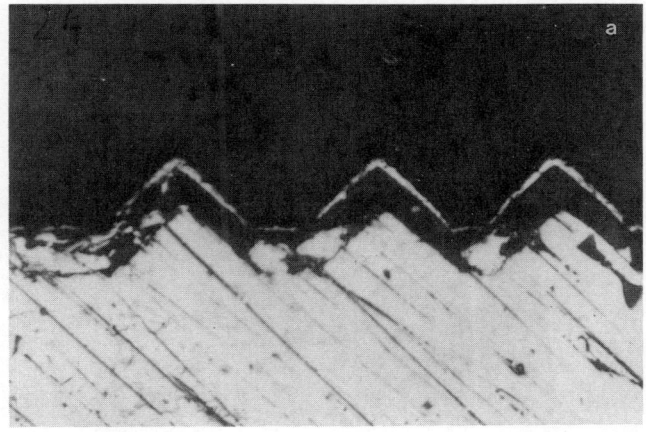

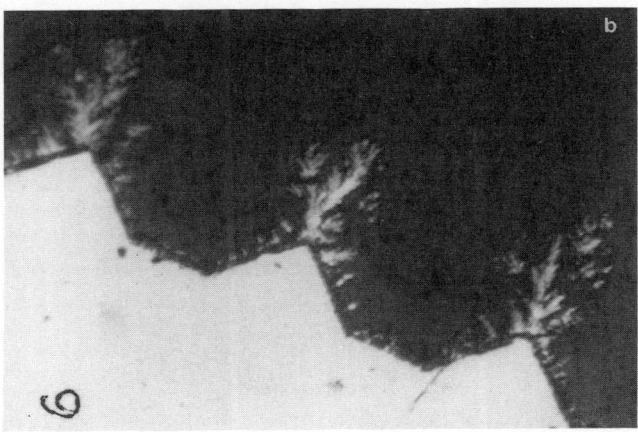

Figure 26. Cross sections of copper deposits obtained in the same manner as those in Fig. 3 from 0.1 mol dm^{-3} CuSO$_4$ in 0.5 mol dm^{-3} Na$_2$SO$_4$. The thickness of the agar diffusion layer was 1.0 mm. Deposition time, 30 min; deposition overpotential: (a) 300 mV, (b) 600 mV; magnification, 450×. (A. R. Despić and K. I. Popov, unpublished data.)

Figure 27. Cross section of copper deposit obtained in the same manner as those in Fig. 3 from 0.5 mol dm^{-3} CuSO$_4$ in 0.5 mol dm^{-3} H$_2$SO$_4$ by pulsating overpotential deposition. The thickness of the agar diffusion layer was 1.0 mm. Pulsation frequency, 5×10^{-4} Hz. Overpotential amplitude, 600 mV. Deposition time, 4 h. Magnification, 750×. (From Ref. 57, with kind permission from Chapman and Hall, London, United Kingdom.)

not formed on the flat part of the electrode. (More information about dendritic growth under conditions of pulsating overpotential can be found in Ref. 58.)

3. Real Systems

It is obvious that further growth of the dendrite precursors in Fig. 28 produces the dendrites shown in Figs. 29–31. Around the tips of dendrite precursors, as well as around the tips of dendrites, spherical or cylindrical diffusion control can take place, which is in good agreement with the requirements of the mathematical model. There is an induction period before initiation of dendritic growth.[1,6,52,53] During this induction period, dendrite precursors are formed and become sufficiently high to satisfy Eq. (58) at a given overpotential,[17] as illustrated by Fig. 32. The crosslike grains seen in Fig. 30a–c further develop into dendrite precursors (Fig.

Figure 28. Precursors of cadmium dendrites obtained by deposition from 0.1 mol dm^{-3} CdSO$_4$ in 0.5 mol dm^{-3} H$_2$SO$_4$ onto a cadmium wire electrode. (a) Deposition overpotential, 50 mV; deposition time, 2 min; magnification, 5000×. (b) Deposition overpotential, 110 mV; deposition time, 2 min; magnification, 3000×. (c) Deposition overpotential, 130 mV; deposition time, 3 min; magnification, 9000×. (From Refs. 50, 59, and 60, with kind permission from Elsevier Science S.A., Lausanne, Switzerland.)

32a,c). The propagation of this structure by branching (Fig. 32d) produces dendrites as in Fig. 32e. The initiation of dendritic growth is followed by a change in the slope of the current density–time curves,[5,52,54,61] indicating a change in the deposit growth mechanism.

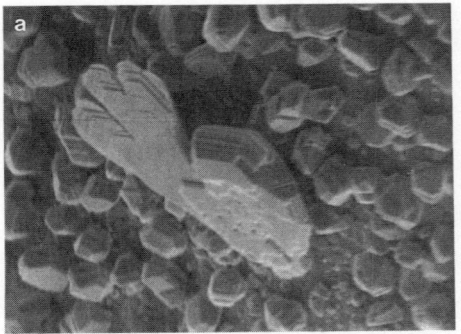

Figure 29. Cadmium dendrites obtained under the conditions specified in the caption to Fig. 28a. (a) Deposition time, 2 min; magnification, 2000×. (b) Deposition time, 10 min; magnification, 200×. (c) The detail from (b); magnification, 1000×. (From Refs. 50 and 60, with kind permission from Elsevier Science S.A., Lausanne, Switzerland.)

The validity of Eqs. (59) and (62) can be qualitatively tested by using the same solutions (1 and 2) as in the examination of spongy deposit formation. In this way, different j_L/j_0 ratios for the same deposition process can be obtained, while the surface energy and the crystallographic

Electrodeposit Formation 303

Figure 30. Cadmium dendrites obtained under the conditions specified in the caption to Fig 28b. Magnification: (a) 3000×, (b) 1000×, (c) 400×. (From Ref. 60.)

properties of the metal are kept the same. As expected, because of the lower j_L/j_0 ratio, dendrites appear at lower overpotentials from the more dilute solution than from the more concentrated one. This is illustrated in Fig. 33.

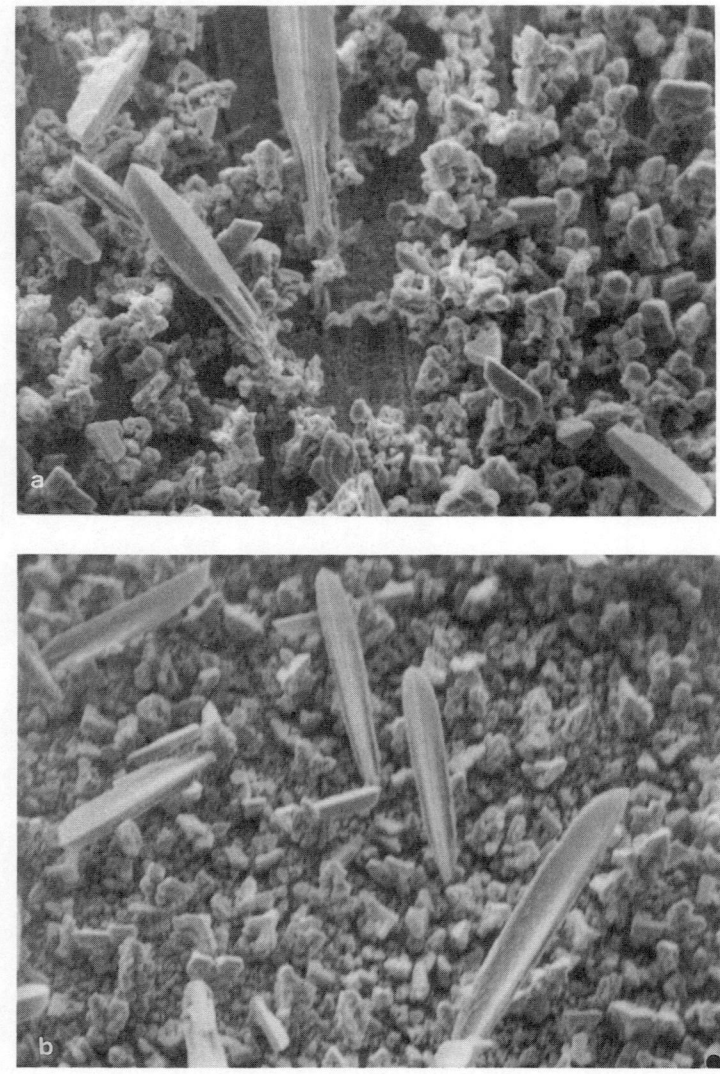

Figure 31. Cadmium dendrites obtained under the conditions specified in the caption to Fig. 28c. Magnification: (a) 2000×, (b) 1000×. (From Ref. 60.)

Figure 32. SEM micrographs of copper deposits obtained by deposition from 0.3 mol dm^{-3} CuSO$_4$ in 0.5 mol dm^{-3} H$_2$SO$_4$ onto a copper wire electrode. Deposition overpotential, 550 mV. (a) Quantity of electricity $q = 2$ mAh cm^{-2}; magnification, 2000×. (b) $q = 2$ mAh cm^{-2}; magnification, 3500×. (c) $q = 5$ mAh cm^{-2}; magnification, 2000×. (d) $q = 10$ mAh cm^{-2}; magnification, 1500×. (e) $q = 10$ mAh cm^{-2}; magnification, 1500×. (From Ref. 17, with kind permission from the Serbian Chemical Society, Belgrade, Yugoslavia.)

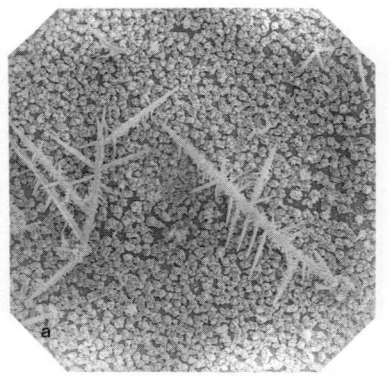
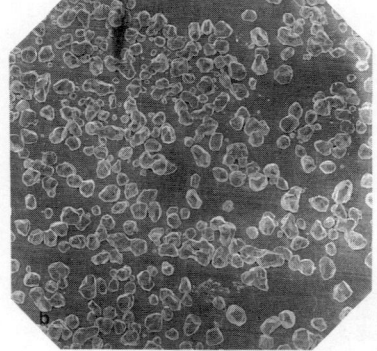

Figure 33. Cadmium deposits. (a) From 0.005 mol dm^{-3} CdSO$_4$ + 0.4 mol dm^{-3} H$_2$SO$_4$ solution. Deposition overpotential, 40 mV; deposition time, 2 h; magnification, 300×. (b) From 1.0 mol dm^{-3} CdSO$_4$ + 0.5 mol dm^{-3} H$_2$SO$_4$ solution. Deposition overpotential, 40 mV; deposition time, 4 min; magnification, 200×. (c) Same conditions as in (b) except the deposition time was 2 h. Magnification, 100×. The substrate is a copper plane electrode. (From Ref. 25, with kind permission from Chapman and Hall, London, United Kingdom.)

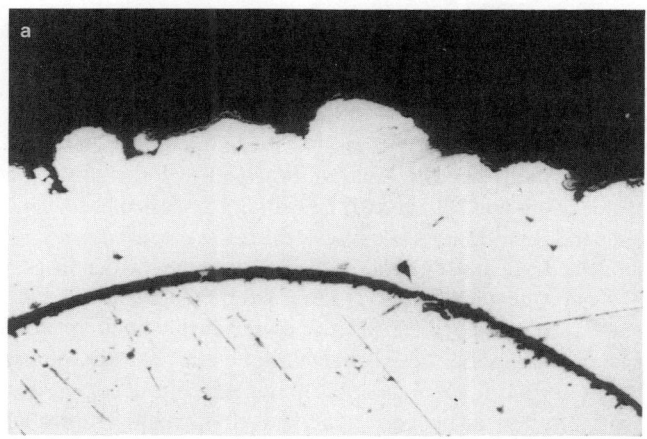

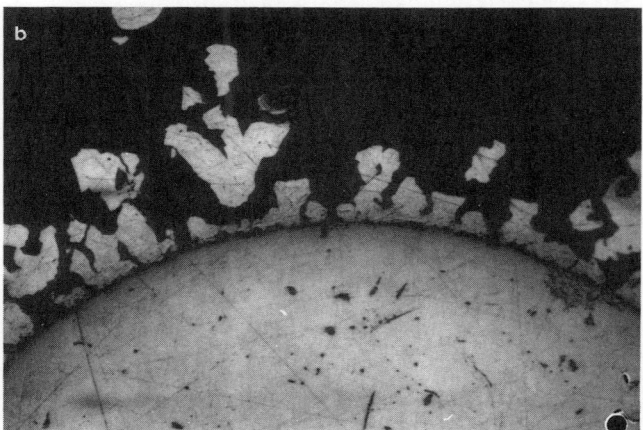

Figure 34. (a) Cross section of copper deposited from 0.2 mol dm^{-3} in 0.5 mol dm^{-3} H$_2$SO$_4$ onto steel wire previously plated with copper from a copper pyrophosphate bath. Deposition overpotential, 200 mV; deposition time, 6 h; magnification, 150×. (b) Cross section of cadmium deposited from 0.1 mol dm^{-3} CdSO$_4$ in 0.5 mol dm^{-3} H$_2$SO$_4$ onto copper wire. Deposition overpotential, 40 mV; deposition time, 2 h; magnification, 150×. (From Ref. 52, with kind permission from Chapman and Hall, London, United Kingdom.)

The above reasoning is also valid for different metals with different exchange current densities, as can be seen from Fig. 34.

It is known[62] that, apart from decreasing the concentration of the depositing ion, the formation of a dendritic powder can also be enhanced by increasing the concentration of the supporting electrolyte, increasing the viscosity of the solution, decreasing the temperature, and decreasing the velocity of motion of the solution. Practically all the above facts can be explained using Eqs. (59) and (62), assuming that a decrease in η_i means enhanced dendrite formation because of the lower electrical work required to produce the dendrites. The possibility of obtaining the dendrites of Pb[55] and Sn[56] from aqueous solutions at lower overpotentials than dendrites of Ag[23] can also be explained by Eq. (66) owing to the much lower melting points of these metals, i.e., their lower surface energy at room temperature. The dendrites of silver are obtained from molten salts at overpotentials of a few millivolts,[6] as in the case of Ph and Sn deposition from aqueous solutions,[55,56] because the difference between the melting point of silver and the working temperature for deposition from molten salts is not very different from the difference between the melting point of lead or tin and room temperature.

Dendrites grow from nuclei of higher indices or twinned ones only.[42] The probability of formation of such nuclei increases with increasing overpotential,[18,63,64] and η_c can also be defined as the overpotential at which they are formed. It is seen from Fig. 35 that different kinds of dendrite precursors are formed at 50 mV, and dendrites like those in Fig. 29 can be obtained.

Deposition at higher overpotentials on the same substrate produces other type of dendrites also. Hence, already existing dendrite precursors develop into dendrites after some critical overpotential is reached. It can be the overpotential at which propagation of the structure by branching produces dendrites like the one in Fig. 35b. If a short overpotential pulse of 130 mV is imposed on the same substrate, further deposition at 95 mV produces needles, as illustrated in Fig. 35c, just as in the case of deposition at 130 mV on cadmium wire. This is due to the new nucleation on top of flat dendrite precursors. In such a case, the critical overpotential of needle growth initiation is that at which the right type of precursor is formed.

Hence, it is difficult to say what η_i really is. Nevertheless, it is possible to estimate it by using Eqs. (59), (62), and (66).

The condition for spongy growth initiation (Eq. 45) can also be satisfied at higher overpotentials[34] where $K_2/\eta^2 \ll 1$, $A \rightarrow K_1 j_0$, and

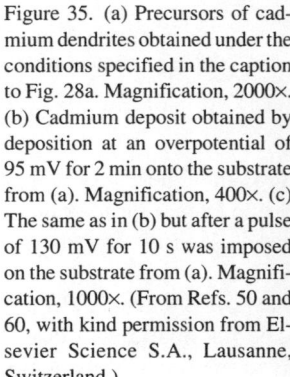

Figure 35. (a) Precursors of cadmium dendrites obtained under the conditions specified in the caption to Fig. 28a. Magnification, 2000×. (b) Cadmium deposit obtained by deposition at an overpotential of 95 mV for 2 min onto the substrate from (a). Magnification, 400×. (c) The same as in (b) but after a pulse of 130 mV for 10 s was imposed on the substrate from (a). Magnification, 1000×. (From Refs. 50 and 60, with kind permission from Elsevier Science S.A., Lausanne, Switzerland.)

At $t_i \approx 0$ if $t_i \to 0$. Hence, spongy deposit formation at high overpotentials starts at very low deposition times when the electrode is not completely covered with deposited metal. This is illustrated in Fig. 36. Deposition is initiated by nucleation (Fig. 36a), which occurs consecutively on practi-

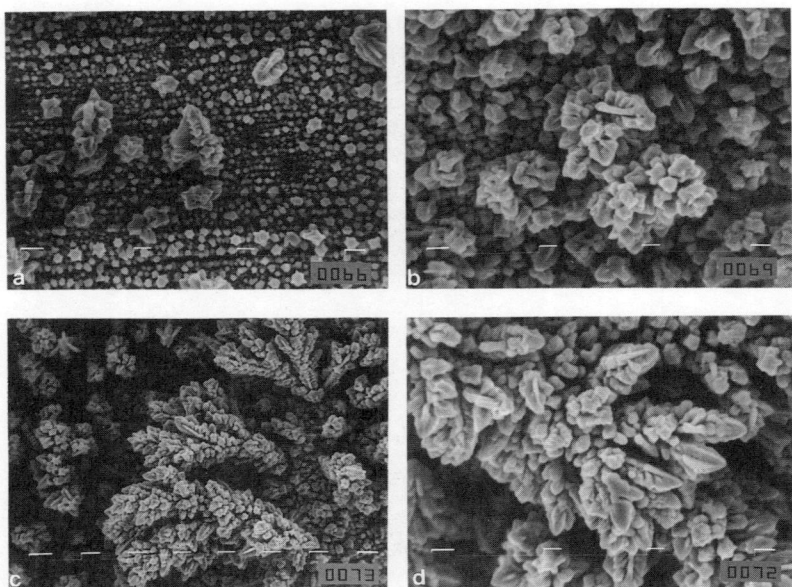

Figure 36. SEM microphotographs of copper deposits obtained from 0.3 mol dm^{-3} CuSO$_4$ in 0.5 mol dm^{-3} H$_2$SO$_4$ solution. Deposition overpotential, 700 mV. (a) Quantity of electricity $q = 0.5$ mAh cm^{-2}; magnification, 3500×. (b) $q = 2$ mAh cm^{-2}; magnification, 3500×. (c) $q = 5$ mAh cm^{-2}; magnification, 1500×. (d) $q = 5$ mAh cm^{-2}; magnification, 3500×. The substrate is a copper wire electrode. (From Ref. 17, with kind permission from the Serbian Chemical Society, Belgrade, Yugoslavia.)

cally all grains formed (Fig. 36b). The further growth of these protrusions in a spherically symmetric fashion (if it is possible from a steric point of view) produces a spongy deposit (Fig. 36c) that is dendritic in structure (Fig. 36d).

The largest grain agglomerations formed at 550 mV (Fig. 32b) and 700 mV (Fig. 36b) are similar in size and shape for small quantities of electrodeposited metal, but a spongy deposit is formed only in the deposition at 700 mV. This is because spongy growth is possible at overpotentials larger than the critical one, which can be determined by substitution of r_c from Eq. (36) and t_i from Eq. (40) into Eq. (45).

It is obvious that the electrochemical conditions, as well as the crystallographic ones, under which dendritic deposits are formed can be precisely determined. One problem that still seems to remain unsolved is the question of what causes the dendrite precursors to appear at regularly spaced locations along the dendrite stem. Further investigations in this direction are necessary.

REFERENCES

[1] A. R. Despić and K. I. Popov, in *Modern Aspects of Electrochemistry*, No. 7, Ed. by B. C. Conway and J. O'M. Bockris, Plenum Press, New York, 1972, Chapter 4.
[2] C. Wagner, *J. Electrochem. Soc.* **101** (1954) 225.
[3] S. I. Krichmar, *Elektrokhimiya* **1** (1969) 609.
[4] A. R. Despić, J. W. Diggle, and J. O'M. Bockris, *J. Electrochem. Soc.* **115** (1968) 507.
[5] J. W. Diggle, A. R. Despić, and J. O'M. Bockris, *J. Electrochem. Soc.* **116** (1969) 1503.
[6] J. L. Barton and J. O'M. Bockris, *Proc. Roy. Soc. London, Ser.* A **268** (1962) 485.
[7] D. R. Hamilton, *Electrochim. Acta* **8** (1963) 731.
[8] K. I. Popov, B. N. Grgur, M. G. Pavlović, and V. Radmilović, *J. Serb. Chem. Soc.* **58** (1993) 1055.
[9] J. O'M. Bockris, in *Modern Aspects of Electrochemistry*, No. 1, Ed. by J. O'M. Bockris and B. E. Conway, Butterworths, London, 1954, Chapter 4.
[10] K. I. Popov and A. R. Despić, *Bull. Soc. Chim. Beograd* **36** (1971) 173.
[11] K. I. Popov, Ph.D. Thesis, University of Belgrade, 1971.
[12] K. I. Popov, Lj. J. Pavlović, M. G. Pavlović, and M. I. Čekerevać, *Surf. Coat. Technol.* **35** (1988) 39.
[13] K. I. Popov, M. G. Pavlović, Lj. J. Pavlović, M. I. Čekerevac, and G. Ž. Remović, *Surf. Coat. Technol.* **34** (1988) 355.
[14] K. I. Popov, M. D. Maksimović, M. G. Pavlović, and D. T. Lukić, *J. Appl. Electrochem.* **10** (1980) 299.
[15] A. Damjanović, *Plating* **52** (1965) 1017.
[16] K. I. Popov, V. Radmilović, B. N. Grgur, and M. G. Pavlović, *J. Serb. Chem. Soc.* **59** (1994) 47.
[17] K. I. Popov, V. Radmilović, B. N. Grgur, and M. G. Pavlović, *J. Serb. Chem. Soc.* **59** (1994) 119.
[18] N. A. Pangarov, *Electrochim. Acta* **9** (1964) 721.
[19] N. A. Pangarov and S. D. Vitkova, *Electrochim. Acta* **11** (1966) 1733.
[20] M. D. Maksimović, K. I. Popov, Lj. M. Jović, and M. G. Pavlović, *Bull. Soc. Chim. Beograd* **44** (1979) 547.
[21] M. D. Maksimović, K. I. Popov, and M. G. Pavlović, *Bull. Soc. Chim. Beograd* **44** (1979) 687.
[22] K. I. Popov and M. D. Maksimović, in *Modern Aspects of Electrochemistry*, No. 19, Ed. by B. E. Conway, J. O'M. Bockris, and R. E. White, Plenum Press, New York, 1989, Chapter 3.
[23] K. I. Popov, N. V. Krstajić, Z. D. Jerotijević, and S. R. Marinković, *Surf. Technol.* **26** (1985) 185.
[24] E. Budevski, V. Bostanov, and T. Vitanov, *Rost Kristallov*, Nauka, Moskow, 1973, p. 230.
[25] K. I. Popov and N. V. Krstajić, *J. Appl. Electrochem.* **13** (1983) 775.
[26] S. Toshev and I. Markov, *Electrochim. Acta* **12** (1967) 281.
[27] J. O'M. Bockris, Z. Nagy, and D. Dražić, *J. Electrochem. Soc.* **120** (1973) 30.
[28] M. Froment and G. Mourin, *Electrodeposition Surf. Treat.* **3** (1975) 245.

[29] K. I. Popov, N. V. Krstajić, S. R. Popov, and M. I. Čekerevac, *J. Appl. Electrochem.* **16** (1986) 771.
[30] I. Markov, *Thin Solid Films* **35** (1972) 11.
[31] I. Markov and D. Kaishew, *J. Cryst. Growth* **16** (1972) 170.
[32] A. Scheludko and M. Todorova, *Commun. Bulg. Acad. Sci. (Phys.)* **3** (1952) 61.
[33] M. Fleishman and H. R. Thirsk, *Electrochim. Acta* **15** (1959) 146.
[34] K. I. Popov, N. V. Krstajić, and S. R. Popov, *J. Appl. Electrochem.* **1** (1985) 151.
[35] N. A. Pangarov and V. Velinov, *Electrochim. Acta* **11** (1966) 1753.
[36] K. I. Popov, N. V. Krstajić, M. V. Simičić, and N. M. Bibić, *J. Serb. Chem. Soc.* **57** (1992) 927.
[37] W. Lorenz, *Z. Elektrochem.* **58** (1954) 912.
[38] N. Ibl, *Electrochim. Acta* **1** (1959) 3.
[39] M. M. Jakšić, *Surf. Technol.* **24** (1985) 193.
[40] I. B. Murashova and A. V. Pomosov, *Itogi Nauki, Elektrokhimiya*, Vol. 30, Moscow, 1989.
[41] G. Wranglen, *Electrochim. Acta* **2** (1960) 130.
[42] I. N. Justinijanović and A. R. Despić, *Electrochim. Acta* **18** (1973) 709.
[43] J. Smit, F. Ogburn, and C. J. Bechtold, *J. Electrochem. Soc.* **115** (1968) 371.
[44] J. W. Faust and H. F. John, *J. Electrochem. Soc.* **110** (1963) 109.
[45] J. W. Faust and H. F. John, *J. Electrochem. Soc.* **108** (1961) 856.
[46] F. Ogburn, C. Bechtold, J. B. Morris, and A. W. Koranyi, *J. Electrochem. Soc.* **112** (1965) 547.
[47] C. Bechtold, F. Ogburn, and J. Smit, *J. Electrochem. Soc.* **115** (1968) 813.
[48] J. B. Kushner, *Met. Prog.* **81** (1962) 88.
[49] H. Fisher, *Elektrolytishe Abscheiung und Elektrokristallization von Metallen*, Springer-Verlag, Berlin, 1954.
[50] K. I. Popov and M. I. Čekerevac, *Surf. Coat. Technol.* **37** (1989) 435.
[51] J. S. Newman, *Electrochemical Systems*, Prentice-Hall, Englewood Cliffs, New Jersey, 1973, p. 177.
[52] K. I. Popov, M. D. Maksimović, J. D. Trnjavčev, and M. G. Pavlović, *J. Appl. Electrochem.* **11** (1981) 239.
[53] K. I. Popov, M. G. Pavlović, and M. D. Maksimović, *J. Appl. Electrochem.* **12** (1982) 525.
[54] A. R. Despić and M. M. Purenović, *J. Electrochem. Soc.* **121** (1974) 329.
[55] K. I. Popov, N. V. Krstajić, R. M. Pantelić, and S. R. Popov, *Surf. Technol.* **26** (1985) 177.
[56] K. I. Popov, M. G. Pavlović, and J. K. Jovićević, *Hydrometallurgy* **23** (1989) 127.
[57] A. R. Despić and K. I. Popov, *J. Appl. Electrochem.* **1** (1971) 275.
[58] K. I. Popov and M. G. Pavlović, in *Modern Aspects of Electrochemistry*, No. 24, Ed. by R. E. White, J. O'M. Bockris, and B. E. Conway, Plenum Press, New York, 1992, Chapter 6.
[59] M. I. Čekerevac and K. I. Popov, *Surf. Coat. Technol.* **37** (1989) 441.
[60] M. I. Čekerevac, Ph.D.Thesis, University of Belgrade, 1988.
[61] N. Ibl and K. Schadegg, *J. Electrochem. Soc.* **114** (1967) 54.
[62] N. Ibl, in *Advances in Electrochemistry and Electrochemical Engineering*, Ed. by P. Delahay and C. W. Tobias, Vol. 2, Interscience, New York. 1962, pp. 50–68.
[63] R. Kaishew, A. Scheludko, and G. Bliznakov, *Izv. Bulg. Akad. Nauk* **1** (1950) 137.
[64] N. A. Pangarov, *Phys. Status Solidi* **20** (1967) 371.

4

The Manganese Dioxide Electrode in Aqueous Solution

T. N. Andersen

Kerr-McGee Chemical Corporation, Oklahoma City, Oklahoma 73125

I. INTRODUCTION

1. Significance of Manganese Dioxide and Its Variations

Ever since 1866, when George Leclanché invented the galvanic cell, manganese dioxide has been the principal cathode constituent in dry cells. Today it is used in alkaline, zinc chloride, Leclanché, magnesium, and aluminum primary batteries as well as alkaline secondary batteries.[1] The zinc–MnO_2 batteries (alkaline, zinc chloride, and Leclanché) account for the majority of the 20 billion dry cells sold annually in the world, and dry cells account for the majority of the small-format consumer cells—a \$10 billion market.[2]

The history of Zn–MnO_2 battery development[3–7] has tied in quite closely with development of battery-grade manganese dioxide.[8,9] This, in turn, has evolved with the advancement of portable devices and the need for increasing drain rates and energy density in batteries.

Tetravalent manganese oxide occurs in more than 15 allotropic forms, of which only a few are useful depolarizers in batteries.[10–12] Some of the marginally active forms contain significant percentages of cations other than manganese and are not strictly manganese dioxides. Manganese dioxides used in batteries are broadly categorized into three groups according to origin—i.e., natural manganese dioxide (NMD), chemical manganese dioxide (CMD), and electrolytic manganese dioxide

Modern Aspects of Electrochemistry, Number 30, edited by Ralph E. White *et al.* Plenum Press, New York, 1996.

(EMD).[8,12,13] NMD is natural ore carefully chosen for high content of battery-active MnO_2. It is not a single stoichiometric MnO_2 but, rather, a mixture of up to 10–20 different manganese oxide minerals[4,11,12] (as well as gangue minerals) that have widely differing battery activities. CMD is synthesized by various processes that include precipitation and chemical oxidation of manganese-containing solutions or compounds. EMD is synthesized by electrochemical deposition at the anode from aqueous solutions of manganese salts, usually the sulfate. NMD is often "activated," which consists of heating it to form Mn_2O_3 and then treating it with dilute sulfuric acid to re-form MnO_2 and $MnSO_4$.

The three battery-active classes of MnO_2 have some crystal similarity. CMD and EMD usually belong to the γ-MnO_2 and/or the closely related ε-MnO_2 crystal class(es). (Several of the manganese dioxides are identified by means of a Greek symbol.) Present NMDs usually contain natural γ-MnO_2, the mineral nsutite, as a major and most important component.[12] The γ- and ε-MnO_2's are nonstoichiometric and contain up to ~4% structural water. They are represented by the chemical formula $MnO_x \cdot mH_2O$, where $x = 1.92$–1.98. Thus, battery manganese dioxides are not strictly "MnO_2," although they are generally given this shorthand designation. Throughout this chapter, the terms MnO_2 and manganese dioxide will be used to represent any and all of the Mn(IV) oxides. When a distinction is to be made between forms of different stoichiometry, the needed terms will be identified. Section IV details the structures of these materials.

Despite crystal similarities, NMD, CMD, and EMD exhibit different physical and electrochemical properties. NMD gives cell performance (energy and power densities) that is greatly inferior to that of the two synthetic forms. EMD exhibits the greatest battery activity for most applications and, particularly, for high drain rates. Nevertheless all three are used extensively, because of substantial cost differences (NMD < CMD < EMD) and because they can be used in different cell chemistries, which are targeted for applications (e.g., drain rates) that are compatible with their performance capabilities.

Zn–MnO_2 cells are distinguished by the type of electrolyte employed, but they also differ in design, construction, and the type of MnO_2 used.[4,5,7,9,12–19] The Leclanché cell employs an electrolyte of aqueous ammonium chloride with or without a lesser amount of zinc chloride. As MnO_2, NMD or activated NMD, often blended with a synthetic MnO_2 (mainly CMD), is employed. The zinc chloride cell (sometimes called the heavy-duty or extra-heavy-duty Leclanché cell) employs an electrolyte of aqueous zinc chloride, with or without a minor fraction of ammonium chloride, coupled with higher quality MnO_2 than employed in the common Leclanché cell—i.e., principally synthetic manganese dioxides with, perhaps, some activated NMD. The Leclanché and zinc chloride cells are collectively referred to as zinc–carbon cells. Alkaline cells employ an electrolyte of aqueous concentrated KOH (usually ~$9M$) and highly developed EMDs.

Both the performance and cost of $Zn-MnO_2$ cells increase in the order Leclanché < $ZnCl_2$ < alkaline.[1,15,17,19] An example of performance difference is shown by the following relative discharge capacities of D-size cells on 2.25-Ω continuous drain to a 0.9-V cutoff[15]:

Standard Leclanché/high-power Leclanché/$ZnCl_2$/alkaline = 0.12/0.24/0.55/1.00

The differences shown by this example, as well as by other similar published examples, are due to several causes (see, e.g., Ref. 16): i.e., (1) packing densities of the various manganese dioxides are very different (EMD being the most dense) and consumer cells are constant-volume devices; (2) cell construction varies, such that MnO_2 is used more efficiently in the costlier cells; and (3) the MnO_2 intrinsically functions better on a unit mass basis with progression from NMD to CMD to EMD.

The modern $Zn-MnO_2$ alkaline cell possesses a collection of advantages that make it the premier primary cell and the most popular primary cell in the industrialized world, with sales of some $6 billion. These advantages include high energy density, high drain capabilities (within its design), stability over a wide temperature range, long shelf life (~4 years), low corrosion-provoked leakage, low-cost and abundant raw materials, and environmental friendliness. This cell is the most costly of the three, not because of the electrolyte or the (most costly) MnO_2, but mainly because of its different design, which serves to maximize the advantages of the alkaline chemical system—i.e., a steel can with an "outside" sleeve-type cathode and inside zinc-slurry anode rather than a zinc can that acts as "outside" anode to an inside "bobbin" cathode.

Current annual production of battery manganese dioxides is about 450,000 tons, comprising 210,000 tons of NMD, 40,000 tons of CMD, and 200,000 tons of EMD.[12,9]

Sorting out the relatively few battery-active forms of MnO_2 has been an important part of battery development.[18] As efforts focused on synthesizing the active structures, the mechanism of electrochemical discharge and its relationship to MnO_2 properties and structure gained a central position in electrochemical studies.

2. Brief Overview of Previous Literature

Despite its complexity, the chemistry of the MnO_2 electrode is a very mature subject, and the number of relevant publications is many thousands, far beyond manageability in a chapter of this size. Comprehensive reviews that are still timely were published in the mid-1970s. These deal with Leclanché cells,[4] alkaline cells,[5] and the electrochemistry of MnO_2.[20,21] Several major reviews have since been published that deal with synthesis and X-ray diffraction studies of manganese dioxides,[22] solid-state and electrochemical properties of manganese dioxides,[23] and structural and electrochemical properties of the proton/γ-MnO_2 system.[24] Preparation and properties of EMD are addressed in two reviews,[20,25] and two smaller reviews deal with MnO_2 rechargeability.[7,26]

Although the value of journal articles cannot be minimized, it seems appropriate to mention the intense concentration of relevant articles in the proceedings of the various symposia on battery materials or MnO_2 sponsored and published by the International Battery Material Association (IBA) and the Electrochemical Society. The respective proceedings volumes bear the titles of the symposia for the period 1975–1985.[27–31] The nine-volume series *Progress in Batteries & Solar Cells*, published by JEC Press in Cleveland, Ohio, between 1978 and 1990, contains numerous articles on MnO_2 electrodes in aqueous solutions as well as other battery systems. This series at first ran concurrently with the IBA symposia, and then the two were merged into the single series *Progress in Batteries and Solar Cells,* starting in 1987. As the IBA symposia continued, the title of the latter publication was changed to *Progress in Batteries & Battery Materials*, which currently consists of Vols. 11–15; this series is published by IBA Incorporated, Brunswick, Ohio.

3. Organization of Chapter

This chapter will continue in Section II with a simplified (two-electron-step) description of MnO_2-electrode behavior in alkaline cells, both because such cells best delineate the chemistry of first- and second-electron discharge for MnO_2 and because alkaline cells are the mainstay of modern aqueous MnO_2 science. Then, descriptions of the MnO_2 electrode in Leclanché and zinc chloride environments will be discussed in Section III as modifications of alkaline discharge. Sections IV and V will then focus on MnO_2 structure and the effect of such structure on battery activity. Section VI will revisit first-electron discharge (proton intercalation) in detail.

Then, the chapter will change focus with a section on the chemistry of recharging MnO_2 in alkaline media, which is currently a technological frontier of intense interest. Finally, the last section will focus on MnO_2 deposition, which, with its parametric variations, is responsible for the varied but controlled structural features of MnO_2.

Manganese oxides are one of the most utilized frameworks or precursors for lithium-battery cathodes and loom very large in next-generation power sources. This aspect of manganese oxides would require a complete review of its own and hence is not included in this chapter. $Mg-MnO_2$ and $Al-MnO_2$ cells are also not covered in this chapter, although these cells employ aqueous electrolytes—i.e., magnesium perchlorate with some magnesium hydroxide (pH = 8.5) in the former case and aluminum or other chlorides in the latter case. Therefore, the MnO_2 behavior is somewhat similar to that in other aqueous solutions. These cells are reviewed in Ref. 32.

II. ALKALINE DISCHARGE FROM TWO-STEP VIEWPOINT

1. Discharge Curves and Steps for $\gamma/\varepsilon\text{-}MnO_2$

When MnO_2 is completely discharged in alkaline solution, the overall reaction is

$$MnO_2 + 2H_2O + 2e^- \rightarrow Mn(OH)_2 + 2OH^- \tag{1}$$

A useful starting point for discussing the mechanism of the reaction in Eq. (1) is examination of the galvanostatic discharge curves shown in Fig. 1. These curves consist of two sections or segments, which describe two discharge steps. In the first, the potential decreases continuously from ~0.3 V to approximately −0.4 V vs. Hg/HgO (or ~1.6 V to ~0.9 V vs. Zn/ZnO_2^{2-}). Then the curve flattens out at −0.4 V for the remainder of the discharge to give stage 2. Finally, when oxide reduction cannot sustain the current further, the potential drops abruptly to ~−1 V, where H_2 evolution occurs; as this is not part of MnO_2 reduction, it will not be addressed further in this chapter. At very low current densities, each segment corresponds to one electron per Mn atom, as first reported by Kozawa and co-workers.[33–36] Correspondingly, these investigators proposed that the

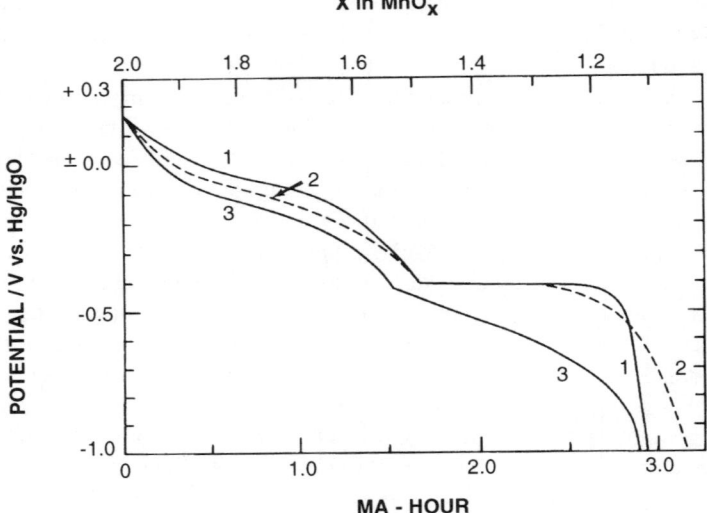

Figure 1. Discharge curves for thin films of EMD on a graphite substrate in 9M KOH at 23°C. 1, $i = 0.01$ mA/cm^2; 2, $i = 0.03$ mA/cm^2; 3, $i = 0.28$ mA/cm^2. (Reprinted from Ref. 33, Fig. 2, with permission from the authors and the Electrochemical Society.)

electrochemical reduction of MnO_2 occurs in a two-stage mechanism according to the following reactions:

$$MnO_2 + H_2O + e^- \rightarrow MnOOH + OH^- \quad \text{Stage A} \quad (2)$$

$$MnOOH + H_2O + e^- \rightarrow Mn(OH)_2 + OH^- \quad \text{Stage B} \quad (3)$$

Kinetic effects cause both segments to decrease with an increase in discharge rate, the second more than the first.

2. Step Mechanisms for $\gamma/\varepsilon\text{-}MnO_2$

(i) First Electron

The first- and second reduction steps differ in mechanism, as one would anticipate from the shape of the discharge curve. First-electron discharge occurs by the so-called proton–electron mechanism, in which the rate-determining step is proton–electron pair diffusion through the solid manganese oxyhydroxide.* The discharging oxide may be described as a mixture of Mn^{4+}, Mn^{3+}, O^{2-}, and OH^- ions (as the early investigators[34] deduced from Vetter's theoretical treatment of nonstoichiometric oxides[37]), through which the protons hop between oxide ions and the electrons hop between Mn^{4+} ions (cf. Fig. 2). The facile proton and electron movement is made possible by (a) the good semiconductivity of the manganese oxide, (b) the Lewis acid strength of the proton and its ability to exchange between neighboring oxygen atoms in O^{2-} ions and water, and (c) the fact that the proton, being so tiny, causes minimum structural change (i.e., $r_{O^{2-}} = 1.40$ Å and $r_{OH^-} = 1.53$ Å).[26] From the Nernstian shape of the potential versus reduction curves (especially at very low rates or at open circuit), the solid phase has been identified as a homogeneous redox system.[34,38] This concept has been accepted and used widely through the 1980s, successfully explaining many observations. However, a homogeneous redox system implies that all proton sites are

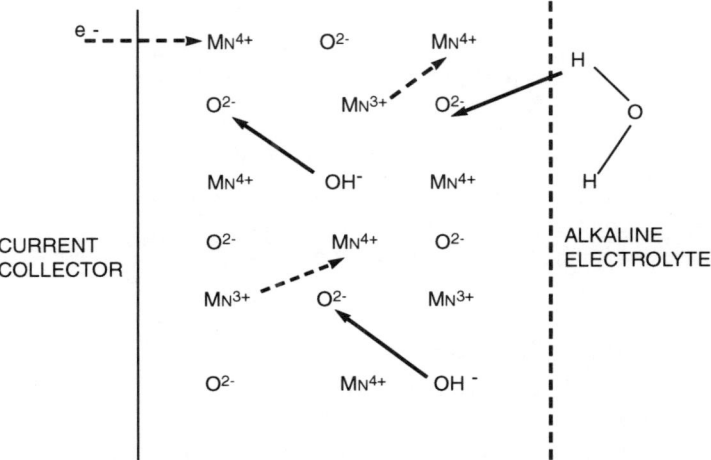

Figure 2. Schematic representation of MnO_2 during first-electron discharge. (Reprinted from Ref. 34, Fig. 1, with permission from the authors and the Electrochemical Society.)

---

*This first-electron step has been referred to in various ways—i.e., proton insertion, intercalation, etc.—which will be used interchangeably in this chapter.

energetically equivalent, and this is at variance with the detailed manner in which various oxide properties change with reduction. Therefore, a full section (Section VI) will address first-electron reduction in detail after MnO_2 structure has been discussed (as necessary background).

(ii) Second Electron

The second MnO_2 discharge step, Eq. (3), is a heterogeneous process, as shown by the presence of two solid phases—i.e., MnOOH and $Mn(OH)_2$—and the corresponding flat equilibrium discharge curve[34,39] (e.g., cf. Fig. 1, curve 1). It is believed to proceed by a dissolution–precipitation type of mechanism, which actively involves the Mn(III) ion in solution. [Mn(III) ions are present during the first step, also, but do not influence the kinetics.[26]] The mechanism may be formulated as the following three steps[26,33,36]:

$$MnOOH + H_2O \rightarrow MnO(OH)_2^- + H^+ \qquad (4)$$

$$MnO(OH)_2^- + e^- + H_2O \rightarrow Mn(OH)_4^{2-} \qquad (5)$$

$$Mn(OH)_4^{2-} \rightarrow Mn(OH)_2 + 2OH^- \qquad (6)$$

3. Experimental Support for Mechanism of $\gamma/\varepsilon\text{-}MnO_2$ Discharge

The first- and second-electron discharge mechanisms were developed from numerous studies of the physical, chemical, and electrochemical properties of manganese dioxide as a function of discharge. These studies are detailed below.

(i) X-Ray Diffraction Studies

Ex situ[20,40,41] as well as *in situ*[42] X-ray diffraction (XRD) studies show that the principal change during the sloping portion of the discharge curve is, at first, dilation of the original lattice without change in crystal structure.* Lattice dilation is necessitated by the increase in ion size during reduction of Mn^{4+} ions ($r = 0.52$–0.60 Å) to Mn^{3+} ions ($r = 0.65$ Å). The

---

*Reduction in which electrons and ions move in the host lattice without changing the host structure, other than through lattice expansion, is termed "topotactic" reduction.

larger Mn^{3+} ion is stabilized in the lattice by Jahn–Teller distortion of the MnO_6 octahedra.[43]

When MnO_2 has been reduced to $\sim MnO_{1.65}$,[†] formation of MnOOH occurs. This is principally α-MnOOH or groutite, but amorphous MnOOH and γ-MnOOH (manganite) also have been observed. Although the MnOOH compounds have separate XRD patterns from MnO_2, they are isostructural with MnO_2, as will be discussed in Section IV, and thereby are compatible with the homogeneous reduction.

During the flat region (second step) of the discharge curve, lower oxides such as Mn_3O_4 and, finally, $Mn(OH)_2$ appear. The XRD and chemical changes are found to be reversible with discharge and recharge during the first reduction stage but not during the second-electron reduction.

(ii) In Situ Microscopy

Kozawa's two-step mechanism was confirmed by Ruetschi[44] using direct visual observation of morphological changes in MnO_2 (EMD) during low-speed cyclic voltammetry (180 mV/h) in both alkaline and neutral (chloride) solutions. Utilizing an ultrathin cell, Ruetschi applied microscopy at 1500× magnification. In 10.2M KOH he found two reduction peaks. During the first peak, he initially observed a change in shape of the original MnO_2 particles. Then, grain dissolution occurred during the descending branch of that peak. Associated with the second peak, small hexagonal platelets of $Mn(OH)_2$ began to crystallize out of solution onto the substrate.

(iii) Equilibrium Potentials as a Function of Discharge

Since equilibrium potentials may be treated rigorously with thermodynamics, equilibrium discharge curves, i.e., plots of equilibrium potential (E_e) versus degree of MnO_2 reduction (r), have played an important role in the elucidation of MnO_2 discharge mechanism. Application of this tool has been particularly intensive for first-electron discharge to rationalize the sloping E–r curve in both alkaline and Leclanché solutions.

[†]Since MnO_2 reduces by the sequential addition of H, i.e., $H^+ + e^-$, the true formula should be $MnO_{2-x}(OH)_x$. However, the shorthand formula MnO_z ($z \leq 2$) is often employed solely to indicate Mn oxidation number, where $z = 2 - x/2$.

The equilibrium potentials are obtained by discharging the electrode to various levels of discharge and then stopping the current and letting the electrode rest on open circuit until the potential decays to a steady state—i.e., the "equilibrium" value (E_e)—which signals equidistribution of the protons throughout the oxide. Most of the potential decay or recovery in alkaline solution occurs in a very few hours, although thorough equilibrium studies usually involve recoveries of a day or more.[20,34,39,45]

The E_e–r curve contains the same two steps as the (dynamic) E–r curve or closed-circuit voltage curve, and, if the current density is not too great, the E_e–r curve is roughly parallel to the E–r curve.[20,38,45]

The sloping E_e–r curve during first-electron reduction is generally explained by treating the manganese oxyhydroxide as a solid redox system in Mn^{4+} and Mn^{3+} ions.[20,38] Thermodynamic derivations result in expressions of the Nernst form; i.e., the potential versus an Hg/HgO reference electrode (which is reversible to OH^- ions) is of the form

$$E_{e,1} \text{ (vs. Hg/HgO)} = E_{e,1}^{\circ} + (ART/F) \ln\{[Mn^{4+}]/[Mn^{3+}]\} \qquad (7)$$

Here $E_{e,1}^{\circ}$ is the equilibrium at the chosen standard state, and $[Mn^{4+}]$ and $[Mn^{3+}]$ are the concentrations or fractions of the Mn in the 4+ and 3+ oxidation states, respectively. When proper consideration is given to the solid-state nature of the oxyhydroxide, largely manifest in the term "A," Eq. (7) fits experimental data through most of the first-electron reduction. Section VI reviews the various treatments of first-electron discharge in detail.

After completion of the first electron discharge, E_e is independent of degree of reduction [i.e., from MnOOH to $Mn(OH)_2$], signaling that the second electron transfer (Eq. 3) is a *heterogeneous* reaction. The corresponding potential is of the form

$$E_{e,2} \text{ (vs. Hg/HgO)} = E_{e,2}^{\circ} + (RT/F) \ln\{(MnOOH)/(Mn(OH)_2)\} \qquad (8)$$

Here (MnOOH) and ($Mn(OH)_2$) represent activities of the respective species, and both activities are 1 throughout the second-electron reduction (which is approximated by curve 1 in Fig. 1). Kozawa and Yeager[33] added insight to the thermodynamics, showing polarographically that the flat potential is halfway between the half-wave potentials for Mn^{2+} oxidation and Mn^{3+} reduction. Also, they showed polarographically that the concentration of Mn^{2+} and Mn^{3+} ions in solution does not change with changes in the relative amounts of the two oxides. This is a reminder that the solubility thermodynamics between the Mn^{2+} and Mn^{3+} ions and their complementary solid form are fundamentally the basis of Eq. (8) and that Eq. (8) could take the same form as Eq. (7).

(i) Modeling of Proton Diffusion for First-Electron Reduction

Modeling studies of first-electron discharge and recovery based on proton diffusion have provided positive evidence for the proton–electron mechanism set down in Section II.2(i). Coleman[47] first recognized, with the Leclanché cell, that MnO_2 discharge involves diffusion of the reaction products in the interior of the oxide. Open solutions of the diffusion equation in one dimension were obtained[48,49] by assuming the reduction to be in equilibrium—i.e., by assuming that the potential was related to the surface oxide composition according to a Nernst-type equation, such as Eq. (7). These solutions were first applied to Leclanché systems during discharge and recovery[48–50] and soon thereafter were shown to describe alkaline systems during both galvanostatic and potentiostatic reduction.[51,52] The diffusion coefficient of protons in the oxide and the electrochemical reaction area, which together appear as a parameter in the solution, appeared to assume reasonable values in fitting of the theory to the results.

Later models[53,54] refined the above treatment to include nonequilibrium of the electrochemical reduction and a diffusion coefficient that changed with degree of reduction, the latter allowing for filling of the diffusion sites. The general agreement between theory and experiment continued to support the proton-diffusion/homogeneous-phase mechanism of the first-electron reduction.

Modeling of MnO_2 discharge will be addressed in more detail in Section VI.

(v) Isotope Effects in First-Electron Reduction

First-electron discharge slows down (gives greater overvoltage) when D_2O is substituted for H_2O.[34,35] When discharge currents for KOH–H_2O and KOD–D_2O systems ($[OH^-] = 9M$) were compared at equal polarization, the ratio of such currents was $I(KOH)/I(KOD) = 1.41$.[36] Polarization was defined as the difference between the IR-free closed-circuit and open-circuit potentials. These investigators considered this ratio to be evidence of rate-determining proton diffusion early in the first-electron transfer, as the theoretical diffusion rates would be expected to be proportional to $(m_{D^+})^{1/2}/(m_{H^+})^{1/2}$ or 1.41.

Gabano et al.[52] studied first-electron reduction in $1M$ KOH in H_2O and $1M$ KOD in D_2O and applied their theoretical diffusion model to the results. They found that the experimental value of $D^{1/2}$ from potentiostatic reduction varied with the hydrogen isotope as $D_H^{1/2}/D_D^{1/2} = 1.4$, in accordance with diffusion theory.

(vi) Parametric Discharge Studies

Kozawa and Yeager[33,55] conducted numerous diagnostic tests in which various discharge parameters were varied. The effects of these parameters on the resulting discharge curves for both first- and second-electron discharge translated into the effects on the rates shown in Table 1.

The second-electron mechanism was further clarified by Kozawa and Yeager[55] through an extension of the galvanostatic curves. MnOOH, prepared by discharge of MnO_2–coke composites, was reduced at various constant currents. Resultant potential–time curves, over the first few minutes, showed polarization that increased slowly with current up to a critical current. At and above this critical current, the potential dropped very rapidly, and it was apparent from the potential–time and current–potential relationships that this critical current is a limiting current for diffusion of Mn(III) ions.

Kozawa and Yeager[55] showed that Mn(III) ions are substantially more soluble than Mn(II) ions ($4.4 \times 10^{-3}M$ vs. $0.4 \times 10^{-3}M$) and that the

Table 1
Effects of Various Parameters on MnO_2 Discharge Curve[a]

Variable	Effect on rate of 1st electron discharge	Effect on rate of 2nd electron discharge
Increase in KOH concentration (0.1 to $9M$)	Very little effect	Increase
Increase in apparent reactant surface area (decrease in MnO_2 particle size)	No significant effect	Increase
Increase in C/oxide ratio [from 1 to 10 (w/w)]	No effect	Increase
Addition of triethanolamine (complexing agent for Mn^{2+} and Mn^{3+} ions)	No effect	Increase
Increase in T ($1m$ NaOH)	Increase	Increase
Substitution of D_2O for H_2O	Decrease	na[b]
(Limited) increase in MnO_2 thickness	No significant effect	na

[a] Electrolyte is $9M$ KOH unless otherwise noted.
[b] na, Not available.

limiting current corresponds to the electrochemical reduction, Eq. (5), which is coupled to the reaction in Eq. (4).

The observations in Table 1 then follow quite understandably: (a) Mn(III)-ion solubility increases with OH⁻-ion concentration and in the presence of the complexing agent; (b) the Mn(III)-ion production rate increases with the MnOOH apparent surface area; and (c) the limiting reduction current at given MnOOH surface area and Mn(III)-ion concentration increases with cathodic area, which is proportional to the amount of carbon present.

In contrast, the electron–proton mechanism, which occurs within the solid MnO_2, does not depend on the above factors to a very significant extent.

Diagnostic tests[33,55] aimed at differentiating first- and second-electron discharge of EMD indicated that first-electron reduction did not depend significantly on solution-related parameters or reasonable changes in the percent carbon in the cathode mix. These tests are discussed further in the next subsection in conjunction with second-electron reduction.

(vii) Diagnostic Chemical Studies

Kozawa[38] conducted several diagnostic tests that helped confirm the one-phase redox concept of the first-electron reduction.

- Acid washing of MnO_2 with 9–10M H_2SO_4 at 80–95°C resulted in a considerable increase in open-circuit potential. Such a potential increase is consistent with Eq. (1) for a one-phase system but not with Eq. (2) for a two-phase system.

- After acid washing MnO_2 and inserting it in solutions of different pH values, Kozawa found permanganate and manganate ions in solution, associated with potentials near the theoretical E–pH line for MnO_2 and MnO_4^- or MnO_4^{2-}.

- H_2 gas was found to be absorbed (oxidized) much faster for MnO_2 that contained a Pd or Ag–Mn catalyst than for various oxides that reduce by the two-phase (heterogeneous) mechanism—i.e., PbO_2, HgO, and CuO.

4. Discharge of β-MnO₂

Kozawa and Powers[39] found that β-MnO_2 discharge in alkaline solution produces a two-step curve, similar in shape to that of γ-MnO_2 but displaced negatively from the latter by ~0.12–0.17 V. From this observation, they concluded that β-MnO$_2$ reduces by the same two-step mechanism as γ-MnO$_2$.

In contrast, however, Bode and Schmier[56] found that the equilibrium potential of MnO$_2$ in KCl, pH 6.7, is independent of the degree of reduction by hydrazine. Also, they found that the β-MnO$_2$ crystal lattice did not dilate in the early reduction stages, but the peaks just weakened in intensity and finally were replaced by those of the product, MnOOH. They further showed that γ-MnO$_2$, under identical chemical reduction, gives a sloping E_e–r plot and crystal lattice dilation. From these observations, they concluded that β-MnO$_2$ cannot intercalate protons and is forced to reduce by a two-electron mechanism (cf. Section III).

McBreen,[57] using slow-scan voltammetry, substantiated Bode and Schmier's conclusion in alkaline solution when he found that the β-MnO$_2$ lattice does not expand upon reduction. McBreen rationalized the sloping E_e–r curve of Kozawa and Powers[39] in terms of E_e being a mixed potential for simultaneous reduction of β-MnO$_2$ to hausmanite (Mn$_3$O$_4$) and of hausmanite to Mn(OH)$_2$. The equilibrium potential plateaus during second-electron reduction of γ-MnO$_2$ and β-MnO$_2$ in alkaline solution were found to superimpose.[39]

III. DISCHARGE IN LECLANCHÉ AND ZINC CHLORIDE ELECTROLYTES

More reactions are needed to describe MnO$_2$ discharge in weakly acidic to near-neutral chloride solutions of the ZnCl$_2$ and Leclanché systems than are portrayed by the two-step MnO$_2$ reduction described in the last section. The mechanism is further complicated because slight differences in experimental conditions, such as electrolyte composition and amount of electrolyte present, change the interplay of these reactions very markedly. The chloride solutions are not as strongly buffered as the strong (~9M) alkaline electrolyte, and so conditions at the oxide surface can change markedly even during a single discharge.

Additionally, whereas discharge of the alkaline cell is dominated by MnO$_2$ reduction, discharge of Leclanché and ZnCl$_2$ cells is impacted significantly by the anode through precipitation of zinc compounds at/in the surface of the cathode. Therefore, this section will start with a review of overall cell phenomenology, after which the individual cathode and anode partial reactions will be delineated. The reaction schemes will then be applied to some of the more recent work, which has focused on the superior performance of the $ZnCl_2$ cell over that of the Leclanché cell, on the role of the precipitates in this difference, and on effects of MnO_2 type on cell performance.

1. Overall Cell Chemistries

The initial open-circuit voltage of the Leclanché cell corresponds closely to the equilibrium

$$MnO_{2(s)} + Zn_{(s)} + 4H^+_{(aq)} \rightleftharpoons Zn^{2+}_{(aq)} + Mn^{2+}_{(aq)} + 2H_2O_{(aq)} \quad (9)$$

where the subscripts s and aq refer to the solid and aqueous phases, respectively. For example, using very precise free energies of formation under carefully defined conditions, DeVries[58] calculated an equilibrium voltage ($V_{e,c}$) of 1.603 V, which compares to a measured voltage ($V_{e,m}$) of 1.598 V. His free energies corresponded to β-MnO_2, which has been quite popular in Leclanché cells.[8]

Inserting ammonia/water equilibria into Eq. (9), such that NH_4^+ ions are the proton donor on the left-hand side and NH_4OH is the product on the right-hand side, leaves $V_{e,c}$ unchanged.

DeVries found that pseudo-equilibrium voltages after various stages of discharge are very difficult to calculate. Problems include unknown concentration gradients in the oxide and in the solution in the pores of the oxide.

On discharge, various products are formed in addition to the Zn^{2+} and Mn^{2+} ions shown by Eq. (9)—i.e., precipitates of zinc, solid hetaerolite ($ZnO \cdot Mn_2O_3$), and lower oxides of manganese. Various overall reactions have been proposed to explain the main aspects of carbon–zinc-cell discharge chemistry while keeping the number of reactions small. Reactions (10)–(14) are such an attempt to show the primary products.[4,18,58–62]

NH_4Cl electrolyte:

$$2MnO_{2(s)} + 2NH_4Cl_{(aq)} + Zn_{(s)} \rightarrow Zn(NH_3)_2Cl_{2(s)} + 2MnOOH_{(s)} \quad (10)$$

$$2MnOOH_{(s)} + 2HCl_{(aq)} \rightarrow MnO_{2(s)} + MnCl_{2(aq)} + 2H_2O_{(aq)} \quad (11)$$

$ZnCl_2$ electrolyte:

$$8MnO_{2(s)} + 4Zn_{(s)} + ZnCl_{2(aq)} + 8H_2O_{(aq)} \rightarrow 8MnOOH_{(s)} +$$
$$[ZnCl_2 \cdot 4Zn(OH)_2]_{(s)} \quad (12)$$

$$8MnOOH_{(s)} + 5ZnCl_{2(aq)} \rightarrow 4MnO_{2(s)} + 4MnCl_{2(aq)} +$$
$$[ZnCl_2 \cdot 4Zn(OH)_2]_{(s)} \qquad (13)$$

Both electrolytes:

$$2MnO_2 + Zn \rightarrow ZnO_{(s)} \cdot Mn_2O_3 \text{ (hetaerolite)} \qquad (14)$$

Some investigators are less specific as to the formula of the zinc-containing precipitate in Eqs. (12) and (13). Also, MnOOH is used as the only solid form of Mn(III) in Eqs. (10)–(13), whereas some investigators also allow for the existence of Mn_2O_3[61] and $Mn(OH)_2$.[62] Other forms of the acid–base reaction given in Eq. (11) are found, such as involvement of NH_4^+ ions and NH_4OH rather than, or in addition to, H^+ and H_2O. Finally, Mn_3O_4 (hausmanite) is formed from MnOOH on prolonged discharge in both types of electrolyte[4,18] but is omitted in the above sequence of reactions, as it is not used in the mechanistic discussion of this section. None of the above differences seriously impacts the MnO_2 electrode fundamentals, as reviewed here.

Equations (10)–(14) summarize the principal processes in these cells, which are as follows:

1. MnOOH is formed initially at the cathode, as in the case of alkaline cells.

2. Dissolution of anodic zinc is followed by precipitation of the diammine zinc chloride and the basic zinc chloride in the NH_4Cl and the $ZnCl_2$ cell, respectively. These salts are formed at the outer edge of the cathode bobbin and have a very significant impact on the discharge capacity of the cell, which will be discussed further in the later part of this section. The alkaline cell does not suffer such severe problems from the zinc oxidation, because hydroxide ions migrate from the cathode to the anode, causing zinc oxide formation to occur in the anolyte away from the cathode. Because the anode is high-surface-area zinc powder, the oxide formation does not cause passivation problems except at very high current densities.

3. MnOOH may be removed from the surface of the MnO_2 by two paths, i.e., transport of protons and electrons into the oxide, as in the alkaline case, and disproportionation to form soluble Mn^{2+} ions. The acidic medium and the complexing power of NH_4^+ ions stabilize soluble Mn^{2+} ions.

4. Hetaerolite, $ZnO \cdot Mn_2O_3$, forms to a much greater extent in the chloride systems than in the alkaline case.

For a given cell, the ratio of products changes during the discharge.

2. Discharge Mechanism

(i) Discharge Curves

Constant-current discharge curves, which are often the starting point for kinetic studies, are typified by those shown in Fig. 3 for γ-MnO_2 (an EMD) in 5M NH_4Cl and 5M $ZnCl_2$ electrolyte, which are taken from Ohzuku et al.[62] The cells in these tests were flooded, and the cathode was in the form of an MnO_2–carbon composite. Significant characteristics of the discharge, which are generally manifest in such discharge data, are:

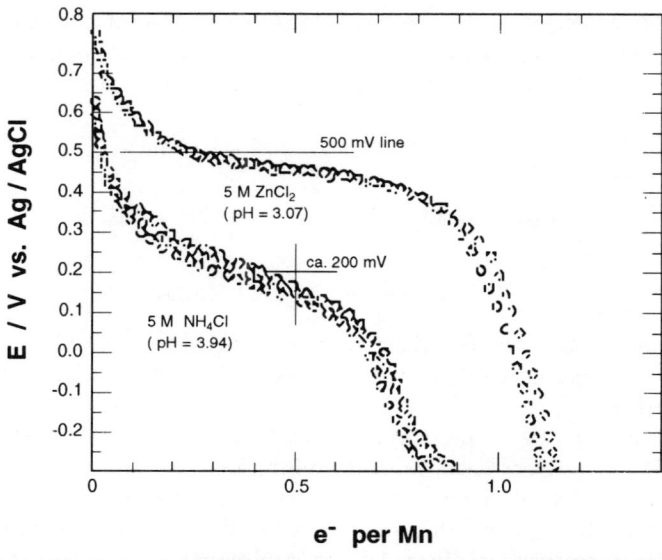

Figure 3. Comparison of EMD discharge curves in 5M NH_4Cl and in 5M $ZnCl_2$ solutions. $i = 10$ mA/g[1] (0.89 mA/cm^2). Cathode mix: MnO_2/acetylene black/electrolyte (0.1 g/0.1 g/2 ml). (Reprinted from Ref. 62, Fig. 2, with permission from the authors and IBA, Inc.)

(a) Both curves are generally S-shaped, much like alkaline discharge curves, although the curve for $ZnCl_2$ solutions usually flattens out for a considerable part of the discharge.

(b) The curve for $ZnCl_2$ occurs at higher potential with respect to a fixed reference electrode than that for NH_4Cl. Also, the discharge capacity to any end point (the author took 0.1 V vs. Ag/AgCl as the "practical" end point) is substantially greater for $ZnCl_2$.

(c) The discharge capacity is greater than 1.0 F per mole of MnO_2 in the case of the $ZnCl_2$ cell.

The above investigators[62] studied numerous MnO_2 materials, including various generic types, and the behavior was the same as (a)–(c) in every case. These results appear general, based on studies by other investigators[60,61,64] employing a variety of NH_4Cl–$ZnCl_2$ solutions and manganese dioxides.

(ii) Discharge Mechanisms

Cahoon, Korver, and Johnson (Ref. 65 and references therein) proposed that manganese dioxide is reduced to Mn^{2+} as the primary cathodic reaction, and the product Mn^{2+} then reacts with the remaining MnO_2 to form MnOOH and, in the presence of Zn^{2+}, $ZnO·Mn_2O_3$; i.e.,

$$MnO_2 + 4H^+ + 2e^- \rightarrow Mn^{2+} + 2H_2O \qquad (15)$$

$$Mn^{2+} + MnO_2 + 2H_2O \rightarrow 2MnOOH + 2H^+ \qquad (16)$$

$$2MnOOH + Zn^{2+} \rightarrow ZnO·Mn_2O_3 + 2H^+ \qquad (17)$$

This mechanism satisfied their data, including the observed products, Mn^{2+} ions and hetaerolite, and the slope of the potential–pH plot (59 mV/pH unit).

In a long series of investigations, Vosburgh and co-workers (Refs. 63 and 66 and references contained therein) found that the initial reaction product was MnOOH and that Mn^{2+} ions occurred only during the later stages of discharge and then built up with time. Mn^{2+} ions appeared in significant amounts only in the latter two-thirds of the discharge in NH_4Cl at pH 7 but appeared much earlier as the pH was decreased below 5. These and other observations led Vosburgh and co-workers to propose the following mechanism:

$$MnO_2 + e^- + H^+ \rightarrow MnOOH \tag{18}$$

$$2MnOOH + 2H^+ \rightarrow MnO_2 + Mn^{2+} + 2H_2O \tag{19}$$

$$MnOOH + H^+ + e^- \rightarrow Mn(OH)_2 \tag{20}$$

$$Mn(OH)_2 + 2H^+ \rightarrow Mn^+ + 2H_2O \tag{21}$$

A reaction such as Eq. (17) would still be necessary to account for hetaerolite formation. Reactions (20) and (21) are only significant in neutral solutions, where reaction (19) cannot depolarize the surface of MnOOH significantly.

As with the total cell reaction, the above reactions can be coupled with the ammonia/ammonium ion equilibrium, which makes NH_4^+ ions the proton donor; i.e.,

$$2MnOOH + 2NH_4^+ \rightarrow MnO_2 + Mn^{2+} + 2NH_4OH \tag{22}$$

Vosburgh's mechanism has found wide support, and reaction (18) is regarded as the primary reaction in the neutral solutions appropriate to Leclanché batteries. Under various equilibrium conditions, the two mechanisms—i.e., reactions (15) + (16) vs. (18)–(21)—reduce to the same overall equation, as is discussed below.

(iii) Equilibrium Discharge Potentials

Key to mechanistic interpretations has been the similarity between the dynamic discharge curves and the corresponding equilibrium discharge curves or open-circuit voltage curves (see Fig. 4) coupled with the thermodynamic interpretation of the potential–discharge curve.

In the absence of Mn^{2+} ions, the equilibrium potential–pH response of Eqs. (15) + (16) is, of course, the same as that of Eq. (18)—i.e., 59 mV/pH unit. This response has been observed for the initial open-circuit cell potential for both NH_4Cl and $ZnCl_2$ electrolytes and for pH values down to 1.[4,64,67] On the other hand, the potential–pH response when Mn^{2+} ions are present is ~118 mV/pH unit and appears to satisfy reaction (15).[68–70] This, of course, is equivalent to reactions (18) + (19) in equilibrium.

When coupled with Vetter's theory for mixed-ion oxides with mobile protons and electrons,[37] reaction (18) also explains the continuous decrease of equilibrium potential with discharge in the early discharge stages (e.g., cf. Ref. 20 and Fig. 4) similar to the alkaline case (Eq. 7).

Analysis of E_e versus R curves during one-electron discharge has been conducted by various investigators and will be detailed in Section VI.

E_e becomes constant during the latter part of the MnO_2 discharge (cf. region C in Fig. 4). This is attributed to reaction (15) predominating at the oxide–solution interface, the dissolution reaction being kinetically much faster than the first-electron transfer. Between regions A and C, both reactions (15) and (18) occur simultaneously.

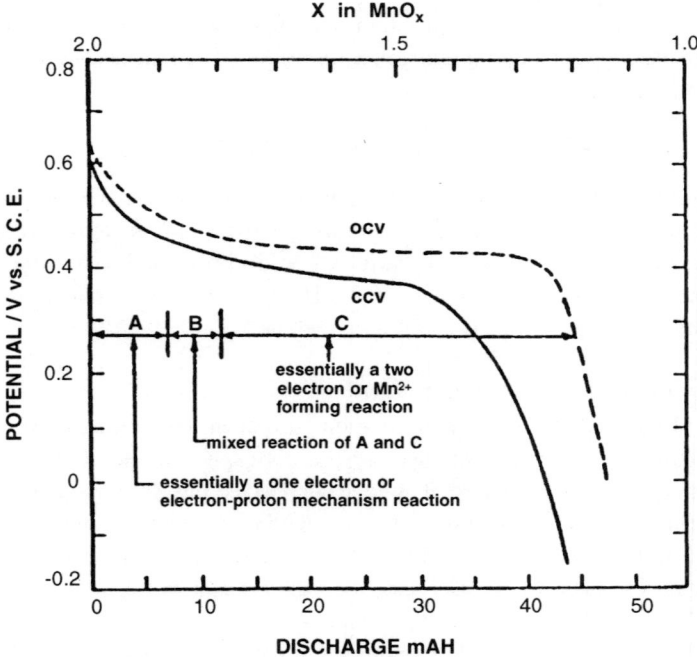

Figure 4. Open-circuit voltage (ocv) and closed-circuit voltage (ccv, *IR*-free) during MnO_2 discharge in 40% $ZnCl_2$, pH 3.5–4.5. The cathode is a compacted mix of 0.10 g EMD + 1 g graphite + 2 g coke + 0.5–0.7 cm^3 electrolyte. Discharge mode: 17 h of discharge (I = 0.35 mA) followed by 7 h of rest each day for 8 days. (Reprinted from Ref. 64, Fig. 10, with permission from the author and IBA, Inc.)

(iv) Discharge and Recovery Phenomena

Reduction usually starts with reaction (18), as in the case of the alkaline cell. As the surface concentration of MnOOH increases, the potential decreases continuously. The MnOOH thus formed is removed from the surface by the parallel paths of (a) proton intercalation in the MnO_2 through proton and electron diffusion and (b) disproportionation of the MnOOH, reaction (19).

In neutral solutions, MnOOH disproportionation is slow enough that proton intercalation and deintercalation control the rates of polarization buildup and decay during the early part of cathodic discharge and open-circuit recovery, respectively. This process was modeled by Scott[48] and Kornfeil,[49] who derived a theoretical relationship between discharge or recovery potential and time based on diffusion of protons between the MnO_2 surface and the bulk MnO_2. (Details of their expression will be reviewed in Section VI.) Using data for a neutral solution, made by adding NH_3 to NH_4Cl solution, these investigators fit both the discharge and recovery curves to their theory over a fairly substantial time range.

MnOOH is much more unstable in weak-acid electrolytes than in neutral or strongly alkaline solution. Therefore, when Mn^{2+} dissolution is favored by electrolyte acidity, by adequate electrolyte volume, and by NH_3 complexation, disproportionation of MnOOH is faster than H^+–e^- diffusion and is the major pathway for depolarization of the surface MnOOH. Such acidity is promoted by concentrated $ZnCl_2$ solutions, either with or without NH_4Cl present, because of hydrolysis, i.e.,

$$Zn^{2+} + 2H_2O \rightarrow ZnO_2^{2-} + 4H^+ \qquad (23)$$

Even NH_4Cl in neutral to weakly acid solutions provides sufficient complexation of Mn^{2+} ions to exhibit a noticeable effect on electrode depolarization, as demonstrated by the frequently quoted diagnostic tests of Era, Takehara, and Yoshizawa.[51,46] These tests showed that electrode polarization upon application of current and the time for such polarization buildup were much greater for weakly acidic to weakly basic KCl electrolyte than for the corresponding solutions that contained ammonium ions. Strong acid and strong alkali, also, gave much smaller polarization than the potassium-salt solutions, and the polarization times in the strong acid and base were much shorter. Likewise, when polarization ceased, the decay time was much greater in the neutral KCl than in the other solutions.

These results are rationalized by the fact that either protons or ammonia molecules depolarize the surface of MnOOH by shifting reaction (19) to the right. In the case of strong KOH solutions, the surface MnOOH is not solubilized substantially, but, rather, the strong base is thought to accelerate the H^+-e^- diffusion into the oxide, perhaps by weakening the $O^{2-}-H^+$ bond.

3. Key Issues in Recent Studies

(i) $ZnCl_2$ versus NH_4Cl Electrolytes

A concentrated effort has been expended to explain the superior discharge performance of zinc chloride cells as compared to that of Leclanché cells. Part of this difference relates to intrinsic electrolyte involvement in the MnO_2 discharge process, as exposed in studies with flooded cells and as exemplified by Fig. 3.[59,61,62,64] Part, however, involves differences in the zinc precipitates formed in the two cases, which is maximized in dry cells.[60–62]

Kozawa[64] studied discharge curves for EMD in flooded cells that contained either 10–70% $ZnCl_2$ or one of several Leclanché formulations. Ohzuku et al.,[62] who also employed flooded cells, studied electrochemical reduction of various manganese oxides in $5M$ or ~44% $ZnCl_2$ (pH = 3.07) and $5M$ or ~25% NH_4Cl (pH = 3.94). The galvanostatic discharge curves of both studies exhibited the three trends generalized in Section III.2(i)—i.e., the curve was at higher potential and yielded greater capacity for $ZnCl_2$ electrolytes than NH_4Cl or Leclanché electrolytes, and, in the case of $ZnCl_2$ solutions, the discharge potential and capacity increased with $ZnCl_2$ concentration. At the higher $ZnCl_2$ concentrations, the discharge capacity was greater than one electron per Mn atom.

MnO_2 discharge in $NH_4Cl-ZnCl_2$ electrolytes was found to be controlled largely by the $ZnCl_2$ concentration, although the addition of 20% NH_4Cl to 25% $ZnCl_2$ solution had a moderately beneficial effect on the discharge capacity.[64] On the other hand, ammonium chloride additions lowered the position of the discharge curve slightly at cathode compositions in the range $MnO_{1.7}$ to $MnO_{1.4}$.

The positive shift of potential with increasing $ZnCl_2$ concentration in the earlier section of the discharge studied by Kozawa correlated favorably with the measured pH of the solution through the equilibrium of reaction (18)—i.e., 59 mV/pH unit. This was taken as evidence in support of the one-electron discharge mechanism. Later in the discharge, when the $E-t$

curve flattened, the pH effect on the discharge potential was somewhat larger, signaling a contribution by the two-electron reduction, reaction (15), or the combination of reactions (18) and (19). Such two-electron involvement is also compatible with the discharge capacities greater than 1 e^-/Mn atom.

Ohzuku et al.[62] showed that discharge in $ZnCl_2$ solutions was superior to that in $ZnSO_4$ or $Zn(NO_3)_2$ solutions and that this superiority paralleled the greater hydrolysis (acidity) in the case of $ZnCl_2$. Kozawa[64] associated the significant benefit of $ZnCl_2$ late in the discharge with the measured solution buffer capacity, which neutralized the basicity created by the discharge reactions (18) and (19). This buffering delays precipitation of manganese and zinc in the MnO_2 pores and these latter processes, according to Kozawa, are the processes that finally shut off the discharge. Discharge capacity in flooded cells (>1 e^-/Mn atom) was shown to be significantly greater than that in electrolyte-starved dry cells.

A small negative effect of ammonium chloride additions to 25% $ZnCl_2$ solutions at $MnO_{1.7}$ to $MnO_{1.4}$ was attributed to their effect of decreasing the buffer capacity, and the small positive effect of 20% NH_4Cl in increasing the overall discharge capacity of 25% $ZnCl_2$ solution was attributed to complexation of Mn^{2+} ions, which retards manganese precipitation.[64]

Studying nonflooded/dry cells, Uetani and co-workers[60,61] also found that $ZnCl_2$ cells gave substantially greater discharge capacities than Leclanché cells. A major part of the cell polarization was shown to be anodic overvoltage, particularly in the case of Leclanché cells or $ZnCl_2$ cells with significant NH_4Cl additions. This overvoltage was shown to be caused by Zn^{2+}-ion buildup on the anode side of the separator. (This is in contrast to the behavior of alkaline cells at moderate current densities, in which most of the polarization is at the MnO_2 electrode.)

The addition of NH_4Cl up to about 5% in the $ZnCl_2$ electrolyte decreases the cathode polarization.[60,61] based on this finding, coupled with the detrimental effect on the anode polarization, the best overall NH_4Cl concentration in the $ZnCl_2$ electrolyte was found to be 5%.

(ii) Role of Zinc Precipitates

Uetani and co-workers[60,61] identified precipitates formed in the outer part of the cathode bobbin as $ZnCl_2 \cdot 4Zn(OH)_2$ in the $ZnCl_2$ cells and as $Zn(NH_3)_2Cl_2$ in the Leclanché cells. They attributed the superior performance of the $ZnCl_2$ cells to the fact that Zn^{2+} ions could penetrate the basic zinc chloride but not the zinc chloroammine—hence the greater zinc-ion buildup near the Leclanché anode. Kozawa[64] added the caveat that the greater conductivity of the ammonium-salt solutions as compared to that of the zinc-chloride solutions causes the chloroammine complex to form over a smaller distance, thus offering more impedance to Zn^{2+} ions.

Ohzuku et al.[62] focused on characterizing the zinc precipitate that formed during discharge in $ZnCl_2$ electrolyte and on explaining how the two-electron reaction to form Mn^{2+} ions could continue in spite of such a precipitate. Using scanning electron microscopy and elemental dot mapping (EPMA), they showed that this precipitate forms as an intermediate layer immediately around the discharging MnO_2 particles. From XRD they identified this solid as $Zn_5(OH)_8Cl_2 \cdot H_2O$.

Crystal-structure considerations led to the conclusion[62] that the basic zinc chloride precipitates with Mn^{2+} ions entrapped in it and that this solid intermediate can exchange Mn^{2+} for Zn^{2+}. Thus, the intermediate transports Mn^{2+} ions out of the discharging MnO_2 while transporting H^+ ions in, according to Eq. (15). This ion transport/migration apparently depends heavily on the presence of the current or field, for α-MnOOH and γ-MnOOH soaked in $5M$ $ZnCl_2$ solution for three months on open circuit did not change. This test, in fact, demonstrates that the intermediate protects the MnOOH core from convection or diffusion of fresh acid. In ammonium chloride electrolyte, the solid formed during discharge $[Zn(NH_3)_2Cl_2]$ does not possess the ion-transporting property shown by basic zinc chloride, which accounts for the great discharge difference between Leclanché and $ZnCl_2$ electrolytes (see Fig. 3). However, NH_4Cl, even without $ZnCl_2$, assists the discharge to some extent through hydrolysis to NH_3 and through the complexation of Zn^{2+} and Mn^{2+} ions by the so-formed NH_3 molecules. This is particularly so in flow cells, as reported in some older literature (cf. Ref. 62).

Figure 5. Proportions of reduction products during MnO_2 discharge in $ZnCl_2$ electrolyte for EMD (a), NMD (b), and CMD (c). Wt. MnO_2 = 31.1 g EMD, 30.7 g NMD, and 23.9 g CMD; △, ▲, 450-mA continuous discharge; ○ and ●, 450-mA intermittent discharge; □ and ■, 45-mA intermittent discharge. (Reprinted from Ref. 61, Fig. 11 with permission from the authors and the Electrochemical Society.)

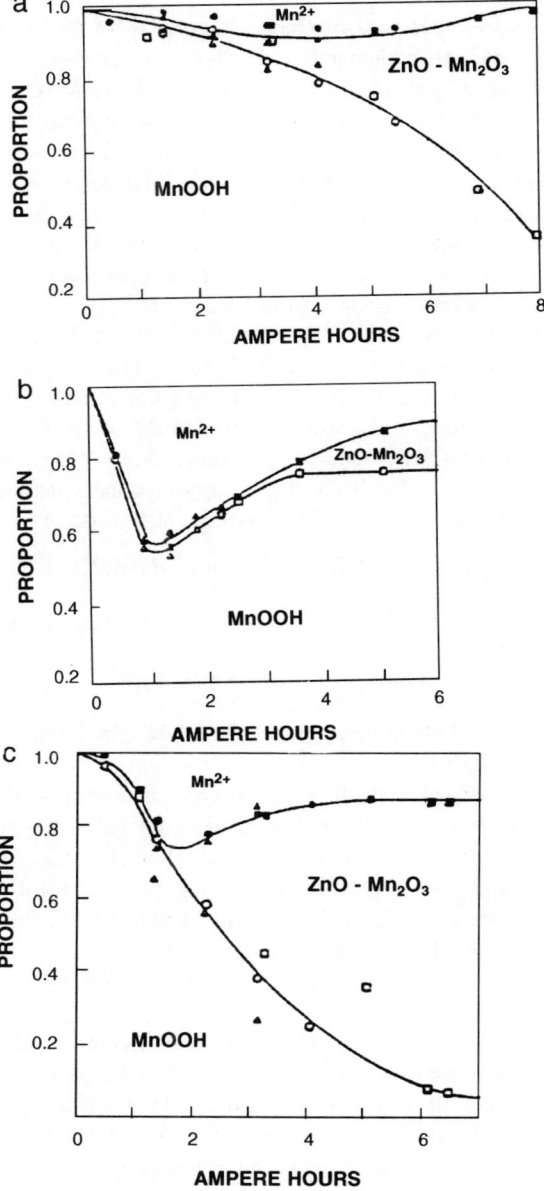

(iii) Effect of Manganese Oxide Type

Various lower manganese oxides, i.e., groutite (α-MnOOH), manganite (γ-MnOOH), and hausmanite (Mn_3O_4), were shown to discharge in $ZnCl_2$ solutions at significantly lower potentials than γ-MnO_2 (both EMDs and CMDs), β-MnO_2, and Mn_2O_3, although substantial discharge was found to occur ultimately in all cases.[62] Thus, (1) the heterogeneous two-electron mechanism, promoted by acidic electrolyte, is important in all these cases, but (2) the rate of ion migration and the free energy of the reaction are structure-specific. In alkaline solution, where depolarization must depend on H^+–e^- diffusion, β-MnO_2 gives only a very tiny discharge capacity, and this only through surface reactions.

Uetani and co-workers[60,61] found that relative discharge capacities of the generic manganese dioxides in $ZnCl_2$ electrolyte increase in the order NMD < CMD < EMD, similar to the findings of other investigators for both $ZnCl_2$ and alkaline electrolyte. Guided by analysis of the various phases for Zn^{2+} and Mn^{2+} ions and characterization of the precipitates as $ZnCl_2 \cdot 4Zn(OH)_2$ or $ZnO \cdot Mn_2O_3$, they interpreted their results in terms of the competition between the following three reactions:

$$8MnO_2 + ZnCl_2 + 4Zn + 8H_2O \to 8MnOOH + ZnCl_2 \cdot 4Zn(OH)_2 \quad (24)$$

$$2MnO_2 + 3ZnCl_2 + 4H_2O + 2Zn \to 2MnCl_2 + \{ZnCl_2 \cdot 4Zn(OH)_2\} \quad (25)$$

$$2MnO_2 + Zn \to ZnO \cdot Mn_2O_3 \quad (26)$$

The proportions of the three products, MnOOH, Mn^{2+}, and $ZnO \cdot Mn_2O_3$, are shown in Fig. 5.

Early discharge for all three MnO_2 types is dominated by MnOOH formation, so intercalation of protons in the unreacted MnO_2 distinguishes the relative performance, as in the case of alkaline discharge. As discharge proceeds, hetaerolite formation and manganese dissolution displace MnOOH accumulation. The inferior performance of NMD correlates with lack of formation of hetaerolite, and the superior performance of EMD over CMD correlates with a continued presence of MnOOH and very little Mn^{2+}-ion buildup.

The beneficial role of hetaerolite formation as compared to Mn^{2+}-ion formation stems from the fact that hetaerolite formation, reaction (26), does not alter the electrolyte composition, whereas Mn^{2+}-ion formation consumes $ZnCl_2$. Mn^{2+}-ion formation lowers electrolyte acidity and, correspondingly, retards depolarization of the MnOOH.

IV. STRUCTURAL CHARACTERIZATION OF MnO_2

One of the main thrusts in the study of battery-grade manganese dioxide has been the search for the battery-active forms, characterization of forms, and synthesis of improved forms. Most interest today focuses on γ-MnO_2 and ε-MnO_2, which are also the focus of this section. (ρ-MnO_2, which has been reported in various studies, now is considered part of the γ-MnO_2 family.) The battery activities of even these active forms vary considerably with synthesis conditions. Thus, as has been noted by Chabre and Pannetier,[24] "a quick examination of the scientific and technological literature . . . gives the impression that there are almost as many MnO_2 forms as there are preparation methods." Coupled with the fact that the different forms are difficult to characterize, the great number of structures provides incentive to review structural characterization as background for kinetic analysis and review of battery activity.

Although physicochemical properties can be addressed in a study of structure, this section will focus on three features: (1) crystal structure, (2) stoichiometry or chemical makeup, and (3) surface properties or mesostructure. Many other properties follow from these, as will be discussed.

1. Crystal Structure

Abundant literature exists on the characterization of manganese dioxides by X-ray diffraction. Detailed bibliography and discussion may be found in reviews by Burns,[81] Fernandes *et al.*,[22] and Chabre and Pannetier.[24]

(i) XRD Patterns

Although X-ray diffraction has been one of the strongest tools in MnO_2-structure elucidation, results have been complicated and difficult to understand. The difficulty stems from the fact that XRD peaks for the battery-active forms are often sparse, broad, and ill-defined and vary somewhat with almost every sample. Some general XRD facts are given here as background for structural definition.

Powder diffraction patterns for a typical commercial EMD, two CMDs, and an NMD are shown in Fig. 6. These materials* have in

---

*These samples are four of a large set of commercial manganese dioxides established by the International Battery Material Association (IBA) as study materials. The IBA has characterized these materials and published the corresponding properties for the benefit of the MnO_2/battery community—e.g., see Ref. 12.

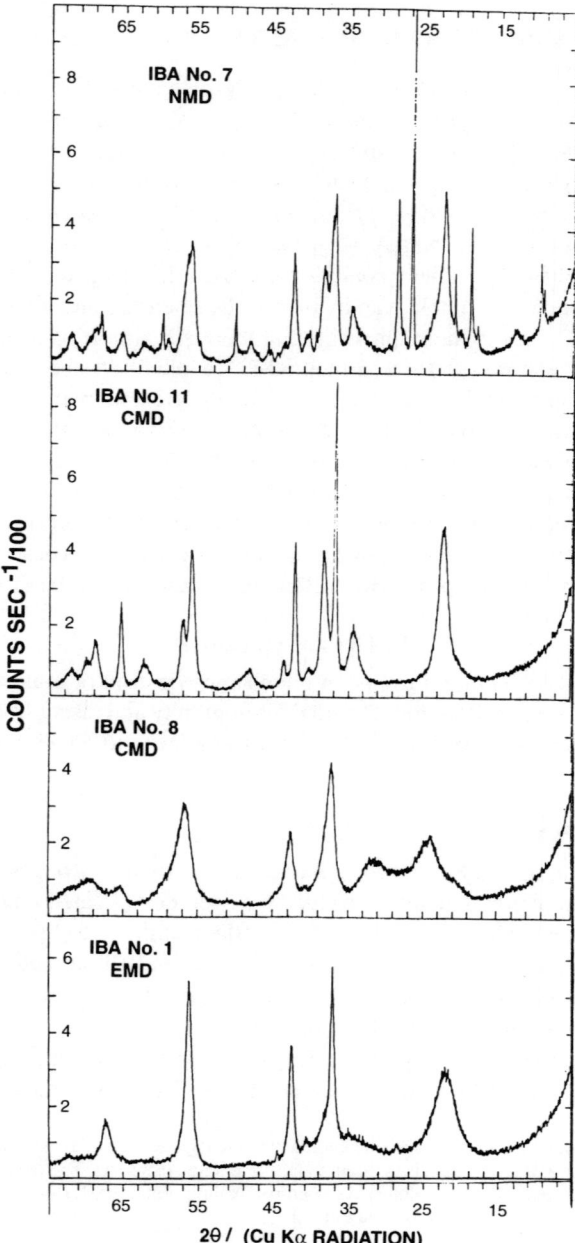

common five major peaks and several other peaks that vary in size with the particular sample. The line-rich pattern of IBA No. 11 (a CMD) closely matches that of the mineral nsutite (MnO_2) as well as that of the sample from which De Wolff deduced the γ-MnO_2 structure.[71,72]

Commercial EMDs are typified by the pattern of IBA No. 1 in Fig. 6, which contains only a few broad peaks.[71] These peaks vary slightly from sample to sample.[20,71,73] Although EMDs have historically been called γ-MnO_2's, because of the peak at 22°, interpretation of the patterns has led recent investigators to identify such materials as ε-MnO_2.[74] Most CMDs and EMDs tend to have patterns between those of IBA No. 1 and IBA No. 11, all the lines being accounted for as γ- or ε-MnO_2.

The NMD pattern represents a mixture of γ-MnO_2 and four other phases, i.e., quartz, pyrolusite, cryptomelane, and todorokite[71]; the last three minerals are MnO_2 species. Such mixtures are not used to determine MnO_2 structure.

When EMD deposition conditions are varied widely, the XRD patterns of the products vary greatly. Commercial EMDs are typically deposited from an acidic $MnSO_4$ + H_2SO_4 electrolyte at current densities of ~6 mA/cm^2 and 95°C.[75] Very low deposition current densities (<1 mA/cm^2) and high temperatures (95°C) yield a product rich in XRD peaks, similar to IBA No. 11 in Fig. 6,[76–78] and the product is most closely identified as γ-MnO_2.[71] As the deposition current density is increased or the temperature is decreased, the crystal pattern tends to that of ε-MnO_2 (the EMD in Fig. 6).[76–78]

All EMDs form a series between γ-MnO_2 and ε-MnO_2. On this basis, EMD has been considered to be a combination of ε-MnO_2 and γ-MnO_2. The γ/ε character is defined[79] by a so-called "Q" ratio, which is the ratio of the peak height for the most characteristic peak of γ-MnO_2 to that of the most characteristic peak of ε-MnO_2, i.e., $2\theta = 22.2°$ and $37.1°$ (Cu $K\alpha$ radiation); i.e.,

$$\gamma/\varepsilon \text{ character} = Q \text{ ratio} = 22.2° \text{ pk ht}/37.1° \text{ pk ht} \tag{27}$$

Q is approximately 1.5 for γ-MnO_2 and 0.35 for ε-MnO_2, and all EMDs fall between these limits.[79,77]

Figure 6. XRD patterns of an EMD, two CMDs, and an NMD. (Reprinted from Ref. 71, Figs. 5.1, 5,8, 5.11, and 5.13, with permission from the authors and IBA, Inc.)

β-MnO$_2$ not only is closely related structurally to γ- and ε-MnO$_2$'s, as will be discussed shortly, but also heat treatment of γ- and ε-MnO$_2$'s to approximately 375°C causes crystallographic changes in which the γ/ε-MnO$_2$ XRD pattern takes on β-MnO$_2$ character; i.e., the γ-MnO$_2$ peak at 2θ(Cu $K\alpha$) = 22° (d = 4.0 Å) is removed and the strong β-MnO$_2$ peak at 2θ(Cu $K\alpha$) = 28.3° (d = 3 Å) appears. On this basis, such heat-treated EMD (HEMD) has often been studied as an example of β-MnO$_2$.[39]

(ii) Simple Polymorphs

The γ- and ε-MnO$_2$ structures, which are nonstoichiometric, are based on two stoichiometric polymorphs, β-MnO$_2$ (pyrolusite) and ramsdellite, a rare MnO$_2$ mineral.[10,11,22,72,80,81] Structures of the above four simple polymorphs of MnO$_2$ are based on approximately hexagonal close-packed layers of oxygen anions. Two types of interstices exist between the close-packed oxygens, defining cation coordination environments with six and four nearest neighbors—i.e., octahedral and tetrahedral sites, respectively. In all four of the simple MnO$_2$ polymorphs, Mn^{4+} cations fill half the octahedral sites of the (a,b) plane, and different arrangements of Mn^{4+} cations in these sites distinguish each polymorph from the others,[81–83] as demonstrated in Fig. 7. The ε-MnO$_2$ phase is said to have a completely disordered distribution of Mn^{4+}, which gives it hexagonal symmetry, in contrast to the orthorhombic symmetry of γ-MnO$_2$.[81,84,85]

Perhaps a more graphic distinguishing feature of the simple polymorphs is given in terms of the manner in which the [MnO$_6$] octahedral group is linked to other [MnO$_6$] octahedra (see Fig. 8).[80,81] In the β-MnO$_2$ structure the [MnO$_6$] octahedra share edges to form single chains running parallel to the c axis; i.e., the half of the close-packed rows of octahedral interstices occupied by Mn^{4+} ions line up in an infinite chain. Each chain is cross-linked with neighboring single chains by corner-sharing of oxygen atoms of adjacent octahedra. In this structure the unoccupied octahedral sites may be considered to form single lines of vacancies parallel to the c axis, and these define "tunnels" with dimensions corresponding to the width of one [MnO$_6$] octahedron, which are designated accordingly as [1 × 1] tunnels.

Pyrolusite or β-MnO$_2$ is described as a rutile structure in which the [MnO$_6$] octahedra are distorted, producing three coplanar 120° Mn^{4+}–O bonds rather than the single 90° and the two 135° angles that would exist if all octahedra were regular.[22,86–88] This results in a lattice expansion in the c direction and a contraction in the other two directions.

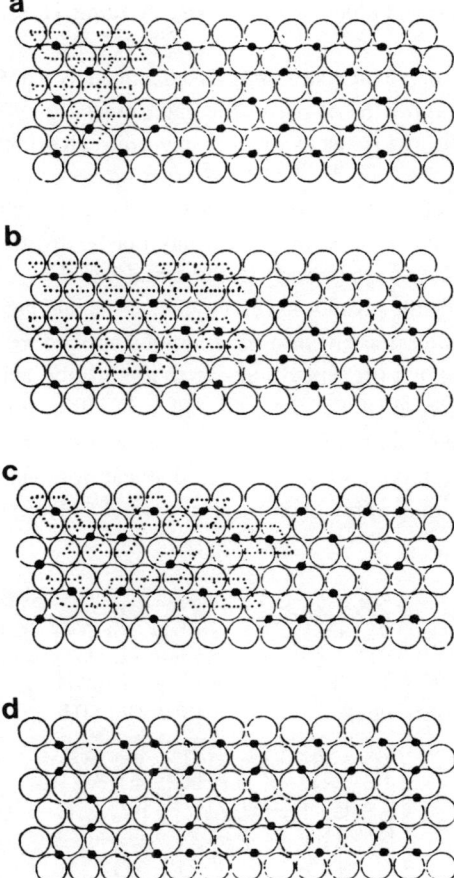

Figure 7. Arrangements of Mn^{4+} cations in octahedral sites of hexagonal close-packed oxide lattice. (a) Pyrolusite; (b) ramsdellite; (c) γ-MnO_2/nsutite; (d) ε-MnO_2. (Reprinted from Ref. 81, Fig. 1, with permission from IBA, Inc.)

Ramsdellite, a dimorph of pyrolusite, contains double chains of linked [MnO$_6$] octahedra. These octahedra are again linked together by edge-sharing, and the double chains run parallel to the c axis. The double chains are further cross-linked to adjacent double chains through corner-sharing of oxygen atoms. The unoccupied octahedral sites form double lines of vacancies parallel to the c axis, defining [1 × 2] dimensional tunnels.

The nsutite or γ-MnO$_2$ structure, first deduced by De Wolff,[72] consists of irregular intergrowths of pyrolusite and ramsdellite. This model considers the pyrolusite to exist as slices of one-unit-cell thickness in a matrix of ramsdellite units (Fig. 8).[10,72] The alternating c-axis chain segments of the basic single and double [MnO$_6$] octahedral chains are random, so that regular periodicity or superstructures may not be apparent. This model claims to account for the poorly defined and variable X-ray diffraction patterns of synthetic γ-MnO$_2$—i.e., the line shifts and broadening. High-resolution transmission electron micrographs applied to a suite of naturally occurring nsutite specimens provide direct evidence for the nonperiodic intergrowths of β-MnO$_2$ and ramsdellite structural units, as well as multiple chains of [MnO$_6$] octahedra that formed other 1 × n tunnels and a variety of stacking faults, dislocations, chain defects, multidimensional tunnels, and other irregular voids.[89,81] Recent X-ray diffrac-

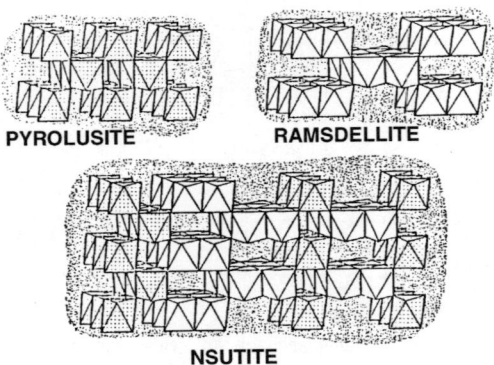

Figure 8. Tetravalent manganese oxides, showing [MnO$_6$] octahedra linkages and tunnel structures. (Reprinted from Ref. 81, Fig. 2, with permission from IBA, Inc.)

tion studies by Charenton and Strobel[90] also confirm De Wolff's intergrowth model.

Recently, De Wolff's model of γ-MnO$_2$ was revisited and extended by Chabre, Pannetier, and co-workers.[92,93,24] These investigators first tried to apply De Wolff's model to differentiate various γ- and ε-MnO$_2$'s. They noted that the possible variables in De Wolff's model are the fraction of ramsdellite versus β-MnO$_2$ layers perpendicular to the c axis and the relative permutations of these alternate types of layers. Using a numerical approach, they calculated the expected evolution of the powder diffraction pattern of γ-MnO$_2$ through the complete structural phase diagram of ramsdellite–pyrolusite. This treatment failed to explain the lack of resolution that characterizes XRD patterns of most EMD and CMD samples.[71] In particular, this model produced only the orthorhombic symmetry of ramsdellite, whereas experimental XRD patterns of ε-MnO$_2$ often exhibit an almost hexagonal symmetry.[73,85,22] Chabre, Pannetier, and co-workers accounted for these results by considering microtwinning on the (021) and/or (061) growth planes. Using methods they developed for determining the β-MnO$_2$–ramsdellite and microtwinning statistics, they analyzed numerous commercial MnO$_2$ samples and classified them. Their results are shown in Fig. 9 as a plot of fraction of twinning (Tw) versus fraction of chain slabs that are pyrolusite, Pr [(1-Pr) = the fraction that are ramsdellite].

The results not only explain XRD features and the corresponding role of structural defects but also serve to relate and categorize, for the first time, all EMDs, CMDs, and NMDs. The diffuse XRD band at a d spacing of approximately 4 Å is caused by the microtwinning and gives the apparent hexagonal symmetry as well as large values of the orthorhombic cell parameter ratio, $b/2c$. Important conclusions from Fig. 9 are as follows: (1) Most available manganese dioxides fall into four families, which span different ranges of Tw and Pr; (2) natural ramsdellite, although rarely occurring as a pure phase, appears to contain very little structural defects; (3) CMDs have Tw in the range of 0.2–0.4 and Pr close to 0.3; (4) EMD has Tw close to 1.0 and Pr in the range of 0.45–0.5; and (5) annealed or heat-treated MnO$_2$ (HTMD) has low Tw (<0.2) but Pr > 0.7. On the basis of the above results, Chabre and Pannetier suggest that heat-treated γ- and ε-MnO$_2$'s would better be described as pyrolusites containing random ramsdellite defects than (as often presently considered) as β-MnO$_2$. The above classification led the investigators to propose

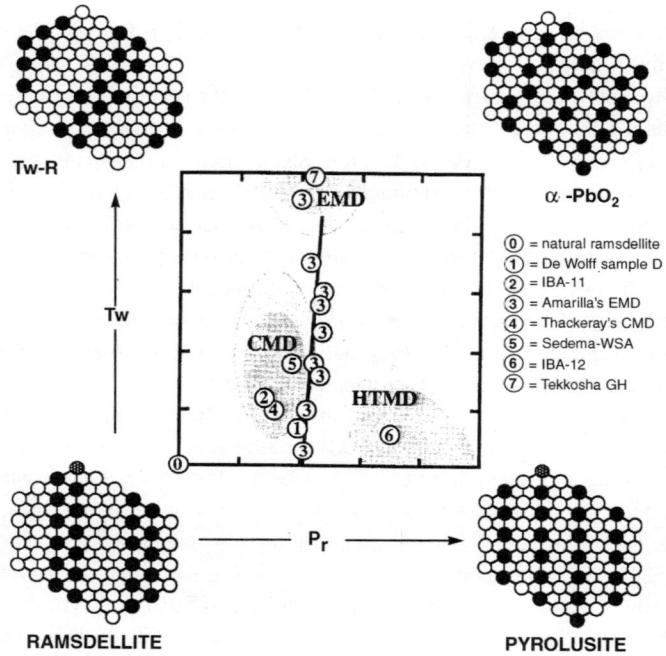

Figure 9. Classification of manganese dioxides based on fraction of twinning (Tw) and of pyrolusite units (Pr). The drawings at each corner of the diagram give a schematic description of the corresponding structure in the lattice gas representation (Mn atoms only). For the sake of simplicity, the twinned structures, at the top part of the diagram, are represented as ordered. Shaded areas are tentative delimitation of the categories introduced in the text. Numbered circles refer to the samples. The line is drawn through points no. 3 and illustrates the regular increase of microtwinning with deposition current density from low at the bottom to high in the upper part of the diagram. (Reprinted from Ref. 24, Fig. 24, with permission from the authors and Elsevier Science Ltd.)

that the denomination ε-MnO_2 be banned and that the CMDs and EMDs instead be designated as γ-MnO_2 followed by values of Tw and Pr.

(iii) Complex Mn(IV) Oxides

The more complex polymorphs of tetravalent manganese oxide also contain chains of $[MnO_6]$ octahedra and tunnels, although the close-packed oxygen lattice no longer exists.[81] Identification of these tunnel structures resulted from high-resolution transmission electron microscopy

(HRTEM) on fibrous minerals by Turner and Buseck.[89,94] The nomenclature for these tunnels is [m, n], where m is the tunnel height in the stacking of chains perpendicular to the c axis and n is the tunnel width, parallel to the chains of octahedra. In α-MnO_2 possessing the hollandite structure, [2×2] tunnels are produced by pairs of ramsdellite-like double chains of edge-shared [MnO_6] octahedra linked at right angles by corner sharing. Larger [2×3] tunnels occur in psilomelane (romanechite), and [3×3] tunnels most commonly occur in todorokite. The HRTEM studies revealed complex intergrowths of these structures and others with tunnel sizes up to [2×7]. The T($2,\infty$) end-member structure represents phyllomanganates such as δ-MnO_2 (vernadite and birnessite). The T($3,\infty$) end-member structure represents the phyllomanganates. These T(a,∞) compounds are infinite two-dimensional sheets of edge-sharing [MnO_6] octahedra, with the layers separated by 7 Å or 10 Å.[11,80,100] These layers are held apart by foreign cations (Na^+, Mn^{2+}, Ca^{2+}, etc.) and H_2O in the intermediate layers.

2. Structural Water and Stoichiometry

(i) Definition

Investigators have long recognized that electrochemically active manganese dioxides, e.g., γ- and ε-MnO_2, always contain a few percent of "structural" or "chemically bonded" water.[95–98,62] This water is differentiated from bulk water adsorbed on the MnO_2 surface by the fact that the latter is removed at ~100–110°C, whereas structural water is removed progressively by heating at temperatures from about 110 to 400–500°C. Much of the structural water is removed at temperatures less than about 350°C without significant change in the XRD pattern.[99,100]

β-MnO_2, whether occurring naturally or synthesized by heat treatment of γ/ε-MnO_2's, contains little or no compositional water. Structural water in EMD varies with deposition conditions, increasing with increasing current density and with decreasing temperature. Thus, it is more prevalent in ε-MnO_2 than in γ-MnO_2.[76–78]

(ii) State of Structural Water; Cation Vacancies

Because protons, through structural water, are an integral part of battery activity, the location and state of such water has created much interest and controversy. It has long been noted that the electrochemical or chemical reduction of battery-active oxides to MnOOH yields less capacity than is predicted by Eq. (2) with the assumption that such oxides are "MnO_2." Thus, several percent of the Mn in γ/ε-MnO_2's is assumed to already be in the Mn(III) state. It was recognized early[63,96,100–107] that at least some of the protons reside on the O^{2-} ions and that each Mn^{3+} ion, which replaces a Mn^{4+} ion, must have one neighboring OH^- ion in place of one O^{2-} ion for charge balance and fixed oxide packing to occur.[105–109] Correspondingly, the formula for the oxide or oxyhydroxide, both before and during discharge, could be written as $(MnO_2)_n(MnOOH)_m$. However, considerably more H^+ ions were liberated as water during heat treatment than this formula could depict for a given average oxidation state of manganese, so additional immobilized water had to be considered. The formula, correspondingly, progressed to the form[105–109]

$$\text{"}MnO_2\text{"} = [MnO_2]_{(2n-3)} \cdot [MnOOH]_{(4-2n)} \cdot mH_2O$$

$$\text{or } MnO_n \cdot (2 - n + m)H_2O \tag{28}$$

The first representation above indicates that m molecules of true lattice (structural) water must exist for each atom of manganese. The second representation is merely for the convenience of representing the average oxidation number of Mn, i.e., $2n$. The structural water has been postulated to be adsorbed, in a high-density state, in microcrevices, in closed pores, in interstices, lattice dislocations, stacking faults, chain defects, dislocations, multidimensional tunnels, and other irregular voids.[72,80–82,85,95,100,110,111]

Ruetschi, in agreement with many earlier studies (see above), showed that structural water must be present as OH^- ions. Moreover, he addressed the question of where these ions reside and the total question of charge balance through his cation-vacancy model.[95] Herein, a fraction of the Mn^{4+} ions is missing, and each of these vacancies is coordinated to four protons for charge balance. Thus, there are two types of protons, those that are associated with vacancies and those that are associated with Mn^{3+} ions. There is no form of Mn other than Mn^{4+} and Mn^{3+}. Mn^{2+} ions are too large to fit in the lattice,[95] and Jahn–Teller distortion of the $[MnO_6]$ octahedron during reduction provides direct evidence of Mn^{3+}.[100] Calling x the fraction of missing Mn^{4+} ions and y the fraction of manganese ions present as Mn^{3+}, Ruetschi's formula for MnO_2 thus becomes

$$MnO_2(\text{cation-vacancy theory}) =$$

$$Mn^{4+}_{(1-x-y)} \cdot Mn^{3+}_y \cdot O^{2-}_{(2-4x-y)} \cdot (OH)^-_{(4x+y)} \tag{29}$$

Values for x and y may be determined experimentally from three assays, i.e., total Mn, Mn oxidation number, and total structural water (which determines total H). Assay methods may be found in Ref. 12 or 20. Three determinations are needed because manganese dioxides contain numerous impurities, the total level being difficult to determine accurately by independent means. These impurities constitute ~1–3% of synthetic manganese dioxides and a much higher fraction of natural ores. Undried samples contain another 1–3% as moisture.

Chemical analyses of various EMD samples along with calculated values of x and y are given in several references[77,78,95,112,113]; for some CMDs, cf. Ref. 149. A commercial EMD is typified in Table 2 with regard to chemical analysis and calculated Ruetschi parameters.[113] The values of x and y for all EMDs are generally found in the ranges $x = 0.03$–0.09 and $y = 0.05$–0.12.

Systematic studies to determine how x and y vary with deposition conditions have not been very definitive. Two studies[77,78] agree that deposition at very low current densities and high temperatures favors lower cation-vacancy population, x, and that lower deposition temperatures favor a higher concentration of Mn(III), i.e., higher y. However, the magnitude of the changes is very different in the two studies.

Very recently, direct evidence has been provided for cation vacancies and proton residence at such cation vacancies through inelastic neutron

Table 2
Chemical Analysis and Cation-Vacancy Parameters
for MnO_2 Sample IBA No. 20 (an EMD)

Chemical analysis
% Mn = 60.82
% MnO_2 = 92.64[a]
% Structural H_2O = 2.94
Impurities = 1.3%

Calculated parameters
n in MnO_n = 1.96 (Eq. 28)
m = 0.105 (Eq. 28)
x = 0.05
y = 0.07

[a] Determined by dissolving the sample in acidic $FeSO_4$ solution and assuming that all oxidizing power is MnO_2.

scattering (INS) studies.[114–116] While most experimental techniques do not provide direct information on proton location, INS is particularly useful because the scattering cross section of H atoms is very large compared to those of Mn and O. Therefore, the INS spectra are dominated by proton signals, the vibrational frequencies providing indication of proton sites or bonds (i.e., the environment), and the intensities indicating the relative amounts of each species.

Changes in INS intensity for EMD heated to various temperatures (which drives off structural water) correlate with changes in the respective thermogravimetric (TGA) measurements.[114] INS spectra of nontreated γ-MnO_2 at various sample temperatures reveal $(H^+)_4$ entities in manganese vacancies.[115] At 100 and 200 K, these entities form freely rotating tetrahedra inside the O^{2-} octahedra with a radius of gyration (~0.5 Å) similar to the ionic radius of the missing Mn^{4+} ion. Two different MnO_2 samples (a CMD and an EMD) yielded INS intensities proportional to the amount of Mn^{4+} vacancies in the lattice.[116]

(iii) Relationship of Structural Water to Other Properties

Ruetschi's cation-vacancy model relates the MnO_2 structure, defined by the stoichiometry of Eq. (29), to various other physical and electrochemical properties—i.e., absolute (pycnometric) density, electronic conductivity, thermodynamic (maximum) electrochemical discharge capacity, and electrode potential.[95,117] Many of these relationships have been confirmed experimentally, although precise determinations of x and y and, therefore, precise correlations are difficult to achieve. This is because of limitations in analytical precision, because the temperature at which water expulsion changes from adsorbed H_2O to structural H_2O (i.e., to OH bonds) is indefinite, and because Mn(II) may be present in MnO_2 as MnO or adsorbed $MnSO_4$ to the extent of a few tenths of a percent.[112]

Nevertheless, theoretical densities[95] agree fairly well with measured densities.[76,118] Also, theory rationalizes observed differences in absolute/pycnometric densities among EMDs deposited at different current densities on the basis of differences in concentration of cation vacancies,[77,78] which are manifest in structural water. These densities range[76,78] from ~4.7 to 4.3 g/cm^3 for deposition at very low to high current densities—i.e., 0.5 to 10 mA/cm^2. The pycnometric density of EMD deposited at very low current densities (i.e., <1 mA/cm^2) approaches the theoretical or crystallographic density of γ-MnO_2.[76]

The gain in density as MnO_2 is heat-treated[119] is explained by the annealing out of cation vacancies with loss of structural water. Thermogravimetric experiments[120] show that manganese dioxide heated in air to 500°C oxidizes quantitatively to MnO_2 [i.e., Mn(IV) oxide] while losing all its structural water; then, between 500 and 600°C, it converts to Mn_2O_3. TGA of α-MnOOH in air also produces MnO_2, although the reaction occurs at 250°C; then it follows the same TGA curve as MnO_2.[120,127] TGA of one mole of manganese dioxide (Eq. 29) in vacuum first yields $1 - x - y$ moles of MnO_2 and $y/2$ moles of Mn_2O_3 between 150 and 300°C. Then the manganese dioxide converts completely to Mn_2O_3. TGA of α-MnOOH in vacuum indicates complete decomposition to Mn_2O_3 by ~200°C.[120]

The electronic conductivity of MnO_2 increases as the structural water is progressively driven off through heat treatment, and the betafied material that results after the last water is removed (at ~375°C) is much more conductive than γ/ε-MnO_2.[121] Ruetschi explained these results[95] through the decrease in cation vacancies that occurs as the water is driven out. Herein, the hopping distance of the electron between Mn^{4+} ions decreases with a decrease in vacancies.

A very useful application of cation-vacancy theory is prediction of the theoretical maximum one-electron discharge capacity per unit weight, C_w, which is given by[95]

$$C_w(A \cdot h/g) = \frac{(1 - x - y)26.80}{MW} \tag{30}$$

where MW is the theoretical molecular weight, i.e.,

$$MW = (1 - x)54.94 + 16(2) + 4x + y \tag{31}$$

On this basis, C_w for IBA No. 20 (Table 2) is 0.279 A·h/g. This value is substantially less than the theoretical capacity of pure MnO_2, which is 0.308 A·h/g. (Both values should be adjusted for the impurities in the EMD if comparison is made to experimental values.)

The theoretical capacity is quite close, but somewhat greater than, the value that would be calculated from the MnO_2 assay (92.64% in Table 2); i.e., $0.308 \times 0.9264 = 0.285$ A·h/g. The high bias from the analytical determination stems from the fact that the $FeSO_4$ titration gives Mn^{3+} ions half value in the oxidizing scheme, whereas in first-electron electrochemical reduction, these ions would not contribute.

Maximum capacities from cation-vacancy theory are in accord with those obtained experimentally at very low discharge current densities. The common knowledge that EMDs deliver 7–10% less capacity on discharge than would be delivered by pure MnO_2 qualitatively attests to cation-vacancy theory. Recent work[78] shows a fair correspondence between the theoretical maximum discharge capacity and the experimental capacity for a suite of EMDs deposited at different current densities. The theoretical capacity is greater, as would be expected from the fact that the experimental capacity would suffer from polarization/overvoltage, which hastens the potential drop to the cutoff voltage.

Many investigations have experimentally shown correlations between structural water and other physicochemical properties—i.e., magnetic susceptibility, infrared absorption, thermogravimetry, differential thermal analysis, and redox properties (e.g., the so-called concentration of surface OH groups), not to mention the obvious compositional features. Many of these correlations may be found in a review by Desai *et al.*,[23] and are detailed further in the references therein. Inasmuch as structural water is vital to electrochemical activity, further review of the above properties will be deferred to Section V, where they will be discussed in conjunction with electrochemical properties.

3. Surface Properties or Mesostructure

(i) Characterization

Surface properties or "mesostructure"[76] includes those features measured by gas sorption, water sorption, mercury intrusion (porosimetry), and ion exchange. The significance of surface involvement is comprehensible from the fact that battery-grade manganese dioxides have surface areas of ~10–100 m^2/g, which corresponds to 10–20% of the atoms occupying a surface position.[38] This contrasts to the superficial area of the particles, i.e., ~0.03 m^2/g for particles that are 20 μm in diameter (a typical mean particle size). Thus, the particle size does not contribute significantly to the surface area.

Surface area, the most studied surface property, is usually determined by applying the Brunauer–Emmett–Teller (BET) isotherm[122,123] to N_2 (or other gas) adsorption data in the form

$$\frac{1}{VP_0/(P-1)} = \frac{1}{V_m C} + \frac{(C-1)P}{V_m C P_0} \tag{32}$$

Here V is the equilibrium volume of N_2 adsorbed at pressure P; P_0 is the N_2 saturation vapor pressure at the temperature of the liquid-nitrogen coolant bath; V_m is the volume of N_2 required to cover the surface with a monolayer; and C is a constant indicative of the N_2–MnO_2 interaction energy.

Ion exchange with zinc ions has been shown to yield surface areas similar to those from adsorption of N_2 or other gases.[124]

The total pore volume is measured by total adsorption of any number of gases or liquids. The sizes of the pores are usually obtained by applying the Kelvin equation to the N_2 (or other gas) desorption isotherm (i.e., to the emptying of pores) above the pressure at which hysteresis starts (see Ref. 123 or instruction briefs provided by manufacturers of sorption instruments); i.e.,

$$\ln(P/P_0) = \frac{-2\gamma V_m}{r_p RT} \cos \phi \qquad (33)$$

Here γ and V_M are the molar volume and surface tension, respectively, of the liquid adsorbate (N_2); ϕ is the contact angle between the liquid and wall of the pore; and r is the radius of the pore.

Most definition of MnO_2 pores has come from application of isotherms to the BET and Kelvin equations (e.g., cf. Refs. 125–131). Measured radii of the pores vary from <20 Å to more than 10^6 Å. These voids are often classified according to their radius as micropores ($r < 20$ Å), transitional pores or mesopores ($r = 20$–200 Å), macropores ($r = 200$–1000 Å), and interstices between grains ($r > 10^3$ Å).[123] Attention has been called to limitations in BET analysis, as micropore data for manganese dioxides do not fit Eq. (32) over the entire applied P/P_0 range[130]; the latter is usually taken as $P/P_0 = 0$–0.3. Also, pore-size-distribution determinations have been questioned[130] on the basis that the Kelvin equation breaks down at relative pressures (P/P_0) below about 0.5 and that this breakdown gives rise to artifacts interpreted as pores of a constant size (about 2-nm radius) independent of the particular sample.

Although large pores, for which the Kelvin equation holds, account for most pore volume, small pores account for most surface area[130] and are the pores strongly implicated in MnO_2 battery activity. Grinding does not appear to strongly affect either surface area[130] or first-electron battery activity at nominal current densities.[20] A method to determine mean pore size was suggested,[130] and results for more than 40 MnO_2 samples show that the mean radius of small pores varies from 12 to 24 Å among these samples. Various studies have concluded that micropores are the surface area of microcrystallites, these crystallites being much smaller than the grains of MnO_2.[20,112]

(ii) Variation of MnO_2 Surface Area with Synthesis/Origin

Pyrolusite (β-MnO_2) and very crystalline γ-MnO_2's, as typified by IBA No. 11 and by natural nsutites, have very small surface areas—i.e., 1–10 m^2/g. EMDs deposited at very low current densities (i.e., <1 mA/cm^2), also are of the γ-MnO_2 form and have low surface areas. EMD surface area increases as the deposition current density is increased and/or as the deposition temperature is decreased,[76-78] the surface areas ranging up to approximately 75 m^2/g. Chemical manganese dioxides synthesized in a variety of ways yield high surface areas, i.e., up to ~100 m^2/g.

When EMD is heat-treated, the surface area at first increases, and then, above about 250°C, the area decreases.[39,131-134] Once adsorbed water is removed, the area increase with heating, especially at $T \geq 150°C$, is associated with the production of micropores[127,133,134] as compositional water is removed and cation vacancies are annealed out. After heating to temperatures above ~250°C, microtwinning is found to decrease.[24] Surface area decreases markedly as γ/ε-MnO_2 is betafied at ~375°C.

(iii) Porosity and Some Other Physical Properties

Because electrochemical behavior and, particularly, first-electron discharge depend on the small pores/surface area, rather than the large pores/pore volume, this chapter focuses but little on properties related to large pores. However, modeling of porous electrodes utilizes the porosity (void space), and this and related properties are technologically important. Therefore, porosity is reviewed briefly.

The porosity (Por), block density (ρ_{block}), and absolute or pycnometric density (ρ_{abs}) are related by the following expression:

$$\text{Por} = 1 - \frac{\rho_{block}}{\rho_{abs}} \tag{34}$$

Equation (34) applies to chunks of material, such as EMD as it is plated. Application of various density methods to EMD samples and resultant porosities are reported in Ref. 135.

The block density of EMD is linearly related to the Vickers hardness[76,136,137] over a significant range, the latter property being related to tool wear in battery-cathode construction. The hardness is linearly related to the reflectance of polished EMD in either chip or powder form.[136] Thus, EMD with the greatest surface area (i.e., ε-MnO_2, deposited at the greatest current densities or lowest temperatures) exhibits the greatest porosity, the least hardness, and the least reflectance.

When powdered manganese dioxides are compressed, as in the construction of cathodes, the corresponding densities correlate with block densities of the respective materials, although the block densities are greater. (The compacted powders usually contain a small amount of carbon to simulate a battery cathode.)

V. EFFECT OF MnO_2 STRUCTURAL FEATURES ON BATTERY ACTIVITY

1. Battery Activity

Useful capacity to a given voltage cutoff is one of the most important figures of merit for a Zn–MnO_2 cell. This capacity depends substantially on the MnO_2, since the decline in cell voltage with discharge often corresponds to the decline in the MnO_2 discharge potential.[59,138,139] Therefore, the utilization of MnO_2, i.e., the fraction of the theoretical discharge capacity realized, has come to be synonymous with the "battery activity" of the MnO_2.

Battery activity depends on the MnO_2 electrode arrangement, the current density, and the potential chosen for cutoff. However, once the test is defined with proper precautions taken to eliminate ohmic drops, anode polarization, and solution concentration changes (i.e., by using excess graphite in the cathode and excess electrolyte), discharge capacities are the standard method of comparing the relative battery activities of various manganese dioxides. Discharge tests used in the industry have been described by Burkhardt[138] and Ohta[140] for alkaline cells and by Schumm[141] for $ZnCl_2$ and Leclanché cells. Of course, battery manufacturers ultimately evaluate cathode materials in commercial dry cells.

Battery-activity tests have often resulted in cell cutoff voltages of 0.9–1.0 V vs. Zn, because of the voltage demands of many electronic devices. With the drain rates chosen, this voltage cutoff allows the best manganese dioxides in alkaline electrolytes to yield most of their theoretical first-electron discharge capacity—i.e., today's EMDs yield 250+ mA·h/g at useful current densities compared with a theoretical maximum of ~280 A·h/g (this value being derived in Section IV). With a cutoff voltage of ~1.0 V vs. Zn, the manganese dioxides with the greatest discharge capacity usually also yield the greatest usable discharge energy.

Battery activities vary as widely as MnO_2 structure, which has prompted numerous studies of correlations between activity and structural features in both natural and synthetic materials. This section summarizes these studies, categorizing them according to the type(s) of Mn(IV) oxide compared, i.e., (a) different crystal classes and different generic types within the γ/ε-class, (b) heat-treated versus non-heat-treated γ/ε-MnO_2's, and (c) different materials within the EMD and CMD classes.

2. Comparison of Different Crystal and Generic Types

(i) Experimental Findings

Studies of natural and synthetic manganese (IV) oxides in KOH, NH_4Cl, and NH_4Cl–$ZnCl_2$ solutions have consistently found the following trends:

1. MnO_2 crystal structure is the major differentiating factor in battery activity (see Refs. 39, 100, 109, 112, 126, and 142–149). When different crystal forms of MnO_2 are studied, the γ- and ε-MnO_2's, represented by EMDs and various CMDs, give the greatest battery activity. β-MnO_2's give much poorer activity, often giving only a small fraction of the discharge capacity realized with the γ- and ε-MnO_2's. The α- and δ-MnO_2's, though variable in behavior, usually give substantially lower capacities than the γ- and ε-MnO_2's. Natural ores are significantly inferior to the synthetic manganese dioxides[112,126,147] because they contain substantial fractions of gangue impurities (10–20% compared to ~1–2% for EMDs and CMDs)[12,71] and nonactive Mn(IV) crystal forms[71,126,148] in addition to the battery-active γ-MnO_2 mineral, nsutite.

2. When the various γ- and ε-MnO_2's are compared, commercial EMDs usually exhibit the greatest discharge capacities, outperforming the CMDs.[52,109,112,138,150–152] EMDs deposited under noncommercial conditions can be vastly inferior to commercial EMDs [cf. Section V.2(ii)]; such nontypical EMDs have shown inferior performance to CMDs, even in alkaline solutions.[147]

3. When CMDs outperform "good" EMDs, it usually is in situations for which proton diffusion in the oxide is not primarily rate-determining—i.e., in $NH_4Cl-ZnCl_2$ electrolytes and at low or intermittent drain rates.[151,152] In this case, dissolution of MnOOH to form Mn^{2+} and MnO_2 is faster than proton diffusion and, in effect, provides a two-electron pathway for discharge, which can yield greater capacities than the one-electron pathway of H^+-e^- insertion (see Section III).

(ii) Features Responsible for Activity Differences between Various MnO_2 Types

(a) γ/ε-MnO_2 versus other forms

Because first-electron discharge involves proton movement through the lattice, and because battery-active forms generally contain much more structural water/mobile protons than nonactive forms, the presence of protons and sites for them has very often been invoked in explaining superior battery activity of γ/ε-MnO_2[95,109,120,148,153] (see also the review by Desai et al.[23]). Although δ-MnO_2 contains bound water, it is in a molecular form and unsuitable for proton movement.[100]

Ruetschi and Giovanoli[95,120] proposed that protons in cation vacancies assist proton diffusion, both by providing initial starting concentration and by making proton bridges for chain-type H^+ transfer. These bridges facilitate H^+ movement in the *a* and *b* directions, whereas the vacancy-free lattice can facilitate H^+ movement only in the *c* direction, along the chains of octahedra.[120] These investigators developed a quantitative relationship for the proton diffusion coefficient.

When lattice diffusion of protons cannot support the current, as in the case of β-MnO_2, then the intrinsically slower "heterogeneous" reduction mechanism must support discharge. Proton insertion in β-MnO_2 should be more difficult than in γ/ε-MnO_2 based on the narrow [1,1] tunnels in β-MnO_2 compared with the larger [1,2] tunnels in the ramsdellite domains of γ/ε-MnO_2.[24]

Voinov[88] concluded, from lattice studies and comparison of MnO_2 with other substances that (a) H^+-ion insertion is into tetrahedral sites of the hexagonally close-packed oxygen layers and (b) proton motion is between such sites. He studied the environments of such tetrahedral sites in terms of the repulsion between protons in these sites and Mn^{4+} ions in adjacent [MnO_6] octahedra which share a face with the tetrahedral site. The impediment to proton insertion or occupancy during transport increases with the average number of octahedra shared with the tetrahedron, and this number increases in the order γ-MnO_2 < β-MnO_2 < α-MnO_2. Voinov noted that this order is opposite to that of battery activity.

Battery-active manganese dioxides have been found to have different (paramagnetic) magnetic susceptibilities than inactive forms,[109,147,149,154] but the relationship to battery activity is not a simple, monotonic one. Also, MnO_2 decomposition temperatures (e.g., to Mn_2O_3) have, in some studies, correlated with battery activity.[23,109]

Chabre and Pannetier[24] studied intercalation of protons into various MnO_2 samples by means of step potential electrochemical spectroscopy (SPECS). They found that intercalation into pyrolusite units occurs at a potential substantially negative of that for intercalation into ramsdellite units. This difference is such that much intercalation into pyrolusite units occurs below the cutoff potential defined for battery activity tests. Since the fraction of pyrolusite units is greater in β-MnO_2 than in γ/ε-MnO_2, the corresponding increasing order of battery activity follows directly.

Factors other than the crystallographic differences also have been considered in accounting for the superior battery activity of γ/ε-MnO_2's over other forms. For example, variation in tunnel cations affects the discharge capacity of α-MnO_2,[147,149] and the absence of any such cations (except for, possibly, protons) has given substantially better battery activity than provided by typical α-MnO_2's which contain alkali and alkaline-earth cations.[147] The cations and H_2O molecules between the layers of $[MnO_6]$ octahedra of δ-MnO_2 are thought to impede battery activity.[100]

Active forms usually have greater packing densities than nonactive forms, which is associated with larger crystallites and less impedance to proton diffusion and electronic conductivity at grain boundaries.[142,148] It has also been associated with greater electrolyte adsorption,[148] but this would imply a rate-determining step other than proton diffusion in the solid.

The fact cannot be overlooked that EMDs and CMDs have considerably greater surface areas than NMDs, activated ores, and inactive manganese dioxides,[126,112] and diffusion-modeling equations predict that this is advantageous.

(b) EMDs versus CMDs

Activity differences between CMDs and EMDs, both of which are γ- or γ/ε-MnO_2's and both of which contain compositional water, are more difficult to explain than activity differences between these collective synthetic forms and the inactive manganese dioxides.

Two diffusion treatments of EMD versus CMD discharge are not clear as to the controlling factor in the superior activity of the EMD. Such treatments assume electrochemical reaction areas and obtain H^+ diffusion coefficients as experimental parameters (as will be detailed in Section VI). Only one sample of each MnO_2 type was studied in each case. In the first study,[52] the EMD had greater surface/reaction area and a slightly greater proton diffusion coefficient than the CMD sample. In the second study,[112] the EMD had smaller surface area and a greater diffusion coefficient. In both studies, the CMDs studied contained less structural water than the EMDs,[109,112] and so the battery-activity differences may be qualitatively explained by the difference in cation vacancies and the corresponding difference in proton diffusion and potential.[95] Kozawa[45] measured the equilibrium potentials of several EMDs and CMDs during intermittent discharge and did not find a clear-cut difference between the EMDs and CMDs. In alkaline solution, he did find a larger polarization for the CMDs, which is consistent with the EMDs having faster proton diffusion.

From recent structural and electrochemical spectroscopy studies, Chabre and Pannetier[24] suggested that greater microtwinning in the case of EMD causes the greater battery activity.

Commercial EMDs and CMDs have different surface areas, which have been considered to be a contributing factor in the activity differences. One complicating factor, however, is that the various CMD study samples have surface areas both greater than and less than those of the EMD study samples.[12,109,112,125] The EMDs generally possess N_2 BET surface areas from about 25 to 50 m^2/g while most CMDs have surface areas from 50 to 100 m^2/g; however, one CMD that has often been studied (IBA No. 11)[109,45] has a surface area of only ~2 m^2/g. Thus, linear regressions (e.g., Ref. 109) are not of high enough order to identify the important but complicated correlation between battery activity and surface area. The linear regression[109] did identify that the majority of pores in the EMD samples had average radii in the range of 15–25 Å (which is at the small end of the pore-size distribution for battery-active manganese dioxides) whereas the (inferior) CMDs had the majority of pores in a range of 25–100 Å. The significance of this is questionable, however, because the actual surface area of pores having radii in the range 15–25 Å (i.e., the surface area multiplied by the fractional pore volume in the given range) was not consistently greater for the EMDs than the CMDs.

When concentration polarization in the pore electrolyte or precipitation of zinc compounds at the entrance to these pores is rate-determining, rather than proton diffusion in the oxide, a predominance of small pores leads to impeded battery activity. This is the case in Leclanché cells at low or intermittent drain. The superior MnO_2 feature in such a case is thought to be the greater porosity or greater average pore size of the CMDs (r = 40 Å) compared to that of the EMDs (r = 20 Å).[109,151,152,157,45]

Porosity is also manifest indirectly in battery discharge performance through the compressed or packed density of the cathode. EMDs possess greater packing densities than CMDs, which allows more active cathode material to be placed in a given volume. Correspondingly, consumer-type cells typically have exhibited greater discharge capacities with EMDs than with CMDs, much of the capacity advantage in medium and hard drain tests in all types of electrolyte being attributed to the fact that a greater amount (e.g., 23%)[151] of EMD than CMD is contained. This difference explained much of the capacity advantage held by the EMD in medium and hard drain tests in alkaline, $ZnCl_2$, and LeClanché electrolytes.[151] The CMD also operated in a 50–100 mV lower voltage range than the EMD. The poorer compactability of the CMD was associated with the larger pore volume (~0.6 versus 0.36 cm^3/g measured with mercury porosimetry) and narrower particle-size distribution. At higher molding pressures (e.g., 3–5 ton/cm^2), the pellets with CMD cracked whereas the pellets with EMD compressed further.

3. Studies of Heat-Treated EMD (HEMD)

Investigators have heated EMDs to temperatures less than 375°C to remove compositional water on a systematic basis, and they have converted EMD to β-like MnO_2 by heating to about 375°C with or without post-treatment with acids. Corresponding changes in battery activity are discussed below.

Heating EMD to about 200–250°C has shown mixed effects on battery activity, causing in some cases an increase[132] and in other cases no change or a decrease.[39,158] In this stage the surface area is increased with successive heating[131,132,159] and Koshiba[132] attributed his observed increased alkaline discharge capacity to the increase in area for occurrence of the electrochemical reaction. Decreases in battery activity were associated with a lowering of the equilibrium potential and an increase of the cathodic overvoltage with removal of compositional water.[158]

From about 250°C to the so-called "betafication" temperature (~375°C), the battery activity has been found to decrease.[39,158] Thermogravimetric studies show that EMD loses its compositional water in this temperature range,[120] the surface area decreases, and the equilibrium potential at all degrees of reduction decreases.[39] Thus, all the factors that affect proton diffusion change in such a direction as to decrease battery activity.

At $T > 375°C$, when the EMD becomes beta-like, a marked decrease in battery activity usually occurs. This β-MnO_2 contains no protons (compositional water) and yields a lower equilibrium potential and surface area than the starting EMD.[39] Kozawa and Powers[39] attributed most of the battery activity decrease to the decreased equilibrium potential (at all stages of reduction) and the decreased surface area.

4. Studies within CMDs and EMDs

(i) General Collections of CMDs and EMDs Produced under Undefined Conditions

Although activity differences among manganese dioxides of a given type are usually smaller than those between different MnO_2 types, studies of similar types have yielded correlations with various MnO_2 properties not otherwise observed. For example, Kanungo et al.[147] observed correlations between discharge capacity and magnetic susceptibility, surface OH groups, and lattice parameters (measured as unit-cell axial ratios, c/b and b/a) for small groups of synthetic samples from several crystal classes. These included α-, β-, γ-, and δ-MnO_2's, where the γ-MnO_2's included both CMDs and EMDs. When all the diverse samples were grouped, the correlations were very poor.

Fernandes et al.[149] observed correlations, for various CMDs, between battery activity (as well as catalytic activity in oxidizing thiosulfate ions), surface area, and the intensity of the (100) peak measured by XRD. Xie et al.[148] regressed magnetic susceptibility, surface OH concentration, and various other EMD properties against the discharge time. From their results, the above investigators gave qualitative explanations for involvement of structural water (or OH groups), magnetic susceptibility, and unit-cell dimensions in facilitation of proton movement in γ/ε-MnO_2, although neither the results nor the explanations from these investigations appeared altogether consistent.

Swinkels[73] epitomized the value of statistical methods and an adequate sample base in demonstrating significant correlation between the XRD patterns of (all) 19 IBA EMD samples and the alkaline discharge capacities measured by Burkhardt.[138] As this work was conducted primarily to demonstrate the value and application of chemometrics, which is a statistical method used to obtain chemical information from data, Swinkels did not go deeply into structure. He did confirm, however, that low levels of β-MnO_2 in the EMD are detrimental. Also, he confirmed and quantified the advantage of a taller 22° peak (Fig. 6), which qualitatively had been noted by prior investigators.[20] More importantly, he identified several other changes in the XRD pattern that accompany changes in the 22° peak. More structural insight into these changes was provided by Williams et al.[78]

One of the focal correlations has been a decrease in battery activity with an increase in surface area for various collections of EMDs.[20,109,112] This contrasts with the opposite activity–surface area correlations often observed in comparisons between different classes of manganese dioxides (cf. Section V.2). The correlation with surface area has perhaps been more visible than that with other structural features because differences in surface area are much larger than differences in, for example, solid-state parameters. Investigators usually associate the increasing surface area with micropores, which represent space/boundaries between MnO_2 crystallites. As these become more numerous, impedance to proton diffusion between crystallites increases.[20,109,112,129]

(ii) Collections of EMDs Deposited under Known Conditions

EMD samples synthesized through systematic variation of deposition parameters have provided an advantageous study base, because (a) causes for the differences are known and controllable and (b) MnO_2 properties can be varied more widely, continuously, and systematically. Such samples have provided a much wider range of crystal morphology and surface area than the collections of available/commercial EMDs or CMDs. In particular, the materials deposited at very low current densities add a dimension not observed with the commercial EMDs, and this has helped clarify some of the conflicting correlations observed with small, available collections as well as some of the phenomenological differences between CMDs and EMDs.

As deposition current density is increased from very low to high values, structural changes include (a) a change from γ- to ε-MnO_2, i.e., a decrease in the unit-cell b parameter and an increase in the c parameter, (b) a decrease in absolute density, (c) an increase in structural-water content and a corresponding decrease in manganese content, (d) an increase in surface area, (e) an increase in porosity, and (f) a decrease in average pore size.[76–78]

Correspondingly, the alkaline discharge capacity of the product maximizes at an intermediate deposition current density—i.e., about 3–4 mA/cm^2 (see Fig. 10).[77,160] Therefore, at least two of the changes in structural features must have opposite effects on battery activity.

Given that a higher deposition temperature has always been found beneficial to the battery activity of the product,[20,77,78,160,161] the two-factor effect has not been manifest in temperature variation studies. Older studies of deposition current density versus product battery activity were restricted to somewhat practical current densities, which covered only the descending capacity versus current density range of Fig. 10.

The ascending branch in Fig. 10 has been ascribed to increases in proton diffusion coefficient, electrochemical reaction area, and oxide free energy (i.e., operating potential) with an increase in deposition current density.[77] Model development is not advanced enough to determine the relative importance of each of these features.

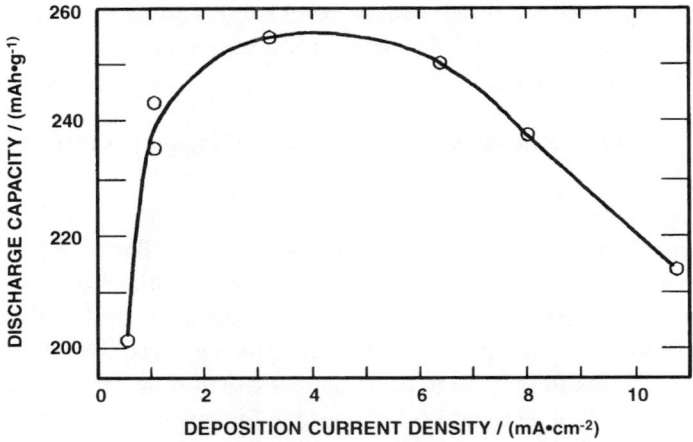

Figure 10. Alkaline discharge capacity of EMD vs. deposition current density. (Reprinted from Ref. 77, Fig. 2, with permission from the author and from IBA, Inc.)

The descending branch has been ascribed to both the increasing surface area and the change in crystallography[20,77,109,160] with increasing deposition current. The average pore size decreases, the population of micropores increases, and the solid crystallite size decreases, all making for more boundaries between solid crystallites.[77,78,112,160] This situation reduces to the general problem of battery activity decreasing with porosity and surface area. It is generally acknowledged that the above boundaries/dislocations/micropores, as they have been described by various investigators, impede the diffusion of protons in the solid.

In high-current-density studies, i.e., at 50 and 100 mA/(g EMD). Williams et al.[182] found that discharge capacities to typical one-electron cutoff potentials (i.e., −0.4 vs. Hg/HgO) increased with a decrease in EMD porosity, typical of EMDs on the descending branch of Fig. 10 and in agreement with most previous investigations. However, the discharge curves for the more porous samples started at higher potentials and then crossed over the curves for the less porous EMDs late in the discharge. These investigators suggested that surface reduction controls the rate (i.e., the potential) during the first part of the discharge, and then proton diffusion in the oxide becomes rate-controlling in the latter part. This investigation also found that coarser particle-size EMD gave lower discharge capacities, for particle sizes in the −45- to 500-μm size range. Some earlier studies had found that EMD particle size has little effect on first-electron discharge.[20,139]

(iii) Secondary Phases/Domains of Inactive Manganese Dioxides in EMD

EMD deposition also affects the battery activity of the product through formation of battery-inactive microdomains. For example, the presence of ammonium or potassium ions in the deposition bath from the processing operation directs some electrocrystallization to the α-MnO_2 morphology[20,162]; the XRD pattern of the product contains some small α-MnO_2 peaks in addition to the major γ/ε-MnO_2 peaks. Deposition at very high current densities is said to lead to formation of some β-MnO_2 in the EMD.[20,161] A weak β-MnO_2 XRD pattern is observed in the γ/ε-MnO_2 pattern of many EMDs[163,164,98]; its origin will be discussed in Section VIII.

VI. DETAILED PROTON–ELECTRON INSERTION MECHANISM

The simplified single-phase, homogeneous-redox mechanism of first-electron discharge presented in Section II did not explicitly address structural effects on the mechanism. MnO_2 structure is key to battery performance, however, as demonstrated in Section V, and a prime focus of MnO_2 development has been identification, characterization, and synthesis of ever-improving battery-active MnO_2's. With Sections IV and V as background, proton–electron insertion will now be reviewed in a more comprehensive manner.

This section will first show how homogeneous reduction is incompatible with the structure of $\gamma\text{-}MnO_2$ and with changes in properties that occur during discharge. Then, detailed mechanisms of proton–electron insertion in $\gamma\text{-}MnO_2$ (or $\gamma/\varepsilon\text{-}MnO_2$) will be reviewed according to some general methods used—i.e., (a) equilibrium potentials as a function of depth of discharge, (b) modeling studies at practical discharge rates, and (c) very slow charging, using various experimental formats. Finally, insertion studies on crystal forms other than $\gamma/\varepsilon\text{-}MnO_2$ will be reviewed.

1. Structural Evidence for Nonhomogeneous Reduction of $\gamma\text{-}MnO_2$

(i) Reduced Forms of Manganese Dioxide

The various solid forms of Mn(III) to which pure MnO_2 forms reduce, as well as the reduction intermediates, have been identified.[81,91,100,165–167] They have been an important factor in defining reduction mechanisms and will be described briefly here as background for the following discussion.

Groutite or $\alpha\text{-}MnOOH$ is the reduced form of ramsdellite, the two forms being essentially isostructural. Low-temperature oxidation/reduction interconverts $\alpha\text{-}MnOOH$ and ramsdellite.

Groutellite has been suggested as a phase intermediate between ramsdellite and groutite. In MnO_2 reduction it is symbolized as Mn_2O_4H. The (rare) mineral form has composition Mn_2O_3OH.

Manganite, $\gamma\text{-}MnOOH$, has a structure similar to that of pyrolusite with a slightly expanded lattice. It results from reduction of pyrolusite.

$\delta\text{-}MnOOH$ is a name sometimes given to the reduction product of $\gamma\text{-}MnO_2$. By analogy with De Wolff's description of $\gamma\text{-}MnO_2$, it appears to be an intergrowth structure of groutite with randomly distributed microdomains of manganite.

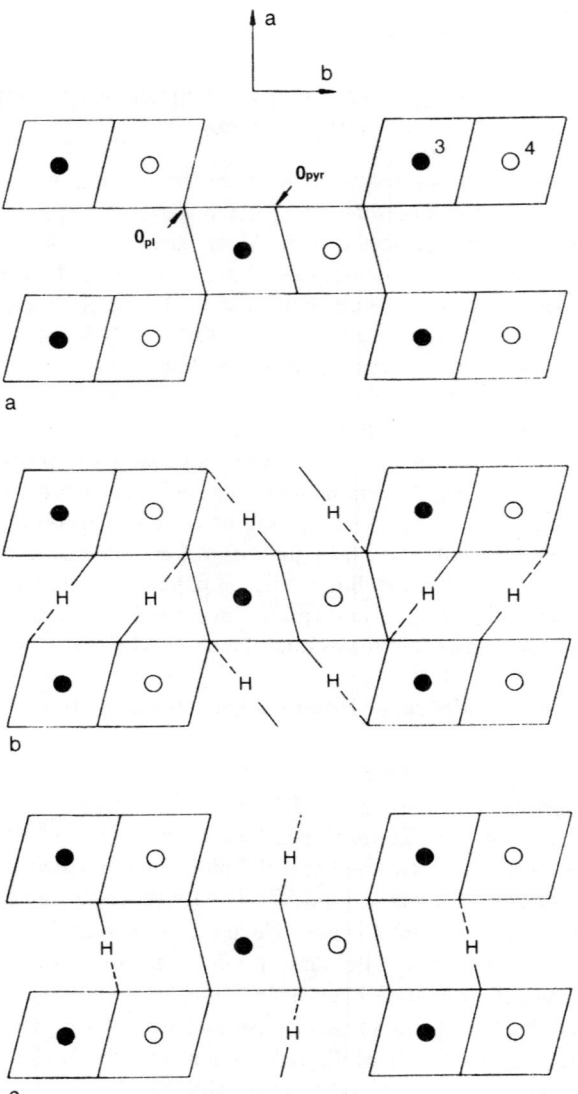

Figure 11. Representation of (001) plane of ramsdellite and related oxyhydroxides, showing different bonding protons: (a) ramsdellite, (b) groutite, and (c) suggested structure of groutellite, $MnO_{1.75}$. Mn atoms are represented by filled and open circles at depths $3c/4$ and $c/4$, respectively. O atoms are at the corners of the octahedra. (Reprinted from Ref. 91, Fig. 6, with permission from the authors and Elsevier Science Ltd.)

(ii) Different Proton Environments

De Wolff's analysis[72] identifies two types of chains in the γ-MnO_2 structure, i.e., ramsdellite and pyrolusite (cf. Section IV). These provide for manganese of two different energies and two different channels through which protons can enter and diffuse.[24] Furthermore, through the contained ramsdellite chains, γ-MnO_2 contains two kinds of oxygen atoms defined by the type of site with respect to neighboring cations (Fig. 11a).[91,168,169] The first kind of atom is located at "planar" sites, at the center of an almost equilateral triangle of Mn^{4+} cations; it is termed O_{pl}. O_{pl} atoms link corner-sharing octahedra and are similar to the oxygen found in rutile and pyrolusite (cf. Section IV). The second kind of atom lies at the apex of a trigonal pyramid of cations; it is termed O_{pyr}.

Maskell et al.[91] suggested that the two types of oxygen sites allow two types of locations for proton bonding: (a) type A or interpyramidal, where a proton is bonded to an oxygen in the O_{pyr} coordination, and (b) type B or pyramidal/planar, where a proton is covalently bonded to an oxygen in the O_{pyr} coordination and hydrogen-bonded to another in the O_{pl} coordination.

(iii) Insertion Changes during Reduction

Discontinuities have been noted in the evolution of various structural properties at about mid-reduction (i.e., at $\sim MnO_{1.75}$). The lattice dilates continuously throughout first-electron electrochemical and chemical reductions.[41,91,170,171] At the same time, changes in the individual lattice parameters exhibit discontinuities at mid-reduction.[91,169,172–174] Corresponding changes are also noted in the development of the activation energy for semiconductivity,[103] in the magnetic moment,[175,154] in discharge curves,[39,40] and in the equilibrium potential.[176]

The different progression of properties before and after mid-reduction suggest two solid solutions, i.e., one between MnO_2 and $MnO_{1.75}$ (e.g., Mn_2O_4H) and another between Mn_2O_4H and $MnOOH$, with the reduction passing through the groutellite stage at $MnO_{1.75}$.[169,176]

Rationale for the different insertion mechanisms before and after mid-reduction has been offered in terms of the two proton locations shown in Fig. 11b and c.[91,169,24] Location "A" between pyramidal oxygens is energetically more favorable for the proton and is filled first; there are only sufficient locations for type A bonds to occur exclusively in the region MnO_2 to Mn_2O_4H (Fig. 11c). Addition of each further proton must result in the location of at least two protons, per H added, in type B positions, or pyramidal planar. In this case a proton is covalently bonded to an oxygen in the pyramidal coordination and hydrogen-bonded to another in the planar coordination as shown in Fig. 11b. Thus, as reduction passes the midpoint (Mn_2O_4H), there must be a switching of existing protons from A to B. At the composition MnOOH, only type B positons may be occupied.

Maskell et al.[91,169] offered another rationale based on electron sharing between neighboring Mn atoms. Between MnO_2 and Mn_2O_4H, added electrons are delocalized between two Mn ions so that Mn^{3+} ions as such are not present. Between Mn_2O_4H and MnOOH, such delocalization can only occur to a limited degree, and some Mn^{3+} ions must be present. The evidence, from variations of lattice parameters and Jahn–Teller distortion (i.e., the ratio of two different "a" distances), is in favor of the model involving electron filling, but both possibilities could occur together.

2. Equilibrium Potentials of γ-MnO_2

(i) Foundations and Limitations of Solution Theory Approaches

Foundations for homogeneous reduction started as early as 1953 when Johnson and Vosburgh[177] observed that an equilibrated mixture of MnO_2 and Mn_2O_3 was stable for months and exhibited a potential that varied continuously with the mole ratio of MnO_2 to Mn_2O_3 in the solid, i.e., $[MnO_2]/[Mn_2O_3]$, as follows*:

$$E'_e = E^o_e + 0.073 \log([MnO_2]/[Mn_2O_3]) \tag{35}$$

The key relationship of this result to homogeneous discharge was not recognized for some years until Vetter put forth his general theory of nonstoichiometric oxides.[37,178]

---

*MnO_2 potentials have been popularly expressed with respect to either a fixed reference electrode, such as the standard hydrogen electrode (SHE), or a reference electrode reversible to H^+ or OH^- ions, such as the Hg/HgO electrode (which is popular in alkaline solutions). When the former is used for the MnO_2–MnOOH couple (i.e., proton insertion), the equilibrium potential must contain a solution-pH term in addition to those shown in Eq. (35), i.e., "(2.3RT/F)pH," whereas when the latter is used, no pH term appears. Often the type of reference electrode is only implied in publications. Here we use the potential expressions given in the original references. Then, to prevent confusion regarding the reference electrode, we will use E to designate the potential with respect to SHE (or a fixed reference electrode) and E' to designate the potential with respect to a reference electrode that is reversible to the H^+ or OH^- ion.

The Manganese Dioxide Electrode in Aqueous Solution

From their own and previous studies, Kozawa and Powers[34,39] proposed that the potential of this system be expressed in a Nernstian type manner, similar to the case of solution redox couples; i.e.,

$$E'_e = E^o_e + (RT/F) \ln([Mn^{4+}]/[Mn^{3+}]) \quad (36)$$

where $[Mn^{4+}]/[Mn^{3+}]$ is the ratio of mole fractions of end products in the oxide. The degree of reduction, r, is usually used as the independent variable, in which case Eq. (36) becomes

$$E'_e = E^o_e + (RT/F) \ln[(1-r)/r] \quad (37)$$

In applying Eq. (37), the initial r value is not zero but can be determined from chemical analysis of the oxide for total manganese and the manganese oxidation state.[179]

Kozawa and Powers found that Eq. (37) fit experimental discharge of γ-MnO_2 over more than half of the first-electron reduction—i.e., to ~$MnO_{1.7}$. Thereafter, the experimental potential decreased more rapidly with potential than the theoretical curve. However, beyond $MnO_{1.7}$ new Mn(III) oxide phases appeared on the electrode surface, which seemed to account for the misfit.

Later workers, focusing on the misfit, refined their experiments and included mild chemical reduction as well as electrochemical reduction[88,164] to allow more uniform reaction throughout the MnO_2 sample over most of the first-electron reduction. Even so, the theoretical potential from Eq. (37) decreased with reduction only about half as much as the experimental potential. The misfit was rationalized by introducing activity coefficients for the MnO_2 and $MnOOH$,[180] but these varied by orders of magnitude with degree of reduction and were different for MnO_2 and $MnOOH$ at all values of reduction except one. Therefore, this empirical approach was not very helpful.

(ii) Solid-State/Statistical-Mechanical Approaches

A major improvement in theory came when solid-state chemistry was applied to the oxide rather than solution chemistry (in the latter, inserted protons and electrons are assigned to discrete chemical compounds). Tye[181] substituted ionic mole fractions of Mn^{4+}, Mn^{3+}, O^{2-}, and OH^- for activities in the logarithmic term of E, arriving at Eq. (38).

$$E'_e = E^o_e + (2RT/F) \ln[(1-r)/r] \quad (38)$$

This expression doubled the potential change during reduction, correcting much of the problem with Eq. (37). Figure 12 compares experimental results in alkaline solution (the solid line) to results calculated by means of Eq. (38), shown as the dotted line. E_e^o depends on the particular oxide, so experimental and theoretical curves are compared by normalizing both to $E_e' = E_e^o$ at $r = 0.5$. Agreement between experimental and theoretical results is reasonably good for $r < 0.65$. Subsequently, Atlung and Jacobson,[179] Tye and co-workers[183,184] put Eq. (38) on a firm statistical basis. They represented the Gibbs free energy of the oxide by the positional entropy of the inserted species, the protons on available sites and electrons in bands. With the condition that the protons and electrons are two thermodynamically independent species, each having its own distribution on the sites/bands, the potential versus reduction relationship became Eq. (38).

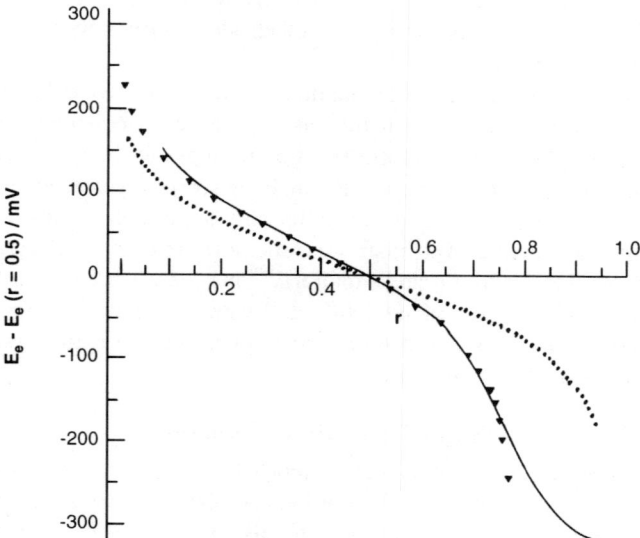

Figure 12. Equilibrium potential vs. degree of MnO_2 reduction (r) in alkaline solution according to experimental results (——), Eq. (38) (• • •), and Eq. (41) (▼▼▼). (Reprinted from Ref. 179, Fig. 5 and 6, with permission from the authors and Elsevier Science Ltd.)

Tye and co-workers found good agreement between Eq. (38) and experimental data only if most of the Mn^{3+} ions originally present in the unreduced oxide (the latter being represented as $MnOOH_{0.1}$) are electrochemically inactive. This was justified on the basis of a Gibbs–Duhem treatment of the potential early in the discharge over varying Mn^{3+} and Mn^{4+} concentrations.[185] This treatment led to the expression[184]

$$E_e = \frac{2(G^\circ_{MnO_2} - G^\circ_{MnOOH_{0.5}})}{F} + \frac{2RT}{F} \ln \frac{(0.5 - r')}{r'} - \frac{2.3RT}{F} pH \quad (39)$$

Here, r' is r corrected for inactive Mn^{3+} and is given by

$$r' = (r - y)/(1 - y) \quad (40)$$

where y is the number of moles of inactive MnOOH per mole of original MnO_2; y was found to be $0.08^{183,185}$ or 0.0865^{184} for material represented as $MnOOH_{0.105}$. The first and last terms in Eq. (39) are, of course, no different from those in other treatments but delineate E° and show the pH term because of the reference electrode. Equation (39) was found to give excellent representation of the reduction of EMD in Leclanché solution over the entire range of r.[184,186]

Recently, Tye and Tye[187] abandoned the "inactive Mn^{3+}" assumption as being contrived and, instead, proposed that the initial oxide contains O^{2-}-ion vacancies. Such anion vacancies do not preclude cation vacancies but must outweigh the latter. The anion vacancies and, with them, their charge-compensating protons are fixed. Thus, these protons do not contribute to the positional entropy, and the mathematics of Eqs. (39) and (40) remains preserved.

These investigators[187] developed a planar random insertion model, through which partial blocking of movements of H^+ and e^- from an inserted pair by an adjacent pair were shown. Good agreement between experimental and calculated values of Gibbs free energy (i.e., the potential) at *all* levels of insertion was achieved by taking into account this blocking.

Both the schools further refined their theories. Atlung and Jacobson proposed that the number of accessible sites (which is initially the number of Mn atoms) diminishes proportionately as insertion proceeds. This led to their final expression,

$$E'_e = E^o_e + \frac{2RT}{F} \ln \frac{[1-(1+\alpha)r]^{(1+\alpha)}}{r(1+\alpha r)^\alpha} \tag{41}$$

This expression fit the alkaline discharge data quite well (Fig. 12), with $\alpha = 0.25$, as determined experimentally.

Maskell et al.[183] refined Eq. (39) even further, proposing from structural considerations that two types of sites are available simultaneously for either protons or electrons. Configurational entropy was maximized between the two types of sites. A further refinement included an enthalpy difference for occupation of the two types of sites. These refinements led to the involved expression

$$E'_e = \frac{1}{F}(G^o_{MnO_2} - G^o_{MnOOH}) - \frac{RT}{F} \ln \frac{(1 - rf_m)^p (rf_m)^p (r - rf_m)^{(1-p)} r}{2^p (1 - r - rf_m)^{(1+p)}(1-r)} - \frac{pd}{F} \tag{42}$$

where

$$p = (1-2r)/[2(1-2r+2r^2)^{1/2}] \tag{43}$$

$$f_m = \{1 - [1 - 2r(1-r)]^{1/2}\}/2r \tag{44}$$

and d is the enthalpy term.

Ruetschi[117] considered two effects of proton insertion on the free energy, i.e., (1) the effect on the distributional partition function for mobile protons on the sites and (2) the effect on the potential energy, caused by change in the lattice binding energy with lattice dilation. He further included the influence of cation vacancies on both terms, arriving at the expression

$$E_e = E^o_e + (RT/F) \ln[(1-r)/r] + (NF\phi)(1-r) - (RT/F)\text{pH} \tag{45}$$

Here the ln term in r represents the statistical term and the linear term in r represents the potential energy term, where ϕ is the $\Delta P\, dV$ work in expanding the lattice from $r = 0$ to $r = 1$ (i.e., the difference in potential energy per proton between the $r = 0$ and $r = 1$ states) and ΔP is what Ruetschi calls the "lattice cohesive pressure." N is Avogadro's number. With values he determined for ϕ and, accounting for the fact that r is a function of the vacancy fraction, x (cf. Section IV), Ruetschi fit the experimental potentials of Tye and co-workers[91,169,183,184,186] quite nicely from $r = 0.1$ to $r = 0.9$.

Desai and co-workers[188] extensively tested the various theories discussed above. Since their work preceded Ruetschi's detailed development,[117] they derived a statistical form of the potential from his 1984 cation-vacancy theory. Open-circuit discharge curves for manganese dioxides belonging to various crystalline phases were measured in both $9M$ KOH and $5M$ $NH_4Cl + 2M$ $ZnCl_2$. The oxide phases studied included various γ-MnO_2's made both electrochemically and chemically as well as δ-MnO_2 (birnessite) and ρ-MnO_2 (now considered a γ-MnO_2).

The results of Desai et al.[188] indicate that all the models successfully predict the behavior of all the samples in the first half of the reduction range, $0 < r < 0.5$, but fail to do so in the latter half to varying degrees. One model of Maskell et al. yielded an excellent match with experimental data in the acidic-neutral electrolyte.

Reasons for discrepancies between the theoretical predictions and experimental data late in the discharge were discussed. Appearance of a new phase was invoked, as it had been originally by Kozawa and Powers.[34] For some of the systems, the equilibrium potential curves flattened out before the first-electron discharge finished, suggesting a potential-controlling second-electron reduction. This situation was more prevalent for the δ-MnO_2 and ρ-MnO_2 and in alkaline solution.

Holton et al.[189] had previously shown that γ-MnO_2 equilibrium potentials in concentrated KOH solutions can be complicated late in the course of reduction by instability (phase changes) and therefore may not be expected to conform accurately to the single phase treatments (see also Ref. 188). For this reason, these investigators restricted the evaluation of their theories to Leclanché solutions. Such phase changes probably are responsible for much of the misfit between Eq. (38) and experimental results in Fig. 12.

3. Modeling of γ-MnO_2 Discharge at Practical Rates

(i) Equilibrium-Diffusion Approaches

The earliest modeling studies were aimed at confirming the idea that solid-state diffusion of protons is rate-determining in MnO_2 discharge and recovery. Therefore, they were not intended to be structure-specific. Current distribution within the bobbin of Leclanché cells and within the MnO_2 particle, itself, was first addressed quantitatively in 1946 by Coleman,[47] who applied Fick's law to study the relationship between discharge capacity and drain rate. Treatments a decade later, by Scott[48] and Kornfeil,[49] solved the diffusion equation for the one-dimensional semi-infinite case, yielding an expression for the concentration of reduction product (protons) as a function of time and depth into the solid. Substituting the concentration–time relation on the surface of the electrode into a Nernst-type equation gave a potential–time function, which could be applied to constant-current discharge and open-circuit recovery experiments; i.e.,

$$E_t = E_0 + K_1 \log \frac{[Mn^{4+}]}{[Mn^{3+}]}$$

$$= E_0 + K_1 \log \frac{1-r}{r}$$

$$= E_0 + K_1 \log \left[\frac{SF(\pi D)^{1/2}}{2IV[t^{1/2} - (t-\tau)^{1/2}]} - 1 \right] \quad (46)$$

Here E_t is the oxide potential at time t, and E_0 is the potential at time zero (the beginning of the discharge). K_1 was taken as a constant determined from the data, rather than $2.3RT/F$; interestingly, it was determined to be 0.068 V at 25°C rather than the 0.059 or 0.12 V determined from first principles. D is the diffusion coefficient of protons in the oxide phase; S is the electrochemical reaction area of the solid; V is the (assumed constant) molar volume of MnO_2; t is the time measured from the beginning of discharge; and τ is the total discharge time. Thus, $t - \tau$ is the recovery time, which is taken as zero for $t < \tau$.

This treatment yields mainly one kinetic parameter that could be considered structure-specific, i.e., $SD^{1/2}$. E_0 naturally varies with the particular MnO_2, as discussed previously, but this is an issue more of lattice energy than of kinetics. Verification that $SD^{1/2}$ was constant, reasonable in magnitude, and invariant for different discharge times and currents constituted verification of the theory.

Further modeling studies have been aimed at (a) further verifying the basis for the treatments; (b) improving the treatments, in terms of both bringing more structural detail into the model and improving the mathematical basis and solutions (e.g., the use of more sophisticated methods); and (c) using the models to distinguish various manganese dioxides. K_1 was later identified with $2.3RT/F$. Correcting Eq. (46) for the fraction of Mn^{3+} in the initial oxide (i.e., Mn that is already reduced[51,52]) brought more structural detail into focus. Gabano et al.[52] verified the theory further by showing the equivalence of results from galvanostatic and potential-step experiments. They also showed that potential steps of different size yield the same value for $SD^{1/2}$ and that $SD^{1/2}$ varies directly with the weight of sample. Importantly, this work demonstrated the expected isotope effect on $SD^{1/2}$ for protonated versus deuterated solutions (cf. Section II).

Gabano et al.[52] also compared results for three different generic manganese dioxides—an EMD, CMD, and NMD—as discussed in Section V. Methods of separating D and S, so as to provide more detailed comparison of different materials, have usually addressed means of estimating the electrochemical reaction area, S. A common assumption has been that S is equivalent to the N_2 BET surface area.[52,53,77,112] Such areas range from 10 to 100 m^2/g and usually result in solid-state proton diffusion coefficients of 10^{-16}–10^{-20} cm^2/s. This assumption finds some justification in the fact that Zn^{2+}-ion adsorption requires the same surface area as gas adsorption.[124] Various investigators believe that the electrochemical reaction area should be much smaller—even as small as the geometric surface area, which is about 10^{-3} of the BET surface area (see Ref. 190, which deals with chemical leaching of MnO_2). On this basis, the D values extracted from diffusion treatments (i.e., from $SD^{1/2}$) would be 10^{-10}–10^{-14} cm^2/s. A value of 6×10^{-10} cm^2/s for D was determined directly from nuclear magnetic resonance[191] but was later claimed to suffer the error of detecting only surface protons in measured jumps.[24]

(ii) Porous Electrode Approaches

The above treatments do not address the porous structure of the electrode in any detail, including the reaction distribution within the electrode. The electrode has a geometric shape within which are porous EMD particles, so diffusion has three regimes—i.e., across the electrode (solution to current collector), in the pores, and in the MnO_2 lattice. Newman[192] and Newman and Tiedemann[193] developed techniques for handling porous electrodes but did not actually model the $MnO_2/MnOOH$ electrode. However, these techniques were used liberally by the subsequent investigators. Atlung and West[194] were perhaps the first to introduce a porous model of the $MnO_2/MnOOH$ alkaline electrode, although they did not use the techniques of the above authors.[192,193]

Wruck and Chapman[53,155] (also see Ref. 156) developed an in-depth porous model based on the superposition of macroscopic solid and liquid continuums and planar, one-dimensional current flow. They considered transport phenomena for all the major species, ohmic resistance in both phases, reaction thermodynamics (appropriate equilibria), electrode kinetics, compositional changes, and electroneutrality. They first applied a preliminary model with simplified descriptions of macroscopic phase properties, (linear) electrode kinetics, equilibrium potential, and current and potential distributions. The relevant equations were solved by computerized/numerical techniques.

Although agreement with the experimental results was fairly good, Wruck and co-workers then expanded their preliminary model by improving representations of electrochemical phenomena and by including additional concentration gradients and transport effects. Model predictions agreed very well with experimental discharge curves for a wide range of electrode thickness (i.e., 0.9–7.5 mm) and current density [3–150 mA/(g EMD)], ranging from near equilibrium to very high current density discharge; theoretical results also predicted electrode utilization successfully.

This work determined the solid-state proton diffusion coefficient to be 4×10^{-17} cm^2/s and the exchange current density for first-electron reduction of MnO_2 to be $i_0 = 6 \times 10^{-7}$ A/cm^2. Dominating factors, which control discharge, were found to be reaction-rate uniformity across the electrode and depletion of solid reactant surface concentration at the electrolyte interface.

Hong et al.[54] modeled EMD reduction over the full first-electron regime of $r = 0$–1, recognizing that the diffusivity of protons changes with r. From fitting their model to potential-step data, they found that the proton diffusion coefficient ranged from ~1.3×10^{-15} to 0.3×10^{-15} cm^2/s, and the D versus r curve showed a discontinuity at $r \approx 0.5$. They also varied temperature and obtained activation energies, which ranged from ~40 to 56 kJ/mol. They considered geometry of the diffusion, comparing planar—i.e., the case usually considered—and spherical diffusion, as had been considered by Valand.[195] Hong et al. obtained the best fit by considering double-direction tunnel diffusion approximated by a double-plane model.

Cheh and co-workers, in a series of papers,[196–198] modeled discharge of cylindrical alkaline cells over a wide range of drain rates [i.e., up to ~100 mA/(g cathode)].[*] Their models showed, as has often been observed,

---

[*]Alkaline cathodes are roughly 85% MnO_2, the remainder being graphite/carbon and electrolyte.

that differences in cathodic overpotential are the primary reason for differences in cell capacities (on a per-cathode-weight basis). These models gained in sophistication with drain rate, as necessitated by observation. Predominant factors in discharge ranged from quasi-equilibrium behavior of the MnO_2 at very low rates (described by $E-r$ functions in Section VI.2) to MnO_2 charge transfer at moderate rates to mixed reactions at high rates. The mixed reactions included effects from diffusion, convection, ohmic resistance, ion migration, and charge transfer in both the solid and liquid phases. These effects were handled by macrohomogeneous porous electrode theory. At high drain rates, zinc precipitation in the separator became a significant factor.

Results[196-198] were applied to commercial cells (which have annular or sleeve-type cathodes) as well as spirally wound cells. Differences in performance of AAA-, AA-, C-, and D-size cells were attributed primarily to differences in cathode thickness and its effect on cathode overpotentials and nonuniform reaction distribution.

Brenet and Picquet[199-201] performed in-depth studies of electrochemical processes in MnO_2 using thermodynamics. They examined the nature of proton movement during MnO_2 discharge and recovery and concluded that faradaic current described the driving force for their calculations better than diffusion. They treated discharge/recovery from the irreversible thermodynamics of a concentration cell.[201]

4. Electrochemical Spectroscopy of γ-MnO_2

Stepwise and low-rate voltammetry studies have decoupled reduction of the ramsdellite and pyrolusite type Mn atoms in γ-MnO_2.

(i) Low-Rate Voltammetry

Using continuous linear-sweep voltammetry at ~25 μV/s, McBreen[57,202] separated the MnO_2 reduction peak in $7M$ KOH into two peaks—i.e., one at -0.19 V and one at -0.45 V vs. Hg/HgO; the latter was the much larger peak after the first cycle. Swinkels et al.[203] galvanostatically discharged MnO_2 as well as MnO_2 that had been heat-treated to various extents in $9M$ KOH, utilizing very low currents (~40-h reduction times). From the differential discharge capacity versus potential curves, they observed shoulders on the main peak, which they attributed to regions of different tunnel widths as well as defects.

(ii) Step Potential Electrochemical Spectroscopy (SPECS)

Using step potential electrochemical spectroscopy (SPECS) (10 mV/2 h), Chabre and co-workers[204,205,24] examined MnO_2 in $1M$ KOH and obtained two reduction peaks similar to those reported by McBreen, as well as a small initial peak at 0.1 V. The SPECS method applies a potential sweep, step by step, while the charge (coulometry) is recorded at each potential level as the reducing current decays to a very small value. The fine peak resolution is obtained by choosing test parameters and instrumentation to give 1-μA·h resolution. An example of the results is shown in Fig. 13, for MnO_2 Sedema WSA mixed with carbon black in $1M$ KOH. A single experiment gives both the quasithermodynamic equilibrium conditions for insertion at definite potentials and the kinetic behavior of the system at each redox state. Each dot in Fig. 13 represents the average current for successive 10-μA·h charge increments at constant potential.

From such studies on various MnO_2 samples in concentrated as well as dilute KOH solutions, Chabre and co-workers[204–206,24] found that the

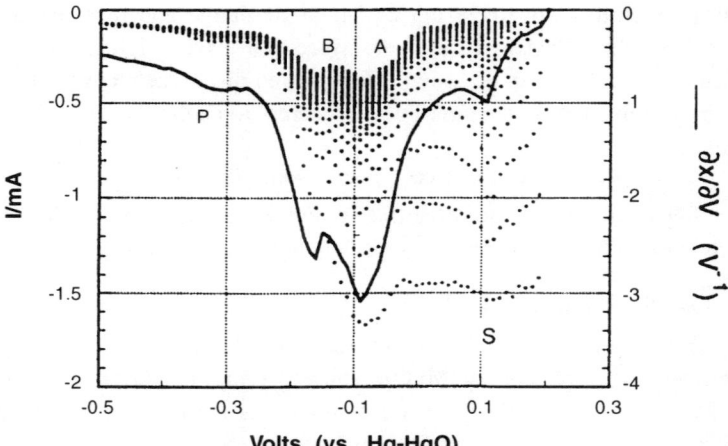

Figure 13. Step potential electrochemical spectroscopy (SPECS) response of MnO_2 (Sedema WSA, a CMD) in $1M$ KOH obtained with 10-mV/2 h potential steps. Each dot at a given potential is the mean reduction current for successive 10-μA·h charge increments. The line gives the incremental capacities, $\partial x/\partial V$. Electrode: 100 mg of MnO_2 + 50 mg of carbon black. (Reprinted from Ref. 206, Fig. 4, with permission from the author and Plenum Press.)

MnO_2 reduction steps are as follows (Fig. 13): (1) reduction at or close to the surface of MnO_2 grains (shown as S); (2) proton insertion into Mn(R) located in ramsdellite units, which occurs in two (A and B) steps, corresponding to 0.5 and 1 proton per manganese; and (3) proton insertion into Mn(r) belonging to pyrolusite units, at $E < -0.3$ V (shown as P).

Slow potential scans (10 mV/h) on pure mineral ramsdellite in $7M$ KOH solutions showed, perhaps surprisingly, a narrow reduction peak centered at −0.47 V vs. Hg/HgO, which is equivalent to the potential of the reduction peak for pure pyrolusite (β-MnO_2) but very different from the potential for reduction of the ramsdellite blocks in γ-MnO_2.[207] In contrast, a CMD (γ-MnO_2) sample gave a reduction band for the ramsdellite blocks centered at ~−0.15 V and a sharp peak for the pyrolusite blocks centered at −0.47 V. Thus, intergrowth of pyrolusite and ramsdellite structural blocks appears to increase substantially the reduction potential of γ-MnO_2 in comparison to the reduction potential of both stoichiometric and ordered forms.

(iii) Potential-Step Voltammetry with In Situ Neutron Powder Diffraction

In situ neutron powder diffraction (NPD) has been used in conjunction with reduction in small potentiostatic steps at long times to confirm the findings of SPECS studies and to complement the findings from open-circuit potentials, XRD, and other measurements made as a function of fractional MnO_2 reduction. Neutrons penetrate solids much deeper than X rays and thus offer the possibilities of studying larger samples and minimizing effects from reduction gradients between the surface and interior of electrodes.

Ripert and co-workers[92,208,209,24] thus studied the electrochemical reduction of two CMDs, IBA No. 11 (cf. Section IV.1) and a commercial sample called Sedema WSA, in $7M$ KOD/D_2O (substitution of the latter solution for KOH/H_2O is necessitated by incoherent scattering from protons). The electrodes were composites of the MnO_2 and carbon black. The two CMD samples are similar in De Wolff defects (Pr = 0.32 and 0.36, respectively, for IBA No. 11 and WSA) but quite different in twinning (Tw = 20 and 36%, respectively).

Both samples reduced similarly in the early stages but differed significantly in the final stages, the more ramsdellite-like IBA No. 11 transforming smoothly into groutite (α-MnO_2) and the more twinned WSA converting into a mixture of groutite and pyrochroite [$Mn(OH)_2$]

phases. Development of the NPD pattern of lattice parameters indicated that reduction proceeded in three stages. In stage 1, ramsdellite slabs transform into groutellite, during which a proton is inserted between each opposing pair of pyramidal O atoms in the (a,b) plane [see Section VI.1(iii) and Fig. 11]. In stage 2, continued insertion of protons into the interior of ramsdellite (now groutellite) slabs occurs with rearrangement of the H^+–O^{2-} bonds to form a network of O_{pyr}–H–O_{pl} bonds, and all pyramidal oxygens in the CMD are hydroxylated. Then the structure may be described as a random stacking of groutite and rutile layers. In stage 3, the planar O atoms that link ramsdellite (now groutite) and rutile layers are reduced. This stage shears the lattice into groutite slabs and layers of Mn^{4+} ions, the latter reducing immediately to form crystallites of pyrochroite, $Mn(OH)_2$.

The above mechanism is in harmony with that derived from SPECS experiments[204–206,24] and, in the case of the ramsdellite slabs, with the mechanism proposed by Maskell et al.[169] It is strengthened by the fact that the charge in each step agrees with that predicted from values of Pr in the structural determinations made by XRD.[24] Chabre and Pannetier[24] compared the above slow, in situ NPD studies with previous faster, galvanostatic, ex situ XRD studies to show that the mechanism of CMD reduction varies according to how near the system remains to equilibrium. Although both types of studies start with a homogeneous-phase reduction, near-equilibrium studies proceed through a groutellite phase (Mn_2O_4H or Mn_2O_3OH) to an intergrowth structure of groutite and manganite—i.e., δ-MnOOH. If the reduction is far from equilibrium (i.e., too rapid), the intermediate is a highly disordered structure formed by an interlayering of ramsdellite- and groutite-like layers with various proton contents and rutile-like slabs.

5. Insertion in Modifications Other than γ/ε-MnO$_2$

Section II reviewed the fact that essentially no proton insertion occurs in β-MnO$_2$. Some other forms of MnO$_2$ appear to reduce by the proton insertion mechanism, although not to as great an extent as γ/ε-MnO$_2$.

α-MnO$_2$ reduces electrochemically in alkaline and weakly acid (NH$_4$Cl–ZnCl$_2$) electrolytes to produce E–r discharge curves that are qualitatively similar to those for γ/ε-MnO$_2$—i.e., the curves slope initially and, in some cases, tend to flatten out.[147,149] XRD studies on α-MnO$_2$, after various levels of electrochemical and chemical reduction, showed lattice expansion up until $r = 0.25-0.5$.[210–212] Then, changes in crystal symmetry occurred, and, finally, new phases appeared.

When δ-MnO$_2$ was chemically reduced by 2-propanol, protons were inserted homogeneously to MnOOH$_{0.6}$ with contraction between the [MnO$_6$] layers and expansion within the layers.[213] Further reduction appeared to be heterogeneous and produced a new phase, either Mn$_3$O$_4$ or γ-Mn$_2$O$_3$.

VII. RECHARGEABLE ALKALINE ELECTRODE

1. Introduction: MnO$_2$ Rechargeability Problems

Rechargeable or secondary battery systems have been studied intensively over the past two decades as the economic and environmental concerns of throwaway, primary cells have been accepted, and Zn–MnO$_2$ has been a leader in development efforts. Several characteristics of the alkaline Zn–MnO$_2$ system make it appealing as a secondary system and, thereby, competitive with the present commercial small-format systems— i.e., nickel–cadmium, nickel–metal hydride, and lithium–cobalt oxide. Advantages include low cost, environmental friendliness, competitive energy and power densities, and long shelf life at elevated temperatures.[214]

However, substantial work has been required to overcome technical barriers and to bring these advantages to fruition. Without very special treatment and precautions, the MnO$_2$ electrode loses capacity rapidly. On deep discharge (more than roughly 1 e^-/Mn atom), new phases form that are nonrechargeable. Though cycling is much better on shallow discharge, the capacity still fades due to mechanical problems.

Electrochemical discharging and charging studies coupled with XRD have characterized the chemical failure mechanisms.[202,215,216,218–228,40,42,55] Reversible behavior of proton insertion and deinsertion in MnO$_2$ occurs for the first ~ 0.8 e^-/Mn atom—i.e., from MnO$_2$ to MnO$_{1.6}$—while the potential decreases continuously to 0.9 V vs. zinc and the γ-MnO$_2$ lattice expands. After reduction to a poorly defined intercalation of 0.3–0.6 H$^+$/Mn (MnO$_{1.85}$ to MnO$_{1.7}$), MnOOH and/or amorphous phases start to form. Proton intercalation is still reversible through this reduction/oxidation regime, since the new phase (α-MnOOH) is isostructural with the rhombic γ-MnO$_2$.

On further reduction, however, new phases form that involve yet lower manganese oxides than MnOOH, which do not fully recharge. On recharge, Mn_3O_4 builds up as cycling progresses and rechargeability diminishes. McBreen[202] used slow cyclic scans and *ex situ* XRD to develop the following scheme for the progression to Mn_3O_4 during cycling:

$$\gamma\text{–}MnO_2 \rightleftharpoons \alpha\text{–}MnOOH \tag{47a}$$

$$\alpha\text{–}MnOOH \to Mn(OH)_2 \tag{47b}$$

$$Mn(OH)_2 \to \{\gamma\text{–}Mn_2O_3 + \beta\text{–}MnOOH + \gamma\text{–}MnOOH\} \tag{47c}$$

$$\{\gamma\text{–}Mn_2O_3 + \beta\text{–}MnOOH + \gamma\text{–}MnOOH\} \to Mn_3O_4 + \delta\text{–}MnO_2 \tag{47d}$$

$$\delta\text{–}MnO_2 \to Mn_3O_4 \tag{47e}$$

$$Mn_3O_4 \to Mn(OH)_2 \tag{47f}$$

The results obtained by Mondoloni *et al.*[42] from cycling with *in situ* XRD (Fig. 14) similarly show that Mn_3O_4 and Mn_2O_3 become bottleneck species in recharging. The irreversibility is manifest in the short charging capacity compared to the discharge capacity.

Ouboumour *et al.*[218] found that overcharging with oxygen evolution during cycling increased the rechargeability of EMD. The increase in cycle life was credited to oxidation of Mn_3O_4 to the initial MnO_2 by oxidizing species generated during O_2 evolution.

Qu *et al.*[219] showed that Mn^{3+} ions are present even during the first-electron discharging and recharging; therefore, some second-electron discharge is possible. This mechanism appears to be in parallel with and slow compared to the first-electron mechanism in those numerous studies for which the proton–electron mechanism is controlling.

After the problems of irreversible phase changes were recognized, R&D of rechargeable alkaline cells took two directions. The first of these employed limitation of the depth of discharge to the reversible range for MnO_2, i.e., $\leq 0.6\ e^-/Mn$ atom. Even before the full-scale research was published, such limitation was attempted in the manufacture of commercial cells in the mid-1960s to the mid-1970s by several U.S. battery companies.[6,7] The cells were withdrawn because they had very short cycle

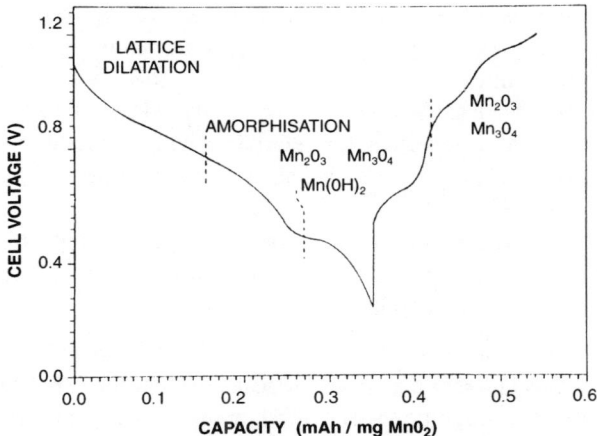

Figure 14. Voltage vs. faradaic capacity for EMD during discharge and recharge. Domains in the curve correspond to structural changes in the MnO_2 measured *in situ*. (Reprinted from Ref. 42, Fig. 2, with permission from the authors and the Electrochemical Society.)

life (<30 cycles), and the depth of discharge had to be closely controlled to prevent total loss of rechargeable properties in very few cycles. The recent flurry of activity has come mainly from the Technical University of Graz (Austria) and Battery Technologies Inc. of Toronto, Canada (see Section VII.2).

The second direction involved modification of the MnO_2 structure from γ-MnO_2 to a stabilized birnessite structure so that the MnO_2 could reversibly deliver more than 1 e^-/Mn atom. This was invented at Ford Motor Company and has been studied intensively at the University of Ottawa (see Section VII.3).

2. Limitation of Depth of Discharge

The first step in preventing MnO_2 phase changes was to limit the MnO_2 depth of discharge to 0.8 e^-/H^+ per Mn atom. This was accomplished by limiting the amount of zinc in the cell.[220] (MnO_2 limits the useful life, above 1 V, in primary alkaline cells.) Even with such anodic limitation, the cells faded fast. Kordesch and co-workers traced this problem to the lattice expansion and contraction that occurs on discharging and charging.[6,221,222] They showed that, without proper electrode forming and support, expansion is not reversed during charging. Therefore, the electrode progressively disintegrates, as manifested by cracking and an increase of electrical resistance.

Measures taken to alleviate this problem included inclusion of appropriate binders and/or fibers in the electrode mix, high-pressure electrode molding with cathode support by the remainder of the cell, and even support by metal cages.[6,223-225]

Once the basic chemical and physical limitations on cycling MnO_2 were alleviated, other secondary modifications were necessary in making the alkaline cell suitably rechargeable, as listed below [6,226]:

1. Zinc migration to the MnO_2 electrode had to be restricted to control the formation of the nonrechargeable haeterolite ($ZnO \cdot Mn_2O_3$) at the cathode. Zinc migration was slowed, but not eliminated, by additives that decrease the solubility of zincate ions in the electrolyte (e.g., Ba^{2+} ions) and by separators that inhibit transport of soluble zinc from the negative to the positive electrode. Even with the above precautions, hetaerolite formation is a limiting factor in cycle life of the cell. When the efficacy of various EMDs in rechargeable cells was compared, the main correlating factor was shown to be hetaerolite formation.[222] Interestingly, the variation of cycle life of numerous IBA study samples of EMD could not be correlated with physical or electrochemical properties of the EMDs, and, particularly, such cycle life did not correlate with performance of these EMDs in primary alkaline cells.[221]

2. Hydrogen and oxygen evolution, which occur during overcharging or cell reversal at the respective electrodes, have to be suppressed to prevent excessive pressure buildup and possible cell rupture. This is accomplished through catalytic conversion of these gases to water at the opposite electrode to which they were formed or at an auxiliary electrode.

Hydrogen is oxidized to water at the MnO_2 electrode or at an auxiliary MnO_2 electrode on a carbon substrate if such an electrode is impregnated with a suitable metal, metal oxide, or metal salt (e.g., Ag or AgO_2).[6,227] Oxygen is reduced at an auxiliary electrode, e.g., porous carbon, the latter being positioned at the top or bottom of the zinc electrode. Alternatively, catalytic particles of graphite or metals may be admixed with the zinc-gel negative electrode. Even with such catalysts, a taper mode of charging had to be developed to limit gas formation to a manageable rate. This setup contrasts to the nickel–cadmium cell, in which the electrodes, themselves, combine O_2 and H_2 gases so rapidly that constant-current charging can be tolerated.

3. Catalysts at the MnO_2 electrode also had to be developed to ensure that O_2, and not manganate ions (MnO_4^{2-}), are formed during battery overcharging. The latter ions not only disproportionate to form battery-inactive Mn(II) but also migrate to the negative electrode, where they interact with and destroy zinc. Various metal oxides (e.g., V_2O_5) and mixed oxides (e.g., $CoAl_2O_4$) are effective catalysts for promoting oxygen evolution and suppressing manganate formation.[228]

4. The rechargeability of the MnO_2, itself, has been shown to be improved by incorporation of titanium oxides, titanium sulfides, or nickel oxides in the MnO_2. This is accomplished by suspending particulates of these compounds in the EMD deposition bath.[224,229,230]

With the above and other extensive design features, secondary alkaline cells based on limited discharge have been developed that can sustain more than 50 deep discharges (~0.5 e^-/Mn atom) and hundreds of shallower discharges.[226] These cells have recently been commercialized and have captured more than 50% of the small-format, rechargeable consumer market in the United States.[214]

3. Modified MnO_2 Structure with Two-Electron Capacity

(i) *Characterization*

A major breakthrough occurred when Wroblowa and co-workers[231-234] discovered that MnO_2, chemically modified by Bi^{3+}-ion and Pb^{2+}-ion incorporation, could be discharged and recharged in strongly alkaline solutions more than 2000 times to the extent of 1.8 electrons per Mn atom. Cyclic voltammetry tests[232,234,235] showed that this extraordinary capacity can be discharged between 0.2 and −0.6 V vs. Hg/HgO, and then the same amount of charge can be delivered back on recharge independent of the depth of discharge between −0.6 and −1.0 V. This contrasts sharply with γ-MnO_2, for which the corresponding cyclic voltammogram shows only partial rechargeability after reduction to −1.00 V and which exhibits no significant recharging current if the cathodic cycle is reversed at −0.6 V (these facts are in harmony with findings of Section VII.1).

The material was synthesized by two methods.[231-234] The first involved precipitation of hydroxides from a mixed nitrate solution of Mn^{2+} and Bi^{3+} or Pb^{2+} ions followed by firing, resulting in what the inventors called chemically modified (CM) MnO_2. The second method involved physically admixing Bi_2O_3 or PbO with various manganese oxides (MnO_2, Mn_3O_4, Mn_2O_3, or MnO). In the latter case, the mixture requires 10–30 electrochemical cycles to develop the extraordinary rechargeability that is exhibited within one or two cycles by the CM MnO_2. The remarkable capacity and rechargeability are realized over the range of bismuth concentration of 1–13 mol % in the MnO_2.

A third method of synthesis appears probable from recent studies[236] in which bismuth was incorporated in EMD during electrodeposition from acid sulfate solutions that contained dissolved Bi^{3+} ions. The EMD deposits showed enhanced rechargeability when compared with EMD deposited without bismuth species.

The remarkable rechargeability appears to have two constraints: (1) the rechargeable capacity rapidly deteriorates when Zn^{2+} ions are introduced into the system, and (2) the rechargeable capacity has only been demonstrated thus far with a high percentage (50–90%) of carbon (Lonza graphite) in the electrode mix.

(ii) Mechanism

The effect of Bi^{3+} and Pb^{2+} ions was intuitively considered by the inventors to be that of stabilizing an open layered lattice structure during cycling. Somewhat supportive of this view is the fact that both the Bi^{3+} and Pb^{2+} ions have a radius of 1.2 Å whereas more than a dozen metal ions of charge +1 to +4 that failed to impart the extraordinary rechargeability to MnO_2 have radii either smaller or larger than 1.2 Å.

However, X-ray diffraction failed to confirm a layered structure, as the material becomes amorphous during cycling. Also, modification of an open solid-state structure is difficult to rationalize at the low levels of 1–2 mol % Bi that still positively influence the rechargeability.

More recently, Qu *et al.*[219] conducted an in-depth study of the discharge and recharge of the bismuth-doped MnO_2 and showed that the mechanism involves a soluble Mn(III) species. They noted that the potential during galvanostatic discharge is nearly constant throughout most of the two-electron electroreduction (Fig. 15), as predicted by thermodynamics for a heterogeneous process (assuming the system stays near equilibrium). This behavior parallels that for the second-electron process proposed by Kozawa and co-workers[33,36] for alkaline discharge [Section II.2(ii)] but applies to most of two-electron discharge, as observed in $ZnCl_2$ discharge (see Section III.2). The galvanostatic discharge for γ-MnO_2,

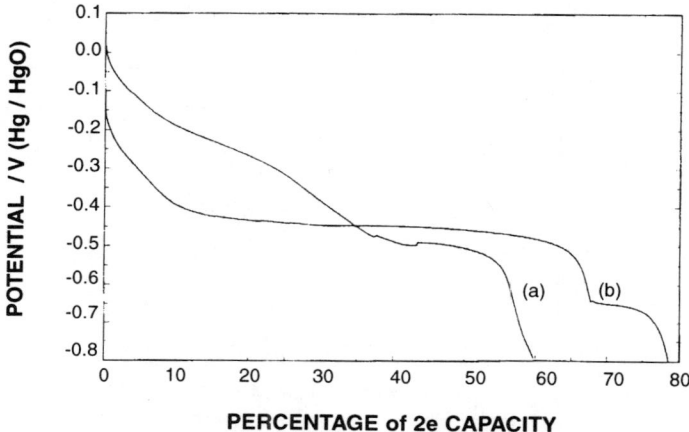

Figure 15. Constant-current discharge of chemically modified MnO_2 (b) and of γ-MnO_2 (a). Current density: 0.2 A/g. (Reprinted from Ref. 219, Fig. 2, with permission from the authors and Chapman and Hall.)

shown in Fig. 15 for contrast, exhibits the proton–electron insertion mechanism for most of the first-electron reduction and then shows the heterogeneous mechanism for just a short part of the second-electron reduction, as is typical at significant current densities.

Qu et al.[219] monitored Mn(III) in solution *in situ* by various means during discharging and recharging of CM MnO_2. Also, they contrasted these results with those obtained for γ-MnO_2 during similar cycling. Slow galvanostatic tests were applied to a stationary MnO_2 electrode while Mn(III) ions were detected with either (a) a stationary Pt detector electrode positioned adjacent to the MnO_2 electrode with potential set (−0.05 V vs. Hg/HgO) to oxidize Mn(III) or (b) spectrochemical means to monitor the 465-nm peak, due to $Mn(OH)_6^{3-}$ ions.[219] During MnO_2 discharge of CM MnO_2, both methods showed an increase in soluble Mn(III), with due account for adsorption, starting at the beginning of the reduction. The Mn(III) concentration maximized at 25% and 40% of two-electron reduction for methods (a) and (b), respectively. Then the concentration decreased with further reduction.

Upon recharge, the concentration of Mn(III) again increased, maximizing at approximately the same level of mean Mn oxidation number as had occurred during reduction, and then decreased again. Figure 16

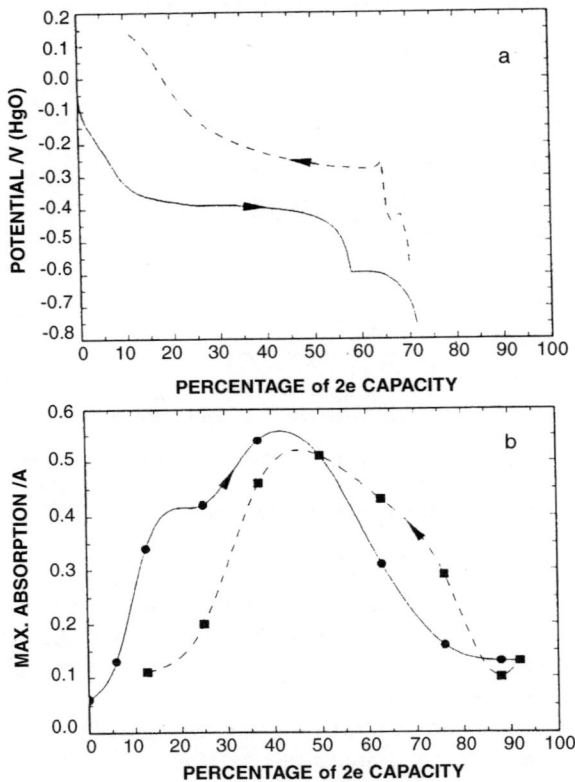

Figure 16. Comparison of constant-current discharge and recharge (a) with maximum absorption at various stages of discharge and recharge (b) for CM MnO_2/Lonza graphite (1:4). Current density: 0.16 A/g. ——, Discharge; ----, recharge. (Reprinted from Ref. 219, Fig. 13, with permission from the authors and Chapman and Hall.)

compares the potential with the spectral absorption maximum during both discharge and recharge at comparable states of MnO_2 reduction. The $Mn(OH)_6^{3-}$ ions were shown to be stable for more than five hours, the time for a single half-cycle of the test, which proved that the decrease in Mn(III) concentration at the extremes of polarization are electrochemically based.

Results from both the stationary electrochemical and the spectroelectrochemical tests indicate that the reaction course is as follows. Mn(III) ions are formed as soon as the discharge begins and then build up to saturation in the MnO_2 electrode. All the while, these ions are diffusing out of the porous MnO_2 electrode into the solution, where they are detected. The second step of the discharge involves reduction of Mn(III) ions to Mn^{2+} ions at surfaces in the composite test electrode. The Mn^{2+}, being less soluble than Mn(III), precipitates as $Mn(OH)_2$. The Mn(III) formation and reduction occur simultaneously, but the decrease in Mn(III) signal occurs as the final Mn(III) ions are reduced on discharge or oxidized on recharge; thus, the Mn(III) signal decreases as the source of Mn(III) becomes denuded, this source being MnO_2 on discharge and $Mn(OH)_2$ on recharge.

For a γ-MnO_2 electrode similarly discharged, the Pt detector electrode sensed a small, although real, concentration of Mn(III) shortly after discharge started. However, a significant increase in the signal arose only when the second-electron reduction started, as manifested by a horizontal plateau in the E versus t discharge curve; this occurred at about 40% of two-electron capacity.

Electrochemical detection of Mn(III) ions was also made in a rotating disk-ring experiment, in which the CM MnO_2 disk electrode was cycled between -1.0 and 0 V vs. Hg/HgO. In this case the disk electrode gave a narrow cathodic peak during reduction, which peaked at about -0.8 V, and the ring, at a potential of -0.1 V, gave a sharp anodic peak due to the Mn(III) ions spun off the disk. Thus, the Mn(III) ions are soluble and are produced over a narrow potential range. Furthermore, the charge at the disk, on reoxidation, was much smaller than that during the cathodic charge, and the ratio of anodic to cathodic charge decreased with increasing speed of the disk. This latter phenomenon follows qualitatively from the fact that the Mn(III) intermediates spun into solution are not available for second-step reduction and subsequent recharge. The investigators believed that the presence of soluble Mn(III) ions plays a role in the rechargeability.

Finally, Qu et al.[219] elucidated more of the discharge and recharge mechanism with potential holding experiments, wherein cathodic linear sweeps were arrested at various potentials and the current decay studied. Then, when the current reached some value (usually zero), the direction of the potential sweep was reversed and the shape of the anodic sweep was studied. Results of these studies showed that soluble Mn(III) ions are oxidized at potentials at and near −0.3 V vs. Hg/HgO whereas Mn(OH)$_2$, once formed, is reoxidized to MnO$_2$ at potentials at and near −0.05 V.

These potential holding studies contrast clearly the behaviors of CM MnO$_2$ and γ-MnO$_2$. The full two-electron capacity of CM MnO$_2$ can be discharged at about −0.43 V vs. Hg/HgO (~0.95 V vs. Zn) and can then be recovered on recharge. Under comparable experimental conditions, γ-MnO$_2$ gives only approximately one-electron capacity on discharge and cannot be recharged unless it is further reduced to Mn(OH)$_2$ at about −1.0 V.[219] Even then, the recharge is less than the discharge.

From all results, then, the following mechanistic facts appear to typify the electrochemical behavior of the bismuth- (and lead-) modified manganese dioxides:

1. A heterogeneous reduction of MnO$_2$ to soluble Mn(III) ions, Mn(OH)$_6^{3-}$, starts immediately upon discharge and circumvents the homogeneous electron–proton mechanism.

2. At −0.43 V (Hg/HgO), where most of the reduction occurs, a second-electron transfer occurs even before the first-electron reduction is complete, and the second-electron transfer converts Mn(III) ions to insoluble Mn(OH)$_2$.

3. The Mn(OH)$_2$, once formed, is recharged to the starting MnO$_2$ at about −0.1 V, whereas Mn(III) ions are recharged to MnO$_2$ at about −0.3 V (Hg/HgO).

Bai et al.[235] studied discharging/charging curves of CM MnO$_2$ as a function of drain rate. Results showed that more than 80% of two-electron capacity could be obtained (a) at drain rates up to C/10* with cathode mixes having equal graphite and MnO$_2$ and (b) at drain rates up to 4C with cathode mixes having a C:Mn ratio of 10:1. The electrode in these cases contained 20 mg of MnO$_2$ per square centimeter. By comparison, electrodes containing γ-MnO$_2$, but identical otherwise, yielded only 1 e^-/Mn atom—i.e., the second-electron reduction yielded virtually no capacity.

Although a high carbon content is needed to support these useful drain rates (remember that the solubility of Mn(III) ions is only ~$10^{-4}M$), the presence of Bi provides a preferred heterogeneous discharge/recharge pathway, involving a soluble Mn(III) intermediate. This pathway provides

---

*In battery notation, C/X indicates that the material is discharged in X hours.

for extraordinarily high rechargeable capacity for MnO_2. How Bi functions is an area of ongoing research.[237]

VIII. MnO_2 DEPOSITION

1. Technology

Electrolytic manganese dioxide (EMD) is rather unique in that it is both produced and utilized electrochemically. However, deposition is conducted in a strongly acidic solution that contains Mn^{2+} ions whereas battery discharge occurs in strongly alkaline to very weakly acidic environments, which, at least initially, contain no dissolved manganese. Correspondingly, the mechanisms of deposition and battery function are quite different. The principal deposition reactions are

$$Anode: Mn^{2+} + 2H_2O \rightarrow MnO_2 + 4H^+ + 2e^- \qquad (48)$$

$$Cathode: 2H^+ + 2e^- \rightarrow H_2 \qquad (49)$$

Commercial EMD production is conducted in a single-compartment cell, and deposition conditions are driven by the demand for maximum battery activity of the product and the sensitivity of such battery performance to deposition conditions (cf. Section V). The rather narrow range of commercial deposition conditions has evolved somewhat with time and is as follows[75,161,217]:

$$T = 90–95°C; \; i_c = 0.5–10 \text{ mA/cm}^2;$$
$$[H_2SO_4] = 0.4–0.7M; \; [Mn^{2+}] = 0.3–0.5M$$

Although commercial anodes have included graphite, lead, and titanium, present-day practice largely utilizes titanium, which is stronger than the other two materials and causes less contamination of the MnO_2 product. Cathodes have included graphite and copper. The cathodic reaction is generally considered to have little influence on the deposition reaction and is not considered further here. The cell is typically covered with a molten wax layer to minimize electrolyte evaporation. Deposition typically is conducted for 2–3 weeks, during which time the MnO_2/EMD plate grows to a thickness of roughly 2 cm.

The anodic current density (typically ~0.6 A/dm² or 6 mA/cm²) is low by most technological deposition standards, which makes such deposition capital-intensive. Higher current densities as well as lower temperatures change the EMD structure from a compact, fine-grained γ/ε-MnO_2 to a more porous ε-MnO_2. The latter structure has a deleterious effect on the battery activity of the product (Section V) and also fosters the formation of undesirable by-products (cf. Section VIII.3) and titanium anode passivation[75,238]; the latter can cause the anodic potential to skyrocket and shut off the process.

An increase in anodic current density by a factor of 2 without the usual product deterioration has been made possible by the ingenious "suspension bath process," in which fine manganese oxide particles are suspended in the deposition bath.[239,79] This process produces a very coarse, cauliflower-like EMD surface, and the product properties (e.g., crystal form and surface area) are much like those produced in conventional deposition at very low current densities.[79,75]

MnO_2 deposition also has been studied (although not commercialized) in aqueous chloride, perchlorate, and nitrate solutions[240] (also, see Ref. 75). The deposits from these baths, although of ε- and γ/ε-MnO_2 morphology, appear shiny and fibrous (very crystalline) in the direction of the electric field and show preferred orientation. This contrasts sharply with the conventional, sulfate-derived EMD. The latter is so fine that individual grains are not observed, even microscopically. The difference is due to the fact that sulfate ions are strongly adsorbed at growth sites and thus foster frequent renucleation, whereas the monocharged anions are much less strongly adsorbed and thus allow crystal growth. The detailed mechanistic work has focused more on the EMD deposited in sulfate baths.

2. Equilibrium Potentials

Experimental studies of MnO_2 potentials have encountered unrepeatability and instability in producing standard potentials for Eq. (48).[241] These difficulties have been associated with the nonstoichiometric nature of EMD and the fact that MnO_2 is very porous and has a strong tendency to undergo ion exchange with the electrolyte, which leads to unstable and undefined local pHs and Mn^{2+}-ion concentrations.

With due care for buffering and equilibration, successful results were obtained[242] for β-MnO_2, which is the most stoichiometric as well as the least porous MnO_2 (cf. Section IV). (Unfortunately, β-MnO_2 is also the least battery active form.) The experimental standard potential for β-MnO_2, 1.233–1.241 V, agrees with that calculated from thermodynamic data; this value has provided the standard by which to compare other manganese dioxides.

Standard-potential determinations for γ/ε-MnO_2 (EMD) were beset with the above problems and were less successful.[242] The equilibrium potentials obtained for EMD in acidic $MnSO_4$ solution were positive of those for β-MnO_2 by 40–99 mV. By comparison, the corresponding difference between EMD and β-MnO_2 potentials in alkaline solutions is about twice as great.[39] Equilibrium-potential differences between various unreduced γ/ε-MnO_2's and heat-treated MnO_2's (including betafied MnO_2's, such as were studied in Ref. 39) in solutions free of Mn^{2+} ions are rationalized by differences in stoichiometry—i.e., cation vacancies—which are manifest in the percentage of structural water.[95,76,117] The controlling potential in this latter case is determined by the MnO_2–MnOOH equilibrium, with the vacancy population determining the free energy of the MnO_2 (cf. Section IV.2).

Vetter and Jaeger[70] concluded that the composition of the deposition electrolyte (i.e., the Mn^{2+}- and H^+-ion activities) completely determine the equilibrium composition and free energy of EMD. Writing the deposition equilibrium as

$$Mn^{z+} + (n + m)H_2O \rightleftharpoons MnO_n \cdot mH_2O + 2nH^+ + (2n - z)e^- \qquad (50a)$$

$$Mn^{2+} + 2H_2O \rightleftharpoons MnO_n \cdot (2 - n)H_2O + 2nH^+ + (2n - 2)e^- \qquad (50b)$$

they reasoned that the free energy of undischarged MnO_2 is completely determined by the Mn oxidation state and the concentration of structural water [n and m, respectively, in Eqs. (50)]. They substantiated this idea experimentally by showing that EMD thin-film electrodes with different initial values of n came to the same equilibrium potential and n value when placed in a solution of given composition. These equilibrium values of n varied with the Mn^{2+}- and H^+-ion concentration of the solution, as Eqs. (50) would predict, but not with anything else. These results allowed development of free energies of EMD. Confirmation of Vetter and Jaeger's theory, i.e., verification that the deposition did, in fact, yield products given by Eqs. (50), required the development of microanalytical/electroanalytical techniques with which to determine n and m. A later theoretical thermodynamic treatment by Atlung and Pohl[243] agreed with the conclusions of Vetter and Jaeger.[70]

In later work,[117] Ruetschi showed that annealing the cation vacancies out of a typical EMD, by heating at various temperatures up to 400°C, caused a lowering of the equilibrium potential in 8.7M KOH by 150 mV but resulted in almost no change in the equilibrium potential in 5.7M H_2SO_4 + 0.01M $MnSO_4$. This finding is in essential harmony with the finding of Vetter and Jaeger. Ruetschi used his cation-vacancy model[95] (cf. Section IV.2) to describe the deposition equilibrium, i.e.,

$$(1-x)Mn^{2+} + 2H_2O \rightleftharpoons Mn^{4+}_{(1-x-y)}Mn^{3+}_y(OH)^-_{(4x+y)}O^{2-}_{(2-4x-y)}$$

$$+ (4-4x-y)H^+ + (2-2x-y)e^- \quad (51)$$

This is, of course, equivalent to Eqs. (50), except that Eq. (51) details the nonstoichiometry. Brenet and Picquet[200] also treated and discussed this equilibrium in terms of modified cation-vacancy theory.

Ruetschi interpreted the behavioral difference between heated EMD in acid and in alkaline solution to mean that the surface of γ/ε-MnO_2 is probably modified in acid solution to become like that of β-MnO_2, which is vacancy-free. His interpretation followed from the fact that the potential change of γ/ε-MnO_2, following heating, in *basic* solution appeared compatible with the energy of removing vacancies.

The conversion of γ/ε-MnO_2 to β-MnO_2 in acidic $MnSO_4$ solutions at open circuit is well documented and has been studied by Kao.[164] Although MnOOH is unstable in acid solution, it is proposed as an intermediate in this transformation; i.e.,

$$\gamma\text{-}MnO_{2(s)} + Mn^{2+}_{(aq)} + 2H_2O \rightarrow 2MnOOH_{(s)} + 2H^+_{(aq)} \quad (52)$$

$$2MnOOH_{(s)} + 2H^+_{(aq)} \rightarrow \beta\text{-}MnO_{2(s)} + Mn^{2+}_{(aq)} + 2H_2O \quad (53)$$

Kao also allowed for the possibility that Mn^{3+} ions are an intermediate and considered additional reactions beyond those given by Eqs. (52) and (53).

3. Side Reactions in Deposition

Before addressing the mechanism of MnO_2 deposition, the role of side reactions is addressed. Oxygen evolution is the first of these; i.e.,

$$2H_2O \rightarrow O_2 + 4H^+ + 4e^- \quad (54)$$

Because dynamic MnO_2 deposition potentials are just negative of the reversible O_2 evolution potential (~1.23 V vs. SHE), some earlier studies associated MnO_2 deposition with initial steps of oxygen evolution—i.e., with the formation of an OH radical, which acted as mediator. However, oxygen evolution at the observed deposition current densities (several milliamperes per square centimeter) requires several hundred millivolts overvoltage on either the bare substrate or the EMD-covered substrate.[20,244] Thus, oxygen evolution plays no role except under special deposition conditions that force the anodic potential positive of 1.23 V; this can occur at localized regions of an electrode that overall is at a potential near 1.23 V.

Such conditions that foster oxygen evolution have been shown to include a decrease in temperature below 75°C, increase in current density above ~10 mA/cm^2, decrease in Mn^{2+}-ion concentration below ~0.5M, and increase in sulfuric acid concentration above 0.5M.[244,20,161] Kano et al.[244] reported that in 90-min tests the potential rose well above 1.23 V and the current efficiency correlated inversely with the oxygen overvoltage. Oxygen evolution may be considered a parallel process to MnO_2 deposition.

A second side reaction is formation of Mn^{3+} ions at the electrode followed by disproportionation of these ions away from the electrode. These Mn^{3+} ions are limited in concentration by the stability constant of the following equilibrium:

$$2Mn^{3+} + 2H_2O \rightleftharpoons Mn^{2+} + MnO_2 + 4H^+ \qquad (55)$$

i.e., at 90°C the equilibrium constant for the above reaction is 1.6×10^7.[245] Nevertheless, this reaction accounts for MnO_2 particulate observed on cell floors and in lines of spent electrolyte. As will be seen in the subsequent subsection on mechanism, Eq. (55) is coupled to the reactions for MnO_2 deposition.

A third side reaction is the formation of β-MnO_2 rather than the principal deposition product, γ/ε-MnO_2. Formation of β-MnO_2, discussed in the last section,[164] is observed in some deposits, probably from conversion of γ/ε-MnO_2 at locations where there is not adequate anodic protection.

4. Deposition on Bare Substrate

Investigators have found that as deposition starts on a bare substrate, the initial rate-determining step is nucleation coupled with three-dimensional growth at the nuclei if the Mn^{2+}-ion concentration is greater than approximately 20mM or if the rotation rate of the disk electrode is fast. This mechanism is manifest by the current, at constant potential, being proportional to (time)^3.[246,247] At low Mn^{2+}-ion concentrations, Mn^{2+}-ion diffusion from the solution to the solid surface is rate-determining. Rotating disk-ring studies[248] indicate that the nucleation on a clean platinum electrode is preceded by Mn^{2+} oxidation to soluble Mn^{3+} ions. Then the Mn^{3+} ions hydrolyze to yield a solid product such, that the mechanism at high Mn^{2+}-ion concentrations is

$$Mn^{2+} \to Mn^{3+} + e^- \qquad (56)$$

$$2Mn^{3+} + 2H_2O \to Mn^{2+} + MnO_2 + 4H^+ \qquad (57)$$

$$Mn^{3+} + 2H_2O \to MnOOH + 3H^+ \qquad (58)$$

5. Deposition on Oxide-Covered Substrate

(i) Phenomenology

Once the surface is covered with nuclei, the current, at constant MnO_2 potential, levels off with deposition time (assuming electrode rotation or solution agitation).[246–250] At high overpotentials, the current may maximize before leveling off. The steady-state current increases with anodic potential near the reversible deposition potential (1.2–1.3 V vs. SHE) but reaches a limiting current at higher overpotentials. This limiting current density is independent of electrode rotation rate/solution agitation and is only of the order of 1–20 mA/cm^2, roughly two orders of magnitude less than the limiting current for solution diffusion with the comparable electrolyte. The Mn^{3+}-ion concentration reaches equilibrium according to Eq. (57) very quickly and remains at this equilibrium value as the oxide thickens.[248]

Cyclic voltammograms yield a peak current akin to the limiting current found in potentiostatic tests. This peak current is independent of electrode rotation rate at nominal Mn^{2+}-ion concentrations and moderate or high rotation rates.

Galvanostatic deposition yields constant MnO_2 potentials, after the seconds required for nucleation to stabilize, which increase with current density. (Although titanium is the favored substrate for commercial MnO_2

deposition, voltage instabilities from titanium passivation make this metal a poor choice for kinetic studies; therefore, kinetic studies have usually been conducted on platinum or carbon.) When deposition proceeds for many days, as in the case of commercial EMD production, the potential across the anode gradually increases. This increase was determined to be ohmic drop through the EMD plate while the charge-transfer/reaction overvoltage remained more or less constant.[251] Interestingly, the resistivity of the EMD is roughly the same as that of the electrolyte, so the cell potential change is nominal unless deposition conditions favor a very resistant EMD (e.g., high current densities create high structural water content; cf. Section IV.2) or cause passivation at the titanium surface adjacent to the inside of the EMD plate.

Although steady-state MnO_2 deposition usually occurs with an overvoltage of less than a few tenths of a volt, the commercial process is very irreversible, as evidenced by the great variation in structure with deposition current density at fixed electrolyte composition. Structural water, cation-vacancy concentration, and, perhaps, relative concentration of Mn(III) in the oxide vary with deposition current density.[76–78]

Proposed mechanisms for continued MnO_2 deposition are dominated by the fact that Mn^{2+}-ion diffusion in solution and initial charge transfer at the oxide/solution interface are not normally the rate-determining steps, as the above facts suggest. Numerous part reactions are possible, i.e., formation and disproportionation of Mn(III) intermediates both in solution* and in the oxide, electrochemical oxidation of these intermediates as well as the reactant, hydrolysis and dehydration of various species, deprotonation of hydrolysis products, and chemical reaction between Mn^{2+} ions and MnO_2. Also, transport of reactants and products through solid products and intermediates increases the number of possible reaction mechanisms. Therefore, because the number of possible mechanisms is very large, support of a unique mechanism becomes very difficult. This complication is manifest in the rather large number of mechanistic investigations reported (see Refs. 248 and 249 for a brief review of such works).

---

*The published standard potential for the Mn^{2+}/Mn^{3+} couple is 1.5 V compared to 1.23 V for Mn^{2+}/MnO_2.[241] Correspondingly, it has been assumed by various investigators that formation of soluble Mn^{3+} is not the first step in the mechanism. However, various *in situ* spectroelectrochemical studies coupled with voltammetry have confirmed the formation of Mn^{3+} ions before the formation of MnO_2. Some investigators (cf. Ref. 251) conclude that the published standard potential of 1.5 V for Mn^{2+}/Mn^{3+} is probably too high.

Here, only a few of the recent mechanisms will be reported, but these mechanisms have rudiments in earlier works and thus seem to represent much of the mainstream thinking.

(ii) Crystallization as Rate-Determining Process

Fleischmann et al.[246] measured the limiting currents [described in Section VIII.5(i)] by means of potentiostatic transients and studied these limiting currents as a function of potential, Mn^{2+}-ion concentration, and H^+-ion concentration. They developed kinetics for crystal growth by counting nuclei and decoupling the nucleation from the growth. Their mechanism is summarized in Fig. 17 as the reaction sequence *not* inside the dashed box. From their results they concluded that the rate-determining step is dehydration of an adsorbed Mn(IV) species to form $\gamma\text{-}MnO_2$—i.e., step 7.

The adsorbed Mn(IV) species, $[Mn(H_2O)_{n-4}(OH)_4]_{ads}$ in Fig. 17, is formed by several preceding steps, which are at semiequilibrium: i.e., (1) the electrochemical oxidation of Mn^{2+} ions to form adsorbed Mn^{3+} ions;

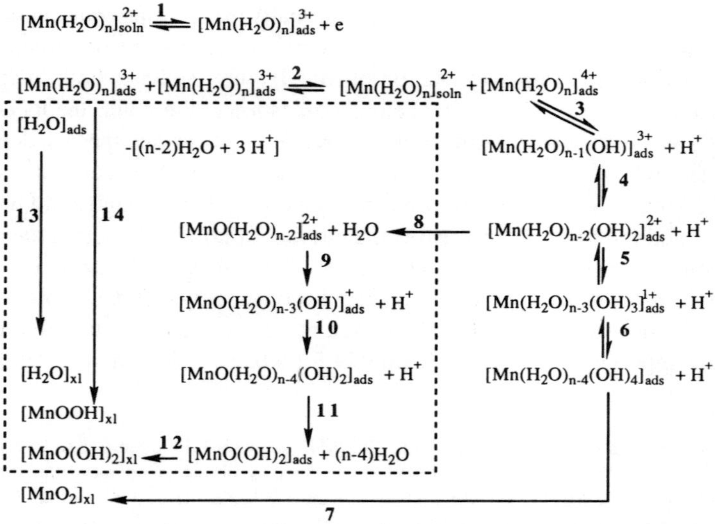

$$[Mn(H_2O)_n]^{2+}_{soln} \xrightleftharpoons{1} [Mn(H_2O)_n]^{3+}_{ads} + e$$

$$[Mn(H_2O)_n]^{3+}_{ads} + [Mn(H_2O)_n]^{3+}_{ads} \xrightleftharpoons{2} [Mn(H_2O)_n]^{2+}_{soln} + [Mn(H_2O)_n]^{4+}_{ads}$$

$$\downarrow 3$$

$[H_2O]_{ads}$ $-[(n-2)H_2O + 3H^+]$ $[Mn(H_2O)_{n-1}(OH)]^{3+}_{ads} + H^+$

$\Updownarrow 4$

$13\downarrow$ $14\downarrow$ $[MnO(H_2O)_{n-2}]^{2+}_{ads} + H_2O \xleftarrow{8} [Mn(H_2O)_{n-2}(OH)_2]^{2+}_{ads} + H^+$

$\downarrow 9$ $\Updownarrow 5$

$[MnO(H_2O)_{n-3}(OH)]^+_{ads} + H^+$ $[Mn(H_2O)_{n-3}(OH)_3]^+_{ads} + H^+$

$\downarrow 10$ $\Updownarrow 6$

$[H_2O]_{xl}$ $[MnO(H_2O)_{n-4}(OH)_2]_{ads} + H^+$ $[Mn(H_2O)_{n-4}(OH)_4]_{ads} + H^+$

$[MnOOH]_{xl}$ $\downarrow 11$

$[MnO(OH)_2]_{xl} \xleftarrow{12} [MnO(OH)_2]_{ads} + (n-4)H_2O$

$[MnO_2]_{xl} \xleftarrow{\quad 7 \quad}$

Figure 17. MnO_2 deposition mechanism, showing steps proposed by Fleischmann et al.[246] (outside the dashed box) and additions by Preisler[76] (inside the dashed box).

(2) disproportionation of adsorbed Mn^{3+} ions to form Mn^{2+} ions and adsorbed Mn^{4+} ions; and (3) deprotonation of the adsorbed Mn^{4+} ions in several sequential steps (i.e., steps 3–6). The above mechanism[246] allowed for dissolution of adsorbed Mn^{4+} ions and reduction of adsorbed Mn(IV) species. These steps are not shown in Fig. 17, as they are not germane to the review given here; however, step 8 in Fig. 17 allowed for dissolution of adsorbed Mn^{4+} ions.

The above mechanism satisfies several diagnostic criteria determined by the investigators:

(A) The growth rate constant, k', increased with anodic potential by an amount $dE/d \log k = 30$ mV at low overpotentials and then leveled off with further increase in potential.

(B) The growth rate increased with Mn^{2+}-ion concentration at low overpotentials.

(C) The growth rate decreased with acidity at low overvoltages, according to the reaction order $[OH^-]^4$.

Facts (A) and (B) were rationalized by simplifying the mechanism so as to consider only one adsorbed Mn(IV) species which adsorbs according to a Langmuir isotherm, the surface coverage being θ. Then, assuming that the rate of deprotonation is proportional to θ, the rate of crystal growth becomes

$$k = \frac{k'\theta K a_{Mn^{2+}} \exp(2EF/RT)}{1 + K a_{Mn^{2+}} \exp(2EF/RT)} \quad (59)$$

where k' and K are constants.

The reaction order with respect to OH^- ions was accounted for in the constant K, which includes four intermediate, successive deprotonation steps, steps 3–6.

Preisler,[76] taking advantage of a wealth of structural data developed subsequent to the work by Fleischmann *et al.*, extended the reaction scheme of the latter group to account for the incorporation of structural water (cation vacancies) and MnOOH in the γ/ε-MnO_2 crystal lattice. His extension is shown by reactions 9–14, which are inside the dashed box of Fig. 17.

Through the complete scheme of Fig. 17, Preisler qualitatively rationalized the changes in structure with deposition current density. At very low current densities, he projected low coverage of the surface with adsorbed

Mn(III), so that growth occurred freely to form the structure with minimum free energy—i.e., vacancy-free "$MnO_{2.00}$." As the (galvanostatic) deposition rate is increased, the surface concentration of Mn(IV) aquo complexes increases, which increases the production of $MnO(OH)_2$. The concentration of adsorbed Mn(III) species also increases, which would result in more reaction via step 14 and an increase in the amount of MnOOH in the solid. The greater surface coverage with all species fosters more nucleation, which in turn leads to more crystal imperfection and randomness of crystal growth. The imperfections allow sites for water molecules to adsorb, where they donate a proton to a lattice surface oxide ion and add to the cation vacancies.

Heavy-metal ions (i.e., Pb^{2+}, Me^{3+}, Cu^{2+}, Co^{2+}, Zn^{2+}, and Ni^{2+}), when present in the deposition bath, adsorb on the oxide surface, suppressing EMD growth and promoting defects in the solid.[253]

(iii) Mechanisms Involving Ionic Diffusion through Thin Films of Solid Intermediates

Cartwright and Paul[249,254] determined the rate-determining step during film growth to be diffusion of Mn^{2+} ions through a porous layer of reaction intermediates, possibly MnOOH. They developed their mechanism from the current response to potentiostatic steps[249,255] as well as impedance measurements at a rotating rod electrode.[249,254] The currents following potential steps decreased with time in the characteristic manner, reaching limiting values that were independent of time, rotation rate, and potential but that increased with temperature and Mn^{2+}-ion concentration. The dominant impedance component was a Warburg response that was independent of rotation rate and increased with overvoltage. The impedance at constant overvoltage was found to decrease significantly with increasing Mn^{2+}-ion concentration but only slightly with increasing H^+-ion concentration.

Two alternative mechanisms were proposed, which are represented by Eqs. (60) and (61).

$$Mn^{2+}_{soln} \rightarrow Mn^{2+}_{MnO_2} \quad \text{(diffusion, rds)} \quad (60a)$$

$$Mn^{2+}_{MnO_2} \rightarrow Mn^{3+} + e^- \quad \text{(fast)} \quad (60b)$$

$$Mn^{3+} + 2H_2O \rightarrow MnOOH + 3H^+ \quad (60c)$$

The Manganese Dioxide Electrode in Aqueous Solution

$$MnOOH \rightarrow MnO_2 + H^+ + e^- \quad (k) \quad (60d)$$

or

$$Mn^{2+}_{soln} \rightarrow Mn^{2+}_{MnO_2} \quad \text{(diffusion, rds)} \quad (61a)$$

$$Mn^{2+}_{MnO_2} \rightarrow Mn^{4+} + 2e^- \quad \text{(fast)} \quad (61b)$$

$$Mn^{4+} + 4H_2O \rightarrow Mn(OH)_4 + 4H^+ \quad (61c)$$

$$Mn(OH)_4 \rightarrow MnO_2 + 2H_2O \quad (k) \quad (61d)$$

The subscripts identify the phase. Both schemes involve an initial step, (a), of diffusion of Mn^{2+} ions through the MnOOH layer to reach the MnO_2 surface. Then two fast reactions, (b) and (c), form the solid diffusion barrier layer. Finally, step (d) converts this layer into MnO_2. The equations were formulated in a standard manner and solved. The final current–time expression for the scheme given by Eqs. (60) is

$$-\frac{i_\infty}{2i - i_\infty} - \ln\left[1 - \frac{i_\infty}{2i - i_\infty}\right] = \frac{i_\infty^2 M}{4F^2 A^2 \rho D C} t \quad (62)$$

where M and ρ are the molecular mass and density of the MnOOH layer, A is the electrode area, D is the diffusion coefficient of Mn^{2+} through the MnOOH layer, C is the Mn^{2+} concentration in the bulk solution, F is the Faraday constant, t is time, and i_∞, the steady-state current at infinite time, is given by

$$i_\infty = 2FAk \exp[(1 - \beta)FE/RT] = 2FADC/\delta_\infty \quad (63)$$

Here δ_∞ is the steady-state thickness of the MnOOH layer at infinite time, β is the electron-transfer symmetry factor, and E is the potential. An expression analogous to Eq. (62) was derived for the scheme given by Eqs. (61).

Cartwright and Paul[249] plotted the left-hand side of Eq. (62) against time, obtaining linearity from 50 to 5000 s. Deposition was controlled by both Mn^{2+} diffusion and charge transfer at $t < 50$ s, and the film was still growing. At $t > 5000$ s the rate-determining step was only Mn^{2+} diffusion in the film. However, direct measurement of i_∞ was not accurate enough for model evaluation, according to the investigators, since it drifted (perhaps from interference of IR). Therefore, they used a computerized technique to find the value of i_∞ that best fit Eq. (62). From this value, linear plots were obtained for Eq. (62) at various deposition conditions, which fit the experimental data at 50–5000 s nicely and produced values for D. Reproduction of i versus t decay curves with these calculated parameters superimposed the experimental results. The values of i_∞ ranged from 0.05 to 16 mA/cm^2, and D values ranged from 8×10^{-9} to 3.8×10^{-7} cm^2/s, depending on the Mn^{2+}- and H$^+$-ion concentrations and the temperature. The D values are about two orders of magnitude less than those for ionic aqueous diffusion. Values of δ ranged from 1 to 34 μm.

From their stationary and rotating platinum/platinum ring-disk electrode studies, Kao and Weibel[248] proposed the following mechanism for EMD growth:

$$Mn^{2+}{}_{(soln)} + MnO_2 + 2H_2O \rightarrow 2MnOOH + 2H^+{}_{(s)} \quad (64a)$$

$$MnOOH \rightarrow MnO_2 + H^+{}_{(s)} + e^- \quad (64b)$$

$$H^+{}_{(s)} \rightarrow H^+{}_{(soln)} \quad (64c)$$

This chemical–electrochemical process starts with Mn^{2+} attack on the MnO$_2$ surface to form an MnOOH intermediate. (This first step was also proposed by various earlier investigators.) As in the case of nucleation, if the Mn^{2+}-ion concentration is very low, i.e., <0.02M, and the electrode rotation rate nominally low (<2000 rpm), then diffusion of Mn^{2+} ions from the bulk electrolyte to the MnO$_2$/electrolyte interface determines the growth rate of MnO$_2$. However, with a sufficient supply of Mn^{2+} ions, as is the case in any practical deposition operation, the surface remains covered with the MnOOH intermediate, and the rate is determined by electrochemical oxidation of the intermediates to MnO$_2$ (Eq. 64b) and the diffusion within the solid of the hydrogen ions that are released simultaneously (Eq. 64c).

Kao and Weibel put forward as support for this mechanism several test results in addition to the independence of the limiting current to disk rotation rate and the decrease of current with time at constant or positively scanning potential. In particular, they demonstrated the presence of both steps (64b) and (64c) by decoupling them as follows: EMD films (0.15

μm thick) were deposited, rinsed, soaked at open circuit in $MnSO_4$–H_2SO_4 solutions for various times, transferred to H_2SO_4 solutions without $MnSO_4$, and there anodized at typical deposition potentials. Finally, the relative amounts of MnO_2 in the treated films were measured by cathodically scanning the electrodes in these same H_2SO_4 solutions. (The chemistry of MnO_2 electrochemical reduction in $MnSO_4$-H_2SO_4 solutions is reported in Refs. 248, 247, and 256.)

Results of the above experiments showed that after the soak–anodization process the MnO_2 film was about 40% thicker than the original film—i.e., this process caused an increase in film thickness of about 0.06 μm. Additional soak time resulted in a decrease in the added film, as acid decomposed MnOOH, but not to less than the original film thickness.

These tests were used not only to demonstrate the part reactions but also to help determine the thickness of the layer of intermediate. By using an independently determined diffusion coefficient of H^+ in MnO_2, together with the limiting current density, and making some assumptions regarding concentrations of protons in the lattice, these investigators calculated the thickness of the MnOOH layer. They obtained values of the order of 0.02 μm, which agree with thicknesses obtained by integrating film growth up to the limiting current during anodic cyclic voltammetric scans and by the soak–anodization experiments. Their values are two to three orders of magnitude less than the thicknesses determined by Cartwright and Paul,[254] but the differences may be partially accounted for by comparably large differences in diffusion coefficients; i.e., Kao and Weibel determined that the diffusion coefficient for H^+ in MnO_2 is 2×10^{-12} cm^2/s at 90°C.

Other evidence has been cited for the existence of a thin layer of intermediate MnOOH, e.g., from surface-enhanced Raman spectroscopy applied to MnO_2 reduction.[256]

(iv) Other Mechanisms

Heusler and Grzegorzewski[250,257] studied MnO_2 deposition by applying pulse polarization techniques at a rotating disk electrode, which was one electrode of a rotating quartz frequency balance, and measuring the flux of hydrogen ions by means of a ring electrode. They considered a mechanism in which Mn^{2+} ions and O^{2-} ions cross the interface in parallel. O^{2-} ions are formed in a rapid dissociation of H_2O molecules.

From their results they concluded that (1) charge transfer of Mn^{2+} ions and that of O^{2-} ions are statistically independent steps, (2) the exchange current of Mn^{2+} ions is much greater than that of O^{2-} ions, (3) Mn^{2+} ions are transferred without water molecules of hydration, (4) H_2O molecules are transferred to the oxide by the O^{2-} ions, and (5) proton transfer between the oxide and the electrolyte proceeds at a negligible rate.

These investigators applied their model to steady-state MnO_2 deposition curves (at current densities, i, less than the limiting current), obtaining the expression

$$i = i_0^o \frac{(r-1)}{r}\left([Mn^{2+}]^{1/r}[OH^-]^2 \cdot \exp\frac{2(r-1)FE}{rRT} - 1\right) \quad (65)$$

Here i_0^o is the exchange current density of O^{2-}-ion transfer; r is the state of Mn oxidation (MnO_r), which, in the steady state, is the ratio of the currents for oxide transfer and manganese transfer; the ion activities are indicated by square brackets, and the other terms have their usual significance. The OH^- activity results from the equilibrium between O^{2-} ions and H_2O/H^+ ions. The deposition conditions in this study[250,257] were quite different from those of commercial EMD production; i.e., $T = 40°C$ rather than ~95°C, pH = 4.1–4.85 rather than ~$0.5M$ H_2SO_4, and Mn^{2+} concentration of 5–250 mM rather than ~500 mM. Nevertheless, Eq. (65) rationalized the Tafel slope of 68 mV (40°C and $r = 1.84$), the reaction order of 2 with respect to OH^- ions, and the reaction order of ~0.54 with respect to Mn^{2+} ions. Heusler and Grzegorzewski attributed the limiting current with high anodic overpotential to the surface reaching limiting coverage with O^{2-} ions, through either mutual repulsion or monolayer coverage. Structural water is built into the lattice through transport from the electrolyte via O^{2-} ions, and MnOOH in the lattice results from the different O^{2-} and Mn^{2+} currents.

ACKNOWLEDGMENT

The painstaking efforts of Marie A. Bufford in preparing a quality manuscript and many of the figures is gratefully acknowledged.

REFERENCES

[1] D. Linden, in *Handbook of Batteries*, 2nd ed., Ed. by D. Linden, McGraw-Hill, New York, 1995, Chapters 6 and 7.
[2] *JEC Battery Newsletter*, No. 6, 1992, p. 20; *JEC Battery Newsletter*, No. 2, 1993; *JEC Battery Newsletter*, No. 3, 1993.

[3] A. J. Salkind, in *Modern Battery Technology*, Ed. by C. D. S. Tuck, Ellis Horwood, New York, 1991, Chapter 1.

[4] R. Huber, in *Batteries, Vol. 1, Manganese Dioxide*, Ed. by K. V. Kordesch, Marcel Dekker, New York, 1974, Chapter 1.

[5] K. V. Kordesch, in *Batteries, Vol. 1, Manganese Dioxide*, Ed. by K. V. Kordesch, Marcel Dekker, New York, 1974, Chapter 2.

[6] K. V. Kordesch, in *Comprehensive Treatise of Electrochemistry*, Vol. 3, Ed. by J. O'M. Bockris, B. E. Conway, E. Yeager, and R. E. White, Plenum Press, New York, 1981, Chapter 6.

[7] H. S. Wroblowa, in *Electrochemistry in Transition from the 20th to the 21st Century*, Ed. by O. J. Murphy, S. Srinivasan, and B. E. Conway, Plenum Press, New York, 1992, p. 147.

[8] K.-J. Euler, *J. Power Sources* **8** (1982) 133.

[9] R. F. Wohletz, E. M. Spore, M. P. Grotheer, and S. F. Burkhardt, in *Proceedings of the Symposium on Quality Management in Industrial Electrochemistry*, Ed. by D. Hall, Y. Kondo, and H. Kawamoto, The Electrochemical Society, Pennington, New Jersey, 1993, p. 49.

[10] R. Giovanoli, *Chimia* **23** (1969) 470.

[11] R. G. Burns and V. M. Burns, in *Manganese Dioxide Symposium, Tokyo, 1980*, Vol. 2, Ed. by B. Schumm, Jr., H. M. Joseph, and A. Kozawa, I.C. MnO_2 Sample Office, Cleveland, Ohio, 1981, p. 97.

[12] D. Glover, B. Schumm, Jr., and A Kozawa, eds., *Handbook of Manganese Dioxides—Battery Grade*, International Battery Material Association, Cleveland, Ohio, 1989.

[13] K. Franzese and N. Bharucha, in *Handbook of Batteries and Fuel Cells*, Ed. by D. Linden, McGraw-Hill, New York, 1984, Chapter 5.

[14] K. V. Kordesch, in *Handbook of Batteries and Fuel Cells*, Ed. by D. Linden, McGraw-Hill, New York, 1984, Chapter 7.

[15] C. A. Vincent, *Modern Batteries*, Edward Arnold, London, 1984.

[16] B. Schumm, Jr., in *Modern Battery Technology*, Ed. by C. D. S. Tuck, Ellis Horwood, New York, 1991, Chapter 3.

[17] J. C. Hunter, in *Modern Battery Technology*, Ed. by C. D. S. Tuck, Ellis Horwood, New York, 1991, Chapter 3.

[18] D. W. McComsey and W. B. Felegyhazi, in *Handbook of Batteries*, 2nd ed., Ed. by D. Linden, McGraw-Hill, New York, 1995, Chapter 9.

[19] R. F. Scarr and J. C. Hunter, in *Handbook of Batteries*, 2nd ed., Ed. by D. Linden, McGraw-Hill, New York, 1995, Chapter 10.

[20] A. Kozawa, in *Batteries, Vol. 1, Manganese Dioxide*, Ed. by K. V. Kordesch, Marcel Dekker, New York, 1974, Chapter 3.

[21] C. C. Liang, in *Encyclopedia of Electrochemistry of the Elements*, Vol. I, Ed. by A. J. Bard, Marcel Dekker, New York, 1973, p. 377.

[22] J. B. Fernandes, B. D. Desai, and V. N. Kamat Dalal, *J. Power Sources* **15** (1985) 209.

[23] B. D. Desai, J. B. Fernandes, and V. N. Kamat Dalal, *J. Power Sources* **16** (1985) 1.

[24] Y. Chabre and J. Pannetier, *Prog. Solid State Chem.* **23** (1995) 1.

[25] J. Prabhaker Rethinaraj and S. Visvanathan, *J. Power Sources* **42** (1993) 335.

[26] B. E. Conway, in *The Fourth International Rechargeable Battery Seminar*, held March 2–4, 1992, sponsored by S. P. Wolsky and N. Marincic, Shawmco, Inc., Tulsa, Oklahoma, 1992.

[27] A. Kozawa and R. J. Brodd, eds., *Manganese Dioxide Symposium, Vol. 1, Cleveland*, I.C. Sample Office, Cleveland, Ohio, 1975.

[28] B. Schumm, Jr., H. M. Joseph, and A. Kozawa, eds., *Manganese Dioxide Symposium, Vol. 2, Tokyo*, I.C. MnO_2 Sample Office, Cleveland, Ohio, 1981.

[29] A. Kozawa and M. Nagayama, eds., *Battery Material Symposium, Vol. 1, Brussels*, International Battery Material Association, Cleveland, Ohio, 1983.

[30] K. V. Kordesch and A. Kozawa, *The 2nd Battery Material Symposium (The 3rd MnO$_2$ Symposium), Vol. 2, Graz, 1985*, International Battery Material Association, Cleveland, Ohio, 1985.
[31] B. Schumm, Jr., M. P. Grotheer, R. L. Middaugh, and J. C. Hunter, eds., *Manganese Dioxide Electrode Theory and Practice for Electrochemical Applications*, The Electrochemical Society, Pennington, New Jersey, 1985.
[32] P. J. Spellman, D. M. Larsen, R. J. Ekern, and J. E. Oxley, in *Handbook of Batteries*, 2nd ed., Ed. by D. Linden, McGraw-Hill, New York, 1995, Chapter 9.
[33] A. Kozawa and J. F. Yeager, *J. Electrochem. Soc.* **112** (1965) 959.
[34] A. Kozawa and R. A. Powers, *J. Electrochem. Soc.* **113** (1966) 870.
[35] A. Kozawa, in *The 2nd Battery Material Symposium (The 3rd MnO$_2$ Symposium), Vol. 2, Graz, 1985*, Ed. by K. V. Kordesch and A. Kozawa, International Battery Material Association, Cleveland, Ohio, 1985, p. 545.
[36] A. Kozawa and R. A. Powers, *J. Electrochem. Soc.* **115** (1968) 122.
[37] K. J. Vetter, *Z. Elektrochem.* **66** (1962) 577.
[38] A. Kozawa, in *Manganese Dioxide Symposium, Vol. 2, Tokyo, 1980*, Ed. by B. Schumm, Jr., Helen N. Joseph, and A. Kozawa, I.C. MnO$_2$ Sample Office, Cleveland, Ohio, 1981, p. 321.
[39] A. Kozawa and R. A. Powers, *Electrochem. Technol.* **5** (1967) 535.
[40] G. S. Bell and R. Huber, *J. Electrochem. Soc.* **111** (1964) 1.
[41] J. P. Brenet, *C. R. Acad. Sci. Paris* **242** (1956) 3064.
[42] C. Mondoloni, M. Laborde, J. Rioux, E. Andoni, and C. Levy-Clement, *J. Electrochem. Soc.* **139** (1992) 954.
[43] J. B. Goodenough, in *Manganese Dioxide Electrode Electrode Theory and Practice for Electrochemical Applications*, Ed. by B. Schumm, Jr., M. P. Grotheer, R. L. Middaugh, and J. C. Hunter, The Electrochemical Society, Pennington, New Jersey, 1985, p. 77.
[44] P. Ruetschi, *J. Electrochem. Soc.* **123** (1976) 495.
[45] A. Kozawa, in *Power Sources 7*, Ed. by J. Thompson, Academic Press, New York, 1979, p. 485.
[46] S. Yoshizawa, in *Manganese Dioxide Symposium, Vol. 2, Tokyo, 1980*, Ed. by B. Schumm, Jr., H. M. Joseph, and A. Kozawa, I.C. Sample Office, Cleveland, Ohio, 1981, p. 1.
[47] J. J. Coleman, *Trans. Electrochem. Soc.* **90** (1946) 545.
[48] A. B. Scott, *J. Electrochem. Soc.* **107** (1960) 941.
[49] F. Kornfeil, *J. Electrochem. Soc.* **109** (1962) 349.
[50] P. Brouillet, A. Grund, F. Jolas, and R. Mellet, in *Batteries 2* (Proceedings of the 4th International Battery Symposium, Brighton, September 1964), Ed. by D. H. Collins, Pergamon, New York, 1965, p. 189.
[51] A. Era, Z. Takehara, and S. Yoshizawa, *Electrochim. Acta* **12** (1967) 1199.
[52] J. P. Gabano, J. Seguret, and J. F. Laurent, *J. Electrochem. Soc.* **117** (1970) 147.
[53] W. J. Wruck, Ph.D. Thesis, University of Wisconsin, Madison, p. 1984.
[54] Z. Hong, C. Zhenhai, and Xia Xi, *J. Electrochem. Soc.* **136** (1989) 2771.
[55] A. Kozawa and J. F. Yeager, *J. Electrochem. Soc.* **115** (1968) 1003.
[56] H. Bode and A. Schmier, in *Batteries*, Pergamon Press, Oxford, 1963, p. 329.
[57] J. McBreen, Electrochim. Acta **20** (1995) 221.
[58] L. E. DeVries, in *Battery Material Symposium, Vol. 1, Brussels 1983*, Ed. by A. Kozawa and M. Nagayama, International Battery Material Association, Cleveland, Ohio, 1983, p. 53.
[59] Eveready Battery Engineering Data, Carbon Zinc, Vol. III, Union Carbide Corporation, Battery Products Division, Danbury, Connecticut, 1984.
[60] Y. Uetani, T. Iwamaru, and Y. Ishikawa, in *Manganese Dioxide Symposium, Vol. 2, Tokyo, 1980*, Ed. by B. Schumm, Jr., H. M. Joseph, and A. Kozawa, I.C. MnO$_2$ Sample Office, Cleveland, Ohio, 1981, p. 505.

[61] Y. Uetani, H. Sasama, and T. Iwamaru, in *Manganese Dioxide Electrode Theory and Practice for Electrochemical Applications*, Ed. by B. Schumm, Jr., M. P. Grotheer, R. L. Middaugh, and J. C. Hunter, The Electrochemical Society, Pennington, New Jersey, 1985, p. 475.
[62] T. Ohzuku, H. Watanabe, and T. Hirai, in *The 2nd Battery Material Symposium (The 3rd MnO_2 Symposium), Vol. 2, Graz, 1985*, Ed. by K. V. Kordesch and A. Kozawa, International Battery Material Association, Cleveland, Ohio, 1985, p. 145.
[63] W. C. Vosburgh, *J. Electrochem. Soc.* **106** (1959) 839.
[64] A. Kozawa, in *The 2nd Battery Material Symposium (The 3rd MnO_2 Symposium), Vol. 2, Graz, 1985*, Ed. by K. V. Kordesch and A. Kozawa, International Battery Material Association, Cleveland, Ohio, 1985, p. 5.
[65] M. P. Korver, R. S. Johnson, and N. C. Cahoon, *J. Electrochem. Soc.* **107** (1960) 587.
[66] H. B. Mark and W. C. Vosburgh, *J. Electrochem. Soc.* **208** (1961) 615.
[67] J. Caudle, K. G. Summer, and F. L. Tye, *J. Chem. Soc., Faraday Trans 1*, **69** (1973) 876.
[68] J. P. Gabano and J. P. Brenet, *Electrochim. Acta* **1** (1959) 242.
[69] P. Benson, W. B. Price, and F. L. Tye, *Electrochem. Technol.* **5** (1967) 517.
[70] K. J. Vetter and N. Jaeger, *Electrochim. Acta* **11** (1966) 401.
[71] D. G. Malpas and F. L. Tye, in *Handbook of Manganese Dioxides, Battery Grade*, Ed. by D. Glover, B. Schumm, Jr., and A Kozawa, The International Battery Material Association, Cleveland, Ohio, 1989, p. 176.
[72] P. M. De Wolff, *Acta Crystallogr.* **12** (1959) 341.
[73] D. A. J. Swinkels, in *Progress in Batteries & Battery Materials*, Vol. 9, Ed. by F. L. Tye, JEC Press, Brunswick, Ohio, 1990, p. 9.
[74] P. M. De Wolff, J. W. Visser, R. Giovanoli, and R. Brutsch, *Chimia* **32** (1978) 257.
[75] E. Preisler, in Progress in *Batteries & Battery Materials*, Vol. 10, Ed. by B. Schumm, Jr. and J. C. Nardi, JEC Press, Brunswick, Ohio, 1991, p. 1.
[76] E. Preisler, in *The 2nd Battery Material Symposium (The 3rd MnO_2 Symposium) Vol. 2, Graz, 1985*, Ed. by K. V. Kordesch and A. Kozawa, International Battery Material Association, Cleveland, Ohio, 1985, p. 247.
[77] T. N. Andersen, in *Progress in Batteries & Battery Materials*, Vol. 11, Ed. by D. A. J. Swinkels, ITE-JEC Press, Brunswick, Ohio, 1992, p. 105.
[78] R. P. Williams, R. A Fredlein, G. A. Lawrance, D. A. J. Swinkels, and C. B. Ward, in *Progress in Batteries & Battery Materials*, Vol. 13, Ed. by J. C. Nardi, ITE-JEC Press, Brunswick, Ohio, 1994, p. 102.
[79] E. Preisler, *J. Appl. Electrochem.* **19** (1989) 540.
[80] R. G. Burns and V. M. Burns, in *Manganese Dioxide Symposium, Vol. 1, Cleveland, 1975*, Ed. by A. Kozawa and R. J. Brodd, I.C. Sample Office, Cleveland, Ohio, 1975, pp. 288, 306; *Manganese Dioxide Symposium, Vol. 2, Tokyo*, Ed. by B. Schumm, Jr., H. M. Joseph, and A. Kozawa, I.C. MnO_2 Sample Office, Cleveland, Ohio, 1980, p. 97.
[81] R. G. Burns, in *Battery Material Symposium, Vol. 1, Brussels, 1983*, Ed. by A. Kozawa and M. Nagayama, International Battery Material Association, Cleveland, Ohio, 1983, p. 341.
[82] A. M. Bystrom, *Acta Chem. Scand.* **3** (1949) 163.
[83] H. Strunz, *Naturwissenschaften* **31** (1943) 89.
[84] F. V. Chukhrov, A. N. Gorchov, A. V. Sivtsov, Y. P. Dikov, and V. V. Berezovskaya, *Izv. Akad. Nauk SSSR Ser. Geol.* **1982** (1) 56.
[85] K. M. Parida, S. B. Kanungo, and B. R. Sant, *Electrochim. Acta* **26** (1981) 435, 1147.
[86] W. F. Nye, S. B. Levin, and H. H. Kedesdy, Proceedings of the 13th Annual Power Sources Conference, Atlantic City, New Jersey, 1959, p. 125.
[87] G. M. Clark, *The Structure of Non-Molecular Solids*, Applied Science, Barking, U.K., 1972.
[88] M. Voinov, *Electrochim. Acta* **27** (1982) 833.

[89] S. Turner and P. R. Buseck, *Nature (London)* **304** (1983) 143.
[90] J. C. Charenton and P. Strobel, *J. Solid State Chem.* **77** (1983) 33.
[91] W. C. Maskell, J. W. A. Shaw, and F. L. Tye, *Electrochim. Acta* **26** (1981) 1403.
[92] M. Ripert, J. Pannetier, Y. Chabre, and C. Poinsignon, *Mater. Res. Soc. Symp. Proc. Ser.* **210** (1991) 359.
[92a] E. Preisler, in *Manganese Dioxide Symposium, Vol. 2, Tokyo, 1980*, Ed. by B. Schumm, Jr., H. M. Joseph, and A. Kozawa, I.C. MnO_2 Sample Office, 1981, p. 184.
[93] J. Pannetier, *Progress in Batteries & Battery Materials*, Vol. 11, Ed. by D. A. J. Swinkels, ITE-JEC Press, Brunswick, Ohio, 1992, p. 51.
[94] S. Turner and P. R. Buseck, *Science* **203** (1979) 456.
[95] P. Ruetschi, *J. Electrochem. Soc.* **131** (1984) 2737.
[96] A. Kozawa, *J. Electrochem. Soc.* **106** (1959) 79.
[97] A. Tvarusko, *J. Electrochem. Soc.* **111** (1964) 125.
[98] D. S. Freeman, P. F. Pelter, F. L. Tye, and L. L. Wood, *J. Appl. Electrochem.* **1** (1971) 127.
[99] H. Ikeda, T. Saito, and H. Tamura, in *Manganese Dioxide Symposium, Vol. 1, Cleveland, 1975*, Ed. by A. Kozawa and R. J. Brodd, I.C. Sample Office, Cleveland, Ohio, 1975, p. 384.
[100] R. Giovanoli, in *Manganese Dioxide Symposium, Vol. 2, Tokyo, 1980*, Ed. by B. Schumm, H. M. Joseph, and A. Kozawa, I.C. Sample Office, Cleveland, Ohio, 1980, p. 113.
[101] W. Feitknecht and W. Marti, *Helv. Chim. Acta* **208** (1945) 129.
[102] O. Glemser and H. Meisiek, *Naturwissenschaften* **44** (1957) 614.
[103] J. P. Gabano, B. Morignat, and J. F. Laurent, *Electrochem. Technol.* **5** (1967) 531.
[104] R. M. Potter and G. R. Rossman, *Am. Mineral.* **64** (1979) 1199.
[105] P. Brouillet, A. Grund, and F. Jolas, *C. R. Acad. Sci. Paris* **257** (1963) 3166.
[106] L. Balevski, J. Brenet, G. Coeffier and P. Lancon, *C. R. Acad. Sci. Paris* **260** (1965) 106.
[107] G. Coeffier and J. Brenet, *Electrochim. Acta* **10** (1965) 1013.
[108] J. P. Gabano, B. Morignat, E. Fialdes, B. Emery, and J. F. Laurent, *Z. Phys. Chem.* **46** (1965) 359.
[109] J. P. Brenet, M. Cyrankowska, G. Ritzler, R. Sada, and K. Traore, in *Manganese Dioxide Symposium, Vol. 1, Cleveland, 1975*, Ed. by A. Kozawa and R. J. Brodd, I.C. Sample Office, Cleveland, Ohio, 1975, pp. 276, 539.
[110] F. Freund, E. Konen, and E. Preisler, in *Manganese Dioxide Symposium, Vol. 1, Cleveland, 1975*, Ed. by A. Kozawa and R. J. Brodd, I.C. Sample Office, Cleveland, Ohio, 1975, p. 328.
[111] J. Koshiba and S. Nishizawa, in *Electrochemistry of Manganese Dioxide and Manganese Dioxide Batteries in Japan*, Vol. II, Ed by K. Takahashi, S. Yoshizawa, and A. Kozawa, U.S. Branch Office of the Electrochemical Society of Japan, Cleveland, Ohio, 1971, p. 85.
[112] T. N. Andersen and J. M. Derby, in *Electrochemistry in Transition from the 20th to the 21st Century*, Ed. by O. J. Murphy, S. Srinivasan *et al.*, Plenum Press, New York, 1992, p. 655.
[113] D. A. J. Swinkels and K. J. Doolan, in *Progress in Battery & Battery Materials*, Vol. 10, Ed. by B. Schumm, Jr. and J. C. Nardi, ITE-JEC Press, Brunswick, Ohio, 1991, p. 176; D. A. J. Swinkels, in *Progress in Battery Materials*, Vol. 11, Ed. by D. A. J. Swinkels, ITE-JEC Press, Brunswick, Ohio, 1992, p. 226.
[114] F. Fillaux, H. Ouboumour, J. Tomkinson, and L. T. Yu, *Chem. Phys.* **149** (1991) 459.
[115] F. Fillaux, H. Ouboumour, C. Cachet, J. Tomkinson, G. J. Kearley, and L. T. Yu, *Chem. Phys.* **164** (1992) 311.
[116] F. Fillaux, C. H. Cachet, H. Ouboumour, J. Tomkinson, C. Le'vy-Cle'ment, and L. T. Yu, *J. Electrochem. Soc.* **140** (1993) 585.
[117] P. Ruetschi, *J. Electrochem. Soc.* **135** (1988) 2657.

[118] D. M. Holton and F. L. Tye, in *Manganese Dioxide Symposium, Vol. 2, Tokyo, 1980*, Ed. by B. Schumm, H. M. Joseph, and A. Kozawa, I.C. Sample Office, Cleveland, Ohio, 1980, p. 244.
[119] A. J. Brown, F. L. Tye, and L. L. Wood, *J. Electroanal. Chem.* **122** (1981) 337.
[120] P. Ruetschi and R. Giovanoli, *J. Electrochem. Soc.* **135** (1988) 2663.
[121] E. Preisler, *J. Appl. Electrochem.* **6** (1976) 311.
[122] S. Brunauer, P. H. Emmett, and E. Teller, *J. Am. Chem. Soc.* **60** (1938) 309.
[123] S. J. Gregg and K. S. W. Sing, *Adsorption, Surface Area and Porosity*, Academic Press, New York, 1967.
[124] A. Kozawa, *J. Electrochem. Soc.* **106** (1959) 552.
[125] T. N. Andersen and R. G. Moody, in *Progress in Batteries & Battery Materials*, Vol. 13, Ed. by J. C. Nardi, ITE-JEC Press, Brunswick, Ohio, 1994, p. 1.
[126] C. St. Claire-Smith, J. A. Lee, and F. L. Tye, in *Manganese Dioxide Symposium Proceedings, Vol. 1, Cleveland, 1975*, Ed. by A. Kozawa and R. J. Brodd, Cleveland, Ohio, 1975, p. 132.
[127] J. A. Lee, C. E. Newnham, F. S. Stone, and F. L. Tye, *J. Colloid Interface Sci.* **45** (1973) 289; J. A. Lee and C. E. Newnham, *J. Colloid Interface Sci.* **56** (1976) 391.
[128] M. A. Dakri, F. L. Tye, and J. L. Whiteman, in *Power Sources 1966*, Ed. by D. H. Collins, Pergamon Press, New York, 1967, p. 65.
[129] M. R. Tarasevich, E. J. Shkol'nikov, V. E. Kazarinov, L. P. Esayan, Z. M. Buzova, B. Gurvitz, and N. I. Hoteeva, in *Progress in Batteries & Solar Cells*, Vol. 7, Ed. by M. Nagayama and A. Kozawa, JEC Press and IBA, Cleveland, Ohio, 1988, p. 49.
[130] D. A. J. Swinkels, N. Bristow, and R. Williams, in *Progress in Batteries & Battery Materials*, Vol. 13, Ed. by J. C. Nardi, ITE-JEC Press, Brunswick, Ohio, 1994, p. 12.
[131] A. Kozawa, in *Manganese Dioxide Symposium, Vol. 1, Cleveland, 1975*, Ed. by A. Kozawa and R. J. Brodd, I.C. Sample Office, Cleveland, Ohio, 1975, p. 470.
[132] J. Koshiba, in *Progress in Batteries & Solar Cells*, Vol. 7, Ed. by M. Nagayama and A. Kozawa, JEC Press and IBA, Cleveland, Ohio, 1988, p. 38.
[133] A. J. Brown, C. R. St. Claire Smith, F. L. Tye, and J. L. Whiteman, *J. Colloid Interface Soc.* **51** (1975) 516.
[134] D. A. J. Swinkels, N. Bristow, and L. Gale, in *Progress in Batteries & Battery Materials*, Vol. 13, Ed. by J. C. Nardi, ITE-JEC Press, Brunswick, Ohio, 1994, p. 22.
[135] R. P. Williams, T. N. Andersen, N. J. Bristow, A. R. Gee, G. A. Lawrance, and D. A. J. Swinkels, Paper presented at the 9th IBA Battery Materials Symposium, Cape Town, South Africa, March 20–22, 1995.
[136] D. A. J. Swinkels and K. Hall, in *Progress in Batteries & Battery Materials*, Vol. 11, Ed. by D. A. J. Swinkels, ITE-JEC Press, Brunswick, Ohio, 1992, p. 16.
[137] E. Preisler, in *Progress in Batteries & Battery Materials*, Vol. 10, Ed. by B. Schumm, Jr. and J. C. Nardi, JEC Press, Brunswick, Ohio, 1991, p. 200.
[138] S. F. Burkhardt, in *Handbook of Manganese Dioxides, Battery Grade*, Ed. by D. Glover, B. Schumm, Jr., and A. Kozawa, IBA, Inc., Cleveland, Ohio, 1989, p. 217.
[139] S. F. Burkhardt, in *Progress in Batteries and Battery Materials*, Vol. 11, Ed. by D. A. J. Swinkels, ITE-JEC Press, Brunswick, Ohio, 1992, p. 136.
[140] A. Ohta in *Handbook of Manganese Dioxides, Battery Grade*, Ed. by D. Glover, B. Schumm, Jr., and A. Kozawa, IBA, Inc., Cleveland, Ohio, 1989, p. 237.
[141] B. Schumm, Jr., in *Handbook of Manganese Dioxides, Battery Grade*, Ed. by D. Glover, B. Schumm, Jr., and A. Kozawa, IBA, Inc., Cleveland, Ohio, 1989, p. 211.
[142] J. P. Brenet, *Chimia* **23** (1969) 444.
[143] T. Tsuchida, in *Manganese Dioxide Symposium, Vol. 1, Cleveland, 1975*, Ed. by A. Kozawa and R. J. Brodd, I.C. Sample Office, Cleveland, Ohio, 1975, p. 230.

[144] Y. Uetani, T. Togo, and T. Iwamaru, in *Manganese Dioxide Symposium, Vol. 1, Cleveland, 1975*, Ed. by A. Kozawa and R. J. Brodd, I.C. Sample Office, Cleveland, Ohio, 1975, p. 183.
[145] J. Watanabe, R. Furumi, and A. Ohta, in *Manganese Dioxide Symposium, Vol. 1, Cleveland, 1975*, Ed. by A. Kozawa and R. J. Brodd, I.C. Sample Office, Cleveland, Ohio, 1975, p. 159.
[146] N. C. Cahoon in *Primary Batteries*, Ed. by N. C. Cahoon and G. W. Heise, Vol. 2, John Wiley & Sons, New York, 1976, p. 68.
[147] S. B. Kanungo, K. M. Parida, and B. R. Sant, *Electrochim. Acta* **26** (1981) 1147.
[148] X. Xie, C. Zhenhai, Z. Peijiang, Z. Haikui, Z. Hong, S. Fang, and L. Hong, in *The 2nd Battery Material Symposium (The 3rd MnO_2 Symposium), Vol. 2, Graz, 1985*, Ed. by K. V. Kordesch and A. Kozawa, International Battery Material Association, Cleveland, Ohio, 1985, p. 501.
[149] J. B. Fernandes, B. D. Desai, and V. N. Kamat Dalal, *Electrochim. Acta* **29** (1984) 181 and 187.
[150] V. M. Burns, R. G. Burns, and W. K. Zwicker, in *Manganese Dioxide Symposium, Vol. 1, Cleveland, 1975*, Ed. by A. Kozawa and R. J. Brodd, I.C. Sample Office, Cleveland, Ohio, 1975, p. 288.
[151] K. Miyazaki and B. Imada, in *The 2nd Battery Material Symposium (The 3rd MnO_2 Symposium), Vol. 2, Graz, 1985*, Ed. by K. V. Kordesch and A. Kozawa, International Battery Material Association, Cleveland, Ohio, 1985, p. 169.
[152] A. Kozawa, G. Kano, K. Horita, and Y. Takeuchi, in *Progress in Batteries & Solar Cells*, Vol. 7, Ed. by M. Nagayama and A. Kozawa, JEC Press, Cleveland, Ohio, 1988, p. 2.
[153] J. P. Brenet and P. Faber, ISE Batteries Symposium, Marcoussis, France, May 1975.
[154] D. Glover, B. Schumm, Jr., and A. Kozawa, eds., *Handbook of Manganese Dioxides, Battery Grade*, IBA, Inc., Cleveland, Ohio, 1989.
[155] W. J. Wruck and T. W. Chapman, The Electrochemical Society Extended Abstracts, Vol. 85-1, Toronto, Ontario, Canada, May 12–17, 1985, p. 759, Abstract 535.
[156] A. J. Wruck, B. Reichman, K. R. Bullock, and W.-H. Kao, *J. Electrochem. Soc.* **138** (1991) 3560.
[157] A. Kozawa, in *Comprehensive Treatise of Electrochemistry*, Vol. 3, Ed. by J. O'M. Bockris, B. E. Conway, E. Yeager, and R. E. White, Plenum Press, New York, 1981, Chapter 5.
[158] E. Preisler, in *Battery Material Symposium, Vol. 1, Brussels, 1983*, Ed. by A. Kozawa and M. Nagayama, International Battery Material Association, Cleveland, Ohio, 1984, p. 143.
[159] A Kozawa, in *Battery Material Symposium, Vol. 1, Brussels, 1983*, Ed. by A. Kozawa and M. Nagayama, International Battery Material Association, Cleveland, Ohio, 1984, p. 430.
[160] E. Preisler and G. Mietens, *DECHEMA Monogr.* **109** (1987) 123.
[161] L. Sen Gupta, in *Proceedings of the International Symposium on Electrometallurgical Plant Practice, Montreal, 1990*, Ed. by P. L. Claessens and G. B. Harris, Pergamon Press, Tarrytown, New York, 1990, p. 141.
[162] W-H. Kao, V. J. Weibel, and M. J. Root, *J. Electrochem. Soc.* **139** (1992) 1223.
[163] A. Agopsowicz and F. L. Tye, *J. Appl. Electrochem.* **1** (1971) 45.
[164] W-H. Kao, *J. Electrochem. Soc.* **135** (1988) 1317; **136** (1989) 13.
[165] C. Klingsberg and R. Roy, *Am. Mineral.* **44** (1959) 819.
[166] L. S. D. Glasser and I. B. Smith, *Mineral. Mag.* **36** (1968) 976.
[167] R. Giovanoli, K. Bernhard, and W. Feitknecht, *Helv. Chim. Acta* **51** (1968) 355.
[168] L. S. D. Glasser and L. Ingraam, *Acta Crystallogr.* **B24** (1968) 1233.
[169] W. C. Maskell, J. E. A. Shaw, and F. L. Tye, *J. Appl. Electrochem.* **12** (1982) 101.
[170] W. Feitknecht, H. R. Oswald, and U. Fetknecht-Steinmann, *Helv. Chim. Acta* **43** (1960) 1947.
[171] J. Fitzpatrick and F. L. Tye, *J. Appl. Electrochem.* **21** (1991) 130.

[172] J. F. Laurent and B. Morignat, in *Batteries, Proceedings of the 3rd International Symposium on Batteries, Bornemouth, October 1962*, Ed. by D. H. Collins, Pergamon Press, London 1963.
[173] J. P. Gabano, B. Morignat, and J. F. Laurent, *Electrochim. Acta* **9** (1964) 1093.
[174] J. P. Gabano, B. Morignat, E. Fialdes, B. Emery, and J. F. Laurent, *Z. Phys. Chem. N. F.* **46** (1965) 359.
[175] J. A. Lee, C. E. Newnham, F. L. Tye, and F. S. Stone, *J. Chem. Soc., Faraday Trans.* **I74** (1978) 237.
[176] S. Atlung, in *Manganese Dioxide Symposium, Vol. 1, Cleveland, 1975*, Ed. by A. Kozawa and R. J. Brodd, I.C. Sample Office, Cleveland, Ohio, 1975, p. 47.
[177] R. S. Johnson and W. C. Vosburgh, *J. Electrochem. Soc.* **100** (1953) 471.
[178] K. J. Vetter, *J. Electrochem. Soc.* **110** (1963) 597.
[179] S. Atlung and T. Jacobson, *Electrochim. Acta* **26** (1981) 1447.
[180] K. Neumann and E. V. Roda, *Z. Elektrochem. Ber. Bunsenges. Phys. Chem.* **69** (1965) 347.
[181] F. L. Tye, *Electrochim. Acta* **21** (1976) 415.
[182] R. P. Williams, A. R. Gee, G. A. Lawrance, and D. A. J. Swinkels, Paper presented at 9th IBA Battery Materials Symposium, Cape Town, South Africa, March 20–22, 1995.
[183] W. C. Maskell, J. E. A. Shaw, and F. L. Tye, *Electrochim. Acta* **28** (1983) 225, 231.
[184] F. L. Tye, *Electrochim. Acta* **30** (1985) 17.
[185] C. Maskell, J. E. A. Shaw, and F. L. Tye, *Electrochim. Acta* **27** (1982) 425.
[186] W. C. Maskell, J. E. A. Shaw, and F. L. Tye, *J. Power Sources* **8** (1982) 113.
[187] F. L. Tye and S. W. Tye, *J. Appl. Electrochem.* **25** (1995) 425.
[188] B. D. Desai, R. A. S. Dhume, and V. N. Kamat Dalal, *J. Appl. Electrochem.* **18** (1988) 62.
[189] D. M. Holton, W. C. Maskell, and F. L. Tye, in *Power Sources 10*, Ed. by L. Pearce, The Paul Press Ltd., London, 1985, p. 247.
[190] M. A. Malati, M. W. Rophael, and I. I. Bhayat, *Electrochim. Acta* **26** (1981) 239.
[191] H. Kahil, F. Dalard, J. Guitton, and J. P. Cogen-Addad, *Surf. Technol.* **16** (1982) 331.
[192] J. S. Newman, *Electrochemical Systems*, Prentice-Hall, Englewood Cliffs, New Jersey, 1973.
[193] J. S. Newman and W. Tiedemann, *AIChE J.* **21** (1975) 25.
[194] S. Atlung and K. West, in *Battery Material Symposium, Vol. 1, Brussels, 1983*, Ed. by A. Kozawa and M. Nagayama, International Battery Material Association, Cleveland, Ohio, 1984, p. 31.
[195] T. Valand, *Electrochim. Acta* **19** (1974) 639.
[196] C. Y. Mak, H. Y. Cheh, G. S. Kelsey, and P. Chalilpoyil, *J. Electrochem. Soc.* **138** (1991) 1607, 1611.
[197] J-S. Chen and H. Y. Cheh, *J. Electrochem. Soc.* **140** (1993) 1205, 1213.
[198] E. J. Podlaha and H. Y. Cheh, *J. Electrochem. Soc.* **141** (1994) 15, 28, 1751.
[199] J. P. Brenet, in *The 2nd Battery Material Symposium (The 3rd MnO_2 Symposium), Vol. 2, Graz, 1985*, Ed. by K. V. Kowdesch and A. Kozawa, International Battery Material Association, Cleveland, Ohio, 1985, p. 45.
[200] J. Brenet and P. C. Picquet, in *Manganese Dioxide Electrode Theory and Practice for Electrochemical Applications*, Ed. by B. Schumm, Jr., M. P. Grotheer, R. L. Middaugh, and J. C. Hunter, The Electrochemical Society, Pennington, New Jersey, 1985, p. 3.
[201] J. Brenet and P. C. Picquet, in *Progress in Batteries & Battery Materials*, Vol. 13, Ed. by J. C. Nardi, ITE-JEC Press, Brunswick, Ohio, 1994, p. 136.
[202] J. McBreen, *Power Sources* **5** (1975) 525.
[203] D. A. J. Swinkels, K. E. Anthony, P. M. Fredericks, and P. R. Osborn, *J. Electroanal. Chem.* **168** (1984) 433.
[204] Y. Chabre, *J. Electrochem. Soc.* **138** (1991) 329.
[205] Y. Chabre and M. Ripert, *Mater. Res. Soc. Symp. Proc.* **210** (1991) 367.

[206] Y. Chabre, in *Chemical Physics of Intercalation II*, Ed. by P. Bernier et al., Plenum Press, New York, 1993, p. 186

[207] C. Poinsignon, M. Amarilla, and F. Tedjar, in *Progress in Batteries & Battery Materials*, Vol. 13, Ed. by J. C. Nardi, ITE-JEC Press, Brunswick, Ohio, 1994, p. 113.

[208] M. Ripert, Thesis, INPG, Grenoble, 1990.

[209] M. Ripert, C. Poinsignon, Y. Chabre, and J. Pannetier, *Phase Transitions* 2/4 (1991) 205.

[210] J. P. Brenet, *Electrochim. Acta* 1 (1959) 231.

[211] H. Bode and A. Schmier, *Chem.-Ing.-Tech.* **38** (1966) 651.

[212] E. D. Gehain, P. C. Picquet, C. J. Spears, and F. L. Tye, in Progress in *Batteries & Battery Materials*, Vol. 13, Ed. by J. C. Nardi, ITE-JEC Press, Brunswick, Ohio, 1994, p. 62.

[213] S. Donne, R. Fredlein, G. Lawrance, D. A. J. Swinkels, and F. L. Tye, in *Progress in Batteries & Battery Materials*, Vol. 13, Ed. by J. C. Nardi, ITE-JEC Press, Brunswick, Ohio, 1994, p. 113.

[214] K. Kordesch and M. Weissenbacher, *J. Powet Sources* **51** (1994) 61.

[215] D. Boden, C. J. Venuto, D. Wisler, and R. B. Wylie, *J. Electrochem. Soc.* **114** (1967) 415; **115** (1968) 333.

[216] H. Y. Kang and Ch. C. Liang, *J. Electrochem. Soc.* **115** (1968) 6.

[217] T. W. Clapper, in *Encyclopedia of Chemical Processing and Design*, Vol. 29, Chief Editor: J. J. McKetta, Marcel Dekker, New York, 1988.

[218] H. Ouboumour, C. Cachet, M. Bode, and L. T. Yr, *J. Electrochem. Soc.* **142** (1995) 1061.

[219] D. T. Qu, B. E. Conway, L. Bai, Y. H. Zhou and W. A. Adams, *J. Appl. Electrochem.* **23** (1993) 693.

[220] Y. Amano et al., U.S. Patent 3,530,496 (1970).

[221] K. Kordesch, J. Gsellmann, M. Peri, K. Tomantschger, and R. Chemelli, *Electrochim. Acta* **26** (1981) 1485.

[222] Y. Sharma and K. Kordesch, in *Progress in Batteries & Battery Materials*, Vol. 11, Ed. by D. A. J. Swinkels, ITE-JEC Press, Brunswick, Ohio, 1992, p. 82.

[223] K. Kordesch, U.S. Patent 3,945,847 (1976).

[224] J. Gsellmann, W. Harer, K. Holzleithner, and K. Kordesch, in *Proceedings of the Symposium on Manganese Dioxide Electrode Theory and Practice for Electrochemical Applications*, Ed. by B. Schumm, Jr., M. P. Grotheer, R. L. Middaugh, and J. C. Hunter, The Electrochemical Society, Pennington, New Jersey, 1985, p. 567.

[225] K. Tomantschger and C. Michalowski, U.S. Patent 5,346,783 (1994).

[226] K. Kordesch and J. Daniel-Ivad, in *Progress in Batteries & Battery Materials*, Vol. 11, Ed. by D. A. J. Swinkels, ITE-JEC Press, Brunswick, Ohio, 1992, p. 70.

[227] K. Tomantschger, E. Oran, and K. Kordesch, U.S. Patent 5,162,169 (1992).

[228] K. Kordesch, L. Binder, and E. Kahraman, PCT Int. Appl. WO 92, 17, 910, Oct. 15, 1992.

[229] S. Fletcher, J. Galia, J. A. Hamilton, T. Tran, and R. Woods, in *Proceedings of the Symposium on Manganese Deoxide Electrode Theory and Practice for Electrochemical Applications*, Ed. by B. Schumm, Jr., M. P.Grotheer, R. L. Middaugh, and J. C. Hunter, The Electrochemical Society, Pennington, New Jersey, 1985, p. 556; *J. Electrochem. Soc.* **133** (1986) 1277.

[230] K. Kordesch, J. Gsellmann, W. Harer, W. Taucher, and K. Tomantschger, in *Progress in Batteries & Solar Cells*, Vol. 7, Ed. by M. Nagayama and A. Kozawa, JEC Press, Cleveland, Ohio, 1988, p. 194.

[231] Y. F. Yao, U.S. Patent 4,520,005 (1985).

[232] Y. F. Yao, N. Gupta, and H. S. Wroblowa, *J. Electroanal. Chem.* **233** (1987) 107; **238** (1989) 98.

[233] H. S. Wroblowa and N. Gupta, *J. Electroanal. Chem.* **238** (1987) 93.

[234] M. A. Dzieciuch, N. Gupta, and H. S.Wroblowa, *J. Electrochem. Soc.* **135** (1988) 2415.

[235] L. Bai, D. Y. Qu, B. E. Conway, Y. H. Zhou, G. Chowdhury, and W. A. Adams, *J. Electrochem. Soc.* **140** (1993) 884.

[236] C. G. Castledine and B. E. Conway, *J. Appl. Electrochem.* **25** (1995) 707.
[237] B. E. Conway, D. Qu, and J. McBreen, in *Synchrotron Techniques in Interfacial Electrochemistry, NATO ASI Ser., Ser. C,* Kluwer Academic Publishers, Amsterdam, 1994, p. 311.
[238] E. Preisler, *J. Appl. Electrochem.* **19** (1989) 559.
[239] M. Misawa, K. Matsuura, T. Okuda (JMC Ltd.), Jpn. Patent 57 1087 (1982); M. Misawa, T. Okuda, and K. Matsuura (JMC Ltd.), Jpn. Patent 57 42711 (1982).
[240] E. Preisler, *J. Appl. Electrochem.* **6** (1976) 301.
[241] J. C. Hunter and A. Kozawa, in *Standard Potentials in Aqueous Solution*, Ed. by A. J. Bard, R. Parsons, and J. Jordan, Marcel Dekker, New York, 1985, p. 429.
[242] A. K. Covington, T. Cressey, B. G. Lever, and H. R. Thirsk, *Trans. Faraday Soc.* **58** (1962) 1975.
[243] S. Atlung and J. P. Pohl, in *The 2nd Battery Material Symposium (The 3rd MnO_2 Symposium), Vol. 2, Graz, 1985,* Ed. by K. V. Kordesch and A. Kozawa, International Battery Material Association, Cleveland, Ohio, 1985, p. 23.
[244] G. Kano, M. Masuda, M. Takashima, and O. Nakamura, *Denki Kagaku* **37** (1969) 356.
[245] J. Y. Welsh, *Electrochem. Technol.* **5** (1967) 504.
[246] M. Fleischmann, H. R. Thirsk, and I. M. Tordesillas, *Trans. Faraday Soc.* **58** (1962) 1865.
[247] Lee, J. A., W. C. Maskell, and F. L. Tye, *J. Electroanal. Chem.* **79** (1977) 79.
[248] W.-H. Kao and V. J. Weibel, *J. Appl. Electrochem.* **113** (1992) 21.
[249] A. Cartwright and R. L. Paul, *Manganese Dioxide Symposium, Vol. 2, Tokyo, 1980,* Ed. by B. Schumm, Jr., H. M. Joseph, and A. Kozawa, I.C. MnO_2 Sample Office, Cleveland, Ohio, 1981, p. 290.
[250] K. E. Heusler and A. Grzegorzewski, *Electrochim. Acta* **35** (1990) 539.
[251] F. R. A. Jorgensen, *J. Electrochem. Soc.* **117** (1970) 275.
[252] H. Zhang, and S. M. Park, in *Electrode Materials and Processes for Energy Conversion and Storage,* Ed. by S. Srinivasan, D. D. Macdonald, and A. C. Khandkar, Proceedings Vol. 94-23, The Electrochemical Society, Pennington, New Jersey, 1994, p. 379.
[253] H. Tamura, K. Ishizeki, M. Nagayama, and R. Furuichi, *J. Electrochem. Soc.* **141** (1994) 2035.
[254] R. L. Paul and A. Cartwright, *J. Electroanal. Chem.* **201** (1986) 113, 123.
[255] G. Ya Slaidin' and A. A. Spritsis, *Elektrokhimiya* **14** (1978) 926.
[256] D. Gosztola and M. J. Weaver, *J. Electroanal. Chem.* **271** (1989) 141.
[257] A. Grzegorzewski and K. E. Heusler, *J. Electroanal. Chem.* **228** (1987) 455.

5

Sorption of Hydrogen on and in Hydrogen-Absorbing Metals in Electrochemical Environments

T. Mizuno

Department of Nuclear Engineering, Faculty of Engineering, Hokkaido University, Sapporo 060, Japan

M. Enyo

Hakodate National College of Technology, Hakodate 042, Japan

I. INTRODUCTION

Hydrogen *ad*sorption on and *ab*sorption in metals in electrochemical environments are of significance in a variety of areas. While adsorption of hydrogen has practical implications in such areas as water electrolysis, electroreduction, metal corrosion, and catalysis, it also has a particular importance in the understanding of basic adsorption phenomena. A typical example in electrochemistry is the hydrogen oxidation/reduction waves observed in current–voltage (i–V) diagrams, or *voltammograms*, which are used to evaluate electrochemically accessible surface area or to probe surface states. Whenever hydrogen plays any role in electrode processes, the hydrogen adsorption behavior is very frequently discussed as a basic property of the electrode.

Hydrogen *ab*sorption has great importance from a practical point of view. For instance, it gives rise to changes in the mechanical properties of the metal, sometimes causing loss of mechanical strength that may lead to a phenomenon called hydrogen embrittlement. Hydrogen-absorbing metals and alloys are important in connection with hydrogen storage, filtration (purification), hydrogen batteries, etc. In recent years, hydrogen (deuterium) absorption into Pd, Ti, and other metals has attracted much attention in connection with the alleged cold nuclear fusion.

Hydrogen may dissolve into many kinds of metals when the metals are placed in an electrochemical environment composed of an aqueous solution (or in a humid atmosphere) and the potential is brought close to or made more negative than the reversible hydrogen electrode (RHE) potential, either directly by cathodization or indirectly by corrosion of the metals, which is often accompanied by hydrogen evolution. In these cases, naturally, the thermodynamic condition (e.g., pressure and temperature) and hence the uppermost level, of the hydrogen sorption is determined by the kinetics and mechanism of the hydrogen electrode reaction (HER): The supply of hydrogen is often much greater and the reducing power is often much stronger than when the metal is just exposed to gaseous hydrogen.

An old concept that relates the hydrogen overpotential and a term representing the reducing characteristics of the hydrogen electrode to the equivalent hydrogen pressure was a Nernst-type equation based on a simple, but not verified, treatment of the kinetic phenomenon in a quasithermodynamic sense. The fact that that equation was not at all applicable was demonstrated by Yoshizawa et al.[1] and DeLuccia et al.[4] on a mild steel electrode. Using a cell like that schematically shown in Fig. 1a, hydrogen gas was introduced into compartment (A), and the rate of hydrogen permeation through a mild steel membrane (M) was followed by observing the potentiostatic anodic stripping current in compartment (B). In addition, analogous measurements were carried out with the use of a cell like that in Fig. 1b in which an electrochemical system was constructed also in compartment (A) to polarize M cathodically. Finally, the hydrogen pressure versus hydrogen overpotential relation was derived by combining these two sets of data. The results are shown in Fig. 2a on a linear scale and in Fig. 2b on a semilogarithmic scale. It was clear that the hydrogen pressure did not obey a Nernst-type equation (broken line), although it certainly increased with the hydrogen overpotential.

Kinetic approaches to the study of the hydrogen *ad*sorption versus *ab*sorption relation have been adopted by very many investigators.[5-23] Typically, polarization behavior in relation to the hydrogen electrode reaction mechanism is considered mainly in terms of surface coverage of hydrogen, as the latter usually enters into the kinetic equations of the

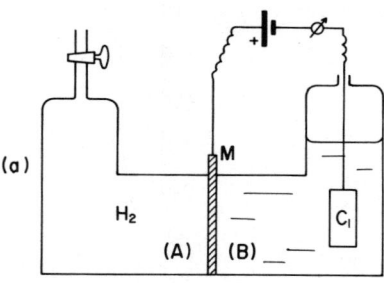

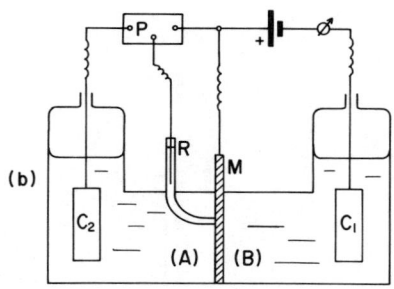

Figure 1. Observation of the hydrogen overpotential vs. hydrogen pressure relation by measuring the hydrogen permeation rate[1–4] through a metal membrane under given hydrogen pressures (a) and under cathodic polarization (b). M, Metal membrane; C, counter electrodes; R, reference electrode; P, potentiostat.

hydrogen electrode reaction, and the role of this reaction in the hydrogen absorption into, for example, the Pd electrode is discussed. To do this, the surface coverage is measured experimentally by using a transient polarization technique. Similarly, Bucur and Bota[5–8,24] carried out a detailed analysis of a kind of equilibrium between adsorbed and absorbed hydrogen at the Pd electrode and at the same time discussed the kinetics of the charge-transfer (Volmer) reaction taking place on the electrode.

As will be seen later in this chapter (Section V.3), the types of behavior represented by curves 1–3 in Fig. 2b can be explained well on the basis of a general kinetic analysis.[25] In this chapter, the conditions that determine the activity of hydrogen adsorbed on the electrode during the HER will be discussed first, but only briefly as this topic has been previously treated in some detail by one of the present authors.[25–29] The term *hydrogen pressure equivalent to hydrogen overpotential* (in short, *equivalent hydrogen pressure*) will be discussed with the use of the *cascade model* of electrode processes. Once the equivalent hydrogen pressure has

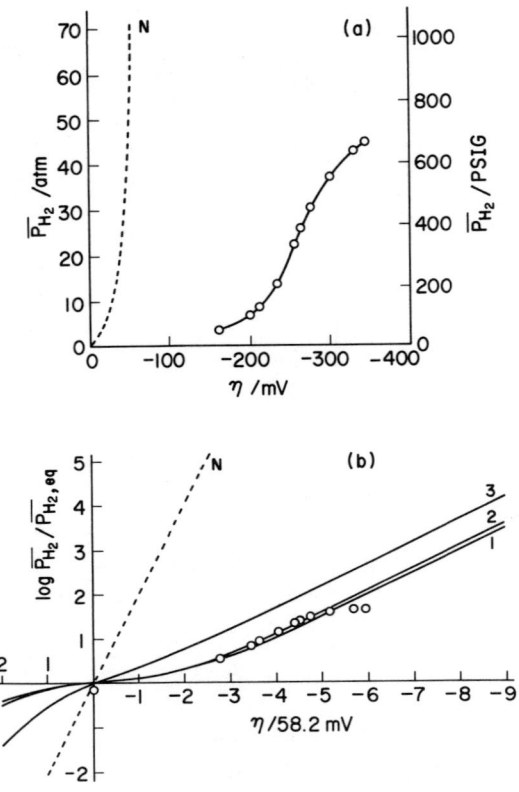

Figure 2. Relationship between hydrogen pressure, $\overline{P_{H_2}}$, and cathodic hydrogen overpotential, η, on a mild steel cathode in 0.05M H$_2$SO$_4$ at 294 K reported by DeLuccia and co-workers,[2-4] (a) Original linear plot; (b) redrawn as a semilogarithmic plot. Dotted lines, labeled N, represent the relations expected from a Nernst-type equation. Solid lines 1, 2, and 3 in (b) represent the relation expected from the analysis based on a model of mixed rate control[26] (see later) in the Volmer–Tafel reaction route with $i_{0V}/i_{0T} = 0.10$, 0.12, and 0.50, respectively.

been determined, it may then be combined with the hydrogen absorption isotherm of any particular metal to give the hydrogen concentration in the bulk of that metal.

A major portion of this chapter will be devoted to a description of the means by which the concentration and distribution of hydrogen in hydrogen-absorbing metals, or possibly metal hydrides, are measured. In subsequent sections, some aspects of hydrogen in metals, such as hydrogen diffusion and hydrogen concentration profiles of metal–hydrogen systems, whose study is made possible through such measurements will be discussed. Loss of mechanical strength of the metal as a result of the inclusion of hydrogen—the phenomenon known as hydrogen embrittlement—will not be a major focus of the present chapter.

II. EXPERIMENTAL ASPECTS OF TRACING HYDROGEN ON AND IN METALS

1. Hydrogen *Ad*sorption on Metal Electrodes

(i) Adsorption Isotherms

For a state of adsorption equilibrium between an adsorbate on a solid substrate and a gas, the basic relation in terms of chemical potentials is

$$\mu(\text{ads}) = \mu(\text{gas}) \tag{1}$$

where $\mu(\text{ads})$ may be given in terms of activity, $a(\text{ads})$, as

$$\mu(\text{ads}) = \mu^0(\text{ads}) + RT \ln a(\text{ads}) \tag{2}$$

The activity is in turn related to surface coverage θ; the form of this relationship depends on the adsorption isotherm adopted. Obviously, the surface coverage is an average quantity although it is known that the amount of adsorbate on the surface generally depends on surface crystalline state and/or microstructure and hence may not be uniform on the solid surface. The following discussion is, however, based on this average quantity, as the "discrete" microscopic model has not been sufficiently developed in the field of reaction kinetics.

A simple form of adsorption isotherm is the Langmuir type, in which the surface coverage θ with the adsorbate at a given gas pressure P is expressed in the form

$$\theta/(1 - \theta) = kP \tag{3}$$

As is well known, this relation is derived by equating the rate of adsorption v_{ads} with that of desorption v_{des}, as is justified at an adsorption equilibrium; these rates are respectively assumed to be given by

$$v_{ads} = k_{ads}P(1 - \theta) \qquad (4)$$

and

$$v_{des} = k_{des}\theta \qquad (5)$$

It is clear from this procedure that the form of the rate expressions for the adsorption and desorption processes must conform to the adsorption isotherm adopted.[29] It is also evident that k (= k_{ads}/k_{des}) and hence θ, but not the activity, at constant pressure of a given gas depend on the nature of the substrate metal.

The Langmuir adsorption isotherm is usually valid only at low θ and is of limited applicability for electrode processes. This is probably because the surface adsorbate on the electrode is densely populated as compared with the gas phase. Alternatively, the Temkin–Frumkin isotherm may be employed with much wider applicability. With the inclusion of an electrode-potential-dependent term $f(\eta)$ in order to make it applicable to electrode systems, it may be expressed as

$$[\theta/(1 - \theta)] \exp(u\theta) = kPf(\eta) \qquad (6)$$

where the term $\exp(u\theta)$ is newly introduced; here, u stands for the heterogeneity of the surface with respect to adsorption energy or, alternatively, U in $u = U/RT$ stands for the (usually repulsive) interaction among the adsorbed species. The Temkin–Frumkin adsorption isotherm will be adopted in later developments in this chapter.

(ii) Evaluation of the Surface Coverage, θ

Although the concept is clear, experimental evaluation of surface coverage θ is not generally easy in electrode systems. No simple method such as the Brunauer–Emmett–Teller (BET) method of surface area measurements in gas-phase adsorption experiments appears to be available, although other techniques such as the use of thin-layer cells[30,31] or radioactive isotopes[32–34] are sometimes successfully employed in adsorption studies.

A technique based on observation of the electrode potential decay (and rise) relaxation is occasionally employed, particularly for the measurement of adsorbed alkali metals,[35] hydrogen,[36,37] and other species to which application of the above-mentioned gas-phase type techniques does not appear to be feasible. A drawback of this technique is that the analysis of the potential decay transients for the evaluation of θ generally requires a degree of conjecture about the reaction mechanism and rate expressions thereof, which frequently involve θ itself. The discussion, therefore, sometimes becomes circular. Nevertheless, the information on θ is important in elucidating reaction mechanisms and rate expressions.

(iii) Significance of θ As Compared with Activity

Surface coverage θ is a very important and physically clear quantity in kinetic expressions. On the other hand, it is unfortunately the term that introduces uncertainties into rate expressions because of ambiguities associated with the adsorption isotherm, etc. In view of this, it seems more reasonable to leave the rate expressions in terms of activity as long as possible and use θ only at the stage where it becomes inevitable. This is to some extent analogous to the use of activity rather than concentration in thermodynamic considerations but is more important because of the high density of the surface adsorbate as compared with ordinary gaseous or solution states.

The hydrogen fugacity at the hydrogen electrode is one of the important topics in this chapter. In order to discuss it in relation to the hydrogen overpotential, and hence in relation to the nature of hydrogen adatoms on the electrode surface, the use of hydrogen activity appears to be more reasonable than the use of θ.

(iv) Measurements of Hydrogen Activity on Pd by a Current Interruption Method

Transient relaxation techniques have been very widely employed in kinetic studies, particularly to obtain information on the amount and nature of reaction intermediates. A number of attempts have been reported to describe, for example, overpotential transients at the moment of application of a constant current or overpotential decay when the flow of current is terminated. However, some of these attempts did not yield correct expressions because it was thought that one has to use the kinetic equations for the rate of an assumed rate-determining step (RDS): Ambiguities were involved in taking into account various terms such as the stoichiometric number of the RDS, electric charges consumed by steps other than the RDS, the treatment to be followed in the case that the RDS is "chemical," etc.

It was shown by one of the present authors[29,36] that the overpotential transient equation should simply be based on the mass-balance requirements for the individual chemical species and for the electric charge, irrespective of any step being rate-determining or not. For example, if one takes the hydrogen electrode reaction (HER), which is assumed to take place in accordance with the Tafel–Volmer mechanism,

$$\text{Tafel} \qquad \text{Volmer}$$
$$H_2 = 2H(ads), \qquad H(ads) = H^+ + e^- \qquad (7)$$

then the following equations must be valid with respect to mass balances for the electric charge on the electrode and for the hydrogen adatoms, respectively:

$$C \frac{d\eta}{dt} = i - (i_{V^+} - i_{V^-}) \qquad (8)$$

$$Q \frac{d\theta}{dt} = (i_{T^+} - i_{T^-}) - (i_{V^+} - i_{V^-}) \qquad (9)$$

where C is the double-layer capacitance, η is the overpotential (taken as positive in the anodic region), Q is the charge corresponding to a monolayer coverage of the electrode surface with H(ads), i is the applied current density, t is time, i_{V^+} and i_{T^+} are the forward rates (measured in units of current) of the Volmer and the Tafel step, respectively, and i_{T^+} and i_{T^-} are defined analogously. No ambiguity should enter into these expressions as they are simply counting the supply/consumption rates of the relevant species. Further developments can be made by giving appropriate expressions for various rates.[29]

A typical circuit for the observation of the overpotential decay transients is shown in Fig. 3.[28] After a steady-state polarization has been established by using a potentiostat or a galvanostat, a potential pulse with a suitable time window is applied from a pulse generator to a solid-state current interruption switch, and the corresponding overpotential decay is recorded by a transient recorder (or the like), which is synchronized by applying the same potential pulse. The polarization current after the interruption period is automatically restored to its original value in order to recover the electrode potential so that subsequent observations with different time windows can readily be carried out. Typical results of such decay transients will be presented later (Fig. 17).

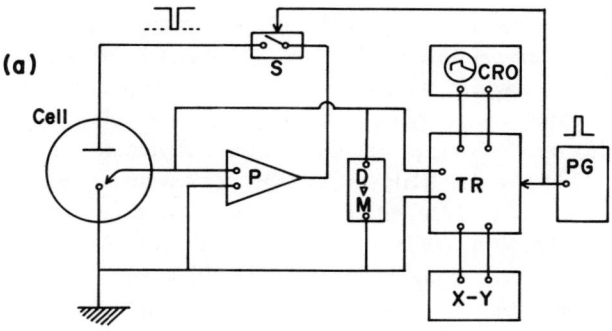

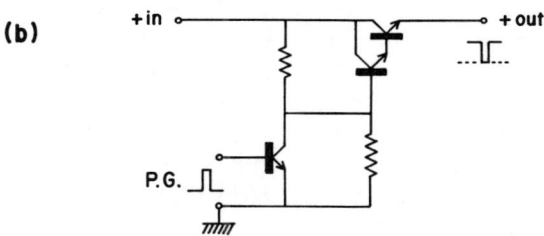

Figure 3. (a) Block diagram of potentiostatic polarization and the overpotential decay measurement circuit.[28] P, potentiostat; DVM, digital voltmeter; TR, transient recorder; CRO, cathode-ray oscilloscope; X–Y, pen recorder, PG, pulse generator; S, transistorized switch. (b) Typical transistorized switching circuit (for positive current).

2. Hydrogen *Ab*sorption into Metals

(i) *Conventional Techniques of Tracing Hydrogen in Metals*

The total amount of hydrogen absorbed in a metal or a metal hydride sample can be estimated by conventional methods such as direct weighing[38–40] such as a quartz crystal microbalance (QCM),[18,19] observation of expansion of the metal specimen[41] or of hydrogen gas released upon heating in vacuum, etc. Hydrogen can be expelled from exothermic hydrogen-absorbing metals, typically Pd, by heating. In the case of Ti, the hydride may be decomposed by heating to 1000 K, and the amount of hydrogen measured precisely by means of, for example, a volumetric technique.[42–45] Results are conveniently summarized in a pressure–composition–temperature (P–C–T) plot, that is, a hydrogen absorption isotherm.[46] In cases in which an electrochemical system is involved, anodic stripping after purging hydrogen out of the solution phase and maintaining the electrode potential at a value sufficient to oxidize hydrogen, e.g., 300 mV vs. RHE, is probably a convenient technique,[47,48] although one that is not often employed. A potentiostatic arrangement, with a current-limiting resistance in the counter electrode circuit to avoid initial current overshooting, may be useful for this purpose.

In the case of Pd, changes in electric resistance as a function of hydrogen content may be applied.[28,42,43,47–50] This technique should have the advantage that the measurement could be carried out without causing significant disturbance to the specimen nor does it require removal of the specimen from the system. Co-conduction by the electrolytic solution during the resistance measurements is usually negligible if reasonable precautions are taken. On the other hand, the resistance–composition relation has an ambiguity at high H/Pd ratios, in particular in the region in which the hydrogen pressure exceeds 1 atm. The relative resistance under 1 atm of hydrogen, where H/Pd at 298 K is 0.69,[28,48] relative to its value at H/Pd = 0, was evaluated to be 1.80 by Flanagan and Lewis[49] and 1.836 by Maoka and Enyo at 303 K.[28]

The resistance as a function of the H/Pd or D/Pd ratio is known to decrease at high loading after reaching a maximum R/R_0 of 1.8 and 2.0 for H and D, respectively, at an H(D)/Pd ratio of 0.72, according to McKubre et al.[51] (Fig. 4). At present, there exists a small degree of numerical discrepancy between these data, and further careful investigations are needed.

The hydrogen concentration profile in the bulk metal near the surface at the time when the hydrogen distribution in the metal has not yet attained a homogeneous distribution because of diffusional limitations is an interesting quantity as it would provide information on the mobility of hydrogen in the metal. In the case of Zr and Ti, the concentration profile could be measured by a differential etching technique, whereby the sample metal is gradually dissolved in an etching solution and the hydrogen gas released is collected and its volume measured. The thickness resolution of the hydrogen concentration profile in this technique is of the order of 10^{-5} cm.[52] The etching solution should provide a relatively mild dissolution rate of the hydride layer, and no additional production of hydrogen should take place during the etching treatment. A 1.2 wt. % HF + 2.0 wt. % HNO_3

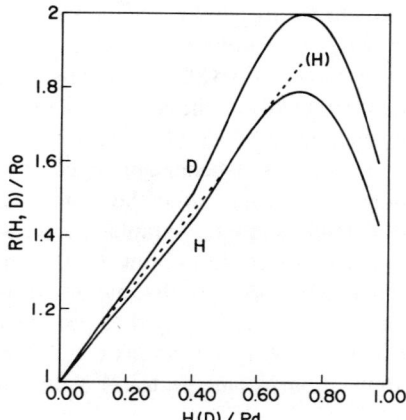

Figure 4. Relative electric resistance of H–Pd system as a function of concentration of hydrogen, including high-concentration region, at 298 K (summarized by McKubre et al.).[51] Dotted line was deduced by Maoka and Enyo[28] at 303 K based on data reported by Flanagan and Lewis.[49]

solution was found suitable for Ti hydride, and a 1.0 wt. % HF + 2.0 wt. % H_2O_2 solution was employed for Zr hydride.[53] This method is, however, not readily applicable to noble metals such as Pd.

(ii) Analysis of Deuterium in Metals by the d–n Reaction Using a Deuteron Accelerator

(a) Principle of the method

It is very important to trace in detail the process of hydrogen absorption into the surface layer of a metal in order to understand the detailed mechanism of hydrogen permeation. Conventional techniques such as the one described in the preceding section yield only an approximate concentration profile and do not allow for a precise evaluation of the rate of hydrogen absorption in a thin layer of hydrogen-absorbing metal.

Butler and Santry[54,55] developed a $d–n$ (d: deuteron, n: neutron) nuclear-chemical method of analysis of deuterium absorbed in metals by deuteron bombardment using an accelerator. Neutrons evolved by the nuclear fusion reaction that took place were measured, and the deuterium concentration in the sample surface layer was estimated therefrom. The method in its original form uses a high-energy beam and thus cannot provide a precise in-depth deuterium concentration profile because deuterons having an energy of the order of 100 keV may penetrate some 10^{-3}

cm into the metal bulk and hence the thickness resolution of the method may be of the order of 10^{-4} cm.

The method was improved by combining it with the chemical etching method. In this modified method, the sample surface is dissolved stepwise by etching and then subjected to the $d-n$ reaction analysis, and the energy of the deuteron beam employed (1–10 keV) is one to two orders of magnitude lower than in the original method so that the penetration depth of deuterons into the sample is decreased to the order of 10^{-4}–10^{-6} cm. A thickness resolution as low as 10^{-7} cm may be achieved. Analysis of the relationship between the rate of neutron evolution and the incident deuteron beam energy may thus provide a detailed profile of the deuterium concentration in the sample metal. Further, this method has an important application in isotope tracer work, as will be presented later [Section VI.4(vi)].

(b) Analysis yielding an in-depth deuterium concentration profile

When a deuteron beam from an accelerator bombards a deuterium-containing metal specimen, the following D + D nuclear reactions take place:

$$^2_1D + ^2_1D \rightarrow ^3_1T + ^1_1H + 4.032 \text{ MeV} \tag{10}$$

$$^2_1D + ^2_1D \rightarrow ^3_2He + ^1_0n + 3.266 \text{ MeV} \tag{11}$$

Neutrons are evolved in the latter reaction [Eq. (11), the $d-n$ reaction]. These two branched reactions are believed to have equal probability when the deuteron energy is on the order of kiloelectron volts.

The penetration depth of deuterons depends on the beam energy and the reaction cross section. The number of neutrons evolved, $N(E_0)$, for an initial value of the deuteron beam energy of E_0 may be given by

$$N(E_0) = \int_0^{R(E_0)} \sigma(x,E) \cdot I \cdot C(x,t) \, dx \tag{12}$$

where $\sigma(x,E)$ is the cross section of the $d-n$ reaction, I is the intensity of the deuteron beam, $C(x,t)$ is deuterium concentration at depth x and time t, and $R(E_0)$ is the upper limit of the penetration depth of deuterons having the initial energy E_0. Changes in the initial deuteron energy would cause changes in $N(E_0)$ through changes in the penetration range $R(E_0)$ and $\sigma(x,E)$.

To obtain the distribution profile $C(x,t)$ of deuterium concentration, $R(E_0)$ and $\sigma(x,E)$ have to be estimated first. There are some data available for the d–n cross section; it was measured by Arnold et al.[56] in the energy range 13–113 keV, by Preston et al.[57] below 500 keV, and by Smith and Perry[58] in the relatively high energy region of 0.5–6 MeV.

The Gamow plot, that is, a plot of the logarithm of the product of σ and the deuteron beam energy E versus $E^{-1/2}$, exhibits good linearity in the energy range between 10 and 100 keV, as shown in Fig. 5.[56,59] The relation may be written as

$$\sigma(E) \times E = (1.56 \times 10^2) \exp(-46.32 E^{-1/2}) \qquad (13)$$

where E is in kiloelectron volts. The value of E after the deuteron beam has passed through a layer of thickness x from the surface is given by

$$E = E_0 - \int_0^x (dE/dx)\, dx \qquad (14)$$

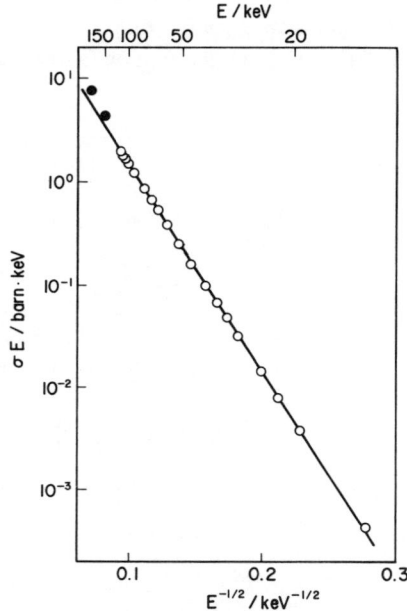

Figure 5. The Gamow plot, σE vs. $E^{-1/2}$, based on Eq. (13), where σ is the d–n reaction cross section and E is the deuteron beam energy. Data are taken from Fowler and Brolley[59] and Arnold et al.[56] (1 barn = 10^{-24} cm^2.)

where dE/dx is called the stopping power. The $d-n$ cross section $\sigma(x,E)$ at depth x is thus obtained from Eqs. (13) and (14), and, having determined these quantities, one can evaluate the distribution of the deuterium concentration $C(x)$ using an experimental relationship between N and E_0. Further developments of the method have been discussed elsewhere.[52]

(c) Experimental arrangement

In this section, a typical experimental setup for the $d-n$ reaction method is described.

Accelerator. A Cock-Croft Walton type accelerator (Toshiba NT-20) was used. The maximum energy is 200 keV, and the maximum current is 1 mA, with the fluctuation stability being within 1%. The beam is 85–90% D^+ ions, together with other components such as D_2^+ and T_3^+.[60–65]

Neutron counting. A BF_3 type detector was fixed near the target at 90° to the direction of the deuteron beam. Paraffin moderators having a thickness of 20 cm were placed between the target and the detector to decelerate and thermalize fast neutrons so that they can be detected by the BF_3 detector, as shown in Fig. 6.[52] The counting efficiency of the system in this geometric configuration was calibrated with a Ra–α–Be standard neutron source and a radiation method using In and Au foils.

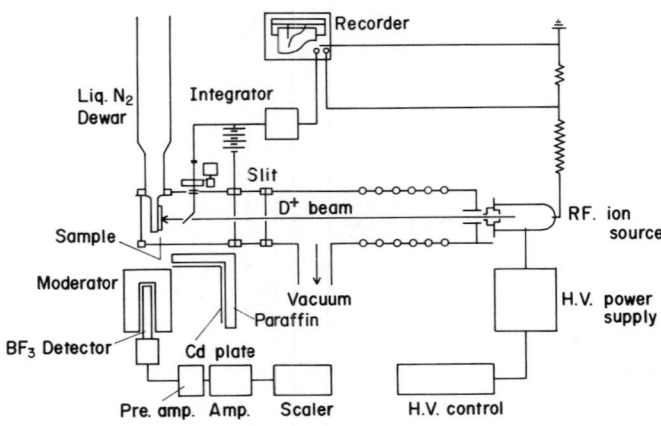

Figure 6. Layout of measuring system for the $d-n$ reaction analysis of deuterium in metal specimens.[52]

Sample. High-purity Ti, Zr, and Pd foils (2 × 2.5 cm and 0.5 mm in thickness) were used to estimate the cross section. They were polished with a buff followed by successive polishing with emery papers down to #600 until a mirror-bright surface was obtained. The samples with various deuterium concentrations were prepared according to the following procedures:

1. Samples were placed into a quartz cell and heat-treated to ~1100 K in the case of Ti and Zr and 600 K in the case of Pd under 10^{-6}–10^{-9} Torr vacuum for 10^4 s.

2. Deuterium gas was introduced into the cell, and the temperature was lowered to 300 K.

3. The above steps were repeated several times. The final hydride compositions obtained were close to those of the stoichiometric deuterides TiD_2, ZrD_2, and PdD.

4. Samples with lower deuterium concentrations were obtained by a heat treatment.

5. A heavy-water ice sample was obtained by deposition of heavy water directly onto a copper plate whose temperature was kept at 77 K using liquid N_2.

III. ABSORPTION OF HYDROGEN IN METALS

1. Light-Hydrogen and Deuterium Absorption Isotherms for Pd

The thermodynamic aspects of hydrogen/deuterium absorption into many kinds of metals have been well summarized[42–45,66–73] and need not be repeated here. As is well known, Pd is a typical metal with an extensive capacity to absorb hydrogen. The phase diagram of the system has two pure phases, the α-phase, in which the hydrogen (deuterium) concentration expressed as the atomic ratio H(D)/Pd, or x in PdH_x, is $x < 0.008$, and the hydrided β-phase, in which $x > 0.607$; the two phases coexist in the region corresponding to intermediate values of x. As shown in the pressure–composition–temperature (P–C–T) diagrams, or the absorption isotherms (Fig. 7),[74] the transition from the α-phase to the $\alpha + \beta$ region in which the two phases coexist is attained at rather low hydrogen pressures, namely, at about 8×10^2 Pa (6 Torr) for hydrogen and 4×10^3 Pa (30 Torr) for deuterium at 303 K; H(D)/Pd reaches 0.69 (0.66) under 1 atm.[28,47] The latter case gives rise to expansion of the lattice by some 3.5%. (A hysteresis between the *ab*sorption and *de*sorption data is generally seen; the data

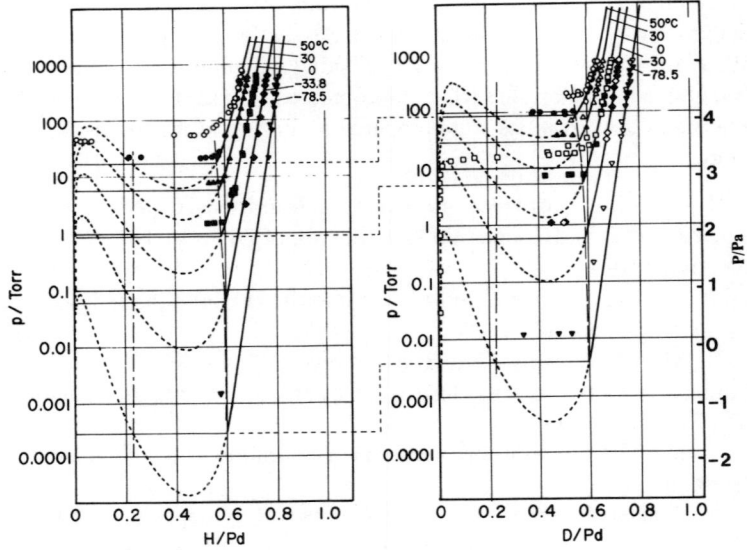

Figure 7. Comparison of hydrogen and deuterium absorption isotherms of Pd.[74]

shown in the figures presented are usually for the desorption branch as they would represent the equilibrium data more closely.[75])

The Pd hydride structure has been widely investigated by neutron diffraction.[54] Hydrogen atoms occupy octahedral sites in the face-centered cubic (fcc) Pd lattice in both the α- and β-phases. The crystal structure of both hydride phases is fcc and has the appearance of two interpenetrating fcc lattices, with Pd centered at (0,0,0) and H at ($\frac{1}{2},\frac{1}{2},\frac{1}{2}$). When x reaches 1.0, all the octahedral sites are occupied.

Deuterium-induced structural changes were studies by an *in situ* technique involving the combination of electrochemical and neutron diffraction methods.[55] A metastable phase of the Pd–D system was found at very high deuterium concentrations, up to a D/Pd ratio of 0.7. At still higher pressures, the value of the D/Pd ratio increases only gradually, and the maximum value in Fig. 7 approaches ca. 0.7 at 303 K. The values are as a rule higher at lower temperatures, approaching 0.8 at 195 K. More data at high pressures, up to 10^2 atm, have been reported; data are presented in Fig. 8[76] for the H/Pd system and in Fig. 9[77] for the D/Pd system.

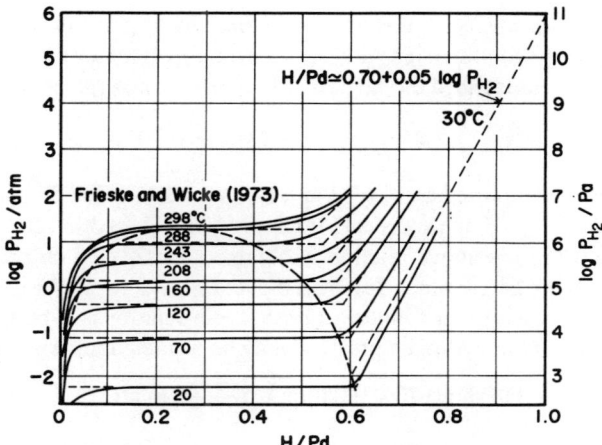

Figure 8. Hydrogen absorption isotherm of Pd in the higher hydrogen pressure region.[76]

The hydrogen absorption condition is much stronger in electrochemical systems; the H/Pd ratio can be estimated by combining the effective hydrogen pressure, obtained by overpotential decay transients as discussed later (Section V.2), with the hydrogen absorption isotherm (such as Fig. 7). The value (strictly, *fugacity*) may reach 10^6 atm at 303 K (cf.

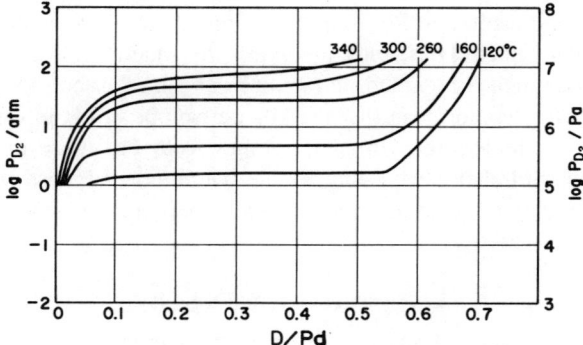

Figure 9. Deuterium absorption isotherm of Pd in the higher deuterium pressure region.[77]

Fig. 24 below); a possible means of evaluating hydrogen concentration(s) in β-H–Pd is to combine this with, e.g., the empirical relation given below, which has been deduced on the basis of reports by several authors:[21,78,79]

$$S_{H(Pd)} = S_{H(Pd)} + K \log P_{H_2} \quad (\text{cm}^3 \text{ H}_2/\text{cm}^3 \text{ Pd}, P_{H_2} \text{ in atm}) \quad (15)$$

where the data reported are $K = 63$ and $S_{H(Pd),1} = 850$ at 298 K[21] or 68 and 956 at 303 K.[78,79] If one accepts H/Pd under 1 atm H$_2$ at 303 K as 0.691,[28] the constant in the above relation is obtained as 972, which practically agrees with the latter value. Based on the latter value for the slope, which almost agrees with that found in Fig. 8,[76] the relationship between the H/Pd ratio and the hydrogen pressure may be written approximately as

$$H/Pd = 0.70 + 0.05 \log P_{H_2} \quad (P \text{ in atm}) \quad (16)$$

If this equation holds, the maximum value of the H/Pd ratio may reach 1.0 at 10^6 atm effective pressure.

There are noticeable differences between the H/Pd and D/Pd systems (and the tritium/Pd system[75,80]). It is seen from the isotherms that both the solubility and the heat of solution (which is exothermic for Pd) are higher for hydrogen than for deuterium: The difference in solubility under equilibrium conditions leads to a noticeable level of isotopic discrimination. Nevertheless, this is not pronounced at ordinary hydrogen pressure as indicated by a smaller difference between β-H–Pd and β-D–Pd phases (e.g., at 1 atm with an equimolar mixture of hydrogen and deuterium, the factor of hydrogen enrichment in Pd as estimated from Fig. 7 is about 1.04). On the other hand, the isotope separation factor when hydrogen and deuterium are taken into Pd from mixed light-/heavy-water solutions by (irreversible) electrolysis, thus involving the kinetic isotope effect, is likely to be much higher, and, in fact, it has been reported to be as large as 13.[81] (This would mean that the kinetic isotope effect in the reverse reaction, i.e., removing hydrogen isotopes from Pd by an irreversible process, should also have a comparable magnitude, because their ratio would be equal to the equilibrium partition coefficient which is, as seen, rather close to unity.)

2. Pressure versus Fugacity

The actual state of gas at high pressures may better be described in terms of *fugacity*. In fact, the term "pressure" used here should be read as "fugacity" at values exceeding 10^3 atm. The deviation of fugacity from

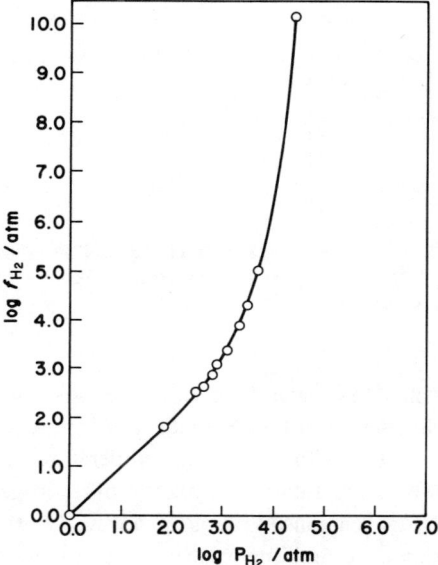

Figure 10. Fugacity vs. pressure relationship for hydrogen at 300 K (summarized by Bockris and Subramanyan[82]).

pressure in the case of hydrogen becomes significant at pressures exceeding 10^2 atm (cf. Fig. 10).[82] Figure 10 may indicate that the apparent hydrogen pressure would hardly exceed $10^{4.5}$ atm.

3. Temperature Dependence of Hydrogen Absorption in Pd

It is clear from Figs. 7–9 that absorption of hydrogen/deuterium in Pd is exothermic within the ordinary range of concentrations. Data on the heat of absorption for Pd as well as some other hydrogen-absorbing metals are summarized in Fig. 11.[83,84] The exothermicity of hydrogen absorption in Pd appears to decrease at higher hydrogen concentrations (β-phase), which seems reasonable in that the entry of hydrogen into the Pd lattice should be less favorable when there is a higher population of hydrogen in the lattice. However, as shown, there is an indication that the process again becomes more exothermic at very high hydrogen concentrations (H/Pd > ca. 0.84). This was explicitly reported by Baranowski,[66] who found that

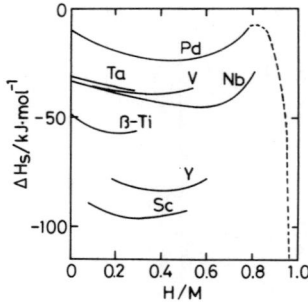

Figure 11. Heat of absorption of hydrogen in some metals as a function of hydrogen concentration.[83]

the exothermic heat of hydrogen absorption increases to some 10^2 kJ mol^{-1} at H/Pd ≈ 0.96. In support of this, Kubota et al.[86] reported a decrease of H/Pd to a value of 0.9 with increasing temperature at high hydrogen concentrations. The temperature dependence of hydrogen absorption at extremely high hydrogen concentrations is not well settled although this kind of information is of vital importance in the discussion of the alleged cold fusion.

4. Effective Charge of Hydrogen in Metals

The effective electric charge of hydrogen dissolved in metals is of some interest. Information as to whether hydrogen is neutral, as usually supposed, or acquires some charge by losing/gaining electrons may be obtained by observing electric migration effects,[87] that is, by following the movement of hydrogen in the host metal upon application of an electric field across it. The value of the H/Pd ratio at 200°C was reported to be ca. +0.55 in Pd or higher (0.6–0.9) in Pd–Ag alloys containing up to 50% Ag. Experimental data for various metals have been summarized.[88] This kind of information would be of great importance in connection with proposed theories of cold fusion.[89]

5. Mechanism of Hydrogen Embrittlement

Several mechanisms have been proposed to explain hydrogen embrittlement phenomena:

(a) The *decohesion mechanism*, in which atomic bonding is considered to be weakened by hydrogen dissolution or hydride formation.

(b) The *stress-induced hydride decomposition mechanism* in which the inner boundary between the precipitant and the substrate is broken by hydride under stressed conditions.

(c) The *hydrogen-enhanced local plasticity mechanism*, in which localized deformation at a point of stress is activated by hydrogen atoms.[67]

An alternative explanation may be considered in the case of Ti and Zr, in which the hydride concentrations have been reported to be low at room temperature.[90-92]

Hydrogen embrittlement is called *reversible* if there is almost complete recovery of ductility after removal of an applied stress. Slip deformation occurs at the (111) plane for Zr and Ti hydrides. Hydrogen atoms are considered to gather and concentrate at cracks where a tensile stress also concentrates and hydrogen precipitates.

Crack propagation and development may be predominant within hydride layers because of their high brittleness. This can explain the crack propagation in Ti and Zr as directly observed at broken surfaces of these metals by means of electron microscopy.[90,91] Stress-induced crack propagation occurs along the site of hydride precipitation transmission electron microscopy (TEM).[67] The third mechanism was also supported by the same observation. The dislocation moved actively under a hydrogen atmosphere, and a deformation developed locally around a crack as also observed in the case of Fe and Ni.

6. Confinement of Hydrogen in Metals and Possible Fracture

It may be interesting in connection with the topic of the alleged cold fusion to look at the highest value of the concentration of D (or H) in Pd that has been attained so far in electrochemical environments. Unfortunately, the literature data are not in good agreement. Thus, Moore[91] reported that the maximum concentration of light hydrogen in β-H–Pd achieved corresponds to an H/Pd ratio of 0.79. An analogous value of 0.89 was reported by Barton *et al.*,[47] and the value at 14 kbar and 298 K is less than 1.0 according to Baranowski.[66] Much higher values were reported in electrochemical systems[93] namely, 1.5 at 50 mA cm^{-2} (Fig. 12) or even an abnormally high value of 2.2 at ~1 A cm^{-2} (not shown) for D/Pd.

Some new observations have been reported recently: Will *et al.*[94] reported that the H/Pd and the D/Pd ratio could reach 1.15 and 1.03, respectively. They used a 2-mm-diameter Pd wire and carried out electrolysis for some 160 h in H_2SO_4 or D_2SO_4 solution. Using a neutron

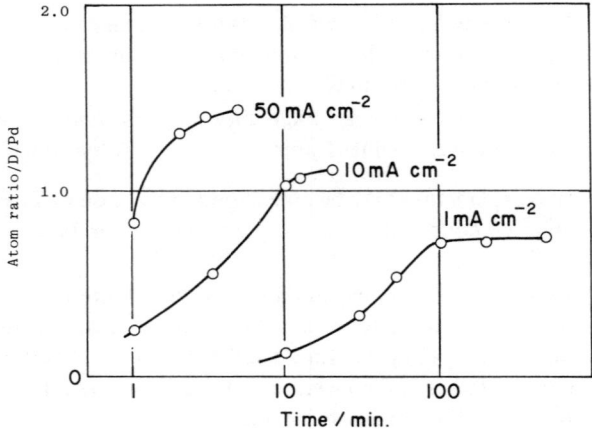

Figure 12. Increase of deuterium concentration, as measured by the neutron evolution rate obtained by the $d-n$ reaction method, in a Pd electrode under cathodic polarization at various c.d.'s.[93] Deuteron beam energy, 100 keV.

diffraction technique, Redey and Myles[95] observed D/Pd = 0.99 after prolonged electrolysis with 500 mA cm^{-2}. Kubota et al.[86] reported similarly high values of H/Pd = 0.88 and D/Pd = 0.79 at 30 mA cm^{-2} and 30°C using a closed-type electrolysis cell. Similarly, Mizuno[96] reported D/Pd = 0.96 at 200 mA cm^{-2} and 110°C.

Recently, Lawson et al.[38] reported that the maximum value (in the deuterium system) was almost independent of solution pH, surface cleanliness, currently density (if it is high enough), etc., this maximum value being D/Pd = 0.73 ± 0.02, which is significantly less than the values cited above. Apart from some numerical disagreements, there is an apparent tendency of approaching toward seemingly limiting values of H(D)/Pd. It may be concluded that there exists a physical limit to confine hydrogen with a high pressure, even if such a condition can be realized in theory in the bulk Pd: There should be no effective way of confining hydrogen in the metal by means of electrode polarization once it becomes gaseous. Thus, cracks, shadow areas with respect to electrode polarization, etc., would facilitate escape of hydrogen from the electrode. Nevertheless, values of η'_2 (see Section V) as negative as –200 mV were observed by Enyo and Biswas[97] in an acidic solution containing a catalytic poison, from which an H/Pd ratio close to unity may be anticipated if one can extrapolate the hydrogen adsorption isotherm in Fig. 8 (shown above) to high hydrogen pressures (up to 10^6 atm).

7. Hydrogen Absorption in Fe and Ni

Hydrogen absorption in Fe is frequently studied because of its particular technological importance in connection with the hydrogen embrittlement of steel.[1–4,98,99] As Fe is an endothermic hydrogen absorber, the solubility increases with temperature; the following relation was reported[100] (P in atm):

$$S_H = 3.7 \times P^{1/2} \exp[-27 \text{ (kJ)}/RT] \quad [\text{cm}^3 \text{ H}_2 \text{ (NTP)}/\text{cm}^3 \text{ Fe}] \quad (17)$$

Further, it is known that the solubility is much higher in molten metal[67] so that the introduction of hydrogen in steel plates during their fabrication by the hot-press technique with the use of water cooling may yield high hydrogen concentrations.

Hydrogen absorption in Ni has been studied extensively. Among others, Smialowski, Baranowski, and co-workers[50,101–103] carried out detailed studies on Ni in an electrochemical environment and showed that the absorption isotherm has a similar shape to that of H-Pd, having α-, β-, and $\alpha + \beta$ phase-transition regions, and that H/Ni can reach a value as high as 0.65[104] or 0.73[101] under cathodic polarization, especially when catalytic poisons (or promoters as they called) are introduced into the system.

8. Hydrogen Absorption in Ti and Zr

Hydrogen dissolved within the α-phase of pure metals exhibits no noticeable effect on the absorption energy. However, there is a marked reduction of the absorption energy in the initial stage of formation of hydride in the α-matrix. The hydrogen absorption behavior of pure Ti and Zr transition was determined to be related to the hydride growth. The hydride phase growth at the α-phase stage shows either a stable δ-phase or a metastable γ-phase,[105] depending on the growth conditions. These phases solubility limits may be different for different stages of nucleation[106] and the strain caused by the deformation.[107]

Hydrogen absorption isotherms for Ti and Zr are shown in Figs. 13 and 14, respectively.[46,108] Ti has three hydride phases —β, δ, and ε—each phase having a different hydrogen content and a different crystal structure. The β-phase persists above 554 K; it has a body-centered cubic (bcc)

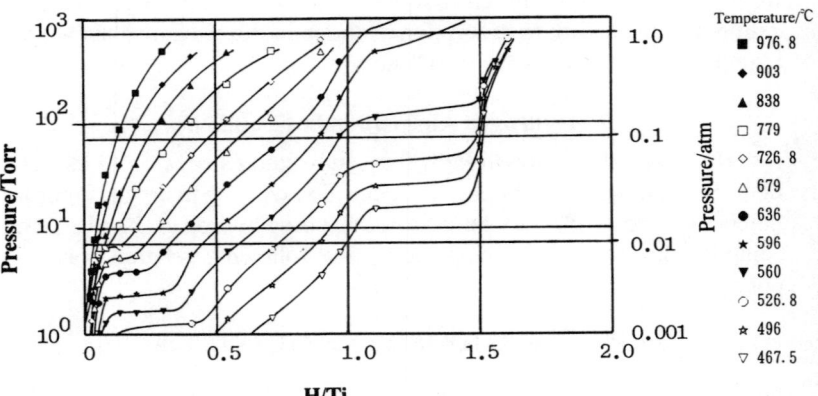

Figure 13. Absorption isotherms of Ti–H system.[46,108]

structure, and its hydrogen concentration ranges from 0 to 50 at. %. The α-phase has a hexagonal close-packed (hcp) crystal structure, and its maximum hydrogen concentration is less than 0.1 at. % (atomic percentage), increasing up to 7.9 at. % at 593 K. The δ-phase appears when the hydrogen content increases up to 7.9 at. %.

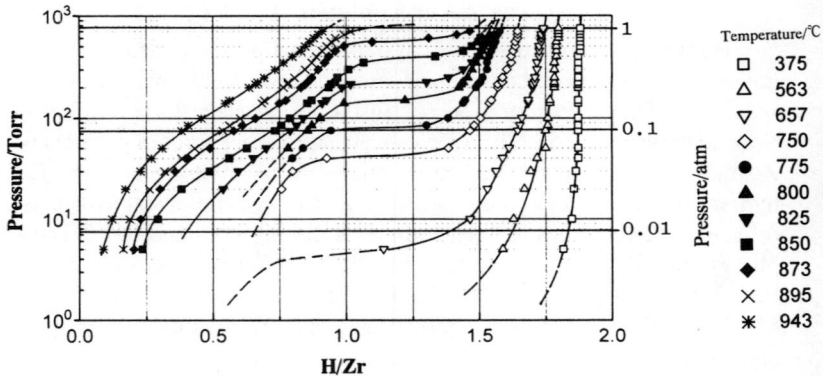

Figure 14. Absorption isotherms of Zr–H system.[109]

The isotherm for H–Zr (Fig. 14) is similar in shape to that for the H–Pd system discussed above, but hydrogen uptake can proceed more extensively, attaining a value close to the composition ZrH_2. Zr has a phase diagram analogous to that of Ti, but the β-phase is stable at temperatures above 837 K. Further, Zr has other phases—δ- and ε-phases—which appear at hydrogen contents from 59.5 to 65.5 at. %. The solubility of hydrogen and the enthalpy of formation of the hydride in Ti and Zr are summarized in Table 1.[109–119]

Table 1
Enthalpy of Hydrogen Dissolution in Ti and Zr

	ΔH_s (kJ mol^{-1})	N_H	Investigator(s)	Reference
α-Ti				
H (706–721°C)	-10.6 ± 0.1	0–0.08	Dantzer et al.	109
D (707°C)	-10.1	0–0.08	Dantzer et al.	109
H and D	-9.6		Ricca	110
H and D	-9.6		Nagasaka and Yamashina	111
βTi				
H (706–721°C)	-13.2	0.36–0.44	Dantzer et al.	109
H (971°C)	-14.2	0.17–0.43	Dantzer et al.	109
D (707°C)	-12.9	0.38–0.44	Dantzer et al.	109
δ-Ti				
TiH$_{1.0-2.0}$	-27.3 to 29.5		Stalinski and Bieganski	112
TiH$_{1.75}$	-31.1		Haag and Shipko	113
ε-Ti				
TiH$_2$	-19.2		Gibbs et al.	114
TiH$_2$	-29.6		McQuillan	115
α-Zr (H, D)	-12.5		Dantzer	116
β-Zr(H)				
ZrH	-15.8		Ells and McQuillan	117
ZrD$_{0.75 \pm 1.25}$	50.3			
δ-Zr and ε-Zr				
ZrH$_{1.86}$	-38.2		Turnbull	118
ZrH$_2$	-39.7		Turnbull	118
ZrD$_2$	-40.2		Fridrickson	119

Hydrogen dissolution reactions in the α- and β-phases of Ti–H and Zr–H are endothermic, and hence the hydrogen solubility increases with temperature, while the decomposition pressure also increases steeply with temperature. On the other hand, the formation of the δ-phase of Ti and the δ- and ε-phases of Zr is exothermic, and hence these phases increase with lowering temperature.

Among the transition metals, those which absorb large quantities of hydrogen are most interesting. At low hydrogen pressure, α-phases, i.e., the solid solutions of hydrogen in the metals, are formed. At higher hydrogen pressure, the β-phase alone or β-, δ-, and ε-phases occur; these phases constitute intrinsic hydrides of transition metals. One of the most important characteristic features of these phases is their variable composition. X-ray, neutron diffraction, and NMR studies have confirmed the model in which hydrogen atoms occupy interstitial positions in the metallic lattice. Unfortunately, the crystal structure of the hydrides has not been settled unambiguously in all cases. A difficulty arises particularly in the case of bcc metals, in which a large number of tetrahedral and octahedral interstices are available for hydrogen atoms.

On the other hand, it can readily be shown (Fig. 15) that the occupation of all octahedral positions in the fcc structure of a metal leads to the composition MeH and an NaCl type structure, whereas the occupation of all tetrahedral sites gives the phase with the composition MeH_2 of the fluoride structure. The latter type is found in the majority of rare-earth

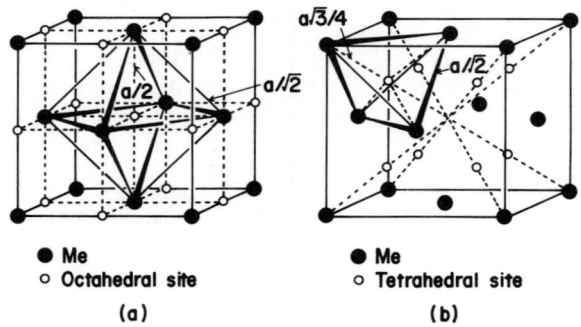

Figure 15. Octahedral and tetrahedral sites of hydrogen atoms in Pd lattice.

dihydrides, as well as in the case of titanium-group dihydrides except for a slight tetragonal distortion.

9. Temperature Hysteresis in Hydrides of Ti and Zr

Hydrogen solubility data for the α-phase of Ti and Zr as evaluated by the electric resistance technique,[120,121] the inner friction technique,[122-124] or direct structural observation[125] may be summarized by the following equations:

$$\alpha\text{–Ti:} \quad S \text{ (at. \%)} = 4.7 \times 10^2 \exp[-2.14(\text{kJ mol}^{-1})/RT] \quad (18)$$

and

$$\alpha\text{–Zr:} \quad S \text{ (at. \%)} = 1.3 \times 10^3 \exp[-3.70(\text{kJ mol}^{-1})/RT] \quad (19)$$

The solubility limit at 303 K is thus evaluated to be 0.09 at. % for Ti and 5 at. % for Zr. The data for hydrogen and deuterium are not significantly different.

A significant hysteresis is seen in repeated dissolution experiments such that the temperature required for redissolution is some 50 degrees higher than that during the preceding cooling process. Volumetric changes also take place. A supercooling phenomenon is also seen during the hydride formation, which is presumably caused by elastic energy due to the strain induced by the transformation. The hysteresis phenomenon may appear because the strain is taking place irreversibly.[126,127]

IV. MODELS OF HYDROGEN ENTRY INTO A HYDROGEN-EVOLVING ELECTRODE

Several models have been proposed in the literature to interpret the kinetics of hydrogen entry into cathodes in relation to the kinetics of the hydrogen evolution reaction. Some of them will be briefly presented in this section.

1. A Concentration Cell Model

A simple case is that in which the hydrogen-evolving electrode is so active that the hydrogen overpotential is practically determined by the concentration overpotential; i.e., it is simply the concentration cell emf established across the Brunner–Nernst diffusion layer. Such a view was taken by Lewis and co-workers,[47,128-131] as the Pd electrodes used by them were sufficiently active: No direct connection with the hydrogen evolution reaction kinetics may then be anticipated.

An analogous case is that in which the hydrogen evolution reaction kinetics yields a Tafel slope of 30 mV (the slow-recombination mechanism presented below), which is mathematically similar to the diffusion-controlled case. This model is likely to be applicable only to active (palladized) Pd (and Pt) electrodes but not to ordinary smooth Pd nor to electrodes of other metals.

2. A Model of Hydrogen Entry Independent of the Hydrogen Evolution Reaction

In proposing a model for hydrogen entry into cathodes, Bagotskaya and Frumkin[132,133] took the view that hydrogen entry into the electrode is possible at a limited number of sites, such as grain boundaries, and is independent of the hydrogen evolution reaction taking place at essentially every site on the surface. Their model may be expressed as

$$H^+ + e^- \xrightarrow{\text{Volmer}} H_{ads} \xrightarrow{\text{Tafel}} \tfrac{1}{2} H_2 \quad (20)$$
$$\phantom{H^+ + e^- \xrightarrow{\text{Volmer}}} \downarrow\uparrow \phantom{H_{ads}} 1 \phantom{\xrightarrow{\text{Tafel}}} 2$$
$$3\downarrow\uparrow -3$$
$$H_{abs,*}$$

where step 3 is assumed to take place on special sites (lines) only, denoted by *, forming $H_{abs,*}$ there. The latter will enter into the metal lattice and diffuse into the metal bulk. The diffusion model may be spherical (or cylindrical) growth of hydrogen-rich regions centered at each site (line), but, unless the entry sites are very scanty on the surface, such sites will soon be smeared out and the diffusion model may become approximately linear or one-dimensional. In any case, one may take the rate of hydrogen entry to be approximately proportional to the concentration/activity of $H_{abs,*}$.

If one assumes[132,133] that the hydrogen permeation rate is controlled by diffusion in the metal bulk and hence that the concentration of $H_{abs,*}$ just beneath the electrode surface is determined by the overpotential η according to a Nerst-type expression, that is, if one assumes quasi-equilibrium conditions for step 3, then it can be shown that the rate J of entry of hydrogen depends on η as

$$d\eta/d \ln J = -RT/F \quad (21)$$

In the case in which the hydrogen evolution rate i depends on η as

$$d\eta/d\ln i = -RT/\beta_1 F \tag{22}$$

which may be anticipated for the rate-determining Volmer mechanism (the slow-discharge mechanism) with the transfer coefficient β_1, then one finds, from these two equations,

$$d\ln J/d\ln i = 1/\beta_1 \tag{23}$$

which is likely to be close to 2. If, on the other hand, the hydrogen evolution rate depends on η as

$$d\eta/d\ln i = -RT/2F \tag{24}$$

which may be anticipated for the rate-determining Tafel mechanism (the slow-recombination mechanism), one finds

$$d\ln J/d\ln i = \frac{1}{2} \tag{25}$$

A difficulty here is that permeation experiments usually support Eq. (25) but the polarization behavior of the hydrogen evolution reaction is consistent with the case which yields Eq. (23). No consistent interpretation of both hydrogen permeation and hydrogen evolution rates appears to be possible on the basis of this model.

3. A Model of Hydrogen Entry via the Intermediate of the Hydrogen Evolution Reaction

A model in which hydrogen entry takes place via the intermediate of the hydrogen evolution reaction was proposed by Bockris, McBreen and Nanis.[22] Experiments on hydrogen permeation through Armco iron membranes conducted by these authors supported Eq. (25), but polarization behavior was in agreement with Eq. (22). In order to interpret such results, they proposed the *coupled discharge–recombination* mechanism. In this model, hydrogen entry takes place via the reaction intermediate H_{ads} of the hydrogen evolution reaction as shown below:

$$H^+ + e^- \xrightarrow{\text{Volmer}} H_{ads} \xrightarrow{\text{Tafel}} \frac{1}{2}H_2$$
$$\quad 1 \quad \downarrow\uparrow \quad\quad 2$$
$$\quad\quad 4\downarrow\uparrow\!-\!4 \tag{26}$$
$$\quad\quad H_{abs}$$

The rate of entry of hydrogen may be given simply as proportional to the activity a_H of H_{ads}, and hence the discussion is reduced to the question of the dependence of a_H on η, which is in turn determined by the operative mechanism of the hydrogen evolution reaction.

In deriving the relation between J and i, the case of the *pure* slow-discharge mechanism is not meaningful, since a_h should be kept constant at the equilibrium level because of the sufficiently rapid Tafel step that is assumed in this mechanism (called the nonembrittling case[82,134]). On the other hand, if one considers the slow-recombination mechanism, one obtains

$$d \ln i / d \ln a_H = 2 \tag{27}$$

or, combining this with the relation that J is proportional to a_H, one obtains Eq. (25) (the embrittling case[82,134]). However, this hydrogen evolution reaction mechanism is not supported by experiments.

To cope with this difficulty, the coupled discharge–recombination mechanism was proposed. This mechanism is based on the assumption that the rate of the hydrogen evolution reaction obeys the slow-discharge mechanism but the rate of the recombination step is finite so that a_H deviates from its equilibrium value $a_{H,eq}$ and is related to i according to

$$i = k_2 (a_H / a_{H,eq})^2 \tag{28}$$

It is obvious that this leads to the relation of Eq. (25); hence it explains the results reported by Bockris *et al.*[22] However, the significance of the finite rate of the recombination step on the kinetics of the hydrogen evolution reaction and further details of the two kinetically combined steps was not discussed.

According to Phillips *et al.*,[135] the experimental value for Ti in $0.05M$ H_2SO_4 is

$$d \ln J / d \ln i = 0.91 \tag{29}$$

Similarly, for Zr in $0.5N$ Na_2SO_4, $1N$ NaOH, and $1N$ H_2SO_4, the value is commonly ~1.0.[136] Thus, experimental results are usually between the values indicated by Eqs. (23) and (25), being often close to unity, and are not well interpreted by any one of these models.

In the present authors' opinion, however, the analyses above are not sufficiently elaborated and a quantitative treatment is required. A typical approach along these lines will be discussed below in Section V.

4. Effects of Catalytic Poisons

It is known that the presence of additives, generally considered as catalytic poisons, gives rise to significant changes in the rate and extent of entry of hydrogen into the electrode metal.[22,132,133,137–140] A widely accepted view about this effect has been that the poisons affect on the $M-H_{ads}$ bond strength, e.g., weakening it, and accelerate hydrogen entry. This interpretation may be questioned, however, on the grounds that weakening the bond strength should also result in a decrease of the steady-state concentration of H_{ads} from its value under otherwise the same conditions, thus contributing to decelerate the hydrogen entry and canceling the above-mentioned effect.

An alternative view may be that catalytic poisons exert effects on the rate of either the discharge or the recombination step, or both. For instance, if a poison retards the recombination step, then the activity of H_{ads} has to be raised in order to maintain a given rate, and hence the rate of hydrogen entry accelerated. In order to proceed with an analysis along these lines, a general treatment of reaction kinetics may be required, as discussed in the following section.

V. ACTIVITY OF HYDROGEN ADATOMS AS THE REACTION INTERMEDIATE OF THE HYDROGEN EVOLUTION REACTION AND THE EQUIVALENT HYDROGEN PRESSURE

1. Hydrogen Electrode Reaction and Hydrogen Permeation in Metals

It seems evident that hydrogen absorption into a metal electrode should be related to the mechanism of the hydrogen evolution reaction (HER) taking place on its surface, as these processes are likely to commonly involve hydrogen adatoms, H(ads), on the surface. Hence, a precise discussion of the mechanism is required. Unfortunately, however, many of the discussions that have appeared in the literature have been qualitative because of a lack of well-founded information on the HER mechanism.

Investigations on the mechanism of the HER have a very long history, dating back to the time of Tafel, as early as 1905, but, in spite of some 60 years of intense investigations since the 1930s, mechanistic studies have not succeeded in providing a final unambiguous picture of the mechanism. Among various theories proposed, the following reaction mechanisms are considered acceptable. These are the Volmer–Heyrovsky mechanism,

Volmer Heyrovsky

$$\text{H}^+ + e^- \xrightarrow{\text{Volmer}} \text{H(ads)}, \quad \text{H(ads)} + \text{H}^+ + e^- \xrightarrow{\text{Heyrovsky}} \text{H}_2(\text{gas}) \quad (30)$$
$$i_{0V} \qquad\qquad\qquad\qquad\qquad i_{0H}$$

and the Volmer–Tafel mechanism,

$$\text{H}^+ + e^- \xrightarrow{\text{Volmer}} \text{H(ads)}, \quad 2\text{H(ads)} \xrightarrow{\text{Tafel}} \text{H}_2(\text{gas}) \quad (31)$$
$$i_{0V} \qquad\qquad\qquad i_{0T}$$

where i_{0V}, i_{0H}, and i_{0T} denote exchange current densities (c.d.'s) for the Volmer, the Heyrovsky, and the Tafel step, respectively. Experimental evidence that seems to support or refute these mechanisms is presented below.

Investigations to correlate the extent of hydrogen absorption with the mechanism of the HER were carried out by several authors.[82,98,135,140–146] Thus, Bockris and Subramanyan[82] concluded that (i) a Nernst-type expression may be derived for the case of the slow Tafel–rapid Volmer mechanism (an embrittling case), (ii) no such increase would occur in the case of the rapid Tafel–slow Volmer mechanism (a nonembrittling case), and (iii) a lower than Nernst-type dependence holds for the coupled (a type of mixed-control) Volmer–Tafel mechanism. It is not possible to accept (i), a slow Tafel mechanism, as the kinetics of the HER, particularly the ~120-mV Tafel slope and the absence of limiting current, do not conform to it, nor (ii), a slow Volmer mechanism, as the effective pressure definitely grows with η; thus (iii) may be a possibility, but this has not been tested quantitatively. A quantitative treatment on the basis of the mixed-control Volmer–Tafel mechanism will be the major focus of the discussion below.[140] On the other hand, a slow Heyrovsky mechanism cannot account for experimental data on the effective hydrogen pressure, at least in the case of Pd and Fe (see below), although it may be acceptable in terms of the Tafel slope.

It thus became clear through these studies that a detailed analysis of the HER mechanism is required for a quantitative understanding of the hydrogen permeation phenomena. As mentioned above, the dependence of the hydrogen permeation rate through mild steel[1–4,147] on the hydrogen overpotential was expressed by no means in a simple fashion,[134] and more detailed analysis was required.

2. Hydrogen Electrode Reaction on Pd

For the HER on Pd, Frumkin and Aladzhalova[20] were the first to report galvanostatic overpotential rise and decay transients, similar to those shown in Fig. 16.[27] They had the view that the overpotential component η_1 (which was thought to be equal to η'_1 in their concept) in Fig. 16 is the genuine hydrogen overpotential and η_2 (hence η'_2) is the shift of the electrode potential due to accumulation of hydrogen in Pd: that is, η_2 could be taken to be a kind of concentration cell emf. This view, which appeared to be adopted by others,[21,148,149] was thought to allow evaluation of the hydrogen concentration in Pd if the value of η_2 (or η'_2) is converted to hydrogen pressure using a Nernst-type equation and then combined with the hydrogen absorption isotherm for Pd.

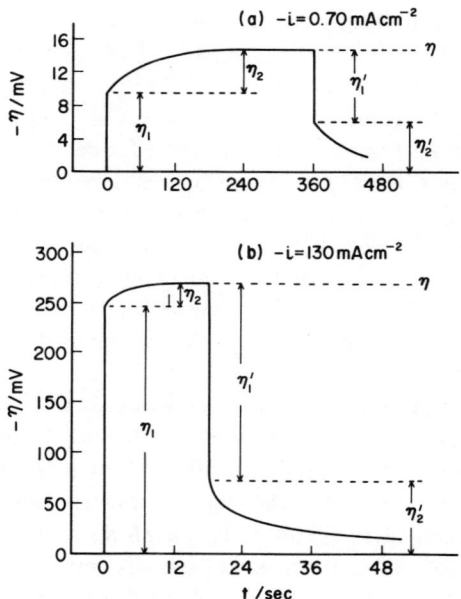

Figure 16. Typical galvanostatic hydrogen overpotential rise and decay transient curves for a Pd wire electrode (0.2-mm diameter) in $0.5M\,H_2SO_4$ at 303 K at low (a) and high c.d.(b).[27] Ohmic pseudo-overpotential has been eliminated.

The model, however, requires amendments in the present authors' opinion. Frumkin and Aladzhalova's concept would imply that the amount of absorbed hydrogen should, in principle, be independent of the (genuine) hydrogen overpotential (i.e., η_1 in their model) in a direct sense. In other words, there is no way of interpreting how η_2 (η'_2 is related to the polarization condition, typically η. Is the equivalence of η'_2 to a concentration cell emf a result of hydrogen accumulation due to the difficulty of removing hydrogen from the electrode by diffusion? A case that presumably is consistent with this view is that in which the electrode is highly active so that the hydrogen overpotential (in particular, the overpotential component for the Tafel step; see below) is negligibly small, and the hydrogen evolution rate is controlled by the limited rates of diffusion of hydrogen away from the electrode: Indeed, such behavior was reported by Barton and co-workers[47,48] for the case of palladized or "flamed," and thus active, Pd electrodes. The highest equivalent hydrogen pressure established on such electrodes was deduced to be $\sim 10^2$ atm.

If this is the case, the amount of absorbed hydrogen would just depend on the applied c.d. and agitation of the solution. On the other hand, the properties of electrode materials should not play any important role in the process of accumulation of hydrogen except for their capacity to absorb hydrogen. At least on ordinary Pd electrodes, however, addition of surface-active substances strongly influences the value of η'_2, as will be discussed later.[28] In fact, there have been a number of reports which indicate strong effects of poisoning additives on hydrogen entry into Pd,[140] Fe,[1-4,22,150-158] Ni,[101,157-161] Ti,[162-164] etc. Also, it was seen that the values of η'_2 were often much higher than those anticipated from a diffusion model.[1-4,36,140]

The overpotential rise and decay curves are almost symmetrical with respect to each other, as in Fig. 16a, if they are recorded at low polarization c.d., thus apparently not conflicting with the view proposed by Frumkin and Aladzhalova,[20] but such symmetry is lost when they are recorded at high c.d., as seen in Fig. 16b. It is, therefore, clear that the abovementioned type of interpretation is not generally applicable, especially for smooth electrodes like the one used to obtain the curves shown in Fig. 16, and one must realize that no common interpretation is possible for η_2 on the rise curve and η'_2 on the decay curve. A revised view of Fig. 16 will be discussed below, with a particular emphasis on the overpotential decay characteristics. In short, it will be shown that η'_2 is the quantity that may

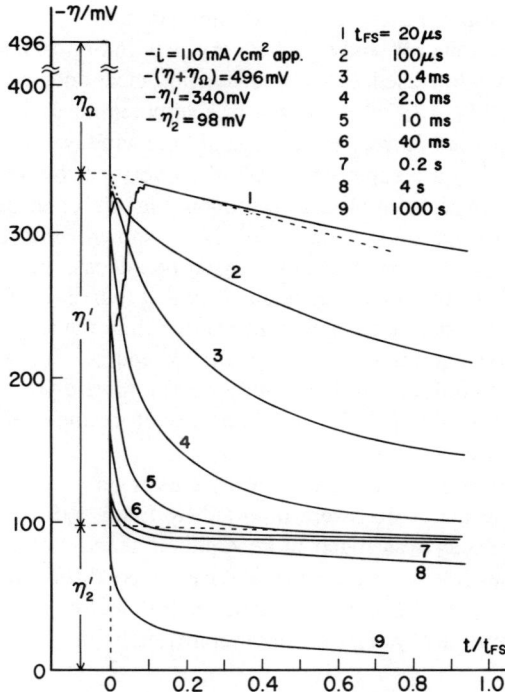

Figure 17. Superposition of overpotential decay curves observed in time windows of various lengths given by the full-scale time, t_{FS} for a Pd foil electrode ($0.012 \times 2 \times 5$ mm) in $0.5M$ H_2SO_4 at 303 K.[27,28] After attaining a steady value of overpotential, a current cutoff pulse from a pulse generator with a prefixed time length was applied to a solid-state switch that was placed in the galvanostatic (or potentiostatic) polarization circuit. The resultant potential decay was recorded in a transient recorder which was synchronized with the pulse generator. The polarization circuit was automatically closed after the pulse. The second (and further) measurement with a different t_{FS} value was carried out after waiting an appropriate time to restore the original overpotential value. A number of decay curves were then stacked on a pen recorder chart. η_Ω represents ohmic pseudo-overpotential.

be assigned to the free-energy decrease *in the Tafel step* and is the quantity that is responsible for the hydrogen accumulation.

The separation of η_1' and η_2' is not very clear in Fig. 16 because of the narrow time window used, but it is sufficiently clear once we expand the time scale. In Fig. 17,[27,28] a number of overpotential decay curves observed with the use of a very wide range of time windows (full-scale time t_{FS} of 10^{-5}–10^3 s) are superposed. Within a very narrow time window, $t_{FS} < 10^{-4}s$, the overpotential decay is monotonous and essentially satisfies a linear relation on a log time scale[165] (not shown) with a slope that generally agrees with the Tafel slope. After this decay, the overpotential does not decay further to any appreciable extent (curves 5–7) and apparently reaches practically a steady value in the time range of 10^{-1} s after the polarization current is switching off: Of course, to satisfy a basic requirement of equilibrium, it certainly decays toward 0 V after a very long time (curve 9)—10^3–10^4 s in the present example of a 0.1-mm-diameter Pd wire.

Rather different overpotential decay curves on a Pd electrode prepared by electrodeposition were reported by Elam and Conway[165]; the overpotential decay was much more rapid. It seems probable that the structure of the electrode was such that the effective diffusion distance for hydrogen in Pd particles was very short, i.e., the electrode was effectively porous, so that the decay of η_2' did not require much time, being completed within seconds.

3. Activity of Hydrogen Adatoms and Hydrogen Pressure Equivalent to Hydrogen Overpotential

If one accepts the Volmer–Tafel reaction mechanism quoted above for the HER, then it follows that the reaction intermediate, H(ads), must be the source of the hydrogen H(M) that enters into the metal electrode.

$$H^+ + e^- \rightarrow H(ads) \rightarrow \frac{1}{2} H_2(gas)$$

$$\downarrow$$

$$H(M) \rightarrow \frac{1}{2} H_2(M) \tag{32}$$

Here, $H_2(M)$ represents the hypothetical existence of molecular hydrogen in the metal, which would indeed be realized if there were a cavity in the

Sorption of Hydrogen

bulk of the electrode: Its pressure after a steady state has been attained is called the *equivalent hydrogen pressure*.

The level of activity of the hydrogen adatoms that is attained at a given value of the hydrogen overpotential depends on the relative rates of steps of the HER. In order to proceed with our analysis, it seems very important to revise our views on the rate-determining characteristics of chemical processes. Thus, it has been, indeed, a firmly established practice in mechanistic studies of electrode (and other) processes to assume the existence of a unique rate-determining step (RDS) and, *simultaneously*, quasi-equilibrium states of all the other steps. However, there is much experimental evidence which does not substantiate these assumptions (particularly the latter one). Furthermore, as discussed below, any detail in the consideration of the activity of the reaction intermediates would be trivial unless these assumptions are abandoned. Typical examples will be presented below.

If one assumes the Volmer–Tafel reaction path, Eq. (31) of the HER and supposes that the Tafel step is the RDS, while the Volmer step is always in quasi-equilibrium (or the free-energy decrease associated with this step is effectively zero), it can be shown readily that the activity of the hydrogen intermediate may be given by a simple, Nernst-type equation.[166] Accordingly, the hydrogen pressure that is hypothetically in equilibrium with such a hydrogen intermediate, or the equivalent hydrogen pressure, may be calculated from the (total) hydrogen overpotential by a relation analogous to the Nernst equation for a concentration cell.

On the other hand, if it is alternatively supposed that the Volmer step is the RDS, then the equivalent pressure should be maintained constant at the level that corresponds to the reversible state before the start of the polarization; that is, no increase of the equivalent hydrogen pressure above that at the equilibrium potential is anticipated: The reasoning is that there must be an *assumed* equilibrium condition between the reaction intermediate hydrogen adatoms and the hydrogen in the real gaseous phase, which is supposed to be maintained constant at, e.g., 1 atm.

Should we accept the rate-determining Tafel mechanism and use a Nernst-type equation in evaluating the equivalent hydrogen pressure from the total overpotential, in view of the fact that there is at least a significant amount of accumulation of hydrogen in Pd[167]? Or should we make other *ad hoc* assumptions so that the theory fits the experimental data[82]? One may think of the case in which the mechanism changes at a certain value of the overpotential for some unknown reason.[23] This type of approach, however, fails in the quantitative interpretation of experimental results.[26]

Experimental investigations of Fe (mild steel)[1–4] (cf. Fig. 2 above) and Pd[25–29] (Fig. 18) hydrogen electrodes indicated that the experimental data conformed to neither of the two above cases. Rather, the actual situation is somewhere between these two extremes; the equivalent hydrogen pressure certainly increases with increase of the cathodic hydrogen overpotential, but not as extensively as a Nernst-type equation requires. It follows, therefore, that in order to interpret quantitatively the increase of the equivalent hydrogen pressure with cathodic hydrogen overpotential, which is, however, lower than that predicted from a Nernst-type equation, it is inevitable that one assume a mechanism that is somewhere between those corresponding to the situations in which the Volmer step is rate-

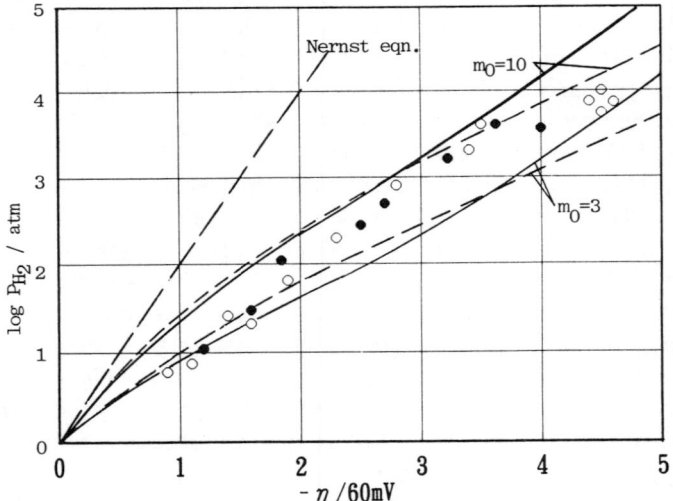

Figure 18. Hydrogen overpotential vs. the equivalent hydrogen pressure for a Pd electrode in $1M$ H_2SO_4 at 303 K.[27,28] The dashed lines were calculated with $m_0 = 3$ or 10 ($m_0 = i_{0V}/i_{0T}$) and $\theta_0 = 0.1$, assuming Langmuir isotherm, the solid lines were similarly calculated assuming a Temkin isotherm (with the interaction factor $u = 5$), and the broken line was obtained by application of the Nernst equation to η.

determining and the Tafel step is rate-determining: that is, one must consider the reaction to be under mixed control.

It has to be stressed here that mixed control is not restricted to exceptional circumstances that may appear in a limited number of situations but rather is ordinarily the case. Conversely, a unique RDS may exist (though it is not very likely) in limiting cases only. Other evidence in support of this view of mixed control has been presented elsewhere.[168]

An interpretation of the overpotential dependence of the equivalent hydrogen pressure is given here on the basis of a general analysis of reaction kinetics in terms of "mixed rate control," that is, eliminating the assumption of any unique RDS and, instead, basing the analysis on a number of steps having rate constants not extremely different from each other.[25,26] It may be stressed that this is done not just to somewhat improve the accuracy of the analysis by treating the case in a general way but to fully interpret the nature of the overpotential dependence.

Thus, the treatment is of essential importance: It has been demonstrated that even, for example, a 100-fold difference in the exchange c.d.'s of the reaction steps is not large enough to justify the approximation of a quasi-equilibrium condition for the rapid step, and, indeed, it is this deviation from the quasi-equilibrium condition that determines the activity of the reaction intermediate.

This type of analysis on the basis of the Volmer–Tafel type reaction mechanism on Pd has been substantiated well in that it can give a quantitative description of the hydrogen absorption phenomena.[140] On the other hand, no reasonable explanation was found possible if one assumes other reaction mechanisms such as the Volmer–Heyrovsky type. Also, the model of direct entry of hydrogen from the gas phase, that is, after it is evolved as molecular hydrogen, was not found to be successful. Accordingly, the state of H(M) may be discussed by considering the behavior of H(ads) within the framework of the Volmer–Tafel HER mechanism. In this case, as will be described later, effects of poisons may also be considered in terms of their effects upon the electrocatalytic activity of the electrode with regard to individual reaction steps.[27,28]

The next task is a quantitative treatment of the rate of hydrogen permeation in the electrode metal, which is often discussed in terms of the hydrogen permeation rate (J) through a metal membrane electrode, in relation to the HER rate (i),[22,23] e.g., $(d \log J)/(d \log i)$. In short, the analysis starts by expressing the concentration, or surface coverage θ_H, of the reaction intermediate H(ads) as a function of applied c.d., in accordance with an HER mechanism, and then combining this expression with a Fick's equation for the permeation rate as a function of θ_H. This approach may not be very convenient as it depends critically on the rate expressions used on the theoretical side and involves rather an ambiguous technique of evaluation of θ_H on the experimental side. It may be worth noting that θ_H at a given value of the activity of H(a) may vary widely among different metals, depending on their hydrogen adsorption isotherms. In the following discussion, therefore, we will instead use the *equivalent hydrogen pressure* as it represents the activity level and hence appears to be a more appropriate and convenient quantity.

More elaborate analysis was carried out later by Bockris et al.[166] An expression for θ was derived on the basis of rate expressions, and then the hydrogen fugacity was evaluated using the adsorption isotherm. This approach is similar to the one developed by Enyo,[29] although the latter seems to be more general and can be used to interpret experimental observations for Pd, or other hydrogen-absorbing metals, without involving any assumption about the hydrogen adsorption isotherm as long as the equivalent hydrogen pressure is discussed. The essence of the analysis is presented below using the *cascade model*; for details of the analysis, the reader is referred to Ref. 29.

The HER on palladium is probably one of the best understood among the reactions on electrocatalytically active metals, owing to the rapid and extensive absorption of hydrogen which gives rise to extremely different time constants of the charge-transfer (Volmer) step and the chemical (Tafel) step, thus permitting a detailed analysis of the polarization behavior after their clear separation. As a consequence, the analysis allows evaluation of overpotential components which are attributable to the individual elementary steps.[25–29,140] The system thus gives information on how the total affinity (overpotential) is distributed among the elementary steps, the kinetics of the individual steps, and how the activity of the reaction intermediate is determined.

4. The Cascade Model for Sequential Reactions

(i) Mixed Rate Control and the Cascade Model

One can readily visualize the dependence of the activity of hydrogen adatoms on the hydrogen overpotential in the *cascade model*, shown in Fig. 19.[169] In Fig. 19a, which represents the case in which there is a unique RDS in a sequential two-step reaction, the chemical potential of the

Sorption of Hydrogen

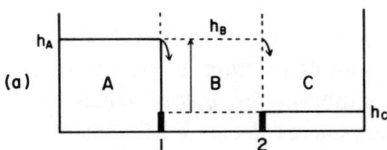

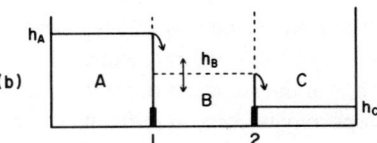

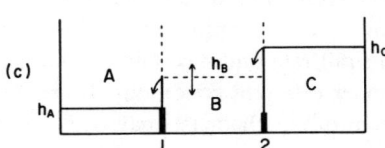

Figure 19. Schematic representation of the "cascade model"[169] for the case of a unique rate-determining step at step 1 (solid line) or 2 (dashed line) (a), mixed rate control with A as the reactant where step 1 behaves as the rate-controlling step (b), and mixed rate control with C as the reactant where step 2 behaves as the rate-controlling step (c).

reaction intermediate (water level h_B) is understood to be equal to either that of A (reactants), h_A, or that of C (products), h_C. In terms of the above-mentioned example, these correspond to either the equivalent pressure given by a Nernst-type equation or the equivalent pressure maintained constant irrespective of the polarization, respectively.

The case of mixed rate control is shown in Fig. 19b for the case of A as reactants. Now, h_B is dependent on the relative rates (hence rate constants) of the two steps and can take any value between h_A and h_C. Further, the level of h_B, and hence the value of, e.g., the equivalent hydrogen pressure and that of the H/Pd ratio (the hydrogen concentration in Pd), which are both connected to h_B, can be formulated with the use of appropriate kinetic rate expressions for the two steps. The net reaction rate is, on the other hand, controlled effectively by step 1. Some additional details have been discussed elsewhere.[29,170]

(ii) Distribution of the Reaction Affinity among the Steps and the Overall Reaction Rate

It may be worthwhile to comment briefly on the reaction affinity distribution among the steps and the overall reaction kinetics.[140] The reaction resistance of the overall reaction may generally be expressed as a sum of reaction resistances of the component elementary steps. Near equilibrium or at low values of the reaction affinity (i.e., the Gibbs free-energy decrease associated with the reaction), the affinity is distributed among the steps in accordance with the (reciprocals of) their exchange c.d.'s, by analogy with the concept of Ohm's law.

At high values of the affinity, however, the relative importance of the steps may change rapidly because of involvement of the electrode potential in some of the rate expressions and, unless the ratio of the exchange c.d.'s of the steps is extremely large (or small), the affinity values for both steps may become significantly large. As a result, even the rapid step may eventually deviate from a quasi-equilibrium condition, because it is not the relative magnitude of the individual affinity value but the absolute magnitude that determines whether the quasi-equilibrium condition obtains. Also, a rapid step that has negligible importance in controlling the overall rate under certain conditions may become kinetically important under different conditions. It can be shown that such a change in rate-controlling character may become important particularly when the step occurs in an early stage of the reaction, thus giving rise to a change in the overall reaction kinetics.[140]

Further, it can be shown that, with increase of the overpotential, the affinity distribution gradually approaches the situation of equipartition among the steps (or just becomes equipartitioned, e.g., in a two-step reaction, if the stoichiometric numbers of the steps are both unity).[140] Indeed, this is a reason why the affinity values for any step may eventually become high so that *all* steps become irreversible.

At sufficiently large overpotential values, h_B should not affect the flow rate of the first step (A → B) as long as the affinity value given to the first step is large enough so that its net rate is effectively given by the forward unidirectional rate, or its reverse unidirectional rate is negligible. This is equivalent, so to speak, to the statement that "the flow rate of water going down in a waterfall is independent of the level of its basin." Further, the net rate of the overall reaction is solely determined by that of the first step because, after all, the rate of the second step (B → C) ought to be equal to that of the first step, provided that the capacity of B (the reaction intermediate) is small, as is usually the case, and thus the reaction (quickly) reaches a steady state. This may be the reason why, in many experimental observations, the reaction kinetics seemingly indicate the existence of a single RDS, even in cases in which a mixed-control mechanism is actually operative.

Figure 19c shows the case in which the sign of the affinity, caused by hydrogen absorption or hydride formation may be reflected in the kinetics of the electrode process taking place on its surface. That is, the electrocatalytic properties of the metal surface may be changed by absorption of hydrogen, e.g., through such changes as an increase in the lattice constant of the metal, new phase formation, etc. For example, in the case of galvanostatic overpotential transients on the Pd electrode, shown above in Fig. 16, the reaction parameters such as the "exchange c.d." of the Volmer step are not necessarily under the condition of steady-state polarization the same as those in the state at 0-V overpotential, i.e., at the very beginning of the application of the cathodic current pulse. Such a difference may influence the decay kinetics in the initial stages.

The "exchange c.d." of the Volmer step at each value of the steady-state polarization (and not at $\eta = 0$) could be evaluated from the time constant involved in the initial portion the overpotential decay transients, namely, from the nature of the overpotential decay starting from η and reaching η'_2. Results obtained for a Pd wire electrode in $0.5M$ H_2SO_4 at 303 K in the cathodic region are shown as $\overline{i_{0V}}$ in Fig. 20[36] as a function of $-\eta_\infty$, where η_∞ is the overpotential at which $\overline{i_{0V}}$ has been evaluated.[36] The Tafel plots of $-\eta$ and $-\eta_2'$ observed at the same time are also given in the same figure. The value of $\overline{i_{0V}}$ was found to be only weakly dependent on the steady-state overpotential, and the dependence was of such an extent that it could be accounted for in terms of changes in the surface coverage with H(ads), which were brought about by the polarization: The change in the kinetic parameters of the Volmer step caused by changes in the surface structure of the substrate electrode is thus not considered appreciable.

By the same token, during long periods of polarization the Pd electrode does not show any appreciable change of overpotential with time, in contrast to the cases of Zr and Ti discussed below. The lack of any appreciable change in the electrocatalytic activity of Pd with respect to the HER as a result of charging of the electrode with hydrogen may be due to the fact that the system under the whole range of cathodic polarization is, after all, within the range of β-phase H/Pd, since the potentials involved are always negative to the $\alpha + \beta$ two-phase region, namely, 50 mV vs. RHE at 303 K.

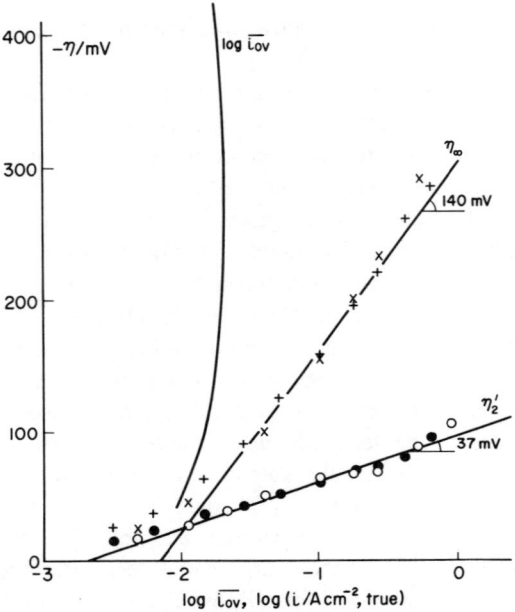

Figure 20. Typical plot of variation of $\overline{i_{0V}}$ with η, together with the Tafel plots of η and η_2', obtained for a Pd wire electrode (0.2-mm diameter) in 0.5M H$_2$SO$_4$ at 303 K.[36]

(ii) Zr and Ti

The variation of the electrocatalytic activity seems to be more significant in the case of Zr and Ti. Figure 21 shows the time dependence of the overpotential on Zr under galvanostatic cathodic polarization conditions.[174] The overpotentials at given c.d. values appeared to have attained their steady-state values soon after the start of the polarization (1 min), but, after prolonged polarizations, they started to decrease and tended to approach new values. It is seen that the time needed to start the decay is certainly far greater than that required to establish a quasi-steady value of the hydrogen-intermediate concentration on the electrode surface, and the start of the decay roughly corresponds to the beginning of the hydride-phase formation, discussed below. Some reduction in the cathodic overpotential occurred after the sample surface was covered with a hydride layer, as will be shown below, and the change was 200–300 mV for Zr in all the solutions studied (50 mV for Ti in sulfuric acid solution[174]).

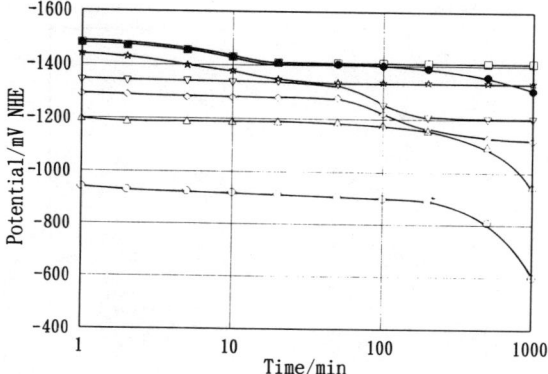

Figure 21. Stability with time of potential on a Zr hydrogen electrode polarized in various solutions.[174] △, ◇, ▽, ☆, and □, In 0.25M Na$_2$SO$_4$ at 0.1, 0.5, 1.0, 5.0, and 10 mA cm^{-2}, respectively; ○, in 0.5M H$_2$SO$_4$ at 10 mA cm^{-2}; ●, in 1M NaOH at 10 mA cm^{-2}.

On the other hand, Bockris *et al.*[166] observed a positive change of potential during a long period (months) of cathodic polarization of Pd in 0.1M LiOD: The change reached 200–300 mV in the c.d. range of 0.1–10 mA cm^{-2}. At the same time, they observed the presence of Pt and Zn by means of X-ray photoelectron spectroscopy (XPS) and energy dispersive spectroscopy (EDS).

The fact that the initial level of electrocatalytic activity was restored once the hydride was decomposed may exclude the possibility that this activity increase could be caused by, e.g., an increase in the surface roughness of the electrode. It may be concluded, therefore, that the overpotential decrease is due to an enhanced electrocatalytic activity of the hydride phase as compared to that of the corresponding metallic phase. (On the other hand, the decrease of electrocatalytic activity by the formation of hydride has also been reported[175].) It is interesting that the phenomenon took a much longer time to take place than one would anticipate; it is not well understood why a thick layer of hydride was needed to cause such a change in the electrocatalytic activity.

The overpotential values observed in the early stage of polarization on Ti and Zr generally show linear Tafel lines over an appreciable range of c.d. in various solutions, as shown in Fig. 22 for Zr.[174] A strong tendency for the overpotential to decrease with time is seen. Incidentally, the slope,

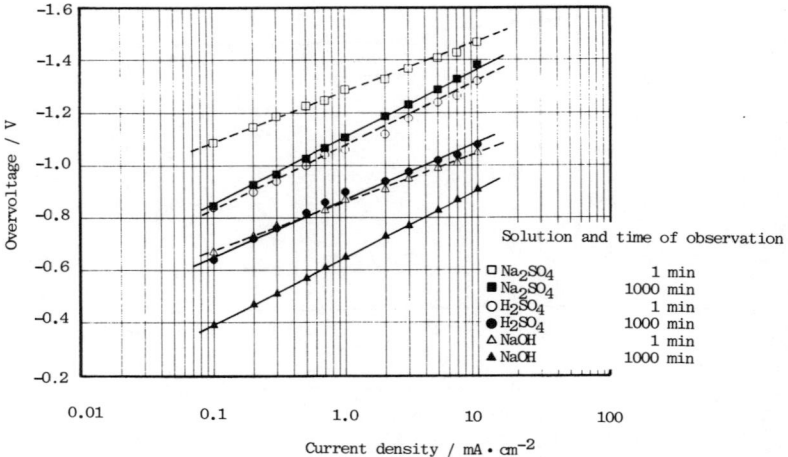

Figure 22. Tafel plots of the hydrogen overpotential on Zr in various solutions measured 1 min (open symbols) and 10^3 min (closed symbols) after the start of polarization.[174]

~140 mV, would indicate that the reaction rate is effectively determined by the charge-transfer step, possibly the Volmer step, but at the same time the latter step, probably the Tafel step, is not in quasi-equilibrium since the polarization gives rise to a definite condition of hydrogen entry into Zr. This may therefore be another example of the mixed-rate-control mechanism discussed in the preceding section.

VI. HYDROGEN PERMEATION IN METALS

Various factors may influence the properties of metals with regard to hydrogen absorption, permeation, hydride formation and of depth of the hydride layer, hydrogen concentration profile, etc. On the metal side, these factors may include surface structure, surface roughness, the presence of an oxide layer, grain boundaries, lattice faults, etc.; on the solution and ambient side, pH, electrolyte, solvent, surface-active substances (surfactants), temperature, and polarization condition are among the factors that may play a role. In what follows, however, only a few limited topics will be touched upon.

1. Permeation of Hydrogen across the Metal Surface Layer and Effects of Poisons

In the process of hydrogen penetration into metals, there must be a process of ingress of adsorbed hydrogen, becoming absorbed hydrogen. No precise data appear to be available for that rate at Pd, but its approximate lower-limit value may be estimated from the observation of the absence of gaseous hydrogen evolution during the initial period of application of a definite strength of cathodic current to a hydrogen-lean electrode. Similarly, the initial (but after removal of *ad*sorbed hydrogen) current upon application of an anodic stripping potential pulse to a hydrogen-charged metal would represent the rate of the process in the reverse direction. On a smooth Pd electrode, this appeared to be at least 0.1 A cm^{-2} in c.d. units at 300 K.[176] Accordingly, it may be assumed that the rate of entry is rapid as compared with the hydrogen diffusion rate in Pd in the majority of cases.

Incidentally, the fact that this value is much larger than the initial rate of entry of gaseous hydrogen into Pd suggests that the rate of dissociative adsorption of H_2 is a slower process.

At Fe, the rate in c.d. units of this ingress/egress process should be much lower than at Pd; it was reported to be 6.7×10^{-5} A cm^{-2} at 298 K.[144]

In connection with this, the effects of surfactants on hydrogen permeation are of interest. Some investigators seem to take the view that some surfactants *promote*[22,146,157,162] the entry of hydrogen into metals by, e.g., weakening the metal–hydrogen bond strength. In the present authors' opinion, this does not appear to be plausible: The motive force, or affinity, of the hydrogen permeation reaction should be dependent on the chemical potential of hydrogen adsorbed on the surface, but the latter should not, in principle, be influenced by the bond strength. For example, in a gas adsorption system, the surface population of adsorbed species under a fixed value of gas pressure may certainly be dependent on the strength of adsorption, but the chemical potential of the adsorbed species is simply determined by the adsorption equilibrium; that is, it may be equated with the chemical potential of the gaseous species. Kinetically also, the lower the bond strength of a hydrogen adatom to the metal surface, the easier it will be to remove, but at the same time the lower will be the adatom concentration under the same conditions.

An analogous view was also presented on the basis of the observation that H_2S, a typical "poison," had the effect of promoting the rate of hydrogen permeation through an Fe membrane when it was added to gaseous hydrogen supplied to the entry side.[157] However, this interpretation may not be free of ambiguity because the hydrogen gas was *atomized* in that work, and hence another explanation of the effect could be that the chemical potential of hydrogen adatoms became higher on the entry side surface as the consumption rate of hydrogen adatoms by $2H(a) \rightarrow H_2$ recombination would have been retarded by H_2S.

By the same token, retardation of hydrogen egress from, e.g., hydrogen-charged Pd by surfactants, as has often been reported, should not be taken as unequivocal evidence of the sluggishness of the hydrogen ingress/egress rate across the surface zone: It may be possible, if not more likely, that the decrease of the desorption rate is due to retardation of the recombination rate of hydrogen adatoms. In support of this view, escape of hydrogen from hydrogen-charged Pd is strongly retarded by a poisoning surfactant present on the surface, but, nevertheless, a large negative value of the concentration cell emf of a hydrogen-charged Pd (after the moment of appearance of η_2'; cf. Fig. 16) is rapidly ($\sim 10^{-3}$ s) restored even after application of an anodic stripping charge pulse that should be sufficient to remove all the hydrogen adatoms.[97] This indicates that the chemical potential of hydrogen adatoms is restored readily after the anodic charge pulse, owing to rapid hydrogen egress from Pd.

It seems likely, therefore, from the point of view of the HER mechanism discussed above, that the retardation of hydrogen recombination caused by the surfactants, giving rise to an increase in the activity of hydrogen adatoms and thus in the equivalent hydrogen pressure during the cathodic polarization, is the cause of the enhanced rate of hydrogen permeation into the electrode metal.[24,154,155,163,177]

The effect of As^{3+} on the process of hydrogen ingress into and egress out of an Fe membrane may also be understood by this line of reasoning.[178] As^{3+} ions were found to promote hydrogen ingress when they were present in the solution on the polarization (ingress) side of the membrane but to retard it when present on the diffusion (egress) side. An attempt to explain this by assuming that the catalytic effects of As^{3+} are opposite in the ingress and egress processes is unreasonable, as it conflicts with a general rule of microscopic balance of catalysis/poisoning, namely, that the active species should be effective in lowering/raising the activation energies of the forward and backward processes of the reaction at the same time. On the other hand, these findings may readily be explained by assuming that As^{3+} retards the recombination process.

2. Hydrogen Permeation into the Pd Electrode

(i) Diffusion Coefficient of Hydrogen in Pd

Data on the diffusion coefficient of hydrogen in α-H–Pd[179–189] have been well summarized by Völkl and Alefeld[72] and are presented in Fig. 23. The equation of the line in the figure is

$$D_{\alpha,H} = 2.90 \times 10^{-3} \exp[-22.19 \text{ (kJ)}/RT] \quad (\text{cm}^2 \text{ s}^{-1}) \tag{33}$$

which yields $D_{303} = 4.3 \times 10^{-7}$ cm^2 s^{-1}.

A comparison of diffusion coefficient data for H and D in Pd is given in Table 2.[179–182,190,191] The mean value of the diffusion coefficient of hydrogen in α-H–Pd, $D_{\alpha,H}$, may be expressed as [in fair agreement with Eq. (33)]

$$D_{\alpha,H} = 3.87 \times 10^{-3} \exp[-23.17 \text{ (kJ)}/RT] \quad (\text{cm}^2 \text{ s}^{-1}) \tag{34}$$

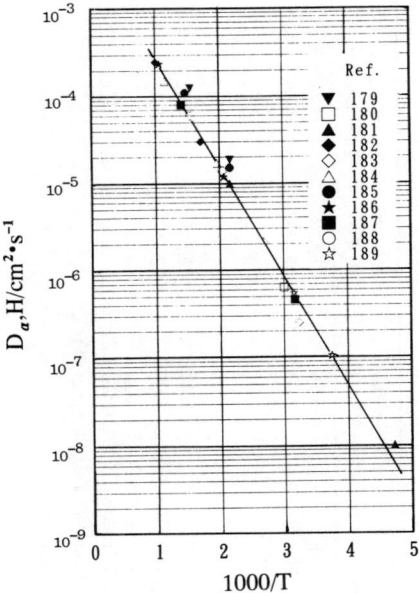

Figure 23. Diffusion coefficient of H in Pd, as summarized by Völkl and Alefeld.[72]

Table 2
Diffusion Coefficients of Light Hydrogen and Deuterium in α-Phase Hydrogen–Palladium System: Reported Values of the Parameters in $D = D_0 \exp(-E_a/RT)$ and of D at 303 K

Investigators	$D_{\alpha,H}$			$D_{\alpha,D}$			Reference
	D_0 (10^{-3} cm^2 s^{-1})	E_a (J)	D_{303} (10^{-7} cm^2 s^{-1})	D_0 (10^{-3} cm^2 s^{-1})	E_a (J)	D_{303} (10^{-7} cm^2 s^{-1})	
Toda (1964)	5.96	23,620	5.08	2.46	20,060	8.61	179
Bohmholdt and Wicke (1967)	3.65	23,430	3.35	2.5	21,550	4.84	180
Völkl et al. (1971)	2.52[a]	17,400[a]	25.3[a]	1.7[a]	19,860[a]	6.47[a]	181
Gol'tzov et al. (1970)	5.25	25,920	1.80	4.46	25,700	1.67	182
Sekine (1977)	3.44	22,530	4.52	2.46	21,230	5.41	190
Fujita (1991)	1.05	20,350	3.27	1.07	19,040	5.61	191
Mean	3.87	23,170	3.6	2.59	21,520	5.1	

[a]Excluded in evaluating the mean values.

or it is about 3.9×10^{-7} cm^2 s^{-1} at 303 K. The data available for deuterium may be summarized as

$$D_{\alpha,D} = 2.59 \times 10^{-3} \exp[-21.72 \text{ (kJ)}/RT] \quad (\text{cm}^2 \text{ s}^{-1}) \quad (35)$$

or $D_{\alpha,D} \approx 5.1 \times 10^{-7}$ cm^2 s^{-1} at 303 K. Thus, the difference in the diffusion coefficient values for hydrogen and deuterium in α-H(D)–Pd is not large.

Hydrogen diffusion in Pd is thus far more rapid than the diffusion of metal in solid metal, such as Na diffusion in Na or Ag diffusion in Ag (diffusion coefficients of 10^{-8} and 10^{-33} cm^2 s^{-1}, respectively at 303 K), but nevertheless is not always rapid enough to be kinetically neglected, process.

The diffusion coefficient of hydrogen in β-H–Pd is believed to be larger than the above-mentioned figure for α-H–Pd. Table 3 gives typical data[141,180,183,192–197] for the comparison of the diffusion coefficients of hydrogen in α- and β-H–Pd, $D_{\alpha,H}$ and $D_{\beta,H}$, respectively.

In connection with this, it is likely that at least some of the data in Table 2 might have been overestimated, because of possible involvement of β-H–Pd. For example, a value of $D = 3 \times 10^{-6}$ cm^2 s^{-1} was obtained with the use of a frequency response technique,[198] but it may be that the value is higher than the one given above for α-H–Pd because of inclusion

Table 3
Diffusion Coefficient of Hydrogen in α- and β-H–Pd (P_{H_2} = 1 atm) at 298 K[a]

Investigators	$D_{\alpha,H}$ (10^{-7} cm^2 s^{-1})	$D_{\beta,H}$ (10^{-7} cm^2 s^{-1})	Reference
Devanathan and Stachursky (1962)	1.3 ± 0.2	—	141
von Stackelberg and Ludwig (1964)	1.6 ± 0.5	30–50	196
Simons and Flanagan (1965)	2.5	—	197
Jewett and Makrides (1965)	3.1 ± 0.6	15	195
Kahrig et al. (1966)	—	27 ± 3	192
Bohmholdt and Wicke (1967)	2.6 ± 0.4	17 ± 3	180
Bohmholdt and Wicke (1967)	2.65 ± 0.25	14.5 ± 1.5	180
Kuballa and Baranowski (1974)		16	193
Bucur (1986)	3.60 ± 0.06[b]	—	194
Mean	2.3	23	

[a]Mainly after Holleck and Wicke (Ref. 183).
[b]At 293 K.

Table 4
Diffusion Coefficient of Hydrogen (and Its Isotopes) in Various Metals: Reported Values of the Parameters in $D = D_0 \exp(-E_a/RT)$ and of D at 303 K[a]

System	Condition	D_0 (cm^{2}s^{-1})	E_a (kJ)	D_{303} (cm^{2}s^{-1})
H/Pd	α-Phase	2.9×10^{-3}	22.19	4.3×10^{-7}
H/Pd	β-Phase			2.3×10^{-6}
D/Pd	α-Phase	2.6×10^{-3}	21.52	5.1×10^{-7}
H/Ni	$T > T_c^b$	6.9×10^{-3}	40.52	7.2×10^{-10}
H/Ni	$T < T_c$	4.8×10^{-3}	39.36	7.9×10^{-10}
H/Fe	(scattered)	7.5×10^{-4}	10.13	1.3×10^{-5}
H/Nb[c]	$T > 273$ K	5.0×10^{-4}	10.23	8.6×10^{-6}
H/Nb	$T < 223$ K	0.9×10^{-4}	6.56	6.6×10^{-6}
D/Nb[d]		5.2×10^{-4}	12.25	4.0×10^{-6}
T/Nb		4.5×10^{-4}	13.02	2.6×10^{-6}
H/Ta[e]	$T > 273$ K	4.4×10^{-4}	13.51	2.1×10^{-6}
H/Ta	$T < 198$ K	2.0×10^{-6}	3.86	4.3×10^{-7}
D/Ta[f]		4.6×10^{-4}	15.44	1.0×10^{-6}
H/V		3.1×10^{-4}	4.34	5.5×10^{-5}
D/V		3.8×10^{-4}	7.04	2.3×10^{-5}
H/Ti[g]	$T = 297$ K		28.8	3.8×10^{-12}
H/Ti[h]				6.1×10^{-12}
H/Ti[i]		5.8×10^{-2}	30.8	2.7×10^{-7}

[a] Mainly after Völkl and Alefeld (Ref. 72).
[b] T_c Curie temperature.
[c] At H/Nb = 0.07.
[d] At D/Nb = 0.048.
[e] At H/Ta = 0.056.
[f] At D/Ta = 0.038.
[g] Ref. 199.
[h] Ref. 200.
[i] Ref. 201.

of β-H–Pd in a significant portion of the sample. Various data on the diffusion coefficient of hydrogen and its isotopes in Pd discussed above are listed in Table 4[72,199–201] together with data for other metals, which are mainly taken from Völkl and Alefeld.[72]

The permeation of light hydrogen/deuterium in Pd is nearly always controlled by diffusion, and hence hydrogen charging or discharging is a relatively slow process. If one takes a specimen (e.g., cylinder) of 4-mm effective thickness (radius), the time required for hydrogen just to travel from the surface to its center will be of the order of 10^5 s at 303 K if its composition is within the α-phase. An even longer time would be required to attain a homogeneous concentration. On the other hand, if the major part of the hydrogen uptake involves the β-phase, as is often the case, then the time required will be roughly one order of magnitude shorter.

Although hydrogen permeation in Pd is believed to be mostly diffusion-controlled, there still remain various aspects yet to be considered, such as how diffusion processes in different H–Pd phases are interrelated, how the concentration profile of hydrogen near the surface is determined (examples will be presented later), etc. There are various phenomena that are difficult to understand on the basis of a simple diffusion model, such as the one discussed in Section VI.4.

Finally, it should be mentioned that there nearly always exists a sizable hysteresis between the hydrogen absorption and desorption curves in the P–C–T diagram at Pd, even if the processes are followed very slowly. Often, the desorption curve is taken to represent an *equilibrium* property, but the cause of the hysteresis is not well understood.

(ii) Light-Hydrogen/Pd System

Hydrogen permeation into the Pd electrode under cathodic polarization and the effects of, e.g., addition of surfactants have been discussed elsewhere.[25–29] In short, the equivalent hydrogen pressure (fugacity), which gives rise to the permeation, increases sharply with cathodic overpotential but not as strongly as a Nernst-type equation would predict (see Section V). The observed dependence of the equivalent hydrogen pressure upon hydrogen overpotential on a Pd electrode in acidic solution has been well explained on the basis of the Volmer–Tafel reaction route under a mixed-control model with $m_0 \equiv i_{0V}/i_{0T} = 3$–$10$ in acidic solution, as shown above (Fig. 18). The maximum pressure estimated was $\sim 10^4$ atm in acidic solutions.[25–29] The highest equivalent hydrogen pressure attained in later experiments with the use of thiourea as an electrocatalytic poison was evaluated to be higher than 10^5 atm at high c.d. (above 0.2 A cm^{-2}) (Fig. 24[97]).

An analogous overpotential–pressure relation is obtained in alkaline solution, as seen in Fig. 25 for the case of LiOH.[202] This is similar to the result in acidic solution (Figs. 18 and 24), and thus no drastic difference in the equivalent hydrogen pressure is noted between acidic and alkaline

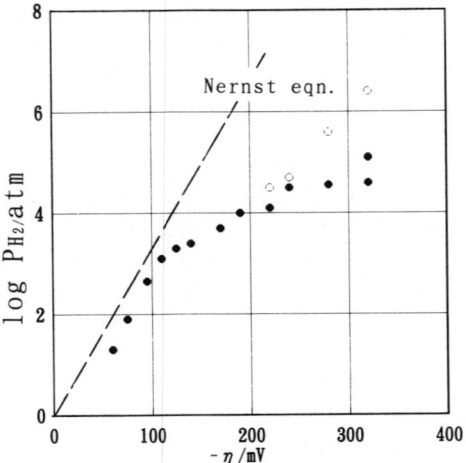

Figure 24. Equivalent hydrogen pressure as a function of cathodic overpotential on Pd foil, 5 μm thick, in 0.5M H_2SO_4 at 303 K with the addition of $1 \times 10^{-5}M$ (○) and $3 \times 10^{-5}M$ thiourea (●).[97] The broken line represents the case of the Nernst equation applied directly to the total overpotential.

solutions, although the total overpotential values are significantly higher in alkaline media. This suggests that the charge-transfer (Volmer) process accounts for a larger portion of the Gibbs free energy decrease (overpotential) in alkaline than in acidic solution.

Although the theory predicts further increase of the equivalent pressure with increase of cathodic overpotential, there may be a practical limit due to various causes, such as vigorous gas evolution, which may disturb the homogeneous current distribution, or deterioration of the electrode surface which leads to an increase of the roughness factor or a decrease of the true c.d. and, furthermore, possible rupture of the Pd specimen.[166] It is interesting that the approximate limiting *pressure* value at very high fugacity in Fig. 10 above, namely, $10^{4.5}$ atm, is roughly the value at which fracture of the Pd metal may occur according to the crack growth model.[166]

(iii) *Deuterium/Pd System*

A lower concentration of deuterium as compared with that of light hydrogen in Pd electrodes under comparable conditions may be antici-

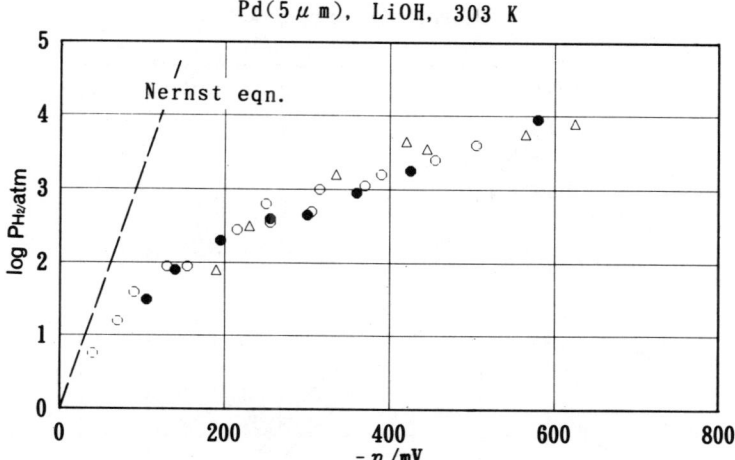

Figure 25. Equivalent hydrogen pressure as a function of cathodic overpotential on Pd foil, 5 μm thick, in 0.085M (open symbols) and 0.83M LiOH (closed symbols) at 303 K. The broken line is for the Nernst equation.[202]

pated if one considers only equilibrium properties, namely, the absorption isotherms (Fig. 7). On the other hand, not much is known about the kinetics of absorption processes, although the difference in the kinetics of light hydrogen and deuterium evolution reactions may play an important role in the process of cathodic hydrogen charging. Further, not much is known about the details of the deuterium evolution reaction, in particular with respect to η_2'.

Figure 26[198,203] shows the Gamow plot (Eq. 13) for a Pd sample; the progress with time of the neutron evolution rate, which is proportional to the deuterium concentration, is plotted against $E^{-1/2}$, where E is the deuteron beam energy. The Pd sample was cathodically polarized in 5 wt. % (1.3M) DCl at a c.d. of 10 mA cm^{-2} at 303 K for 1–20 min. The rate at which a steady value of neutron counts was reached is seen to be more rapid at low energy values, i.e., at shallow depths in the specimen, as indicated by the less extensive variation of the neutron evolution rate with time at low E; this fact indicates that the permeation of deuterium in Pd is not sufficiently rapid to attain a steady concentration of deuterium in a given time of polarization.

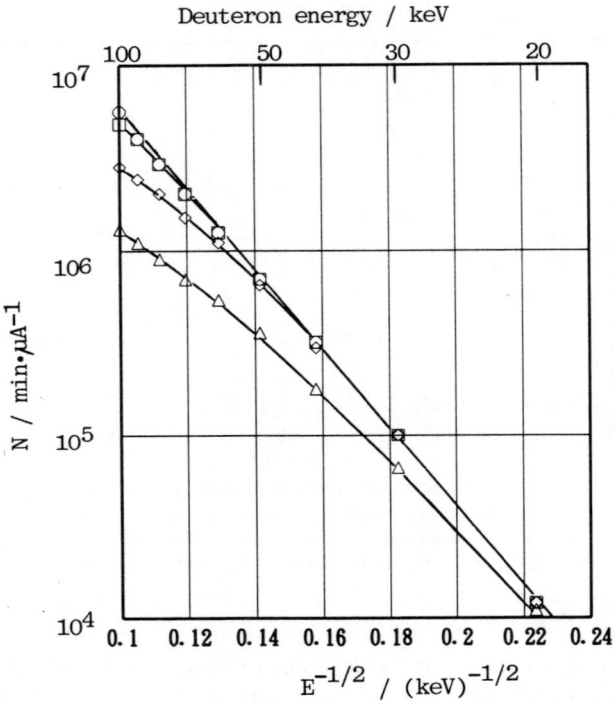

Figure 26. Gamow plot of the rate of neutron evolution from a Pd electrode in which deuterium absorbed was by cathodic polarization for various periods of time: △, 1 min; ◇, 3.5 min; □, 10 min; ○, 20 min. Electrolysis conditions: 5 wt. % DCl/D_2O solution at 303 K, 10 mA cm^{-2}.[93]

Furthermore, the rate of permeation of deuterium in Pd is strongly dependent upon the c.d. of the cathodization. Figure 12 above shows a semilog plot of the deuterium concentration in a Pd electrode, as measured by the neutron evolution rate, at various cathodic c.d.'s as a function of polarization time.[92,93] The time required for the curve to reach saturation is much shorter at higher c.d.'s. Each curve shows an approximately linearize before the stage at which the deuterium concentration finally reaches a saturation value, especially at low polarization c.d.'s. The slope of the rising portions of the curves, ca. 0.5, suggests that the process is essentially diffusion-controlled.

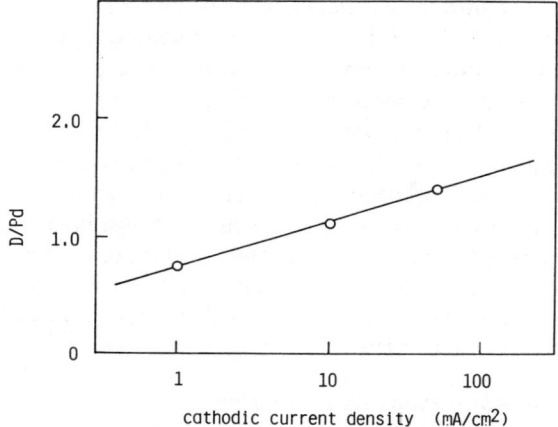

Figure 27. Saturation values of deuterium concentration in Pd plotted against the logarithm of the current density of cathodic polarization.[92,93] Deuteron beam energy, 100 keV; 0.3-mm Pd foil, 303 K, 5 wt. % DCl/D$_2$O.

The saturation values of D/Pd are plotted in Fig. 27[92,93] as a function of log c.d.(i); these may be summarized in an empirical linear relation,

$$\text{D/Pd} = 0.74 + 0.39 \log[i \, (\text{mA cm}^{-2})] \quad (36)$$

The highest value observed, D/Pd = 1.44 at 50 mA cm^{-2}, was rather high by comparison with the literature data. The reason for this high value, although it refers only to the very thin surface layer, is not well understood.

(iv) Absorption and Desorption Behavior of Deuterium

The absorption/desorption behavior of deuterium in Pd during cathodic polarization at very high c.d. is of interest in view of its importance in the alleged cold fusion phenomena. Most of the experiments on hydrogen permeation in the past were carried out in relation to purification of hydrogen gas, hydrogenation catalysts, etc., and hence under relatively mild conditions. Likewise, in electrochemical systems, studies on the Pd–hydrogen system have been mostly carried out at low c.d.'s, e.g., up to 10^{-2} A cm^{-2}. However, there is now interest in the behavior of this system under much more severe conditions.

Newer experimental techniques are being used to measure deuterium absorption in Pd. A closed electrolysis cell containing an H_2/O_2 recombination catalyst is often employed for *in situ* measurements of hydrogen absorption/desorption behavior: The catalyst converts the hydrogen and oxygen gases evolved into water.[203] In this cell, the oxygen gas pressure developed should indicate the amount of hydrogen that was absorbed into the Pd cathode. An improved type of cell, using a hydrogen ionization anode, has also been developed; in this cell, no change in the gas pressure should be involved if there is no absorption, and the gas volume additionally supplied to maintain the gas pressure at a constant level directly gives the amount of absorption. Results indicate that it is very difficult to attain concentrations exceeding H/Pd = 1.

(v) *Models of Absorption and Desorption*

Typical absorption/desorption behavior of deuterium in a Pd rod is shown in Fig. 28.[93] The deuterium loading increased steeply at the beginning of the electrolysis but then slowed down and reached a steady value. Similarly, the deuterium release from Pd after termination of the electrolysis was rapid at the beginning, but again the rate slowed down

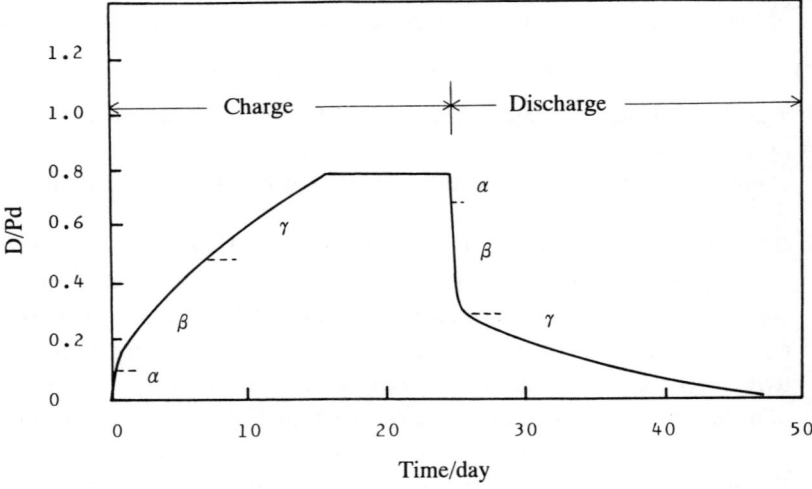

Figure 28. Behavior of deuterium absorption and desorption process in the D/Pd system [Pd rod (1-cm diameter), 44 mA cm^{-2}, 353 K].[93]

when the loading decreased to a certain value. Thus, we see three stages during both hydrogen absorption and desorption.

Three models may be considered in order to interpret the behavior of the rate of hydrogen absorption (Fig. 29): (a) a simple (cylindrical) diffusion model, assuming a single (α) H/Pd phase with a smooth distribution of the hydrogen concentration; (b) a protective film model, in which it is postulated that a film is formed on the metal surface and suppresses the hydrogen absorption; and (c) a new-phase-formation model, in which a constant rate of growth of a hydrogen-rich phase (named the γ-phase and having a low D value[93]) with a moving front toward the hydrogen-lean bulk Pd is postulated.

(a) Diffusion model (Fig. 29a)

A cylindrical diffusion equation, with abscissa $r = 0$ at the center of a cylinder of radius a,

$$\frac{\partial C}{\partial t} = D\frac{\partial^2 C}{\partial r^2} + \frac{1}{r}\frac{\partial C}{\partial r} \tag{37}$$

may be solved with the boundary conditions $C = 0$ at $0 \le r \le a$ and $t = 0$ and $C = C_s$ at $r = a$ and $t > 0$, where C is the concentration, C_s is the surface concentration, t is time, and D is the diffusion coefficient. Thus,

$$C(r, t) = C_s \left[1 - \frac{1}{a}\sum_{n=1}^{\infty} \exp(-D\alpha_n^2 t) \frac{J_0(\alpha_n r)}{\alpha_n J_1(\alpha_n a)} \right] \tag{38}$$

where J is the zeroth-order Bessel function and α_n are the roots of the equation $\alpha J_1(\alpha) - CJ_0(\alpha) = 0$. The total amount of deuterium is obtained by integration as

$$C(t) = C_s \left[1 - 2\sum_{n=1}^{\infty} \frac{1}{\alpha_n} \exp(-D\alpha_n^2 t) \right] \tag{39}$$

where the α_n are given in the table of Carslaw and Jaeger[204] as 2.4048 ($n = 1$), 5.5201 ($n = 2$), 8.6537 ($n = 3$), etc.

(b) Protective film model (Fig. 29b)

Difficulties of penetration of hydrogen through a phase boundary such as at a gas/solid or liquid/solid interface are often encountered at Ti

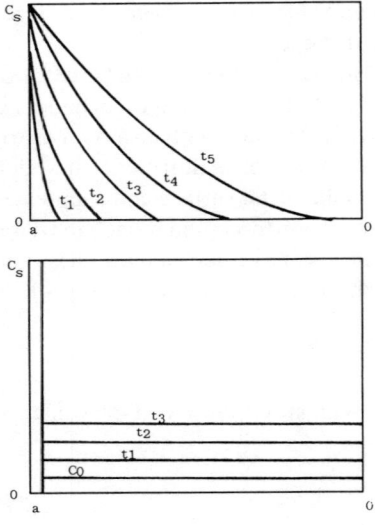
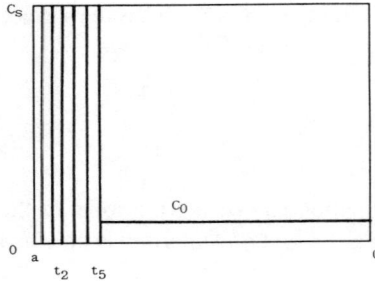

Figure 29. Schematic representation of models of hydrogen absorption in metals: (a) a diffusion model (Eq. 39); (b) a protective film model (Eq. 40); (c) a new-phase-formation (moving boundary) model (Eq. 41).

and Zr. It appears thus possible that hydrogen permeation is retarded by a film having a low permeability to hydrogen.[93] In this case, the rate of increase of the hydrogen concentration C (which is proportional to H/Pd) may be simply assumed to obey the relation

$$C/C_s = [(C_0/C_s) + kt]/(1 + kt) \qquad (40)$$

where k is the permeation rate constant and C_0 is the initial concentration.

(c) New-phase-formation model (Fig. 29c)

One may suppose the formation of a hydrogen-rich new Pd–H phase near the surface, with the front boundary moving with time toward the bulk. For a rod of radius a in which a cylindrical new phase layer of thickness r is formed,

$$C = C_s [1 - (1 - \xi)^2] + C_0(1 - \xi)^2 \tag{41}$$

where $\xi \equiv a/r$ ($0 \leq \xi \leq 1$). If $C_s \gg C_0$,

$$C/C_s = (2\xi - \xi^2) \tag{42}$$

or, assuming that the growth of the hydrogen-rich layer is proportional to time, namely, $\xi = kt$,

$$C/C_s = [2kt - (kt)^2] \tag{43}$$

Figure 30 shows typical plots of the increase of deuterium concentration with time expected for these three models. The question of which of these models is suitable to explain the experimental data may be discussed

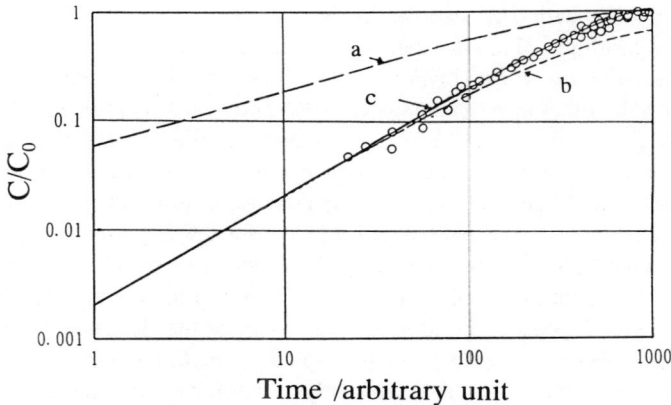

Figure 30. Relationship between amount of deuterium absorption in Pd and time, as anticipated for various models of absorption: (a) a diffusion model; (b) a protective film model; (c) a new-phase-formation model. Experimental points are also plotted.

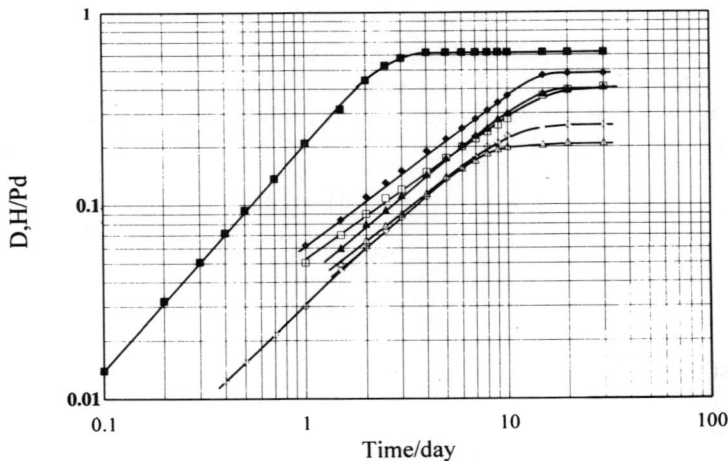

Figure 31. Increase of deuterium absorption in Pd with time under various conditions, as reported by Mizuno,[205] Kunimatsu et al.,[206] and Will et al.[207] ■, 200 mA cm^{-2}, 105°C, Ref. 205; ◆, 44 mA cm^{-2}, 140°C, Ref. 205; □, 44 mA cm^{-2}, 105°C, Ref. 205; ◇, 44 mA cm^{-2}, 80°C, Ref. 205; ▲, 30°C, Ref. 206; △, 30°C, Ref. 207.

by comparing the expected hydrogen absorption behavior according to these models with experimental results.

Logarithmic plots of experimental data of deuterium concentration against time are shown in Fig. 31.[205-207] All the data, including some obtained by other investigators, are similar and generally indicate proportionality with time for the most part and exhibit a relatively sudden approach to saturation. This behavior is consistent with the new-phase-formation model [Fig. 29c or Eq. (43)]. Accordingly, the electrode surface may be assumed to be covered with a layer of a new hypothetical hydride (named the γ-phase) with a high deuterium concentration which is realized under strong cathodic polarization. The inner region is composed of the α-phase (or β-phase, depending on the history of the electrode, temperature, etc.). We may suppose that the γ-phase is metastable and is formed only during high-c.d. cathodic electrolysis and further that the growth of the γ-phase at the interface between the γ-phase layer and the inside bulk phase will be rate-determining (i.e., we assume that the rates of all the other processes are sufficiently rapid). In this way, the model can interpret, for example, the experimental observation that the rate of deuterium absorption against time was linear during electrolysis on a large-size Pd rod reported above (Fig. 31).

The rate of growth of the hydride layer thickness appears to be linearly dependent on c.d. as shown in Fig. 32;[205-207] the data points in this figure are estimated from the data obtained under various conditions that are plotted in Fig. 31, and the evaluation by the d–n reaction method is concerned with the initial region of hydrogen absorption. This linearity would suggest that the process here is not diffusion-limited but is simply determined by the rate of hydrogen supply by electrolysis.

On the other hand, as shown in Fig. 28, the deuterium desorption process is composed of two steps: deuterium gas is released rapidly in the first step within several hours after termination of the electrolysis and then very slowly in the second step, which lasts close to several weeks. The fast step seems to be the gas release from the β-phase layer because the desorption rate and its kinetics obey a diffusion model: The diffusion coefficient is estimated to be of the order of 10^{-6} cm^2 s^{-1}. Conversely, no suitable diffusion model was found that explains the abnormally slow second step, suggesting that the kinetics follow another model such as a phase transformation process.

The hydrogen absorption into Pd cathode under strong cathodization may be summarized as follows:

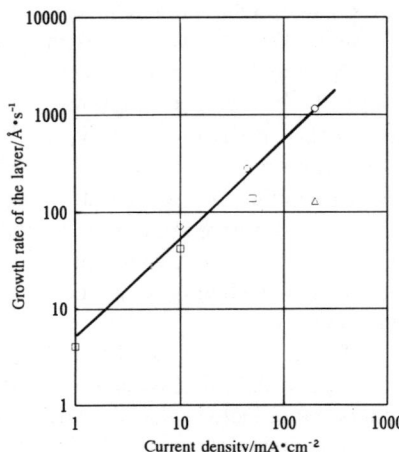

Figure 32. Dependence of the growth rate of hydride layer in Pd on current density of cathodization, as reported by Mizuno[205] (○, □), by Kunimatsu et al.[206] (△), and by Will et al.[207] (◇).

(a) The initial rapid step is the absorption of hydrogen into the β-phase, and this is then followed by the formation of the β-phase. It is known that the current efficiency of the hydrogen absorption often reaches 100% during cathodization under mild conditions. As the hydrogen entry is rapid, a layer with very high deuterium concentration is not likely to be formed at the Pd surface.

(b) The slower absorption stage sets in after the Pd specimen is practically filled up with β-phase so that no further hydrogen can be readily absorbed. Accumulation of hydrogen on the surface leads to a highly concentrated zone, and the latter may be transformed into a hypothetical metastable γ-phase hydride layer. This transformation process is considered to be rate-determining in the later stage of the hydrogen absorption process.

(c) The initial rapid desorption step occurs immediately after electrolysis is terminated; deuterium gas is released rapidly from the α- and β-phase.

(d) The slower desorption step sets in: The release of hydrogen from the γ-phase is postulated to be very slow, being entirely different from the release processes from the β- and the α-phase.

3. Hydrogen Permeation into Fe and Ni Electrodes

The diffusion coefficient of hydrogen in Fe is reported to be about 3×10^{-6} cm^2 s^{-1} for Armco Fe at 30–40°C,[157] 6.25×10^{-6} cm^2 s^{-1} at 25°C for α-Fe,[143] 3×10^{-6} cm^2 s^{-1} at 30°C,[208] etc. This magnitude of D is comparable to that in Pd. Consequently, permeation of hydrogen through Fe membranes may well be discernible,[1-4] as mentioned already.

The permeation of hydrogen through steel membranes is well recognized.[178,209,210] A convenient electrochemical means of evaluating the hydrogen permeation rate using an Fe membrane was demonstrated by Devanathan and Stachurski.[98] This sensitive and rapid way of measuring the permeation rate by means of observation of electric current makes it possible to observe not only the steady-state rate of permeation but also its transient behavior, which provides information on the diffusion coefficient. This technique was accordingly employed by many later investigators.

The dependence of the hydrogen permeation rate on the surface coverage with hydrogen adatoms led Devanathan and Stachurski[98] and others[99] to conclude that the HER mechanism on Fe in acidic solution involves a rate-determining Volmer step followed by a rapid Tafel step at low overpotential and then changes to a slow Heyrovsky (electrochemical) desorption mechanism at high overpotentials (>600 mV cathodic). On the other hand, other investigators[22,208] concluded that the coupled Volmer–Tafel mechanism was the operative mechanism in alkaline solution except at very high overpotential. The rate-determining step in alkaline media was identified by Beck et al.[154] as the Tafel step through investigations of impurity (CN^-) effects on the permeation rate: The view adopted was that the poison retards the Tafel reaction and thus leads to an increase in the activity of hydrogen adatoms on the electrode surface and hence an increase in the permeation rate. The latter type of mechanism of poisoning appears to be supported by other investigators.[153,195]

Investigations by DeLuccia and co-workers[1-4] of the permeation rate through an Fe membrane and evaluation of the equivalent hydrogen pressure, mentioned in Section I, indicated that the equivalent hydrogen pressure is far lower than the value anticipated from a Nernst-type relation directly applied to the total overpotential. This discrepancy was later explained by Enyo[25,26,140] through the use of the concept of mixed rate control within the framework of the Volmer–Tafel reaction route, as presented above in Section II. The value adopted for $m_0 \equiv i_{0V}/i_{0T}$ was ~0.12, and the fit of the calculated relation of the equivalent hydrogen pressure versus total overpotential to the experimental one is acceptable, as demonstrated in Fig. 2. No direct and independent experimental support for this m_0 value seems to be available, but it appears to be reasonable as the corresponding value on Ni is of the order of magnitude of unity,[29] as compared with ~50 on Pt and ~10 on Pd.

Through comparison of permeation rates in various forms of Fe specimens, e.g., polycrystalline versus single crystalline, the conclusion was drawn that grain boundaries do not contribute significantly to the hydrogen permeation process.[22,143] On the other hand, strain in the bulk seems to have significant effects on the absorption/diffusion kinetics.[210–212]

Intensive studies on hydrogen entry in Ni were carried out by Smialowski, Baranowski and co-workers.[50,101–103] The H/Ni value was demonstrated to reach ~0.7 at 298 K, with the use of various catalytic poisons (thiourea, SeO_2, As_2O_3, Na_2S, etc.). The depth of the hydride formed was dependent on the kind of poison. Like the H–Pd system, the H–Ni system exhibits an $\alpha + \beta$ two-phase region at –0.13 V vs. RHE after cessation of cathodic polarization. The kinetics of hydrogen desorption was first-order in hydrogen content in Ni, and thus the rate was considered to be controlled by decomposition of the hydride and not by a bulk diffusion process.

4. Hydrogen Permeation into Ti and Zr Electrodes and Hydride Formation

(i) Diffusion Coefficient

Diffusion coefficient data for Ti and their temperature dependencies are summarized in Table 5.[213-223] The value for α-Ti–H at 303 K reported by Brauer et al.,[216] 2.8×10^{-7} cm^2 s^{-1}, is far larger than the data from other reports and is closer to the values for Nb (5×10^{-6} cm^2 s^{-1}),[224] Ta (2×10^{-6} cm^2 s^{-1}),[225] or V (2×10^{-5} cm^2 s^{-1}).[226]

The data for β-Ti–H are rather few, and the difference between the value reported by Someno and Nagasaki[219] and those reported by others[214,215] is significant. In γ-Ti–H, Korn and Zamir[221] evaluated the value using a pulsed NMR technique in the temperature range 25–500°C. The heat of activation of the diffusion process was 48.9 kJ mol^{-1}. Stalinski et al.[220] determined the value in the γ-phase at 24°C; the heat of activation was 39.3 kJ mol^{-1} for TiH$_{1.6}$ and 42.7 kJ mol^{-1} for TiH$_{1.9}$. According to Brauer and Nann,[216] who measured the diffusion coefficient in the temperature range 24–66°C, the value was dependent on the polarization current, being 2.2×10^{-12}, 5.6×10^{-12}, and 7×10^{-12} cm^2 s^{-1} at 24°C at 5 mA cm^{-2}, 30 mA cm^{-2}, and higher c.d., respectively, and the heat of activation was 28.9 kJ mol^{-1} for TiH$_{0.9}$.

Diffusion coefficient data for Zr are mostly values at higher temperatures. Such data are summarized in Table 6.[223,227-236] The heat of activation of the diffusion process in the α-phase is usually 40–50 kJ mol^{-1}, while it is somewhat lower in the β-phase, and conversely somewhat higher in the δ-phase.

(ii) Kinetics of Hydrogen Absorption

A hydride layer beneath the electrode surface having very high hydrogen concentration is likely to be formed during hydrogen absorption into Ti and Zr, and the layer may retard the rate of hydrogen absorption, especially during electrolysis at high c.d. or for long charging times. The kinetics of hydrogen absorption into Zr in 0.5M Na$_2$SO$_4$ shown in Fig. 33[174] obeys a t^n ($n > 1$) rate law at the beginning of cathodic polarization at various c.d.'s but changes to a linear rate law (namely, the slope is unity) or to a parabolic law (slope is $\frac{1}{2}$) after the sample has absorbed roughly 0.03 and 0.3 cm^3 of hydrogen, respectively, per unit area (the latter case may correspond to H/Zr = 2.0 near the surface).

Table 5
Diffusion Coefficient of Hydrogen in α-, β-, γ-, and δ-Phase Ti–H: Reported Values of the Parameters in $D = D_0 \exp(-E_a/RT)$ and of D at 303 K

Investigators	D_0 (cm^2 s^{-1})	E_a (kJ mol^{-1})	Temperature (°C)	D_{303} (cm^2 s^{-1})	Reference
α-Phase					
Philips et al.	6×10^{-2}	60.3	25–100	2.56×10^{-12}	213
Papazoglou and Hepworth	3×10^{-2}	61.4	610–900	—	214
Wasilewski and Kehl	1.8×10^{-2}	51.8	500–824	—	215
Brauer and Nann	6×10^{-2}	31.0	20–80	2.85×10^{-7}	216
Williams	0.9×10^{-2}	51.8	300–700	—	217
Covington	6×10^{-2}	60.2	21–104	2.56×10^{-12}	218
Someno and Nagasaki	0.44×10^{-2}	42.2	300–800	—	219
β-Phase					
Papazoglou and Hepworth	1.95×10^{-3}	27.8	610–1200	—	214
Wasilewski and Kehl	1.95×10^{-3}	27.6	520–840	—	215
Someno and Nagasaki	5.7×10^{-3}	36.4	900–1200	—	219
Someno and Nagasaki	4.38×10^{-3}	42.2	300–750	—	219
γ-Phase					
Brauer and Nann[a]	3.6×10^{-7}	28.9	24–66	$(2.2–5.7) \times 10^{-12b}$	216
δ-Phase					
Stalinski et al.[c]	—	39.3–42.7	−196–200	—	220
Korn and Zamir	—	48.9	25–500	—	221
Bustard et al.[d]	1.45×10^{-3}	50–52.4	450–550	—	222
Pope and Narang[e]	—	49.0	−13–327	—	223

[a] TiH$_{0.9}$.
[b] D at 297 K.
[c] TiH$_{1.6-1.9}$.
[d] TiH$_{1.55-1.71}$.
[e] TiH$_{1.98}$.

Table 6
Diffusion Coefficient of Hydrogen in α-, β-, and δ-Phase Zr–H: Reported Values of the Parameters in $D = D_0 \exp(-E_a/RT)$ and of D at 303 K

	D_0 (cm^2 s^{-1})	E_a (kJ mol^{-1})	Temperature range (°C)	Reference
α-Phase				
Schwarz and Mallett	1.15×10^{-3}	31.4	—	227
Mallett and Albrecht	7×10^{-4}	32.8–41.5	305–600	228
Gulbransen and Andrew	1.09×10^{-3}	47.6	60–250	229
Someno	4.15×10^{-3}	39.6	450–700	230
Kearns	7.00×10^{-3}	44.6	275–700	231
Mazzolai and Ryll-Nardzewski	7.14×10^{-3}	49.3	245–636	232
Naito	1.44×10^{-2}	48.5	600–814	233
β-Phase				
Someno	7.37×10^{-3}	35.7	870–1100	230
Gelezunas and Conn	5.32×10^{-3}	34.8	760–1010	234
Schwarz and Mallett	5.3×10^{-3}	34.7	—	227
δ-Phase				
Hon	—	52.3	20–200	235
Pope	—	80.0	127–527	223
Korn and Goren	—	55.7	97–496	236

^{a}ZrH$_{1.684}$.

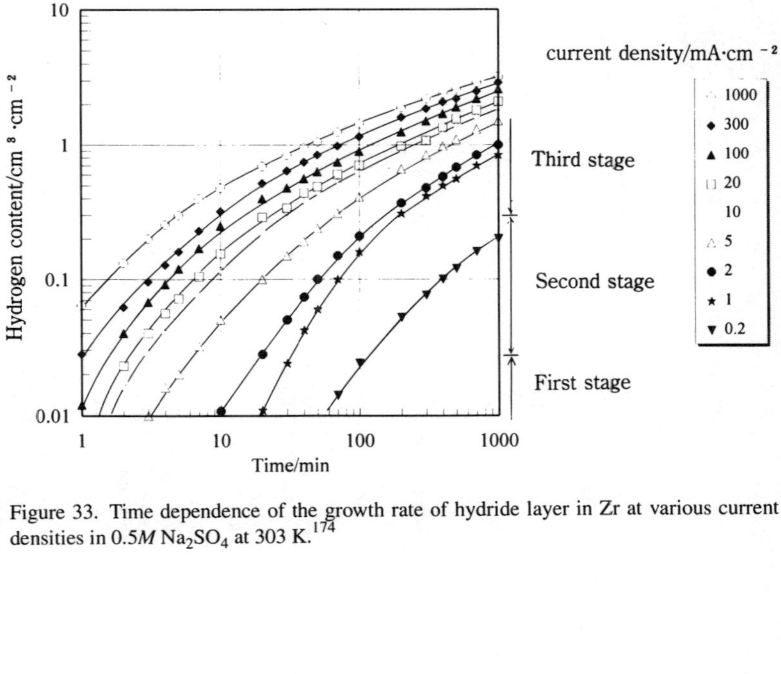

Figure 33. Time dependence of the growth rate of hydride layer in Zr at various current densities in 0.5M Na$_2$SO$_4$ at 303 K.[174]

Figure 34[174] shows the temperature dependence of hydrogen absorption observed in Ti at two c.d. values. Hydrogen absorption obeys a linear rate law at lower c.d.'s (0.1–0.2 mA cm^{-2}) but follows a parabolic rate law at 10 mA cm^{-2} except for the first stage of hydrogen absorption. Taken together, these data indicate that the hydrogen absorption process may be composed of three stages depending on the hydrogen content in the sample and/or on the hydrogen supply rate.[237]

The hydrogen absorption rate is, naturally, accelerated at high temperature. Although different rate laws are obtained for Ti and Zr—one is linear and the other is parabolic (cf. Fig. 34)—the activation energy values for hydrogen absorption are nevertheless nearly the same, being ~17 kJ mol^{-1} for Ti and ~15 kJ mol^{-1} for Zr.

The composition at a very thin surface layer can be obtained by the d–n method; typical results for Ti and Zr are shown in Fig. 35.[174] It is seen that the absorption of hydrogen (deuterium) increases almost linearly with time initially. This may indicate that formation of a hydride having a constant concentration (D/Ti ≈ 1.45) proceeds with time until the δ-phase covers the surface. For Ti, the hydrogen concentration in the surface layer finally reaches D/Ti ≈ 2.0 but it continues to increase at Zr.

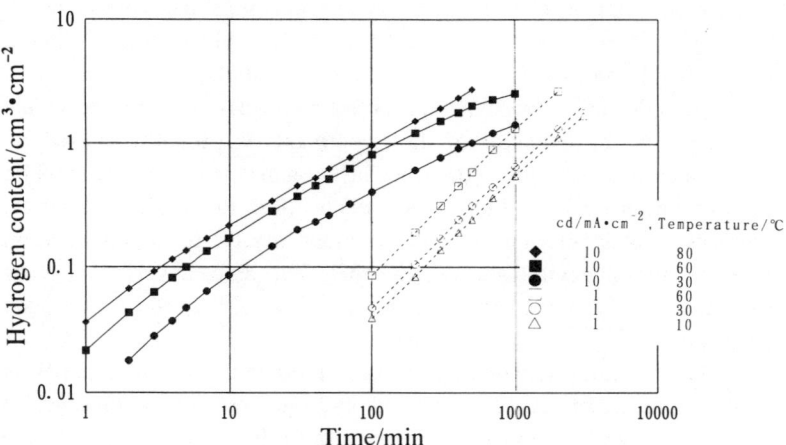

Figure 34. Temperature dependencies of the hydrogen absorption into Ti in 0.5M H$_2$SO$_4$ at two different cathodic current densities.[174]

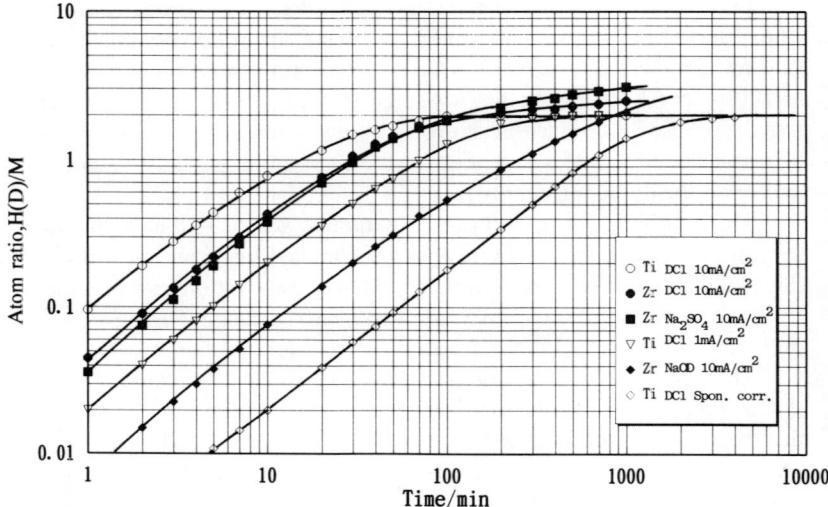

Figure 35. Time dependence of surface deuterium concentration in Ti or in Zr measured in various electrolytes at 303 K.[174]

Changes in the pH give rise to significant changes in the hydrogen absorption behavior. Thus, hydrogen absorption into Ti takes place readily at low pH values but not significantly at high pH (Fig. 36). On the other hand, absorption into Zr takes place over the whole range of pH (Fig. 36[136]), and the rates show a maximum around the neutral-pH region. This difference in the behavior of the two metals is probably caused by differences in the stability of the oxide layer on the metal surface; Zr oxide is easily reduced by cathodic electrolysis over the whole pH range, whereas Ti oxide is removed at low pH under strong cathodic polarization but is stable at neutral and alkaline pH. The oxide layer, even if its thickness is only on the order of nanometers, would retard the hydrogen permeation.

The decrease of the hydrogen absorption rate at low and high pH on Zr may indicate increased rates of the hydrogen recombination reaction at such pHs, which provides an easier path for hydrogen escape. On the other hand, the increase of the absorption rate seen in the presence of Na^+ and K^+, as compared with the rate in the presence of NH_4^+, may indicate

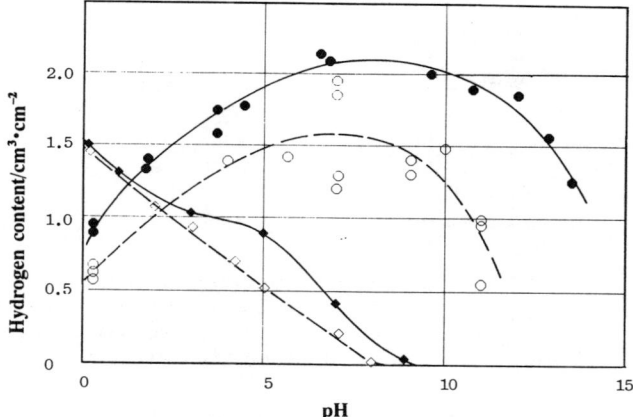

Figure 36. Effect of pH and alkaline ions on the rate of hydrogen absorption in Zr and Ti. Electrolysis conditions: 10 mA cm^{-2}, 10^3 min, 303 K.[136] ●, Zr in [H$_2$SO$_4$ or NaOH] + [Na$_2$SO$_4$ or K$_2$SO$_4$]; ○, Zr in CH$_3$COONH$_4$ or in (NH$_4$)$_2$SO$_4$ + [H$_2$SO$_4$ or NH$_4$OH]; ◆, Ti in (NH$_4$)$_2$SO$_4$ + [H$_2$SO$_4$ or NH$_4$OH] + Na$_2$SO$_4$; ◇, Ti in (NH$_4$)$_2$SO$_4$ + [H$_2$SO$_4$ or NH$_4$OH].

suppression of the rate of recombination, thereby raising the activity of hydrogen adatoms.

(iii) Current Efficiency of Hydrogen Absorption into Ti and Zr

The current efficiency of hydrogen absorption during cathodization is readily obtained. The value of the activation energy as obtained from Fig. 37[237] is ~24 kJ mol^{-1} for the second, linear stage and ~33 kJ mol^{-1} for the third, parabolic stage. This would mean that the hydrogen absorption is controlled by the surface reaction and the diffusion process in the linear and the parabolic stage, respectively.

The dependency of the current efficiency on c.d. is shown in Fig. 38[237] for Ti at various stages of hydride formation. The efficiency, as estimated by the d–n and etching methods, is almost constant, $\varepsilon = 0.15$–0.3, during the first stage of hydrogen absorption. In the second stage, it is reduced to $\varepsilon = 0.15$–0.09 and shows a weak dependence on c.d. in the higher c.d. region after the hydrogen absorption has reached 0.15 cm^3 cm^{-2} (i.e., when the metal surface should be completely covered with a hydride layer), but it stays constant at $\varepsilon = 0.1$ in the lower c.d. region. In the third stage, the

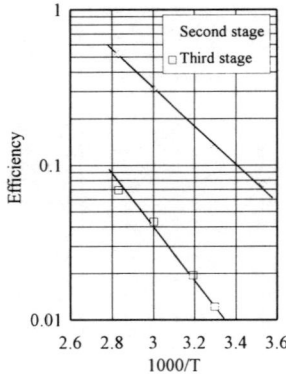

Figure 37. Arrhenius representation of hydrogen absorption efficiencies during the second and third stages for Ti. Electrolysis conditions: 10 mA cm^{-2} in 0.5M H$_2$SO$_4$ and 0.5M D$_2$SO$_4$ solutions.[237]

efficiency becomes inversely proportional to the c.d. in the higher c.d. region.

The characteristics of hydrogen absorption effected by the hydride layer in Zr are nearly the same as for Ti in the absence of Na$^+$ or K$^+$ ions in the solution. Of course, there are some differences between Ti and Zr; hydrogen absorption readily takes place in the low-pH region only at Ti but it occurs at any pH at Zr. On the other hand, the kinetics of hydrogen permeation are different between Ti and Zr in solutions that contain

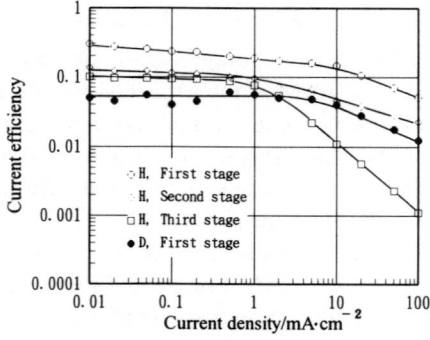

Figure 38. Dependency of efficiencies of hydrogen absorption into Ti on the cathodic c.d. during the early, the second, and the third stage and during the early stage of deuterium absorption. 0.5M H$_2$SO$_4$, 303 K.[237]

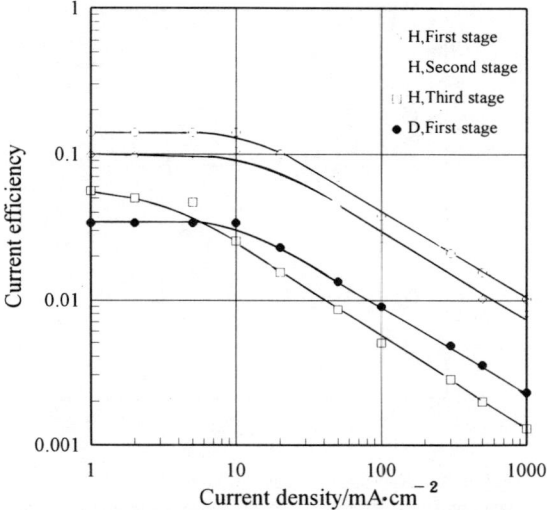

Figure 39. Dependency of efficiencies of hydrogen absorption into Zr on the cathodic c.d. during the early, the second, the third, and the fourth stage and during the early stage of deuterium absorption. 0.5M Na$_2$SO$_4$, 303 K.[238]

halogen ions. The current efficiency ε at Zr is shown in Fig. 39[238]; ε has a constant value of around 0.15–0.20 at the first stage on Zr under 10 mA cm^{-2} of c.d. and then becomes proportional to $I^{-1/2}$ at high c.d. until the surface is covered by a hydride layer of high hydrogen concentration, such as H/Zr = 2.0, whereas at lower c.d. ε is constant. Even if the surface is completely covered by a hydride layer of very high hydrogen concentration, such as H/Zr > 2.5, at high c.d., still ε becomes proportional to $i^{-1/2}$ and hence the permeation rate $J \propto i^{1/2}$.

This phenomenon is the hydrogen absorption process in solutions that contain alkaline ions such as Na$^+$ and K$^+$. This means that diffusion control is not in effect for the case of Zr in solutions that contain the alkaline ions: the surface reaction thus plays a role in the hydrogen absorption process on Zr.

(iv) *Hydrogen Concentration Profile in Ti and Zr Hydride Layers*

The hydrogen concentration profile in Ti obtained by a volumetric technique has a sigmoidal shape as shown by the solid line in Fig. 40,[239]

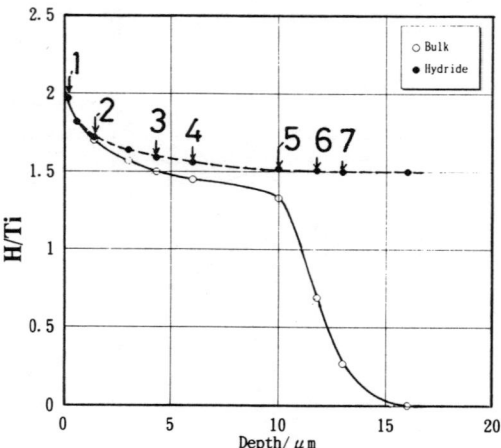

Figure 40. Hydrogen concentration profile in the hydride layer and bulk of Ti. ———, Hydrogen concentration profile in the bulk metal; – – –, hydrogen concentration profile of the hydride phase in the mixed phase.[239] 0.5M H_2SO_4, 10 mA cm^{-2}, $t = 6 \times 10^4$ s.

with the precise shape of the curve detail being dependent on the cathodic c.d. It exhibits a clear plateau region after a long time of cathodic polarization, e.g., 6×10^4 s at a c.d. of 10 mA cm^{-2} in 0.5M H_2SO_4. The H/Ti ratio in the plateau region increases with the c.d. The sigmoidal shape of the concentration profile clearly indicates that the concentration of hydrogen in Ti is not simply controlled by diffusion but that a hydride layer (H/Ti ≈ 1.5) is growing with time toward the bulk of the Ti specimen.

Formation of a similar hydride layer in Zr is also indicated (Fig. 41[239]). The hydrogen concentration profiles generally have a sigmoidal shape, with a plateau region that becomes thicker and clearer with increasing c.d. and duration of cathodic polarization.

Structural analysis at each of the numbered points in these figures indicates that the structures are changing with change in the hydride composition, as given in Table 7.[241] It was seen on Ti that both α- and δ-Ti–H are formed from the beginning of the hydride formation. The composition of the δ-phase is almost constant, being TiH$_{1.5-1.54}$, but the hydrogen concentration increases further after disappearance of the α-

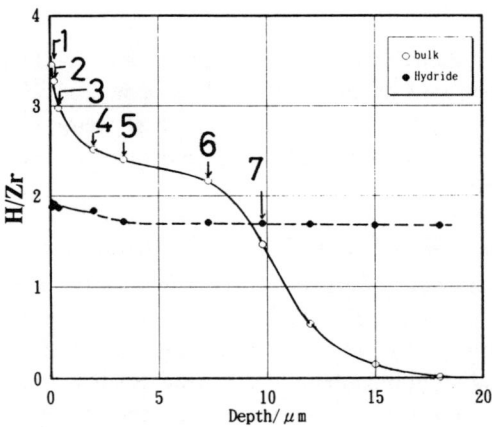

Figure 41. Hydrogen concentration profile in the hydride layer and bulk of Zr. ———, Hydrogen concentration profile in the bulk metal; – – –, hydrogen concentration profile of the hydride phase in the mixed phase.[239] $0.5M$ Na_2SO_4, 10 mA cm^{-2}, $t = 6 \times 10^4$ s.

Table 7
Composition, Structure, and Lattice Parameters of Ti and Zr Hydride at the Numbered Positions in Figs. 40 and 41[a]

Position	x in MH$_x$	Structure	Lattice Parameter(s) (Å)
M = Ti[b]			
1	2.0	fct(ε)	$a = 4.44$
2	1.71	fcc(δ)	$a_0 = 4.42$
3	1.54	fcc(δ)	$a_0 = 4.41$
4	1.45	fcc + hcp(a)	$a_0 = 4.41$
5	1.33	fcc + hcp(a)	$a_0 = 4.41$
6	0.72	fcc + hcp(a)	$a_0 = 4.41$
7	0.30	fcc + hcp(a)	$a_0 = 4.41$
M = Zr[c]			
1	3.45	fct(ε)	$a_0 = 4.96$, $c = 4.47$
2	3.27	fct(ε)	$a_0 = 4.98$, $c = 4.48$
3	2.98	ε + fcc(δ) + hcp(a)	$a_0 = 4.96$, $c = 4.48$, $a_0 = 4.80$
4	2.52	ε + fcc(δ) + hcp(a)	$a_0 = 4.95$, $c = 4.48$, $a_0 = 4.80$
5	2.42	ε + fcc(δ) + hcp(a)	$a_0 = 4.92$, $c = 4.53$, $a_0 = 4.78$
6	2.16	ε + fcc(δ) + hcp(a)	$a_0 = 4.90$, $c = 4.70$, $a_0 = 4.78$
7	1.47	$\delta + a$	$a_0 = 4.78$

[a]Ref. 241.
[b]Refer to Fig. 40.
[c]Ref. to Fig. 41.

phase. The hydrogen concentration profiles in the δ-Ti–H phase only near the surface, measured independently by the d–n method,[239] are also shown in Figs. 40 and 41. These data agree with the data obtained by the volumetric method just beneath the surface but tend to deviate from them for deeper positions.

(v) Microscopic Observation of Growth of Hydride Layer

SEM observations of the surface hydride layer formed by cathodization of Ti are shown in Fig. 42.[242] The hydride layer consists of two portions, a uniform hydride layer and a layer of needlelike/wedgelike structure intruding into the bulk of the metal. Such a structure is formed initially with a thin needlelike shape, but the needles gradually grow both in length and width, which may reach 2–3 μm. A cross-sectional view of the wedgelike portion is shown in Fig. 43,[239] which shows the hydrided portion (white) and the α-Ti phase (black).

The progress of hydride formation in metals may be summarized as follows: (1) At the beginning, a thin needlelike hydride structure is formed,

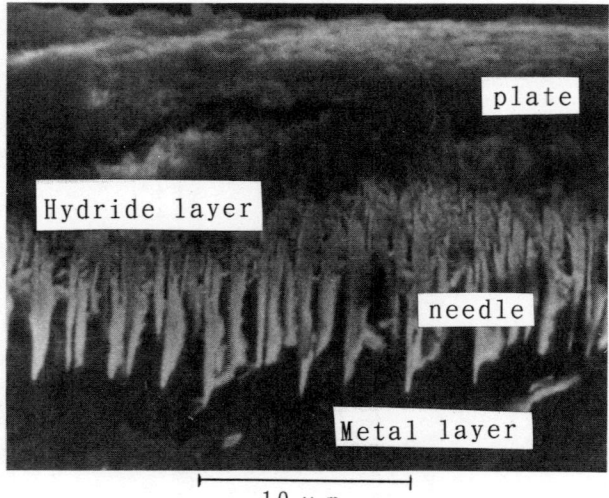

Figure 42. Typical SEM photograph of a surface hydride layer on Ti formed in 0.5M H$_2$SO$_4$ after cathodic polarization at a current density of 10 mA cm^{-2} for 1000 min at 303 K.[240]

Sorption of Hydrogen

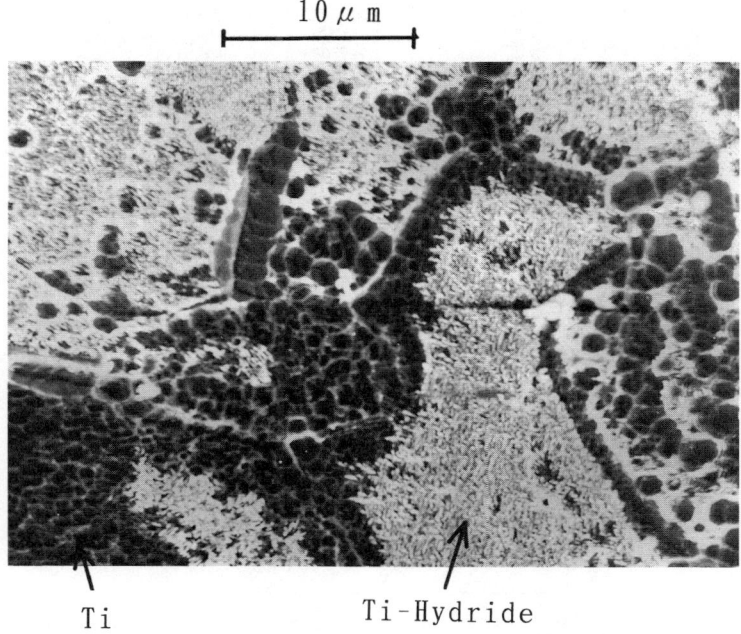

Figure 43. SEM image of mixed phases of Ti hydride and α-Ti, obtained by etching the specimen until dissolution of the hydride layer reached at position no. 7 in Fig. 40.[239]

which consists of δ-phase in the case of Ti and δ- and ε-phase in the case of Zr; (2) the needles grow thicker and tend to merge with each other; the phase gradually becomes uniformly δ-phase in the case of Ti and ε-phase in the case of Zr; (3) the surface layer becomes of a platelike uniform structure; and (4) the hydrogen concentration in the platelike hydride gradually increases and finally reaches the stoichiometric composition. These stages are schematically represented in Fig. 44.[239]

(vi) Further Details of Hydrogen Absorption Behavior as Studied by the d–n Reaction Method

A more detailed picture of the hydrogen absorption behavior can be obtained by combining the etching/volumetric method described above with the *d–n* reaction method. The specimen is first polarized in a light-water solution followed by electrolysis in a heavy-water solution, or this procedure may be carried out in the reverse order, and both the (total)

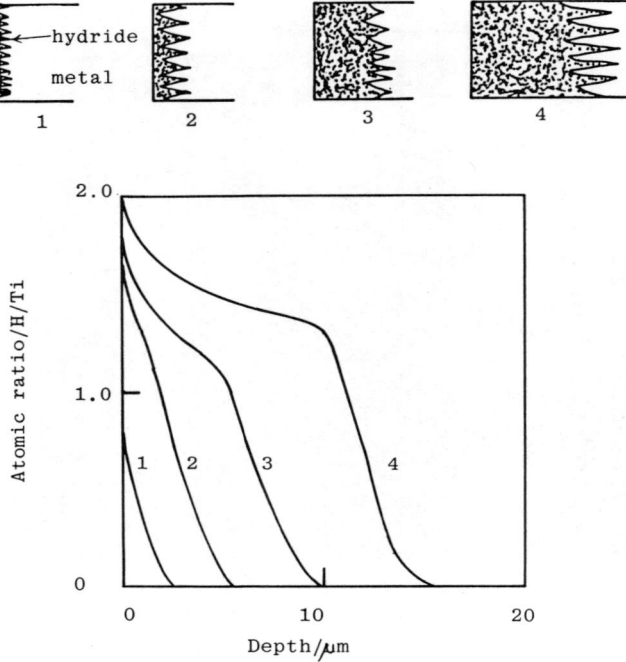

Figure 44. A model of progressive changes of the hydrogen concentration profile during the hydride growth.[239] 1, Formation of needlelike structure; 2, progression of needlelike structure; 3, needlelike plus platelike structure; 4, progression of platelike plus needlelike structure.

hydrogen content and deuterium concentration are evaluated at various depths from the surface. Typical results are shown in Fig. 45[242] for Zr polarized in 0.5M Na$_2$SO$_4$ at a c.d. of 100 mA cm^{-2} at 303 K. The concentration profiles shown in Fig. 45a were obtained from an experiment in which H was introduced first by electrolysis in light-water solution to form a hydride layer of needlelike structure and then D was introduced. It can be seen that H was "pushed in" by D, which was introduced later. Figure 45b shows the results of a similar experiment but conducted in the reverse order: The push-in effect is still seen although the extent is less pronounced than in Fig. 45a. Figure 45c shows the results of an experiment similar to that in Fig. 45a but for the case of platelike structure of the

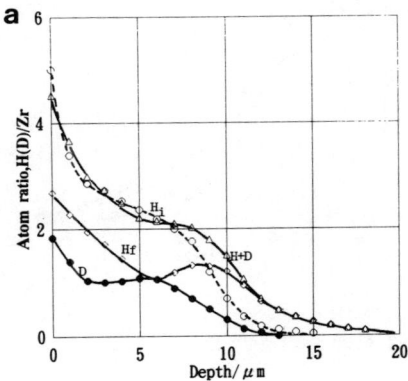

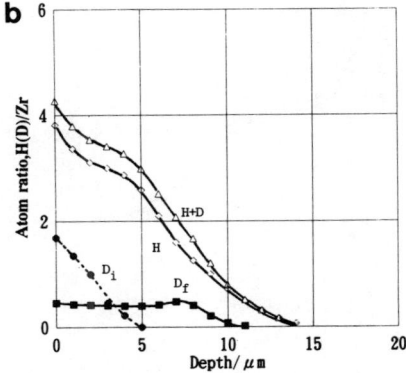

Figure 45. Changes of hydrogen and deuterium concentration profile in Zr after successive isotope absorption in needlelike (a and b) and platelike (c and d) hydride layers.: (a) and (c) light hydrogen followed by deuterium absorption; (b) and (d) in the reverse order.[242]

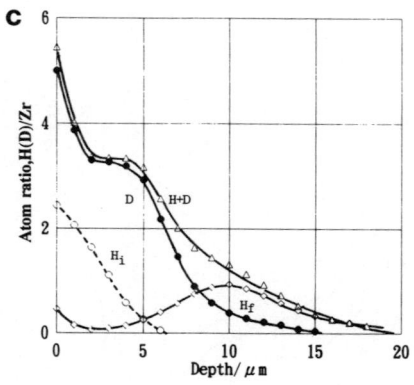

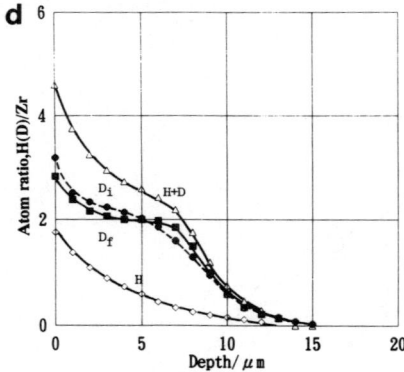

Figure 45. Continued

hydride (deuterium introduced later), and Fig. 45d is for a similar experiment but in the reverse order (i.e., light hydrogen introduced into deuteride).

The results of these experiments show that the hydrogen and deuterium atoms strongly interact with each other, or are sharing the same sites in the metal lattice and hence colliding with each other, during the diffusion; otherwise, the final distribution of the two isotopes should be the same. In this case, the hydrogen atoms introduced first will be kicked by hydrogen atoms introduced later through collisions and hence will be "pushed in" in a statistical sense. The effect will be more pronounced if the former species are H atoms and the latter are D atoms because of the heavier mass of the latter. Conversely, the pushing-in effect will be less pronounced if the heavier species are introduced first.

The diffusion coefficients of deuterium and hydrogen atoms during their absorption may be estimated by using the displacement distances and diffused amounts in a given time,[242] i.e.,

$$J = \Delta P/\Delta t = -D(\Delta c/\Delta x) \quad (44)$$

where J is the permeation of hydrogen atoms, ΔP is the number of atoms that were present at a certain depth, Δt is the time, and D is the diffusion coefficient. One can hence estimate D from the average displacement of atoms and the total number of diffused atoms. The diffusion coefficients for the cases shown in Fig. 45a and c and Fig. 45b and d were shown to be:

$$D_{HD} = 8.9 \times 10^{-4} \exp(-1170/RT) \quad (cm^2\ s^{-1}) \quad (45)$$

$$D_{DH} = 6.2 \times 10^{-5} \exp(-1170/RT) \quad (cm^2\ s^{-1}) \quad (46)$$

Such a large difference (a factor of 14.4) can hardly be explained by conventional diffusion processes; it can therefore be concluded that a strong interaction exists between the hydrogen atoms during the absorption process.

VII. CONCLUDING REMARKS

The phenomena of *ad*sorption and *ab*sorption of hydrogen are intimately related in the cases of many hydrogen-absorbing metals. When a metal is placed in a protonic environment such as water, hydrogen adatoms are formed on the surface and may go into the metal as well. In this chapter, several views of the electrochemistry of hydrogen entry into metal electrodes metal were reviewed, and then the most plausible view was presented. Thus, the surface activity of hydrogen adatoms was discussed first on the basis of the mechanism of the electrochemical hydrogen evolution reaction, and it was argued that the use of activity rather than surface coverage is preferable, as it is a quantity that can be analyzed in terms of electrode kinetics.

The rest of the chapter was devoted to the entry of hydrogen into metal electrodes. Typical findings with Pd, Fe, Ni, Ti, and Zr were discussed by combining electrochemical concepts and the results of analysis of deuterium in metals obtained with the use of a deuteron accelerator. The absorption behavior depends on the metals and the conditions of cathodic electrolysis. Pd metal absorbs hydrogen atoms as soon as the hydrogen evolution reaction takes place on the surface and forms the α- and then the β-phase. During a long period of polarization or under comparatively high c.d., however, a new phase seems to be formed, which may be called the γ-phase. This process seems to differ from the α- and β-phase formation, which follows a diffusion model, in that the γ-phase is formed following a model of a moving front of the phase having a constant concentration of hydrogen. The rates of formation and decomposition of the phase are rather slow and depend on the c.d. For example, formation of a 1-cm-thick layer requires as long as 46 days. The surface layer can be considered as a phase in which the hydrogen atoms dissolve in excess into the β-phase.

The process of phase formation is also similar in the case of Ti and Zr hydrides. First, a low-hydrogen-concentration α-phase was formed during cathodic electrolysis, followed by formation of a $TiH_{1.5}$ δ-phase in the case of Ti or a $ZrH_{1.7}$ δ-phase in the case of Zr. Transformation to the ε-phase occurred during further hydrogen absorption. Interesting results from studies of hydrogen entry into Ti or Zr using light hydrogen followed by deuterium, or the reverse, have been presented; these results seem to indicate that hydrogen isotopes are distinguishable from each other during the permeation process.

In conclusion, further electrochemical investigations combined with other techniques, such as the use of deuteron beams, are likely to provide a rather detailed picture of hydrogen adsorption on and hydrogen absorption into hydrogen-absorbing metals.

REFERENCES

[1] S. Yoshizawa and K. Yamakawa, *Denki Kagaku (J. Electrochem. Soc., Japan)* **39** (1971) 845 [in Japanese].
[2] J. J. DeLuccia, K. Yamakawa, and L. Nanis, private communication.
[3] L. Nanis and J. J. DeLuccia, in *Materials Performance and the Deep Sea*, ASTM STP 445, American Society for Testing and Materials, Philadelphia, 1969, p. 155.
[4] J. J. DeLuccia, K. Yamakawa, and L. Nanis, Technical Report, UPH 2-001, NR 036-077, October 1969.
[5] R. V. Bucur, *J. Electroanal. Chem.* **10** (1965) 8.

[6] R. V. Bucur, *Electrochim. Acta* **11** (1981) 1653.
[7] R. V. Bucur and F. Bota, *Electrochim. Acta* **27** (1982) 521.
[8] R. V. Bucur and F. Bota, *Electrochim. Acta* **28** (1983) 1373.
[9] V. Breger and E. Gileadi, *Electrochim. Acta* **16** (1971) 177.
[10] E. Wicke and G. H. Nernst, *Ber. Bunsenges. Phys. Chem.* **68** (1964) 224.
[11] J. C. Barton and F. A. Lewis, *Z. Phys. Chem. (N.F.)* **33** (1962) 1.
[12] G. W. Castellan, *J. Electrochem. Soc.* **111** (1964) 1273.
[13] R. A. LaPietra and G. W. Castellan, *J. Electrochem. Soc.* **111** (1964) 1276.
[14] J. McBreen, L. Nanis, and W. Beck, *J. Electrochem. Soc.* **113** (1966) 1218.
[15] J. O'M. Bockris, M. A. Genshaw, and M. Fullenwider, *Electrochim. Acta* **15** (1970) 47.
[16] J. P. Chevillot, J. Farcy, C. Hinnen, and A. Rousseau, *J. Electroanal. Chem.* **64** (1975) 39.
[17] R. Carta, M. S. Dernini, A. M. Polcaro, P. F. Ricci, and G. Tola, *Surf. Coat. Technol.* **30** (1987) 375, 389.
[18] N. Yamamoto, T. Ohsaka, T. Terashima, and N. Oyama, *J. Electroanal. Chem.* **296** (1990) 463.
[19] L. Gräsjo and M. Seo, *J. Electroanal. Chem.* **296** (1990) 233.
[20] A. Frumkin and N. Aladzhalova, *Acta Physicochim. URSS* **19** (1944) 1; *Zh. Fiz. Khim.* **18** (1944) 498.
[21] M. von Stackelberg and H. Bischoff, *Z. Elektrochem.* **59** (1955) 467.
[22] J. O'M. Bockris, J. McBreen, and L. Nanis, *J. Electrochem. Soc.* **112** (1965) 1025.
[23] J. O'M. Bockris and P. K. Subramanyan, *J. Electrochem. Soc.* **118** (1971) 1114.
[24] R. V. Bucur, *J. Catal.* **70** (1981) 92.
[25] M. Enyo, *Electrochim. Acta* **18** (1973) 163.
[26] M. Enyo, *Electrochim. Acta* **18** (1973) 155.
[27] T. Maoka and M. Enyo, *Electrochim. Acta* **26** (1981) 607.
[28] T. Maoka and M. Enyo, *Electrochim. Acta* **26** (1981) 615.
[29] M. Enyo, in *Modern Aspects of Electrochemistry*, No. 11, Ed. by B. E. Conway and J. O'M. Bockris, Plenum Press, New York, 1975, pp. 251–314.
[30] M. P. Soriaga and A. T. Hubbard, *J. Am. Chem. Soc.* **104** (1982) 2735.
[31] M. P. Soriaga and A. T. Hubbard, *J. Electroanal. Chem.* **167** (1984) 79.
[32] G. Horanyi and A. Wieckowski, *J. Electroanal. Chem.* **294** (1990) 267.
[33] A. V. Shepakov, V. N. Andreev, and V. E. Kazarinov, *Elektrokhimiva* **25** (1989) 78.
[34] A. Wieckowski, in *Modern Aspects of Electrochemistry*, No. 21, Ed. by R. E. White, J. O'M. Bockris, and B. E. Conway, Plenum Press, New York, 1990, pp. 65–119.
[35] R. Notoya and A. Matsuda, *J. Res. Inst. Catal. Hokkaido Univ.* **14** (1966) 198; *J. Phys. Chem.* **93** (1989) 5521.
[36] M. Enyo and T. Maoka, *J. Electroanal. Chem.* **108** (1980) 277.
[37] M. A. V. Devanathan, J. O'M. Bockris, and W. Mehl, *J. Electroanal. Chem.* **1** (1959/60) 143.
[38] D. R. Lawson, M. J. Tierney, I. F. Cheng, L. S. VanDyke, M. W. Espensheid, and C. R. Martin, *Electrochim. Acta* **36** (1991) 1515.
[39] K. Ohta, H. Yoshitake, O. Yamazaki, M. Kuratsuka, K. Yamaki, K. Ando, Y. Iida, and N. Kamiya, Fourth International Conference on Cold Fusion, Maui, Hawaii, December 6–9, 1993, C2.7; O. Yamazaki, H. Yoshitake, N. Kamiya, and K. Ohta, *J. Electroanal. Chem.* **390** (1995) 127.
[40] P. K. Iyengar, in *Proceedings of the Fifth International Conference on Emerging Nuclear Energy Systems*, Karlsruhe, July 1989 FRG World Scientific Singapore, 1989, p. 2.
[41] H. Numapta and I. Ohno, 183rd Electrochemical Society Meeting, Honolulu, Hawaii, May 16–21, 1993, Paper No. 1772.
[42] T. B. Flanagan and F. A. Lewis, *Trans. Faraday Soc.* **55** (1959) 1400, 1409.
[43] J. C. Barton and F. A. Lewis, *Talanta* **10** (1963) 237.
[44] J. Schirber and C. Northrup, Jr., *Phys. Rev. B* **10** (1974) 3818.

[45] N. Lewis, C. Barnes, M. Heben, A. Kumar, S. Lunt, G. McManis, G. Miskelly, R. M. Penner, P. Santangelo, G. Shreve, B. Tufts, M. Youngquist, R. Kavanagh, S. Kellogg, R. Vogelaar, T. Wang, R. Kondrat, and R. New, *Nature* **340** (1989) 525.
[46] A. D. McQuillan, *Proc. R. Soc. London, Ser. A* **204** (1950–51) 309.
[47] J. C. Barton, F. A. Lewis, and I. Woodward, *Trans. Faraday Soc.* **59** (1963) 1201.
[48] J. C. Barton and F. A. Lewis, *Z. Phys. Chem. (N.F.)* **33** (1962) 99.
[49] T. B. Flanagan and F. A. Lewis, *Z. Phys. Chem. (N.F.)* **27** (1961) 104; F. A. Lewis, *The Palladium–Hydrogen System*, Academic Press, New York, 1967.
[50] B. Baranowski, Z. Szklarska-Smialowska, and M. Smialowski, *Bull. Acad. Pol. Sci., Ser. Sci. Chim.* **6** (1958) 179.
[51] M. C. H. McKubre, S. Crouch-Baker, A. M. Riley, R. C. Rocha-Filho, M. Schreiber, S. I. Smedley, and F. L. Tanzella, in *Hydrogen Storage Batteries and Electrochemistry*, Ed. by D. A. Corrigan and S. Srinivasan, Proceedings Vol. 92-5, The Electrochemical Society, Pennington, New Jersey, 1992, p. 269.
[52] T. Mizuno and T. Morozumi, *Bull. Fac. Eng., Hokkaido Univ.* **93** (1979) 23.
[53] P. F. A. Bijlmar, *Metal Finish.* **12** (1970) 64.
[54] J. P. Butler and D. C. Santry, *Can. J. Chem.* **39** (1961) 689.
[55] J. P. Butler, in *Radiochemical Methods of Analysis*, Vol. 1, 1965, p. 393; Proceedings of the Symposium on the Salzburg Act, International Atomic Energy Agency, Vienna, 1964.
[56] W. R. Arnold, J. A. Phillips, G. A. Sawyer, E. J. Stovall, and J. L. Tuck, *Phys. Rev.* **93** (1954) 483; Los Alamos Scientific Laboratory Report LA-2014.
[57] G. Preston, P. F. D. Shaw, and S. A. Young, *Proc. R. Soc. London, Ser. A* **226** (1954) 206.
[58] R. K. Smith and J. E. Perry, Los Alamos Scientific Laboratory, private communication, 1958.
[59] J. L. Fowler and J. E. Brolley, *Rev. Mod. Phys* **28** (1956) 103.
[60] D. S. Stark and A. W. J. Menley, *Nucl. Instrum. Methods* **91** (1971) 301.
[61] Y. Okamoto and H. Tamagawa, *Jpn. J. Appl. Phys.* **38** (1969) 1114.
[62] H. G. Poole, *Proc. R. Soc. London* **163** (1937) 404.
[63] W. V. Smith, *J. Chem. Phys.* **11** (1943) 110.
[64] B. J. Wood and H. Wise, *J. Chem. Phys.* **29** (1958) 1416.
[65] A. D. Hansen, R. F. Taschex, and J. H. Williams, *Rev. Mod. Phys.* **21** (1949) 635.
[66] B. Baranowski, in *Hydrogen in Metals*, Ed. by G. Alefeld and J. Völkl, *Topics in Applied Physics*, Vol. 29, Springer-Verlag, Berlin, 1978, pp. 157–200.
[67] D. P. Smith, *Hydrogen in Metals*, University of Chicago Press, Chicago, 1948.
[68] I. M. Bernstein and A. W. Thompson, eds. *Hydrogen in Metals*, Proceedings of the International Conference of the American Society Metals, 1974.
[69] G. Alefeld and J. Völkl, eds., *Hydrogen in Metals, Topics in Applied Physics*, Vol. 28, Springer-Verlag, Berlin, New York, 1978.
[70] J. Worsham, M. Wilkinson, and C. Shull, *J. Phys. Chem. Solids* **3** (1957) 303.
[71] F. A. Lewis, *The Palladium–Hydrogen System*, Academic Press, New York, 1967.
[72] J. Völkl and G. Alefeld, in *Hydrogen in Metals*, Ed. by G. Alefeld and J. Völkl, *Topics in Applied Physics*, Vol. 28, Springer-Verlag, Berlin, 1978, pp. 321–348.
[73] T. B. Flanagan, W. Luo, and J. D. Clewley, *J. Less-Common Met.* **172** (1991) 42.
[74] H. Brodowsky, *Z. Phys. Chem. (N.F.)* **44** (1965) 129.
[75] R. Lässer and T. Schöber, *Mater. Sci. Forum* **31** (1988) 39.
[76] H. Frieske and E. Wicke, *Ber. Bunsenges. Phys. Chem.* **77** (1973) 48.
[77] E. Wicke and G. Nernst, *Ber. Bunsenges. Phys. Chem.* **68** (1964) 224.
[78] P. S. Perminov, A. A. Orlov, and A. N. Frumkin, *Dokl. Akad. Nauk SSSR* **84** (1952) 749.
[79] A. I. Fedorova and A. N. Frumkin, *Zh. Fiz. Khim.* **27** (1953) 247.
[80] R. Lässer and K. H. Klatt, *Phys. Rev. B* **28** (1983) 748.
[81] B. Dandapani and M. Fleischmann, *J. Electroanal. Chem.* **39** (1972) 315, 323.

[82] J. O'M. Bockris and P. K. Subramanyan, *Electrochim. Acta* **16** (1971) 2169.
[83] Y. Fukai, The 3rd International Conference on Cold Fusion (ICCF3), Nagoya, Japan, October 21–25, 1992, paper 22PII-3; *Frontiers of Cold Fusion*, Ed. by H. Ikegami, Universal Academy Press, Tokyo, 1993, p. 265.
[84] Y. Fukai, *The Metal–Hydrogen Systems—Basic Bulk Properties*, Springer Series in Materials Science, Vol. 21, Springer-Verlag, Heidelberg, 1993.
[85] Y. Sakamoto, Fundamental Study of New Hydrogen Energy, Interim report, The Institute of Applied Energy of Japan, 1994, p. 60 [in Japanese].
[86] A. Kubota, H. Akita, Y. Tsuchida, T. Saito, A. Kubota, N. Hasegawa, N. Imai, N. Hayakawa, and K. Kunimatsu, The 3rd International Conference on Cold Fusion (ICCF3), Nagoya, Japan, October 21–25, 1992, paper 22PII-8; *Frontiers of Cold Fusion*, Ed. by H. Ikegami, Universal Academy Press, Tokyo, 1993, p. 565.
[87] J. F. Mareché, J. C. Rat, and A. Hérold, *Proceedings of the 2nd International Congress on Hydrogen in Metals, Paris, June 6–10, 1977*, Vol. 1, Pergamon Press, Oxford, 1977, IC9.
[88] H. Wiph, in G. Alefeld and J. Völkl, *Hydrogen in Metals, II, Topics in Applied Physics*, Vol. 29, Springer-Verlag, Berlin, 1978, pp. 273–304.
[89] T. A. Chubb and S. R. Chubb, *Fusion Technol.* **20** (1991) 93.
[90] A. Mitchel, *C. R. Sci. Acad.* **221** (1944) 218.
[91] G. A. Moore, *Trans. Electrochem. Soc.* **75** (1939) 237.
[92] M. v. Stackelberg and P. Zudwig, *Z. Naturforsch.* **19a** (1964) 93.
[93] T. Mizuno, Dissertation, Hokkaido University, 1976: T. Mizuno, T. Akimoto, K. Azumi, and M. Enyo, *Denki Kagaku (J. Electrochem. Soc., Japan)* **60** (1992) 405 [in Japanese].
[94] F. G. Will, K. Cedzynska, and D. C. Linton, National Cold Fusion Institute Report, University of Utah, 1-131, 1991.
[95] L. Redey and K. M. Myles, Electrochemical Society Fall Meeting, Phoenix, Arizona, October 13–17, 1991, Extended Abstracts Vol. 91.2, No. 118.
[96] T. Mizuno, The 4th International Symposium, Catalysis Research Center, Hokkaido University, Sapporo, September 18–19, 1991; T. Mizuno, T. Akimoto, K. Azumi, and M. Enyo, The 3rd International Conference on Cold Fusion (ICCF3), Nagoya, Japan, October 21–25, 1992, paper 22PII-10; *Frontiers of Cold Fusion*, Ed. by H. Ikegami, Universal Academy Press, Tokyo, 1993, p. 373.
[97] M. Enyo and P. C. Biswas, Spring Meeting of the Japan Electrochemical Society, 1992, Abstract 2A21 [in Japanese]; M. Enyo and P. C. Biswas, *J. Electroanal. Chem.* **357** (1993) 67.
[98] M. A. V. Devanathan and Z. Stachursky, *J. Electrochem. Soc.* **111** (1964) 619.
[99] A. Kawashima, K. Hashimoto, and S. Shimodaira, *Nippon Kinzoku Gakkaishi (J. Metal Soc., Japan)* **38** (1974) 553 [in Japanese].
[100] H. H. Johnson, in *Hydrogen in Metals*, Proceedings of the International Conference on the Effects of Hydrogen on Materials Properties and Selection and Structural Design, September 23–27, 1973; *Am. Soc. Metals* **2** (1973) 35.
[101] B. Baranowski, Z. Szklarska-Smialowska, and M. Smialowski, Proc. 2nd Congress on Catalysis, ACTES. Paris, Eds. Tec. Nip. 1960, No. 115, pp. 2269–2283.
[102] B. Baranowski, *Bull. Acad. Pol. Sci., Ser. Sci. Chim.* **7** (1959) 743, 881, 897, 907.
[103] B. Baranowski and Z. Szklarska-Smialowska, *Electrochim. Acta* **9** (1964) 1497.
[104] Z. Szklarska-Smialowska and M. Smialowski, *J. Electrochem. Soc.* **110** (1963) 444.
[105] J. S. Bradbrook, G. W. Lorimer, and N. Ridley, *J. Nucl. Mater.* **42** (1972) 142.
[106] B. Nath, G. W. Lorimer, and N. Ridley, *J. Nucl. Mater.* **58** (1975) 153.
[107] M. P. Puls, *Acta Metall.* **32** (1984) 1259.
[108] M. N. A. Hall, S. L. H. Martin, and A. L. G. Rees, *Trans. Faraday Soc.* **41** (1945) 306.
[109] P. Dantzer, O. J. Kleppa, and M. E. Melnichak, *J. Chem. Phys.* **64** (1976) 139.
[110] F. Ricca, *J. Phys. Chem.* **71** (1967) 3632.

[111] M. Nagasaka and T. Yamashina, *J. Less-Common Met.* **45** (1976) 53.
[112] B. Stalinski and Z. Bieganski, *Bull. Acad. Pol. Sci., Ser. Sci. Chim.* **10** (1962) 247.
[113] R. M. Haag and F. J. Shipko, *J. Am. Chem. Soc.* **78** (1956) 5155.
[114] T. R. P. Gibbs, J. J. McSharry, and R. W. Bragdon, *J. Am. Chem. Soc.* **73** (1951) 1751.
[115] A. D. McQuillan, *Proc. R. Soc. London, Ser. A* **204** (1950–51) 309.
[116] P. Dantzer, *J. Chem. Soc. Solids* **44** (1983) 913.
[117] C. E. Ells and A. D. McQuillan, *J. Inst. Met.* **85** (1956–57) 89.
[118] A. G. Turnbull, *Aust. J. Chem.* **17** (1964) 1063.
[119] D. R. Fridrickson, *J. Phys. Chem.* **67** (1963) 1506.
[120] R. S. Vitt and K. Ono, *Metall. Trans.* **2** (1971) 608.
[121] N. E. Paton, B. S. Hickman, and D. H. Leslie, *Metall. Trans.* **2** (1971) 2791.
[122] L. G. Ritehie and K. W. Sprungmann, *J. Phys. (Paris)* **44** (1983) C9–C313.
[123] H. Numakura and M. Koiwa, *Trans. Jpn. Inst. Met.* **26** (1985) 653.
[124] H. Numakura, T. Ito, and M. Koiwa, *J. Less-Common Met.* **141** (1988) 285.
[125] C. D. Cann and A. Atrens, *J. Nucl. Mater.* **88** (1980) 42.
[126] H. Numakura and M. Koiwa, *Sci. Rep. Res. Inst. Tohoku Univ., Ser. A* **32** (1984) 46.
[127] G. J. C. Carpenter and J. F. Watters, *J. Nucl. Mater.* **73** (1978) 190.
[128] F. A. Lewis, R. C. Johnson, M. C. Witherspoon, and A. Obermann, *Surf. Technol.* **18** (1983) 147.
[129] F. A. Lewis, M. N. Hull, R. C. Johnson, and M. C. Witherspoon, *Surf. Technol.* **18** (1983) 167.
[130] T. B. Flanagan and F. A. Lewis, *J. Electrochem. Soc.* **108** (1961) 473.
[131] J. A. Green and F. A. Lewis, *Trans. Faraday Soc.* **60** (1964) 2234.
[132] Z. A. Bagotskaya, *Zh. Fiz. Khim.* **36** (1962) 2667.
[133] A. N. Frumkin, in *Advances in Electrochemistry and Electrochemical Engineering*, Vol. 3, Ed. by P. Delahay, Interscience, New York, 1963, p. 63.
[134] P. K. Subramanyan, in *Comprehensive Treatise of Electrochemistry*, Vol. 4, Ed. by J. O'M. Bockris, B. E. Conway, E. Yeager, and R. E. White, Plenum Press, New York, 1981, pp. 411–462.
[135] J. J. Phillips, P. Poole, and L. L. Shreir, *Corros. Sci.* **14** (1974) 533.
[136] T. Morozumi, T. Mizuno, and T. Kurachi, *Boshoku Gijyutu* **28** (1979) 285 [in Japanese].
[137] W. Kobozev and V. Montblanova, *Zh. Fiz. Khim.* **6** (1935) 38.
[138] W. Kobozev and V. Motblanova, *Acta Physicochim. URSS* **1** (1936) 611.
[139] M. Smialowski and B. Baranowski, Paper presented at the Electrochemistry Conference, Moscow, 1956.
[140] M. Enyo, in *Comprehensive Treatise of Electrochemistry*, Vol.7, Ed. by B. E. Conway, J. O'M. Bockris, E. Yeager, S. U. M. Khan, and R. E. White, Plenum Press, New York, 1983, pp. 241–300.
[141] M. A. V. Devanathan and Z. Stachursky, *Proc. Roy. Soc. London, Ser. A* **270** (1962) 90.
[142] M. A. V. Devanathan, Z. Stachursky, and W. Beck, *J. Electrochem. Soc.* **110** (1963) 886.
[143] W. Beck, J. O'M. Bockris, J. McBreen, and L. Nanis, *Proc. Roy. Soc. London, Ser. A* **290** (1965) 220.
[144] C. D. Kim and B. E. Wilde, *J. Electrochem. Soc.* **118** (1971) 202.
[145] T. C. Franklin and A. W. Bayerlein, *Denki Kagaku (J. Electrochem. Soc., Japan)* **41** (1973) 186.
[146] R. N. Iyer, H. W. Pickering, and M. Zamanzadeh, *J. Electrochem. Soc.* **136** (1989) 2463.
[147] J. Chene, J. Galland, and P. Azoa, in *Hydrogen in Metals*, Proceedings of the 2nd International Congress, Paris, June 6–10, 1977, Pergamon Press, Oxford, Vol. 1, p. 4A4.
[148] M. Hitzler, C. Knorr, and F. Mertens, *Z. Elektrochem.* **53** (1949) 228.
[149] R. Clamroth and C. Knorr, *Z. Elektrochem.* **57** (1953) 399.
[150] R. D. McCright and R. W. Staehle, *J. Electrochem. Soc.* **121** (1974) 609.

[151] M. Zamanzadeh, A. Allam, C. Kato, B. Ateya, and H. W. Pickering, *J. Electrochem. Soc.* **129** (1982) 284.
[152] A. Kawashima, K. Hashimoto, and S. Shimodaira, *Nippon Kinzoku Gakkaishi (J. Jpn. Soc. Met.)* **38** (1974) 1046 [in Japanese].
[153] B. J. Berkowitz and H. H. Horowitz, *J. Electrochem. Soc.* **129** (1982) 468.
[154] W. Beck, A. L. Glass, and E. Taylor, *J. Electrochem. Soc.* **112** (1965) 53.
[155] H. Angerstein-Kozlowska, *Bull. Acad. Pol. Sci., Ser. Sci. Chim.* **7** (1959) 881.
[156] H. Jarnolowicz and M. Smialowski, *J. Catal.* **1** (1962) 165.
[157] W. Palczewska and I. Ratajczyk, *Bull. Acad. Pol. Sci., Ser. Sci. Chim.* **9** (1961) 267.
[158] W. Palczewska, *Bull. Acad. Pol. Sci., Ser. Sci. Chim.* **7** (1959) 743.
[159] B. Baranowski and Z. Szklarska-Smialowska, *Electrochim. Acta* **9** (1964) 1497.
[160] B. Baranowski and M. Smialowski, *Bull. Acad. Pol. Sci., Ser. Sci. Chim.* **7** (1959) 663.
[161] B. Baranowski, *Bull. Acad. Pol. Sci., Ser. Sci. Chim.* **7** (1959) 897.
[162] Z. A. Foroulis, *J. Electrochem. Soc.* **128** (1981) 219.
[163] T. Okada, *Electrochim. Acta* **27** (1982) 1273; **28** (1983) 1113.
[164] H. Saito and M. Someno, *Nippon Kinzoku Gakkaishi (J. Jpn. Soc. Met.)* **36** (1972) 791 [in Japanese].
[165] M. Elam and B. E. Conway, *J. Electrochem. Soc.* **135** (1988) 1678.
[166] J. O'M. Bockris, D. Hodko, and Z. Minevski, in *Proceedings of the Symposium on Hydrogen Storage Materials, Batteries and Electrochemistry*, Ed. by D. A. Corrigan, Proceedings Vol. 92-5 The Electrochemical Society, Pennington, New Jersey, 1977, pp. 223–247.
[167] S. A. Semiletov, R. V. Baranova, Yu. P. Khodyrev, and R. M. Imamov *Sov. Phys. Crystallogr.* **25**(6) (1980) 665.
[168] M. Enyo, *J. Res. Inst. Catal. Hokkaido Univ.* **27** (1979) 63: **25** (1977) 17.
[169] M. Enyo, *Denki Kagaku (J. Electrochem. Soc., Japan)* **47** (1979) 2 [in Japanese].
[170] M. Enyo, *Int. J. Chem. Kinet.* **7** (1975) 463.
[171] A. N. Frumkin and G. Tedoradze, *Z. Elektrochem.* **62** (1958) 251.
[172] S. Toshima and H. Okaniwa, *Denki Kagaku (J. Electrochem. Soc., Japan)* **34** (1966) 641 [in Japanese]; Extended Abstract, 17th CITCE Meeting, Tokyo, 1966, p. 211.
[173] T. Yokoyama and M. Enyo, *Electrochim. Acta* **15** (1970) 1921.
[174] T. Morozumi, T. Mizuno, and T. Kurachi, *Boshoku Gijyutsu* **28** (1979) 285 [in Japanese].
[175] W. Wilhelmsen and A. R. Grande, *Electrochim. Acta* **35** (1990) 1913.
[176] M. Enyo, unpublished results.
[177] P. Millenbach, M. Givon and A. Aladjem, *J. Appl. Electrochem.* **13** (1983) 169.
[178] P. Kedzierzawski, Z. Szklarska-Smialowska, and M. Smialowski, *J. Electrochem. Soc.* **127** (1980) 2550.
[179] G. Toda, *J. Res. Inst. Catal. Hokkaido Univ.* **12** (1964) 39.
[180] G. Bohmholdt and E. Wicke, *Z. Phys. Chem. (N.F.)* **56** (1967) 133.
[181] J. Völkl, G. Wollenweger, K. H. Klatt, and G. Alefeld, *Z. Naturforsch.* **26a** (1971) 92.
[182] V. A. Gol'tzov, V. B. Vykhodets, G. Ye. Kagan, and P. V. Gel'd, *Phys. Met. Metallogr.* **29** (1970) 195.
[183] G. Holleck and E. Wicke, *Z. Phys. Chem. (N.F.)* **56** (1967) 155.
[184] G. Holleck, *J. Phys. Chem.* **74** (1970) 503.
[185] W. Jost and A. Widman, *Z. Phys. Chem.* **B45** (1940) 285.
[186] W. D. Davis, Atomic Energy Commission Report, Knolls Atomic Power Lab. 1954, p. 1227.
[187] S. A. Koffler, J. B. Hudson, and G. S. Ansell, *J. Met.* **20**(8) (1968) 75A.
[188] O. M. Katz and E. A. Gulbransen, *Rev. Sci. Instrum.* **31** (1960) 615.
[189] H. Zuckner, *Z. Naturforsch.* **25A** (1970) 1490.
[190] K. Sekine, *J. Res. Inst. Catal. Hokkaido Univ.* **25** (1977) 73.
[191] T. Fujita, M. Eng. Thesis, Hokkaido University, 1991.

[192] E. Kahrig, D. Kirstein, and F. Lange, *Ber. Bunsenges. Phys. Chem.* **70** (1966) 592.
[193] M. Kuballa and B. Baranowski, *Ber. Bunsenges. Phys. Chem.* **78** (1974) 335.
[194] R. V. Bucur, *Electrochim. Acta* **31** (1986) 385.
[195] D. N. Jewett and A. C. Makrides, *Trans. Faraday Soc.* **61** (1965) 932.
[196] M. von Stackelberg and P. Ludwig, *Z. Naturforsch.* **19a** (1964) 93.
[197] J. W. Simons and T. B. Flanagan, *J. Phys. Chem.* **69** (1965) 3581.
[198] K. Azumi, T. Fujita, T. Ito, T. Mizuno, and M. Seo, *Denki Kagaku (J. Electrochem. Soc., Japan)* **61** (1993) 576.
[199] E. Brauer and E. Nann, *Werkst. Korros.* **25** (1974) 309.
[200] M. Tsutsui and N. Fujise, *Nippon Kinzoku Gakkaishi (J. Jpn. Met. Soc.)* **39** (1975) 460 [in Japanese].
[201] E. Brauer, R. Dörr, and H. Zöchner, *Z. Phys. Chem. (N.F.)* **100** (1976) 109.
[202] M. Enyo and P. C. Biswas, *J. Electroanal. Chem.* **335** (1992) 309.
[203] T. Mizuno, T. Akimoto, K. Azumi, and N. Sato, *J. Electrochem. Soc. Jpn.* **59** (1991) 798.
[204] H. S. Carelaw and J. C. Jaeger, *Conduction of Heat in Solids*, Oxford University Press, Fair Lawn, New Jersey, 1959, p. 150.
[205] T. Mizuno, *J. Electrochem. Soc. Jpn.* **60** (1992) 405.
[206] K. Kunimatsu, N. Hasegawa, A. Kubota, N. Imai, M. Ishikawa, H. Akita, and Y. Tsuchida, *Frontiers of Cold Fusion*, No. 4, Ed. by H. Ikegami, Universal Academy Press, Tokyo, 1993, p. 31.
[207] F. G. Will, K. Cedzynska, and D. E. Linton, National Cold Fusion Institute Report, 1-131, 1991.
[208] M. Zamanzadeh, A. Allam, and H. W. Pickering, *J. Electrochem. Soc.* **127** (1980) 1688.
[209] R. T. Davis, Jr. and T. J. Butler, *J. Electrochem. Soc.* **105** (1958) 563.
[210] J. N. Andrews and A. R. Ubbelohde, *Proc. R. Soc. London, Ser. A* **253** (1959) 6.
[211] J. O'M. Bockris, W. Beck, M. A. Genshaw, P. K. Subramanyan, and F. S. Williams, *Acta Metall.* **19** (1971) 1209.
[212] R. F. Blundy, R. Royce, P. Poole, and L. L. Shreir, International Conference on Stress Corrosion Cracking and Hydrogen Embrittlement of Iron Base Alloys, Unieux Prminy, France, 1973, paper E-9.
[213] I. I. Phillips, P. Poole, and L. L. Shreir, *Corros. Sci.* **14** (1974) 553.
[214] T. P. Papazoglou and M. T. Hepworth, *Trans. Metall. Soc. AIME* **242** (1968) 684.
[215] R. J. Wasilewski and G. L. Kehl, *Mettalurgia* **1954** (November) 225.
[216] E. Brauer and E. Nann, *Werkst. Korros.* **5** (1974) 309.
[217] D. N. Williams, Hydrogen in Titanium and Titanium Alloys, TML Reports, No. 100, May 16, 1958.
[218] L. C. Covington, Corrosion (Houston) **35**(8) (1979) 378.
[219] M. Someno and K. Nagasaki, *Shinku Kagaku (Vacuum Science)* **8**(4) (1960) 145.
[220] B. Stalinski, C. K. Coogan, and H. S. Gutowsky, *J. Chem. Phys.* **33** (1960) 933.
[221] C. Korn and D. Zamir, *J. Phys. Chem. Solids* **31** (1970) 489.
[222] L. D. Bustard, R. M. Cotts, and E. F. W. Seymour, *Phys. Rev.* **22** (1980) 12.
[223] J. M. Pope and P. P. Narang, *J. Phys. Chem. Solids* **42** (1981) 519.
[224] M. Nagano, Y. Hayashi, N. Ohtani, M. Isshiki, and K. Igaki, *Scr. Metall.* **16** (1982) 973.
[225] Zh. Qi, J. Völkl, R. Lässer, and H. Wenzl, *J. Phys. F* **13** (1983) 2053.
[226] T. Eguchi and S. Morozumi, *Nippon Kinzoku Gakkau Kaiho* **41** (1977) 795 [in Japanese].
[227] C. M. Schwarz and M. W. Mallett, *Trans. Am. Soc. Met.* **41** (1945) 306.
[228] M. W. Mallett and M. W. Albrecht, *J. Electrochem. Soc.* **104** (1957) 142.
[229] E. A. Gulbransen and K. F. Andrew, *J. Electrochem. Soc.* **101**, (1954) 474.
[230] D. Someno, *Nippon Kinzoku Gakkaishi* **24** (1960) 249 [in Japanese].
[231] J. J. Kearns, *J. Nucl. Mater.* **43** (1972) 330.
[232] F. M. Mazzolai and J. Ryll-Nardzewski, *J. Less-Common Met.* **49** (1976) 323.
[233] S. Naito, *J. Chem. Phys.* **79** (1983) 3113.

[234] V. L. Gelezunas and P. K. Conn, *J. Electrochem. Soc.* **110** (1963) 799.
[235] J. F. Hon, *J. Chem. Phys.* **36** (1962) 759.
[236] C. Korn and S. Goren, *J. Less-Common Met.* **104** (1984) 113.
[237] T. Mizuno and M. Enyo, *J. Electrochem. Soc., Jpn.* **63** (1995) 719.
[238] T. Morozumi and T. Mizuno, *Bull. Jpn. Inst. Met.* **16** (1977) 119.
[239] T. Mizuno and M. Enyo, *Mem. Fac. Eng., Hokkaido Univ.* **18**(4) (1994) 17.
[240] T. Shindo, Dissertation, Hokkaido University, 1981.
[241] T. Mizuno, T. Shindou, and T. Morozumi, *Boshoku Gijyutsu* **26** (1977) 185 [in Japanese].
[242] T. Mizuno, *Bull. Jpn. Inst. Met.* **55** (1991) 553.

Cumulative Author Index for Numbers 1–30

Author	Title	Number
Abruña, H. D.	X Rays as Probes of Electrochemical Interfaces	20
Adžić R.	Reaction Kinetics and Mechanisms on Metal Single Crystal Electrode Surfaces	21
Agarwal, H. P.	Recent Developments in Faradaic Rectification Studies	20
Albella, J. M.	Electric Breakdown in Anodic Oxide Films	23
Allongue, P.	Physics and Applications of Semiconductor Electrodes Covered with Metal Clusters	23
Amokrane, S.	Analysis of the Capacitance of the Metal-Solution Interface. Role of the Metal and the Metal-Solvent Coupling	22
Andersen, H. C.	Improvements upon the Debye-Huckel Theory of Ionic Solutions	11
Andersen, T. N.	The Manganese Dioxide Electrode in Aqueous Solution	30
Andersen, T. N.	Potentials of Zero Charge of Electrodes	5
Appleby, A. J.	Electrocatalysis	9
Arvia, A. J.	Transport Phenomena in Electrochemical Kinetics	6
Arvia, A. J.	A Modern Approach to Surface Roughness Applied to Electrochemical Systems	28
Augustynski, J.	Application of Auger and Photoelectron Spectroscopy of Electrochemical Problems	13
Badawy, W. A.	Photovoltaic and Photoelectrochemical Cells Based on Schottky Barrier Heterojunctions	30
Badiali, J. P.	Analysis of the Capacitance of the Metal–Solution Interface. Role of the Metal and the Metal–Solvent Coupling	22
Baker, B. G.	Surface Analysis by Electron Spectroscopy	10
Balsene, L.	Application of Auger and Photoelectron Spectroscopy to Electrochemical Problems	13

Author	Title	Number
Barthel, J.	Temperature Dependence of Conductance of Electrolytes in Nonaqueous Solutions	13
Batchelor, R. A.	Surface States on Semiconductors	22
Bauer, H. H.	Critical Observations on the Measurement of Adsorption at Electrodes	7
Bebelis, S. I.	The Electrochemical Activation of Catalytic Reactions	29
Becker, R. O.	Electrochemical Mechanisms and the Control of Biological Growth Processes	10
Beden, B.	Electrocatalytic Oxidation of Oxygenated Aliphatic Organic Compounds at Noble Metal Electrodes	22
Benderskii, V. A.	Phase Transitions in the Double Layer at Electrodes	26
Berg, H.	Bioelectrochemical Field Effects: Electrostimulation of Biological Cells by Low Frequencies	24
Berwick, A.	The Study of Simple Consecutive Processes in Electrochemical Reactions	5
Blank, M.	Electrochemistry in Nerve Excitation	24
Bloom, H.	Models for Molten Salts	9
Bloom, H.	Molten Electrolytes	2
Blyholder, G.	Quantum Chemical Treatment of Adsorbed Species	8
Bockris, J. O'M.	Electrode Kinetics	1
Bockris, J. O'M.	Ionic Solvation	1
Bockris, J. O'M.	The Mechanism of Charge Transfer from Metal Electrodes to Ions in Solution	6
Bockris, J. O'M.	The Mechanism of the Electrode Position of Metals	3
Bockris, J. O'M.	Molten Electrolytes	2
Bockris, J. O'M.	Photoelectrochemical Kinetics and Related Devices	14
Boguslavsky, L. I.	Electron Transfer Effects and the Mechanism of the Membrane Potential	18
Breiter, M. W.	Adsorption of Organic Species on Platinum Metal Electrodes	10
Breiter, M. W.	Low-Temperature Electrochemistry at High-T_2 Superconductor/Ionic Conductor Interfaces	28
Brodskii, A. N.	Phase Transitions in the Double Layer at Electrodes	26

Author	Title	Number
Burke, L. D.	Electrochemistry of Hydrous Oxide Films	18
Burney, H. S.	Membrane Chlor-Alkali Process	24
Čekerevac, M. I.	The Mechanism of Formation of Coarse and Disperse Electrodeposits	30
Charle, K. P.	Spin-Dependent Kinetics in Dye-Sensitized Charge-Carrier Injection into Organic Crystal Electrodes	19
Cheh, H. Y.	Theory and Applications of Periodic Electrolysis	19
Christov, S. G.	Quantum Theory of Charge-Transfer Processes in Condensed Media	28
Conway, B. E.	The Behavior of Intermediates in Electrochemical Catalysis	3
Conway, B. E.	Electroanalytical Methods for Determination of Al_2O_3 in Molten Cryolite	26
Conway, B. E.	Fundamental and Applied Aspects of Anodic Chlorine Production	14
Conway, B. E.	Ionic Solvation	1
Conway, B. E.	Proton Solvation and Proton Transfer Processes in Solution	3
Conway, B. E.	Solvated Electrons in Field- and Photo-assisted Processes at Electrodes	7
Conway, B. E.	The Temperature and Potential Dependence of Electrochemical Reaction Rates, and the Real Form of the Tafel Equation	16
Covington, A. K.	NMR Studies of the Structure of Electrolyte Solutions	12
Daikhin, L. I.	Phase Transitions in the Double Layer at Electrodes	26
Damaskin, B. B.	Adsorption of Organic Compounds at Electrodes	3
Damjanovic, A.	The Mechanism of the Electrodeposition of Metals	3
Damjanovic, A.	Mechanistic Analysis of Oxygen Electrode Reactions	5
Desnoyers, J. B.	Hydration Effects and Thermodynamic Properties of Ions	5
Despić, A.	Electrochemistry of Aluminum in Aqueous Solutions and Physics of Its Anodic Oxide	20

Author	Title	Number
Despić, A. R.	Electrochemical Deposition and Dissolution of Alloys and Metal Components—Fundamental Aspects	27
Despić, A. R.	Transport-Controlled Deposition and Dissolution of Metals	7
Djokić, S. S.	Electrodeposition of Nickel-Iron Alloys	22
Djokić, S. S.	Electroanalytical Methods for Determination of Al_2O_3 in Molten Cryolite	26
Drazic, D. M.	Iron and Its Electrochemistry in an Active State	19
Efrima, S.	Surface-Enhanced Raman Scattering (SERS)	16
Eisenberg, H.	Physical Chemistry of Synthetic Polyelectrolytes	1
Elving, P. J.	Critical Observations on the Measurement of Adsorption at Electrodes	7
Enyo, M.	Mechanism of the Hydrogen Electrode Reaction as Studied by Means of Deuterium as a Tracer	11
Enyo, M.	Sorption of Hydrogen on and in Hydrogen-Absorbing Metals in Electrochemical Environments	30
Erdey-Grúz, T.	Proton Transfer in Solution	12
Fahidy, T. Z.	Recent Advance in the Study of the Dynamics of Electrode Processes	27
Falkenhagen, H.	The Present State of the Theory of Electrolytic Solutions	2
Farges, J.-P.	Charge-Transfer Complexes in Electrochemistry	12
Farges, J.-P.	An Introduction to the Electrochemistry of Charge Transfer Complexes II	13
Findl, E.	Bioelectrochemistry-Electrophysiology-Electrobiology	14
Floyd, W. F.	Electrochemical Properties of Nerve and Muscle	1
Foley, J. K.	Interfacial Infrared Vibrational Spectroscopy	17
Friedman, H. L.	Computed Thermodynamic Properties and Distribution Functions for Simple Models of Ionic Solutions	6
Frumkin, A. A. N.	Adsorption of Organic Compounds at Electrodes	3
Fuller, T. F.	Metal Hydride Electrodes	27

Author	Title	Number
Fuoss, R. M.	Physical Chemistry of Synthetic Polyelectrolytes	1
Galvele, I. R.	Electrochemical Aspects of Stress Corrosion Cracking	27
German, E. D.	The Role of the Electronic Factor in the Kinetics of Charge-Transfer Reactions	24
Gileadi, E.	The Behavior of Intermediates in Electrochemical Catalysis	3
Gileadi, E.	The Mechanism of Oxidation of Organic Fuels	4
Girault, H. H.	Charge Transfer across Liquid-Liquid Interfaces	25
Goddard, E. D.	Electrochemical Aspects of Adsorption on Mineral Solids	13
Goodisman, J.	Theories for the Metal in the Metal–Electrolyte Interface	20
Gores, H.-J.	Temperature Dependence of Conductance of Electrolytes in Nonaqueous Solutions	13
Goruk, W. S.	Anodic and Electronic Currents at High Fields in Oxide Films	4
Grätzel, M.	Interfacial Charge Transfer Reactions in Colloidal Dispersions and Their Application to Water Cleavage by Visible Light	15
Green, M.	Electrochemistry of the Semiconductor–Electrolyte Interface	2
Gregory, D. P.	Electrochemistry and the Hydrogen Economy	10
Gu, Z. H.	Recent Advance in the Study of the Dynamics of Electrode Processes	27
Gurevich, Y. Y.	Electrochemistry of Semiconductors: New Problems and Prospects	16
Gutiérrez, C.	Potential-Modulated Reflectance Spectroscopy Studies of the Electronic Transitions of Chemisorbed Carbon Monoxide	28
Gutmann, F.	Charge-Transfer Complexes in Electrochemistry	12
Gutmann, F.	The Electrochemical Splitting of Water	15
Gutmann, F.	An Introduction to the Electrochemistry of Charge Transfer Complexes II	13

Author	Title	Number
Habib, M. A.	Solvent Dipoles at the Electrode-Solution Interface	12
Haering, R. R.	Physical Mechanisms of Intercalation	15
Hamann, S. D.	Electrolyte Solutions at High Pressure	9
Hamelin, A.	Double-Layer Properties at sp and sd Metal Single-Crystal Electrodes	16
Hamnett, A.	Surface States on Semiconductors	22
Hansma, P. K.	Scanning Tunneling Microscopy: A Natural for Electrochemistry	21
Harrington, D. A.	Ultrahigh-Vacuum Surface Analytical Methods in Electrochemical Studies of Single-Crystal Surfaces	28
Heiland, W.	The Structure of the Metal-Vacuum Interface	11
Herman, P. J.	Critical Observations on the Measurement of Adsorption at Electrodes	7
Hickling, A.	Electrochemical Processes in Glow Discharge at the Gas–Solution Interface	6
Hine, F.	Chemistry and Chemical Engineering in the Chlor-Alkali Industry	18
Hoar, T. R.	The Anodic Behavior of Metals	2
Hopfinger, A. J.	Structural Properties of Membrane Ionomers	14
Humffray, A. A.	Methods and Mechanisms in Electroorganic Chemistry	8
Hunter, R. J.	Electrochemical Aspects of Colloid Chemistry	11
Jaegermann, W.	The Semiconductor/Electrolyte Interface: A Surface Science Approach	30
Jaksic, M. M.	The Electrochemical Activation of Catalytic Reactions	29
Johnson, C. A.	The Metal–Gas Interface	5
Jolieoeur, C.	Hydration Effects and Thermodynamic Properties of Ions	5
Jović, V. D.	Electrochemical Deposition and Dissolution of Alloys and Metal Components—Fundamental Aspects	27
Kahn, S. U. M.	Photoelectrochemical Kinetics and Related Devices	14
Kahn, S. U. M.	Some Fundamental Aspects of Electrode Processes	15
Kebarle, P.	Gas-Phase Ion Equilibria and Ion Solvation	9

Author	Title	Number
Kelbg, G.	The Present State of the Theory of Electrolytic Solutions	2
Kelly, E. I.	Electrochemical Behavior of Titanium	14
Krstajić, N. V.	The Mechanism of Coarse and Disperse Electrodeposits	30
Lyklema, J.	Interfacial Electrostatics and Electrodynamics in Disperse Systems	17
Lynn, K. G.	The Nickel Oxide Electrode	21
Lyons, M. E. G.	Electrochemistry of Hydrous Oxide Films	18
MacDonald, D. D.	The Electrochemistry of Metals in Aqueous Systems at Elevated Temperatures	11
MacDonald, D. D.	Impedance Measurements in Electrochemical Systems	14
Maksimović, M. D.	Theory of the Effect of Electrodeposition at a Periodically Changing Rate on the Morphology of Metal Deposits	19
Mandel, L. J.	Electrochemical Processes at Biological Interfaces	8
Marchiano, S. L.	Transport Phenomena in Electrochemical Kinetics	6
Marincic, N.	Lithium Batteries with Liquid Depolarizers	15
Markin, V. S.	Thermodynamics of Membrane Energy Transduction in an Oscillating Field	24
Martinez-Duart, J. M.	Electric Breakdown in Anodic Oxide Films	23
Matthews, D. B.	The Mechanism of Charge Transfer from Metal Electrodes to Ions in Solution	6
Mauritz, K. A.	Structural Properties of Membrane Ionomers	14
McBreen, J.	The Nickel Oxide Electrode	21
McKinnon, W. R.	Physical Mechanisms of Intercalation	15
McKubre, M. C. H.	Impedance Measurements in Electrochemical Systems	14
Mizuno, T.	Sorption of Hydrogen on and in Hydrogen-Absorbing Metals in Electrochemical Environments	30
Murphy, O. J.	The Electrochemical Splitting of Water	15
Nagarkan, P. V.	Electrochemistry of Metallic Glasses	21
Nágy, Z.	DC Electrochemical Techniques for the Measurement of Corrosion Rates	25

Author	Title	Number
Nágy, Z.	DC Relaxation Techniques for the Investigation of Fast Electrode Reactions	21
Neophytides, S. G.	The Electrochemical Activation of Catalytic Reactions	29
Newman, J.	Determination of Current Distributions Governed by Laplace's Equation	23
Newman, J.	Metal Hydride Electrodes	27
Newman, J.	Photoelectrochemical Devices for Solar Energy Conversion	18
Newman, K. E.	NMR Studies of the Structure of Electrolyte Solutions	12
Nişanciağlu, K.	Design Techniques in Cathodic Protection Engineering	23
Novak, D. M.	Fundamental and Applied Aspects of Anodic Chlorine Production	14
O'Keefe, T. J.	Electrogalvanizing	26
Orazem, M. E.	Photoelectrochemical Devices for Solar Energy Conversion	18
Oriani, R. A.	The Metal–Gas Interface	5
Padova, J. I.	Ionic Solvation in Nonaqueous and Mixed Solvents	7
Paik, Woon-kie	Ellipsometry in Electrochemistry	25
Parkhutik, V.	Electrochemistry of Aluminum in Aqueous Solutions and Physics of Its Anodic Oxide	20
Parkhutik, V. P.	Electric Breakdown in Anodic Oxide Films	23
Parsons, R.	Equilibrium Properties of Electrified Interphases	1
Pavlovic, M. G.	Electrodeposition of Metal Powders with Controlled Particle Grain Size and Morphology	24
Perkins, R. S.	Potentials of Zero Charge of Electrodes	5
Pesco, A. M.	Theory and Applications of Periodic Electrolysis	19
Piersma, B.	The Mechanism of Oxidation of Organic Fuels	4
Pilla, A. A.	Electrochemical Mechanisms and the Control of Biological Growth Processes	10
Pintauro, P. N.	Transport Models for Ion-Exchange Membranes	19
Pleskov, Y. V.	Electrochemistry of Semiconductors: New Problems and Prospects	16

Author	Title	Number
Plonski, I.-H.	Effects of Surface Structure and Adsorption Phenomena on the Active Dissolution of Iron in Acid Media	29
Plzak, V.	Advanced Electrochemical Hydrogen Technologies: Water Electrolyzers and Fuel Cells	26
Pons, S.	Interfacial Infrared Vibrational Spectroscopy	17
Popov, K. I.	Electrodeposition of Metal Powders with Controlled Particle Grain Size and Morphology	24
Popov, K. I.	The Mechanism of Formation of Coarse and Disperse Electrodeposits	30
Popov, K. I.	Theory of the Effect of Electrodeposition at a Periodically Changing Rate on the Morphology of Metal Deposits	19
Popov, K. I.	Transport-Controlled Deposition and Dissolution of Metals	7
Pound, B. G.	Electrochemical Techniques to Study Hydrogen Ingress in Metals	25
Power, G. P.	Metal Displacement Reactions	11
Reeves, R. M.	The Electrical Double Layer: The Current States of Data and Models, with Particular Emphasis on the Solvent	9
Revie, R. W.	Environmental Cracking of Metals: Electrochemical Aspects	26
Ritchie, I. M.	Metal Displacement Reactions	11
Rohland, B.	Advanced Electrochemical Hydrogen Technologies: Water Electrolyzers and Fuel Cells	26
Roscoe, S. G.	Electrochemical Investigations of the Interfacial Behavior of Proteins	29
Rusling, J. F.	Electrochemistry and Electrochemical Catalysis in Microemulsions	26
Russell, J.	Interfacial Infrared Vibrational Spectroscopy	17
Rysselberghe, P. Van	Some Aspects of the Thermodynamic Structure of Electrochemistry	4
Sacher, E.	Theories of Elementary Homogeneous Electron-Transfer Reactions	3
Saemann-Ischenko, G.	Low-Temperature Electrochemistry at High-T_2 Superconductor/Ionic Conductor Interfaces	28

Author	Title	Number
Salvarezza, R. C.	A Modern Approach to Surface Roughness Applied to Electrochemical Systems	28
Sandstede, G. S.	Water Electrolysis and Solar Hydrogen Demonstration Projects	27
Savenko, V. I.	Electric Surface Effects in Solid Plasticity and Strength	24
Scharifker, B. R.	Microelectrode Techniques in Electrochemistry	22
Schmickler, W.	Electron Transfer Reactions on Oxide-Covered Metal Electrodes	17
Schneir, J.	Scanning Tunneling Microscopy: A Natural for Electrochemistry	21
Schultze, J. W.	Electron Transfer Reactions on Oxide-Covered Metal Electrodes	17
Scott, K.	Reaction Engineering and Digital Simulation in Electrochemical Processes	27
Searson, P. C.	Electrochemistry of Metallic Glasses	21
Šepa, D. B.	Energies of Activation of Electrode Reactions: A Revisited Problem	29
Seversen, M.	Interfacial Infrared Vibrational Spectroscopy	17
Shchukin, E. D.	Electric Surface Effects in Solid Plasticity and Strength	24
Sides, P. J.	Phenomena and Effects of Electrolytic Gas Evolution	18
Snook, I. K.	Models for Molten Salts	9
Somasundaran, P.	Electrochemical Aspects of Adsorption on Mineral Solids	13
Sonnenfeld, R.	Scanning Tunneling Microscopy: A Natural for Electrochemistry	21
Soriaga, M. P.	Ultrahigh-Vacuum Surface Analytical Methods in Electrochemical Studies of Single-Crystal Surfaces	28
Stickney, J. L.	Ultrahigh-Vacuum Surface Analytical Methods in Electrochemical Studies of Single-Crystal Surfaces	28
Stonehart, P.	Preparation and Characterization of Highly Dispersed Electrocatalytic Materials	12
Szklarczyk, M.	Electrical Breakdown of Liquids	25
Taniguchi, I.	Electrochemical and Photoelectrochemical Reduction of Carbon Dioxide	20

Author	Title	Number
Tarasevich, M. R.	Electrocatalytic Properties of Carbon Materials	19
Thirsk, H. R.	The Study of Simple Consecutive Processes in Electrochemical Reactions	5
Tilak, B. V.	Chemistry and Chemical Engineering in the Chlor-Alkali Industry	18
Tilak, B. V.	Fundamental and Applied Aspects of Anodic Chlorine Production	14
Trasatti, S.	Solvent Adsorption and Double-Layer Potential Drop at Electrodes	13
Tributsch, H.	Photoelectrolysis and Photoelectrochemical Catalysis	17
Tsong, T. Y.	Thermodynamics of Membrane Energy Transduction in an Oscillating Field	24
Uosaki, K.	Theoretical Aspects of Semiconductor Electrochemistry	18
Van Leeuwen, H. P.	Interfacial Electrostatics and Electrodynamics in Disperse Systems	17
Vayenas, C. G.	The Electrochemical Activation of Catalytic Reactions	29
Velichko, G. I.	Phase Transitions in the Double Layer at Electrodes	26
Verbrugge, M. W.	Transport Models for Ion-Exchange Membranes	19
Vijh, A. K.	Perspectives in Electrochemical Physics	17
Viswanathan, K.	Chemistry and Chemical Engineering in the Chlor-Alkali Industry	18
Von Goldammer, E.	NMR Studies of Electrolyte Solutions	10
Vorotyntsev, M. A.	Modern State of Double Layer Study of Solid Metals	17
Wachter, R.	Temperature Dependence of Conductance of Electrolytes in Nonaqueous Solutions	13
Wendt, H.	Advanced Electrochemical Hydrogen Technologies: Water Electrolyzers and Fuel Cells	26
Wenglowski, G.	An Economic Study of Electrochemical Industry in the United States	4
West, A. C.	Determination of Current Distributions Governed by Laplace's Equation	23

Author	Title	Number
Wieckowski, A.	Ultrahigh-Vacuum Surface Analytical Methods in Electrochemical Studies of Single-Crystal Surfaces	28
Wiekowski, A.	*In Situ* Surface Electrochemistry: Radioactive Labeling	21
Willig, F.	Spin-Dependent Kinetics in Dye-Sensitized Charge-Carrier Injection into Organic Crystal Electrodes	19
Wojtowicz, J.	Oscillatory Behavior in Electrochemical Systems	8
Woods, R.	Chemisorption of Thiols on Metals and Metal Sulfides	29
Wroblowa, H. S.	Batteries for Vehicular Propulsion	16
Wurster, R.	Water Electrolysis and Solar Hydrogen Demonstration Projects	27
Yeager, E. B.	Ultrasonic Vibration Potentials	14
Yeager, H. L.	Structural and Transport Properties of Perfluorinated Ion-Exchange Membranes	16
Yeo, R. S.	Structural and Transport Properties of Perfluorinated Ion-Exchange Membranes	16
Young, L.	Anodic and Electronic Currents at High Fields in Oxide Films	4
Zana, R.	Ultrasonic Vibration Potentials	14
Zobel, F. G. R.	Anodic and Electronic Currents at High Fields in Oxide Films	4

Cumulative Title Index for Numbers 1–30

Title	Author	Number
Adsorption of Organic Compounds at Electrodes	Frumkin, A. A. N. Damaskin, B. B.	3
Adsorption of Organic Species on Platinum Metal Electrodes	Breiter, M. W.	10
Advanced Electrochemical Hydrogen Technologies: Water Electrolyzers and Fuel Cells	Plzak, V. Rohland, B. Wendt, H.	26
Analysis of the Capacitance of the Metal–Solution Interface. Role of the Metal and the Metal-Solvent Coupling	Amokrane, S. Badiali, J. P.	22
The Anodic Behavior of Metals	Hoar, T. P.	2
Anodic and Electronic Currents at High Fields in Oxide Films	Young, L. Goruk, W. S. Zobel, F. G. R.	4
Application of Auger and Photoelectron Spectroscopy to Electrochemical Problems	Augustynski, J. Balsenc, L.	13
Batteries for Vehicular Propulsion	Wroblowa, H. S.	16
The Behavior of Intermediates in Electrochemical Catalysis	Gileadi, E. Conway, B. E.	3
Bioelectrochemical Field Effects: Electrostimulation of Biological Cells by Low Frequencies	Berg, H.	24
Bioelectrochemistry–Electrophysiology– Electrobiology	Findl, E.	14
Charge Transfer across Liquid–Liquid Interfaces	Girault, H. H.	25
Charge-Transfer Complexes in Electrochemistry	Farges, J.-P. Gutmann, F.	12
Chemisorption of Thiols on Metals and Metal Sulfides	Woods, R.	29

Title	Author	Number
Chemistry and Chemical Engineering in the Chlor-Alkali Industry	Hine, F. Tilak, B. V. Viswanathan, K.	18
Computed Thermodynamic Properties and Distribution Functions for Simple Models of Ionic Solutions	Friedman, H. L.	6
Critical Observations on the Measurement of Adsorption at Electrodes	Bauer, H. H. Herman, P. J. Elving, P. J.	7
DC Relaxation Techniques for the Investigation of Fast Electrode Reactions	Nagy, Z.	21
DC Electrochemical Techniques for the Measurement of Corrosion Rates	Nagy, Z.	25
Design Techniques in Cathodic Protection Engineering	Nişancioğlu, K.	23
Determination of Current Distributions Governed by Laplace's Equation	West, A. C. Neuman, J.	23
Double-Layer Properties at sp and sd Metal Single-Crystal Electrodes	Hamelin, A.	16
An Economic Study of Electrochemical Industry in the United States	Wenglowski, G.	4
Effect of Surface Structure and Adsorption Phenomena on the Active Dissolution of Iron in Acid Media	Plonski, I.-H.	29
Electrical Breakdown of Liquids	Szklarczyk, M.	25
The Electrical Double Layer: The Current Status of Data and Models, with Particular Emphasis on the Solvent	Reeves, R. M.	9
Electric Breakdown in Anodic Oxide Films	Parkhutik, V. P. Albella, J. M. Martinez-Duart, J. M.	23
Electric Surface Effects in Solid Plasticity and Strength	Shchukin, E. D. Kochanova, L. A. Savenko, V. I.	24
Electroanalytical Methods for Determination of Al_2O_3 in Molten Cryolite	Djokić, S. S. Conway, B. E.	26
Electrocatalysis	Appleby, A. I.	9
Electrocatalytic Oxidation of Oxygenated Aliphatic Organic Compounds at Noble Metal Electrodes	Beden, B. Léger, J.-M. Lamy, C.	22

Title	Author	Number
Electrocatalytic Properties of Carbon Materials	Tarasevich, M. R. Khrushcheva, E. I.	19
The Electrochemical Activation of Catalytic Reactions	Vayenas, C. G. Jaksic, M. M. Bebelis, S. I. Neophytides, S. G.	29
Electrochemical Aspects of Adsorption on Mineral Solids	Somasundaran, P. Goddart, E. D.	13
Electrochemical Aspects of Colloid Chemistry	Hunter, R. J.	11
Electrochemical Behavior of Titanium	Kelly, E. J.	14
Electrochemical Investigations of the Interfacial Behavior of Proteins	Roscoe, S. G.	29
Electrochemical Mechanisms and the Control of Biological Growth Processes	Becker, R. O. Pilla, A. A.	10
Electrochemical and Photoelectrochemical Reduction of Carbon Dioxide	Taniguchi, I.	20
Electrochemical Processes at Biological Interfaces	Mandel, L. J.	8
Electrochemical Processes in Glow Discharge at the Gas–Solution Interface	Hickling, A.	6
Electrochemical Properties of Nerve and Muscle	Floyd, W. F.	1
The Electrochemical Splitting of Water	Gutmann, F. Murphy, O. J.	15
Electrochemical Techniques to Study Hydrogen Ingress in Metals	Pound, B. G.	25
Electrochemistry of Aluminum in Aqueous Solutions and Physics of its Anodic Oxide	Despić, A. Parkhutik, V.	20
Electrochemistry and Electrochemical Catalysis in Microemulsions	Rusling, J. F.	26
Electrochemistry and the Hydrogen Economy	Gregory, D. P.	10
Electrochemistry of Hydrous Oxide Films	Burke, L. D. Lyons, M. E. G.	18
Electrochemistry of Metallic Glasses	Searson, P. C. Nagarkan, P. V. Latanision, R. M.	21
The Electrochemistry of Metals in Aqueous Systems at Elevated Temperatures	Macdonald, D. D.	11
Electrochemistry of Nerve Excitation	Blank, M.	24
Electrochemistry of Semiconductors: New Problems and Prospects	Pleskov, Y. V. Gurevich, Y. Y.	16

Title	Author	Number
Electrochemistry of the Semiconductor-Electrolyte Interface	Green, M.	2
Electrochemistry of Sulfide Minerals	Koch, D. F. A.	10
Electrochemical Aspects of Stress Corrosion Cracking	Galvele, J. R.	27
Electrochemical Deposition and Dissolution of Alloys and Metal Components—Fundamental Aspects	Despić, A. R. Jović, V. D.	27
Electrode Kinetics	Bockris, J. O'M.	1
Electrodeposition of Metal Powders with Controlled Particle Grain Size and Morphology	Popov, K. I. Pavlovic, M. G.	24
Electrodeposition of Nickel–Iron Alloys	Djokic, S. S. Maksimovic, M. D.	22
Electrogalvanizing	Lindsay, J. H. O'Keefe, T. J.	26
Electrolyte Solutions at High Pressure	Hamann, S. D.	9
Electron Transfer Effects and the Mechanism of the Membrane Potential	Boguslavsky, L. I.	18
Electron Transfer Reactions on Oxide-Covered Metal Electrodes	Schmickler, W. Schultze, J. W.	17
Ellipsometry in Electrochemistry	Paik, Woon-kie	25
Energies of Activation of Electrode Reactions: A Revisited Problem	Šepa, D. B.	29
Environmental Cracking of Metals: Electrochemical Aspects	Revie, R. W.	26
Equilibrium Properties of Electrified Interphases	Parsons, R.	1
Fundamental and Applied Aspects of Anodic Chlorine Production	Novak, D. M. Tilak, B. V. Conway, B. E.	14
Gas-Phase Ion Equilibria and Ion Solvation	Kebarle, P.	9
Hydration Effects and Thermodynamic Properties of Ions	Desnoyers, J. B. Jolieoeur, C.	5
Impedance Measurements in Electrochemical Systems	Macdonald, D. D. McKubre, M. C. H.	14
Improvements upon the Debye–Hückel Theory of Ionic Solutions	Andersen, H. C.	11

Title	Author	Number
In Situ Surface Electrochemistry: Radioactive Labeling	Wiekowski, A.	21
Interfacial Charge Transfer Reactions in Colloidal Dispersions and Their Application to Water Cleavage by Visible Light	Grätzel, M.	15
Interfacial Electrostatics and Electrodynamics in Disperse Systems	Van Leeuwen, H. P. Lyklema, J.	17
Interfacial Infrared Vibrational Spectroscopy	Pons, S. Foley, J. K. Russell, J. Seversen, M.	17
An Introduction to the Electrochemistry of Charge Transfer Complexes II	Gutmann, F. Farges, J.-P.	13
Ion and Electron Transfer across Monolayers of Organic Surfactants	Lipkowski, J.	23
Ionic Solvation	Conway, B. E. Bockris, J. O'M.	1
Ionic Solvation in Nonaqueous and Mixed Solvents	Padova, J. I.	7
Iron and Its Electrochemistry in an Active State	Drazic, D. M.	19
Lithium Batteries with Liquid Depolarizers	Marincic, N.	15
Low-Temperature Electrochemistry at High-T_2 Superconductor/Ionic Conductor Interfaces	Lorenz, W. J. Saemann-Ischenko, G. Breiter, M. W.	28
The Manganese Dioxide Electrode in Aqueous Solution	Andersen, T. N.	30
The Mechanism of Charge Transfer from Metal Electrodes to Ions in Solution	Matthews, D. B. Bockris, J. O'M.	6
The Mechanism of the Electrodeposition of Metals	Bockris, J. O'M. Damjanovic, A.	3
The Mechanism of Formation of Coarse and Disperse Electrodeposits	Popov, K. I. Krstajić, N. V. Čekerevac, M. I.	30
Mechanism of the Hydrogen Electrode Reaction as Studied by Means of Deuterium as a Tracer	Enyo, M.	11
The Mechanism of Oxidation of Organic Fuels	Gileadi, E. Piersma, B.	4

Title	Author	Number
Mechanisms of Stepwise Electrode Processes on Amalgams	Losev, V. V.	7
Mechanistic Analysis of Oxygen Electrode Reactions	Damjanovic, A.	5
Membrane Chlor-Alkali Process	Burney, H. S.	24
Metal Displacement Reactions	Power, G. P. Ritchie, I. M.	11
The Metal–Gas Interface	Oriani, R. A. Johnson, C. A.	5
Metal Hydride Electrodes	Fuller, T. H. Newman, J.	27
Methods and Mechanisms in Electroorganic Chemistry	Humffray, A. A.	8
Microelectrode Techniques in Electrochemistry	Scharifker, B. R.	22
Models for Molten Salts	Bloom, H. Snook, I. K.	9
A Modern Approach to Surface Roughness Applied to Electrochemical Systems	Salvarezza, R. C. Arvia, A. J.	28
Modern State of Double Layer Study of Solide Metals	Vorotyntsev, M. A.	17
Molten Electrolytes	Bloom, H. Bockris, J. O'M.	2
The Nickel Oxide Electrode	McBreen, J. Lynn, K. G.	21
NMR Studies of Electrolyte Solutions	von Goldammer, E.	10
NMR Studies of the Structure of Electrolyte Solutions	Covington, A. K. Newman, K. E.	12
Oscillatory Behavior in Electrochemical Systems	Wojtowicz, J.	8
Perspectives in Electrochemical Physics	Vijh, A. K.	17
Phase Transitions in the Double Layer at Electrodes	Benderskii, V. A. Brodskii, A. N. Daikhin, L. I. Velichko G. I.	26
Phenomena and Effects of Electrolytic Gas Evolution	Sides, R. J.	18
Photoelectrochemical Devices for Solar Energy Conversion	Orazem, M. E. Newman, J.	18

Title	Author	Number
Photoelectrochemical Kinetics and Related Devices	Khan, S. U. M. Bockris, J. O'M.	14
Photoelectrolysis and Photoelectrochemical Catalysis	Tributsch, H.	17
Photovoltaic and Photoelectrochemical Cells Based on Schottky Barrier Heterojunctions	Badawy, W. A.	30
Physical Chemistry of Ion-Exchange Resins	Kitchener, J. A.	2
Physical Chemistry of Synthetic Polyelectrolytes	Eisenberg, H. Fuoss, R. M.	1
Physical Mechanisms of Intercalation	McKinnon, W. R. Haering, R. R.	15
Physics and Applications of Semiconductor Electrodes Covered with Metal Clusters	Allongue, P.	23
Potential-Modulated Reflectance Spectroscopy Studies of the Electronic Transitions of Chemisorbed Carbon Monoxide	Gutiérrez, C.	28
Potentials of Zero Charge Electrodes	Perkins, R. S. Andersen, T. N.	5
Power Sources for Electric Vehicles	Kordesch, K. V.	10
Preparation and Characterization of Highly Dispersed Electrocatalytic Materials	Kinoshita, K. Stonehart, R.	12
The Present State of the Theory of Electrolytic Solutions	Falkenhagen, H. Kelbg, G.	2
Proton Solvation and Proton Transfer Processes in Solution	Conway, B. E.	3
Proton Transfer in Solution	Erdey-Grúz, T. Lengyel, S.	12
Quantum Chemical Treatment of Adsorbed Species	Blyholder, G.	8
Quantum Theory of Charge-Transfer Processes in Condensed Media	Christov, S. G.	28
Reaction Engineering and Digital Simulation in Electrochemical Processes	Scott, K.	27
Reaction Kinetics and Mechanism on Metal Single Crystal Electrode Surfaces	Adžić, R.	21
Recent Advances in the Study of the Dynamics of Electrode Processes	Fahidy, T. Z. Gu, Z. H.	27
Recent Advances in the Theory of Charge Transfer	Kuznetsov, A. M.	20

Title	Author	Number
Recent Developments in Faradaic Rectification Studies	Agarwal, H. P.	20
The Role of Electrochemistry in Environmental Control	Kuhn, A. T.	8
The Role of the Electronic Factor in the Kinetics of Charge-Transfer Reactions	German, E. D. Kuznetsov, A. M.	24
Scanning Tunneling Microscopy: A Natural for Electrochemistry	Sonnenfeld, R. Schneir, J. Hansma, P. K.	21
The Semiconductor/Electrolyte Interface: A Surface Science Approach	Jaegermann, W.	30
Small-Particle Effects and Structural Considerations for Electrocatalysis	Kinoshita, K.	14
Solvated Electrons in Field- and Photo-Assisted Processes at Electrodes	Conway, B. E.	7
Solvent Adsorption and Double-Layer Potential Drop at Electrodes	Trasatti, S.	13
Solvent Dipoles at the Electrode-Solution Interface	Habib, M. A.	12
Some Aspects of the Thermodynamic Structure of Electrochemistry	Rysselberghe, P. van	4
Some Fundamental Aspects of Electrode Processes	Kahn, S. U. M.	15
Sorption of Hydrogen on and in Hydrogen-Absorbing Metals in Electrochemical Environments	Mizuno, T. Enyo, M.	30
Spin-Dependent Kinetics in Dye-Sensitized Charge-Carrier Injection into Organic Crystal Electrodes	Charle, K.-P. Willig, F.	19
Structural and Transport Properties of Perfluorinated Ion-Exchange Membranes	Yeo, R. S. Yeager, H. L.	16
Structural Properties of Membrane Ionomers	Mauritz, K. A. Hopfinger, A. J.	14
The Structure of the Metal-Vacuum Interface	Heiland, W.	11
The Study of Simple Consecutive Processes in Electrochemical Reactions	Bewick, A. Thirsk, H. R.	5
Surface Analysis by Electron Spectroscopy	Baker, B. G.	10
Surface-Enhanced Raman Scattering (SERS)	Efrima, S.	16
Surface Potential at Liquid Interfaces	Llopis, J.	6
Surface States on Semiconductors	Batchelor, R. A. Hamnett, A.	22

Title	Author	Number
Temperature Dependence of Conductance of Electrolytes in Nonaqueous Solutions	Barthel, J. Wachter, R. Gores, H.-J.	13
The Temperature and Potential Dependence of Electrochemical Reaction Rates, and the Real Form of the Tafel Equation	Conway, B. E.	16
Theoretical Aspects of Semiconductor Electrochemistry	Uosaki, K. Kita, H.	18
Theories for the Metal in the Metal-Electrolyte Interface	Goodisman, J.	20
Theories of Elementary Homogeneous Electron-Transfer Reactions	Sacher, E. Laidler, K. J.	3
Theory and Applications of Periodic Electrolysis	Pesco, A. M. Cheh, H. Y. Popov, K. I.	19
Theory of the Effect of Electrodeposition at a Periodically Changing Rate on the Morphology of Metal Deposits	Maksimovic, M. D.	19
Thermodynamics of Membrane Energy Transduction in an Oscillating Field	Markin, V. S. Tsong, T. Y.	24
Transport-Controlled Deposition and Dissolution of Metals	Despić, A. R. Popov, K. I.	7
Transport Models for Ion-Exchange Membranes	Verbrugge, M. W. Pintauro, P. N.	19
Transport Phenomena in Electrochemical Kinetics	Arvia, A. J. Marchiano, S. L.	6
Ultrahigh-Vacuum Surface Analytical Methods in Electrochemical Studies of Single-Crystal Surfaces	Soriaga, M. P. Harrington, D. A. Stickney, J. L. Wieckowski, A.	28
Ultrasonic Vibration Potentials	Zana, R. Yeager, E. B.	14
Water Electrolysis and Solar Hydrogen Demonstration Projects	Sandstede, G. Wurster, R.	27
X-Rays as Probes of Electrochemical Interfaces	Abruña, H. D.	20

Index

Absorption
 of Br on tungsten selenide, 155
 of D
  in palladium, 471
  into Pd, and time, 475
 of halogens, on silicon, 101
 of H
  in metals, 419, 429
  in titanium, 437
  in zirconium, 437
 of hydrogen
  studied by the d–n reaction, 490
  on titanium, at different current densities, 483
 of ions at interfaces, 34
Absorption coefficient, as a function of photo energy for tin oxide, 228
Accelerator, its use in detecting H as a function of depth of penetration, 428
Activity
 and coverage, 421
 determined by current interrupter method, 421
 of hydrogen atoms, and equivalent hydrogen pressure, 450
Activity differences, between manganese dioxide types, 357
Adsorbates, monovalent, on silicon surfaces, 98

Adsorption and desorption on palladium, diagrammated, 472
Adsorption behavior, of molybdic sulfide, 146
Adsorption energies, on semiconductors, 52
Adsorption rate, of hydrogen into zirconium, 484
Affinity, in hydrogen evolution, and overpotential, 456
Aggregation-diffusion control, in copper deposits, 274
Ammonium chloride electrolyte, and manganese dioxide, 327
Analysis, of hydride layers in zirconium, 488
Anodic oxidation, of chloride on tin oxide, 251
Arrhenius plot, of hydrogen absorption onto titanium, 486
Arsenic, its effect on permeation, 462
Atlung and Pohl, thermodynamic treatment of manganese dioxide, 393

Bagotskaya and Frumkin, hydrogen entry into palladium, 443
Bai, and the discharge–charge curves for chemically modified manganese dioxide, 390

527

Band bending
  involving water and semiconductors, 143
  and semiconductors, 148
Band configuration, for Schottky barrier diode, 200
Band formation, in an LCAO model, 49
Band-edges, 30
Band gaps
  in gallium arsenide, 114
  of semiconductors, 204
Band spectra, in semiconductors, 190
Band structure, of three-dimensional semiconductor, 46
Baranowski
  high pressure in palladium, 435
  hydrogen absorption, 437
  intensive studies of H entry into nickel, 479
Bard, his acceptance of the Schottky barrier model, 202
Barton
  and high pressure within palladium, 435
  flamed palladium electrodes, entry of hydrogen therein, 448
Barton–Bockris theory, of morphology, 261
Batchelor and Hamnett, on surface states, and solar photovoltaics, 203
Battery activity
  as a function of manganese dioxide structure, 355
  and magnetic susceptibility, with manganese dioxide, 358
  and surface area, 362
Beck, and Tafel state in hydrogen permeation through iron, 479
Bending, of energy bands, 73
BET isotherm, of manganese dioxide, 352
Binding energies
  for bromine on tungsten selenide, 162
  in XPS diagrams for oxygen adsorption, 94
Birnesite, 347

Birnesite structure, its use in Wroblowa's ideas on manganese dioxide reduction, 383
Bismuth, its effect on the rechargeability of manganese dioxide, an unfinished story, 390
Bloch states, 50
Bockris
  and changes of potential during polarization on palladium, 459
  and Khan, photoelectrochemistry with high surface state concentration, 196
  and Subramanyan
    fugacity and pressure, 433
    the Volmer–Tafel relationship, 446
  and Utosaki, and surface states affecting kinetics at the semiconductor solution interface, 203
Bockris et al.
  a model for hydrogen entry, 443
  the relation of hydrogen entry rates to the hydrogen evolution rate, 444
Bode and Schmier, discharge on manganese dioxide, 326
Bonds, at the gallium arsenide–solution interface, 118
Born, 16
Brauer, hydrogen into titanium, 480
Bromine adsorption, on indium selenide, reversible, 152
Burkhart, his work on alkaline cells of manganese dioxide, 355
Burns, his work on crystal structure, of manganese dioxide, 339
Butler and Sentry, nuclear chemical method, for analysis of deuterium adsorbed in metals, 425

Cachet, surface potentials at gallium arsenide, 139
Cadmium
  in dendritic form, and diagrams, 302
  and dendritic growth, 301
  and photographs of deposits, 307

Index

Cadmium dendrites, 304
  precursors of, 309
Cadmium deposits, 306
  and growth of grains, 293
  spongy, 286
  from sulfate solution, 292
Capacity, and deposition current density, 363
Carbon content, of chemically modified manganese dioxide, 390
Cartwright and Poul
  diffusion in deposition on manganese dioxide, 401
  rate determining step during growth on manganese dioxide electrodes, 400
Cascade model, in hydrogen evolution, 454
Catalytic poisons, their effect upon the rate of hydrogen evolution, 445
Cation vacancies,
  in manganese dioxide, 347
  their parameters, 349
Cation vacancy theory, and its prediction of maximum capacity, 351
Cauliflower deposits, under normal conditions, 272
Cell chemistries, for manganese dioxide, 326
Chabre et al., their work on γ-manganese dioxide, 345
Chalcogenide semiconductors, 139
  and the monoadsorption phase, 140
Charge, in the interior of a semiconductor, 24
Charge transfer involving surface states, in titanium oxide films, 239
Cheh et al., and modeling of cylindrical cells, 376
Chemically modified manganese dioxide, discharge curve, 387
Chemically treated gallium arsenide, and its interface with the solution, 130

Chemical studies, diagnostic, for manganese dioxide, 325
Chemical vapor deposition on tin oxide films, 213
Chloride evolution,
  on tin oxide, 251
  on tin oxide films, 250
Cleaning, of a silicon-solution interfaces, 108
Clean surfaces, and adsorption at electrochemical interfaces, 86
Cleavage planes, with gallium arsenide, 116
Co-adsorption, of water with bromine, 160
Co-adsorption phases, 153
Concentration profile, for hydrogen in zirconium, 490
  of hydrogen in titanium, zirconium, 487
  near the surface of palladium, 424
Constant current discharge and recharge for manganese dioxide, 388
Contact potential difference, at semiconductor-solution interfaces, 31
Conway, his work on rechargeable manganese dioxide, 383
Copper deposits
  micrographed, 305
  with mixed activation-diffusion control, 280
  with cauliflower-like appearance, 277
  dendritic, 299
  granulated, diagram of, 269
  under mixed activation-diffusion control, 274
Copper electrodeposit, photograph of, 267
Covalent single bonds, for oxygen on silicon, 99
Coverage,
  and activity, 421
  with hydrogen, 421
Cowley and Sze, mathematical expression for pinned metal–semiconductor states, 56

Crack promulgation, and hydride layers, 435
Cross section, of copper, deposited on cross section, 307
Crystallization, as a rate determining process, 398
Crystal structure
  of manganese dioxide, 339
  and manganese dioxide electrodes, 336
Crystal types
  of manganese dioxide, 314
  in manganese dioxide activity, 356
Current density–potential characteristics for tin oxide silicon systems, 217
Current density, its value as a function of crystal change in manganese dioxide, 363
Current efficiencies, of hydrogen absorption into titanium and zirconium, 485
Current transport, in solar cells, 199
Current voltage curve for photoelectrochemical cells involving tin oxide, 249
Cyclic voltammetry, test of Wroblowa's manganese dioxide electrodes, 386

Dark current, involving in oxide, 242
Decohesion mechanism, for hydrogen in palladium, 434
Defects, formed by disorder of oxygen adsorption, 97
Defined surfaces, 65
Deluccia, Permeation of hydrogen into iron, 479
Dendrites, and dependence on concentration, 308
  of cadmium, 300
    diagrammated, 303
    diagrams of, 302
    with cadmium, 300
    their growth, 294
    and twinning, 308

Dendritic deposits, of copper, 299
Dendritic growth, 294
  and exchange current density, 297
  and its physical simulation, 298
  induction theory, 300
Deposition current density, in capacitative discharge, 363
Deposition potentials, for copper deposits, with various promentaries, 275
Deposition, theory of roughness produced therein, 265
  of manganese dioxide, 391
  on oxide covered substrate, 396
Deposits
  coarse, theory of, 262
  spongy, 283
Desai, prediction of behavior of manganese dioxide according to various models, 373
Desorption, of F from palladium, 471
Despic and Popov, experiments on well defined triangular profiles, 266
Despic, Diggle, and Bockris, theory of coarseness in electrodeposition, 263
Deuterium,
  in metals, detected using a deuteron accelerator, 425
  its concentration profile in palladium, 426
Deuterium
  concentration, in palladium, as a function of time, 436
  absorption in palladium, as a function of time, 476
Deuterium-induced structural changes, in palladium, 430
Deuterium-palladium system, 468
Devanathan and Stachurski, diffusion of hydrogen in palladium, 478
DeWolff, his deduction of the γ-manganese dioxide structure, 341
Diffusion coefficient data, hydrogen into zirconium, 482

Index

Diffusion coefficients of light hydrogen,
  various values, 464
  data, hydrogen into palladium, 481
  and hydride formation, 480
  of hydrogen
    in α and β, palladium, values of different authors, 465
    during absorption process, 494
    as given by Volkl and Alefeld, 463
    in palladium, 463
Diffusion, cylindrical, 296
Diffusion model, for hydrogen into palladium, 473
Diggle–Despic–Bockris theory, of morphology, 261
Dipole contributions, at the adsorbate interface, 145
Discharge curves
  for manganese dioxide, effective variables, 325
  for manganese dioxide reduction, 317
Discharge, limiting, in MnO_2 discharge, work by Kordesch, 383
  of β-manganese dioxide, 325
  from manganese dioxide, 329
  for manganese dioxide, and equilibrium, 321
  rates, and modeling, 373
  surface, of manganese dioxide, 324
Dissolution
  of anodic zinc, 328
  of semiconductors, 206
Distribution function, for electrons at interfaces, 17
Distribution, of electronic states, in tin oxide films, 237
Doping, and its effect on tin oxide, 246
Dye molecules, adsorbed on semiconductor solution interfaces, 210

Electrocatalysis, of hydrogen
  on titanium, 458
  on zirconium, 458
Electrochemical behavior of tin oxide films, 230

Electrochemical spectroscopy, for manganese dioxide, 378
Electrode kinetics, involving tin oxide, 232
Electrodeposition,
  on coarse surfaces, 268
    diagrammated, 270
  and diffusion control: photographs, 271
Electron energy, and contribution to the work function, 86
Electronic band structure, of three dimensional semiconductor, 46
Electronic states, their distribution, 244
Electron transfer, to adsorbed bromine molecules, 156
Electropolishing, and gallium arsenide, 132
Embrittlement, by hydrogen, 434
Energy band diagram for tungsten selenide–bromine interfaces, 158
Energy bands, in semiconductors, 72
Energy diagram
  for electronic states in the band gap, 238
  from molybdic sulfide interfaces, 150
  involving semiconductors, 195
  involving tin oxide, 245
  for photoelectrolytic cells in equilibrium, 208
Energy levels, in electrolytes, 194
Energy, of mixed phases in titanium hydride, 491
Enthalpy, of hydrogen dissolution, in metals, 439
Enyo, his reinterpretation of Deluccia's work, in terms of mixed control, 479
  his work
    on hydrogen in palladium, 447
    on the sorption of hydrogen, 415
Equilibrium potential
  as a function of discharge, 321
  manganese dioxide, 322
Equilibrium potentials
  for manganese dioxide, 331, 392
  versus degree of manganese dioxide reduction, 370

Equivalent hydrogen pressure, and overpotential, 445
Era, Takuhara and Yoshizawa, their work on depolarization, for manganese dioxide, 333
Etching
 of gallium arsenide, 131
 of silicon, 109
Exchange current density,
 and affect on dendritic growth, 297
 in hydrogen evolution, and the palladium electrode, 457
 for manganese dioxide electrodes, work of Heusler and Grzegorzewski, 405
Extrinsic semiconductors, 189
Extrinsic surfaces, and an LCAO model, 50

Fermi level
 in a semiconductor, 23
 and semiconductors, 42
Fermi level, pinning, 53
 at gallium arsenide, 134
 with gallium arsenide, 117
 due to surface dipoles, 60
Fernandez, correlation between battery activity and crystal type, 361
Flat band potential, with gallium arsenide, 136
 in electrolytic solutions, 32
 show Nernstian shift, 113
 and work functions, 137
Fleischmann
 his collaboration with Preisler, 399
 limiting current in manganese dioxide, 398
 manganese dioxide deposition, 398
Fluctuation model, for electrons at the interface, 15
Fluoride doping, and the crystallinity of tin oxide films, 227
Fluorine incorporation, its effect on tin oxide films, 224

Foreign atoms and effect on optical properties of tin oxide films, 220
Frumkin and Aladzhalova, and hydrogen entry into palladium, classic work, 448
Fugacity
 and a million atmospheres, 431
 versus pressure, 432

Gallium arsenide, 113
 exposure of its surface to various entities from solution, 120
 and electropolishing, 132
 its surface, 115
 on photo emission spectra, 125
 a typical semiconductor, 51
 its flat band potentials, 136
Galvani, P. D., 27
Gamow plot, for palladium, 469
Gavano, and the various manganese dioxide types, compared, 375
Gerischer, his acceptance of the Schottky barrier model, 202
Grains, growth of, 293
Grouthellite, 365
Groutite, 365
Growth forms,
 irregular shape, 282
 and surface coarseness, 281

Halogen adsorption, 147
Halogens, their adsorption in silicon, 103
Harrison, and two center resonant integral of atomic orbitals in surface states, 48
Heat treatment, of manganese dioxide, 360
Helium, on surfaces, 90
Heusler and Grzegorzewski, their study of manganese dioxide deposition, 403
Heyrovsky (electrochemical) mechanism, 479
History, of zinc–manganese dioxide, 313
HOMO level, for the solvent, 14

Index

Hydride
  formation, in zirconium, as a function of depth, 493
  growth of, on zirconium, 482
Hydride layer
  growth thereof, 490
  on palladium during cathodization, diagrammated, 477
  in titanium, 490
  in zirconium, 488
Hydrogen
  absorption
    on Fe and Ni, 436
    in solutions containing alkaline ions, 487
  adsorption, 100
    on Fe and Ni, 436
    kinetic approaches thereto, 416
    and diffusion coefficient into palladium, 466
  content, in palladium, and its effect on resistance, 424
  diffusion into palladium, temperature dependence, 483
  dissolution, 416
    its enthalpy, 439
    in metals, 417
  entry
    independent of hydrogen evolution, 442
    via an intermediate in the hydrogen evolution reaction, 443
    into palladium, work of Frumkin and Bagotskaya, 443
  evolution
    and catalytic poisons, 445
    its effect on hydrogen entry, 442
    its mechanisms, and the pressure developed inside palladium, 451
    on manganese dioxide, 384
    on palladium, best understood reaction, in electrochemistry?, 454
  in palladium, overpotential and the hydrogen pressure relation, 467
  its absorption, 415

Hydrogen (*cont.*)
  its adsorption on gallium arsenide, 126
  its sorption on metals in electrochemical environments, 415
  in metals, and possible fracture, 435
  into palladium, conclusions, 494
  overpotential
    as a function of pressure, 450
    related to equivalent hydrogen pressure, diagrammatic, 452
  and palladium, and Devanathan and Stachurski, 478
  permeation
    in metals, 460
    through steel, as a function of overpotential, 446
  pressure
    and cathodic potential, 418
    equivalent as a function of cathodic overpotential, in palladium, 469
    equivalent to hydrogen overpotential, 417
Hydrogen electrode reactions, in palladium, 446

Inactive phases, in manganese dioxide, 364
Induction theory, of dendritic growth, 300
Initiation, of dendritic growth, 294
Insertion mechanisms, in Groutite-like layers for manganese dioxide, 380
Intercalation, in manganese dioxide cells, 319
Interface potentials
  at different layered semiconductor solution interfaces, 142
  at the semiconductor-metal contact, 26
Intrinsic semiconductors, 189
Intrinsic surface states, by an LCAO approach, 47
Ionic diffusion, in manganese dioxide, 400
Ionization potentials, affected by the adsorption, 106

Iron, and hydrogen absorption therein, 436
Iron electrodes and hydrogen diffusion, therein, 479
Isotherms
  for hydrogen
    on palladium, 429
    and zirconium, 438
  for hydrogen absorption, methods, 419
Isotope effect, for first electron reduction, 323

Johnson and Vosburgh, and their observation of the stability of mixed manganese oxides, 368
Junction characters, of tin oxide, 216
Junction properties, investigated by UHV, 62

Kao and Weibel, limiting current at manganese dioxide electrodes, 402
Kelvin equation, and the pores in manganese dioxide, 353
Kelvin overpotential, and dendritic growth, 298
Kelvin probe, used in semiconductor–solution interfaces, 75
Kordesch, and rechargeable manganese dioxide battery, 383
Korn and Zamir, hydrogen permeation into palladium, 480
Koshiba
  and alkaline discharge capacity, effect of area, 360
  and discharge of flooded cells involving manganese dioxide, 334
  and his study of the second electron, 386
  and powers, and the discharge of manganese dioxide, 325
  and the two step mechanism for manganese dioxide, 321
  and Yeager
    their study of the second electron mechanism, 324
    their work on manganese dioxide mechanisms, 322
  his work on manganese dioxide, 325

Krishmar, his theory of coarseness and deposition, 262

Lattice parameters, for titanium, and hydrides, 489
LCAO approach, for intrinsic surface states, 47
LCAO diagrams
  for bonding to silicon, 100
  for donor states in tungsten selenide, 144
LCAO energy diagrams, and surface states, 51
Léclanché cells, with zinc chloride electrolyte, 326
LEED and its use for semiconductor solution interfaces, 77
Leverence et al., their work on silicon planes, 63
Lewis, and his ideas on hydrogen entry, 441
Low energy ion scattering spectroscopy, 78
LUMO level, for the solvent, 14

M'hamedi, and his unchanged ionization potential, 127
Mangalite, 365
Manganese dioxide
  cations, in oxyhedral sites, 343
  classifications of, 346
  deposition
    on bare substrate, 395
    as proposed by Fleischmann, 398
  and electrochemical spectroscopy, 378
  electrodes
    and crystal structure, 336
    and zinc precipitates, 335
  effect of those crystal types on battery activity, 357
  effect of heat treatment, 351
  in rechargeability problems, 382
  and in-situ microscopy, 321
  its deposition, 391
  its discharge mechanism, 329

Index

Manganese dioxide (*cont.*)
  and its inactive phase, 364
  and the Léclanché cell, 313
  modification of its structure, by Wroblowa et al., 383
  modified, its two electron capacity, 385
  reduction and the equilibrium potential, 370
  schematic representation of the first discharge, 319
  in spectroscopic studies, 359
  and statistical mechanical approach, 369
  and structural characteristics, 339
  structure, its effect on battery activity, 355
  and the structure of water, 348
  variation of surface area, 354
  the various crystal types, 314
Manganese dioxides, crystal structure, 339
Manganese oxides, complex, 346
Manganese oxide types, effect of, on manganese dioxide electrodes, 338
Maskell
  and distinction of sites, in manganese dioxide, 372
  and electron sharing in manganese dioxide reduction, 368
Mathematical model, for spongy deposits, 283
McBreen
  and rechargeability, for manganese dioxide, 382
  his use of slow scan voltammetry, in manganese dioxide mechanisms, 326
  his use of slow scan voltammetry, with manganese dioxide, 377
  sweep voltammetry, with manganese dioxide, 377
Mechanical strength, and hydrogen absorption, 415
Mechanism
  for continued manganese dioxide deposition, 397
  of hydrogen evolution, 451

Mechanism (*cont.*)
  involving ionic diffusion in manganese dioxide, 400
  proton and electron insertion, for manganese dioxide, 365
  of second electron step, for manganese dioxide, 320
  in steps, for manganese dioxide reduction, 318
  of Wroblowa's manganese dioxide, 386
Metallicity, of semiconductors, measured by photoemission studies, 91
Metal oxide films, 211
Metals
  and hydrogen detection therein, 423
  and hydrogen permeation, 460
Micrographs, of copper deposits, 305
Microscopic observations, 490
Mixed control of hydrogen evolution reaction, its effect on pressure, 453
Mizuno, his work on hydrogen sorption, 415
Model(s)
  basic, for coarseness and deposition, 263
  for hydrogen absorption in metals, 474
  for hydrogen and palladium, protective films, 473
  for hydrogen entry into an electrode, 441
  mathematical, for spongy deposits, 283
Model electrolyte, and information obtainable, 83
Model experiments
  the surface reactions, 151
  with actual semiconductors, their comparison with UHV results, 81
Modeling
  of γ-MnO_2 and practical discharge rates, 373
  of manganese dioxide, 374
Molecular beams, and the investigation of semiconductor interfaces, 66
Molybdic sulfide, and adsorption behavior, 146

Monch, and models for the 111 plane, 89
Morphology, and electrode deposits, 261

Nucleation, 287
Nuclei, their presence, in spongy deposits, 284

Ohta, his work on alkaline cells, 355
Ore, natural, chosen for battery active MnO_2, 314
Overlap, spherical diffusion layers, 278
Overpotential
 and activity of hydrogen inside palladium, 450
 decay curves, under various conditions, for hydrogen on palladium, 449
 and dendritic formation, 308
 at early stages, on titanium zirconium, 459
 of hydrogen
  and reduction of materials at electrodes, 416
  and the equivalent hydrogen pressure, 417
  rise and decay, in hydrogen electrodes, 448
Overview, of manganese dioxide batteries, 316
Oxides
 complex manganese, 346
 on gallium arsenide, 124
Oxygen
 adsorption, on semiconductors, 93
 evolution, on manganese dioxide, 384, 395
 interaction, with gallium arsenide, 121
Palladium
 changes in electrical resistance, as a function of hydrogen content, 424
 diffusion coefficient of hydrogen through it, 463
 and hydrides formed therein, 430
 and hydrogen absorption therein, 433
 and hydrogen electrode reaction, 446

Pandey, his chain model for silicon, 89
Parameters of surfaces, and spherical diffusion, 273
Permeation
 of hydrogen
  into iron and nickel, 478
  in metals, 460
  through palladium, effect of poisons, 461
  into titanium, as a function of hydride formation, 480
 rate of, for deuterium in palladium, 470
Phase changes, irreversible, in manganese dioxide rechargeability, 382
Photoelectric characteristics, of tin oxide, 247
Photoelectrocatalysis, described by Bockris and Khan, 196
Photoelectrochemical and photovoltaic systems, 240
Photoelectrochemical devices, 193
Photoelectrochemical processes, 203
Photoelectrochemical reactions, and interface potentials, 147
Photoelectrolytic cells, 207
 under an applied potential, 209
Photoelectron spectroscopy, 68
Photoemission studies, of gallium arsenide, 123
Photoemission techniques, their sensitivity, 70
Photovoltaic and photoelectrochemical systems, 340
Photovoltaic junction, 191
Photovoltaic parameters involving tin oxide, 253
Photovoltaics, and the conversion of light to electrical energy, 188
Physical simulation
 of coarseness developed in deposition, 266
 of dendritic growth, 298
Pinning effects, originally considered by Bardeen, 57
Planes, from molybdic sulfide, 150

Index

pn photovoltaics, 192
Poisons, effect on permeation of hydrogen through palladium, 461
Polarization, of palladium in hydrogen evolution, changes of overpotential on a month's scale, 459
Polymorphs, 342
Polyselenide solutions, and gallium arsenide, 135
Pores, in manganese dioxide, 353
Porosity, in manganese dioxide, 354
  effect on battery performance, 360
Porous electrode approaches, to manganese dioxide, work of Tiedemann, 375
Potential-discharge, for manganese dioxide, diagrammated, 332
Potentials, equilibrium
  of γ-manganese dioxide, 368
  with manganese dioxide, 321
  in the semiconductor-electrolyte interface, 33
Potential step voltammetry, 379
Potentiostatic polarization, instrumentation thereto, 423
Power characteristics involving tin oxide, 248
Precursors
  of dendrites, and appearance at regular intervals, 311
  of dendritic growth, 300
Preisler, his application of Fleischmann's work, 399
Pressure
  versus fugacity, 3432
  due to hydrogen, as a function of cathodic overpotential, 468
  inside palladium, 451
Production, of manganese dioxide batteries, 315
Production products, during manganese dioxide electrode production, 337
Profile, for hydrogen in zirconium, as a function of depth, 493

Protective films, and a model for hydrogen and palladium, 473
Proton and electron insertion mechanisms, 365
Proton diffusion coefficient, and manganese dioxide, 376
Proton effect, in manganese dioxide electrodes, 323
Proton insertion, for manganese dioxide, 319
Protons, their different environments in manganese dioxide batteries, 367
Protrusions, in electrodeposits, 276
Pyrolusite, 342

Qu
  and the bismuth-manganese dioxide, mechanism, 386
  and Mn_3^+ ions, present during first electron discharge, for manganese dioxide, 382
  his monitoring of chemically modified manganese dioxide, 387
Quantum mechanical theory, of surface states, 45
Qu et al., and the discharge and recharge mechanism, for manganese dioxide, 389

Rate determining process, in crystallization, 398
Rechargeability problems, for manganese dioxide, 382
Redox potentials
  involving bromine and semiconductors, 157
  of semiconductors, 205
Redox reactions, and tin oxide films, 231
Reduction of manganese dioxide, 368
Refractive index, of tin oxide films, 223
Relaxation, at the gallium arsenide solution interface, 114
Ripert, and the electrochemical reduction of manganese dioxide, 379

Ruëtschi
  and cation vacancies, in manganese dioxide, 393
  and his theory of proton insertion in manganese dioxide, 372
  morphological changes, in manganese dioxide, 321
  and the structure of water in manganese dioxide, 348
Ruëtschi's cation vacancy model, in manganese dioxide, 350
Ruëtschi and Giovanoli, proton and cation vacancies, in the mechanism for manganese dioxide activity, 357
Ruthenium dioxide, a catalyst in photoelectrochemistry, 254

Santoni, the ionization potential, as affected by hydrogen adsorption, 127
Schlesinger and Xjanietz, dipole potential contributions, at the semiconductor solution interface, 138
Schottky, 20
Schottky barrier-like contacts, 29
Schottky barrier
  and application to solar cells, 197
  solar cell, diagrammated, 198
Schumm, his work on zinc chloride cells, 355
Second electron step, in manganese dioxide, mechanism of, 320
Semiconductor–electrolyte contact, 28
  and pinning, 58
Semiconductor–electrolyte interface, 84
  and the Cowley–Sze formalism, 59
Semiconductor electrochemistry, and surface science, 61
Semiconductor electrolyte interface, model experiments, 80
Semiconductor metal barrier, with Fermi level pinning (Bardeen limit), 54
Semiconductor–metal interface, 21

Semiconductor metal contacts, in terms of Fermi level pinning, 53
Semiconductors, and the Schottky limit, 20
Semiconductors, extrinsic, intrinsic, 189
Semiconductor–solution interface, a summary, 166
Shockley states, 39
Short circuit current for tin oxide-silicon photocells, 218
Side reactions, in manganese dioxide, 394
Silicon, 87
  the 111 plane, and Pandey's chain model, 89
Silicon–solution interface, Lewis and Bocarsly study, 107
Silicon surfaces, and oxygen adsorption thereon, 96
Sites, for hydrogen atoms in palladium, 440
Smialowski
  and hydrogen adsorption, 437
  his intensive studies of H entry into nickel, 479
Solar cells, and current transport, 199
Solar energy and energy conversion, 203
Solomon, his anodic treatment of gallium arsenide, 133
Someno and Nagasaki, hydrogen permeation into palladium, 480
Space charge layer
  in a semiconductor, 22
  various types, 23
Spectroscopic studies, of manganese dioxide, 359
Spectroscopy
  electrochemical
    of γ-MnO_2, 377
    and manganese dioxide, 378
  of surface states, 61
Spherical diffusion, around promontories on surfaces, 273
Spherical diffusion layers, with overlap, 278

Index

Spongy deposits, 283
  their initiation, 288
  their physical model, 285
Spongy growth, 308
  with zinc deposits, 290
Spongy nucleus, an ideal type, 201
Standard potential, for manganese dioxide, 393
Statistical mechanical approach, to manganese dioxide, 369
Steps, in manganese dioxide reduction, 317
Structural characteristics, of manganese dioxides, 339
Structural parameters, of tin oxide films, 226
Structural water, and Ruëtschi's cation vacancy model for manganese dioxide, 350
Structured investigation, of tin oxide films, 225
Surface coverage, for hydrogen on metals, 420
Surface dipoles
  and the absolute redox potential, 18
  at the interface electrode– electrolyte, 19
  at the metal–semiconductor interface, 25
  at semiconductor solution interfaces, 85
Surface fermi level, as a function of water adsorption, 161
Surface photovoltage, 76
Surface potentials
  at the electrolyte–vacuum interface, 15
  for gallium arsenide, 127
  of halogen covered gallium arsenide, 128
  in UHV adsorption, 149
Surface properties, of manganese dioxide, 352
Surface reconstruction, of gallium arsenide, 117
Surfaces
  clean, involving gallium arsenide, 113
  involving hydrogen in silicon, 111
  and monovalent adatoms, 98
  and their parameters, with spherical diffusion, 273

Surface science, in semiconductor electrochemistry, 61
Surface science techniques, review of, 67
Surface states
  diagrammatic, 38
  extrinsic, 41
  Gaertner, and the Schottky barrier model without a solution, 201
  "good" and "bad," defined, 43
  intrinsic, 37
  limit of 10^{13} for applicability of Schottky barrier, 57
  for real semiconductors, 40
  in terms of quantum mechanics, 44
  their effect upon the semiconductor-solution interface, 36
  transient, 43
Swinkels, his correlation of x-ray patterns, with battery activity in manganese dioxide, 362
Synthesis, of the Wroblowa manganese dioxide, 386

Tafel–Volmer mechanism, 422
Tafel mechanism for hydrogen evolution, and the pressure developed inside palladium, 451
Tafel plots
  as a function of current density for various i_0 values, 458
  with titanium dioxide, 252
Tafel slopes
  and the hydrogen evolution reaction, 445
  on tin oxide, 235
Temperature dependence, for hydrogen absorption in palladium, 433
Theories
  of inactive manganese in manganese dioxide, 371
  of irregularity in electrodeposition, 263
Thomas–Fermi potential, 79
Tiedemann, and his work on the porous electrode approach to manganese dioxide, 375

Time dependence of absorption of D in palladium, 475
Tin oxide
  and electrode kinetics, 232
  and improvements of its characteristics, 219
  and its junction characteristics, 216
  method for its preparation, 212
  as a semiconductor, 211
Tin oxide films, 221
  affected by the presence of foreign atoms, 220
  and photoelectron spectroscopy, 233
  and redox reaction, 231
  their conductivity, 215
  their electrochemical behavior, 229
  their homogeneity, 214
  their structural parameters, 226
  their transparency, 222
Tin oxide layer, involved in heterojunctions, 241
Tip of dendrite, condition at, 295
Titanium
  and hydrogen absorption, 437
  its electrocatalytic activity, 458
Transparency, of tin oxide films, 222
Tufts, his work on the etching of gallium arsenide, 133
Tungsten selenide–solution interfaces, 163
Tunneling, through films, 243
Turner and Bufseck, and their work on fibrous minerals, 347
Twinning
  and dendrites, 308
  in manganese dioxide, 346
Tye, and the amount of Mn_3^+ ions in unreduced oxide, 371

Uetani, and precipitates formed in manganese dioxide electrodes, 335
UHV analysis of silicon, 110
UHV studies, of water, on semiconductors, 104
UHV techniques, and interfaces involving semiconductors, 62
UHV work, tabulated, 64
Ultraviolet photoelectron spectroscopy, and tin oxide films, 233
University of Ottawa, work threat on the rechargeable manganese dioxide electrode, 383
UPS spectra, of tin oxide, 234

Valency band spectra
  for oxygen-exposed silicon surfaces, 96
  in semiconductors, 71
  for tungsten selenide, 154
  for tungsten selenide–water interfaces, 141
Vetter and Jaeger, the equilibrium properties of manganese dioxide, 393
Voinov, and proton insertion into manganese dioxide, 357
Volkl and Alefeld
  their diffusion coefficient of H on Pd, 463
  their listings of diffusion coefficients of hydrogen and palladium, 466
Volmer–Heyrovsky mechanism, 445
Volmer–Tafel mechanism, and pressure developed inside palladium, 451
Voltammetry
  for manganese dioxide, work of McBreen, 377
  potential step, for manganese dioxide, 379
Vosburgh, polarization mechanism, for manganese dioxide batteries, 331

Water adsorption, on silicon, 104
Water
  its dissociated adsorption, 105
  its interaction with gallium arsenide, 129
  structural, in manganese dioxide, 347
  structural, state of, in manganese dioxide, 347

Index

Will, and a D/Pd ratio of 1.15, 435
Williams, discharge capacities for related cut off potentials, 364
Work functions, at semiconductor solution interfaces, 69
Wranglen, his definition of dendrites, 294
Wroblowa
  her major breakthrough at Ford, 385
  her manganese dioxide, mechanism of rechargeability, 386
  her seminal work on rechargeable manganese dioxide, 383
  and the insertion of bismuth, 386
Wruck and Chapman, porous model for manganese dioxide, 375

X-ray diffraction, of manganese dioxide, 320
XPS data, for gallium surfaces, 122
XPS diagrams for two peak levels of fluorine attached to silicon, 102
XPS spectra
  for adsorption of bromine, sodium, and water, 164
  for co-adsorption of bromine and water on tungsten selenide, 159
XPS studies, of oxygen adsorption on silicon, 95

Zinc-manganese dioxide cells, alkaline, 315
Zinc
  anodic dissolution thereof, 328
  and cadmium, real system and their electroposition at low HOMO potentials, 287
  deposits, on copper, their spongy nature, 289
  migration, in manganese dioxide, 384
  precipitates, their role in manganese dioxide electrodes, 335
Zirconium
  and absorption of H therein, 474
  and hydrogen absorption, 437
  its electrocatalytic activity, 458